Routledge Revivals

Ireland

At the time of the publication of this book in its fourth edition in 1969, Ireland was alone globally in having experienced a decline of population for more than a century. National movements in Wales and Scotland made the story of the Irish Republic's first fifty years increasingly interesting. Traditional and conservative as Ireland's life may have seemed in the late 20th Century it had changed considerably since 1921. Like much of Western Europe it continues to share the experience of a declining number of agricultural workers and its government, like that of many other countries is concerned with the problem of industrial growth. The book analyses the physical environment and the life of 20th century Ireland whilst it was in the throes of an economic revolution.

Ireland

A General and Regional Geography

T.W. Freeman

Routledge
Taylor & Francis Group

First published in 1950 and as a fourth edition in 1969 by Methuen & Co Ltd.

This edition first published in 2025 by Routledge
4 Park Square, Milton Park, Abingdon, Oxon, OX14 4RN
and by Routledge
605 Third Avenue, New York, NY 10158.

Routledge is an imprint of the Taylor & Francis Group, an informa business

ISBN 13: 978-1-041-04078-1 (hbk)
ISBN 13: 978-1-003-62675-6 (ebk)
ISBN 13: 978-1-041-04142-9 (pbk)
Book DOI 10.4324/9781003626756

Ireland

IRELAND

A GENERAL AND REGIONAL GEOGRAPHY

T. W. FREEMAN

Reader in Economic Geography in the University of Manchester

LONDON: METHUEN & CO LTD

11 New Fetter Lane, E.C.4

First published June 8 1950
Second edition, revised and reset 1960
Third edition 1965
Fourth edition 1969
Printed in Great Britain
by Butler and Tanner Ltd, Frome and London

4.1
SBN 416 12740 1

Distributed in the U.S.A.
by Barnes & Noble, Inc.

TO THE MEMORY OF

MARGARET JOYCE FREEMAN

'Ceramic'

December 7, 1942

Preface

Several years have passed since the writing of this book was suggested by Dr Lloyd Praeger and by the late Professor Edmund Curtis, of Dublin University, and by many more men and women in Ireland and elsewhere. The need for a new geography of Ireland is acknowledged by all those who have any appreciation of what the subject teaches, and, to some extent, by a wider public. In Ireland there is now a lively interest in natural science, history and economics, not only in universities, schools and learned societies, but also among men and women of inquiring mind elsewhere; and Dublin is now, as in past years, a city in which the researcher may find constant help and inspiration. This book was written mainly within the walls of Trinity College, but many of its pages were first conceived on the roads and lanes of Ireland and in the remote recesses of its countryside. From writing this work the author has learned beyond doubt that Ireland is a fundamentally different country from either England, Wales or Scotland, in each of which he lived for periods until 1935: to some, this book may be of interest in relation to Great Britain; to others, as a treatment of the western island. If one begins by recognizing the individuality of a country separated by sea from even its nearest neighbours, the geographical way may be made plainer both to the author and to the reader.

It is not the purpose of this book to be polemic, and those who search its pages for political judgements will seek in vain, but on the other hand it is not the intention to be so 'safe' that desiccation results, for much of the interest of Ireland lies in the social life of its people; and problems such as emigration are a subject of constant comment in the Press and in conversation. Population distribution, housing, agriculture, industry, transport and many other subjects discussed here are all social questions, and not infrequently the hidden springs of a political life that is vigorous if at times acutely personal. No longer can the woes of Ireland be ascribed solely to Whitehall or to Dublin Castle, for Dail Eireann and Stormont rule the destinies of twenty-six counties and of six, and out of this change a new Ireland is emerging. The shock of war has been less acute than in Britain, yet the whole country, neutral or belligerent, has changed markedly between 1938 and 1948; and the changes between 1848 and the present are of vast scope and importance. A century ago the Industrial Revolution was only just beginning in Ireland, and even now its imprint is less marked than in Great Britain, though at the same time indelible and pervasive. However great the industrialization of the future, Ireland will still remain distinct in its quality of life, having its own kind of social geography, even though the older customs may pass away.

The physical geography of Ireland is so closely related to that of Great Britain that to the imaginative eye the intervening seas exist only as a geological creation of yesterday. Climatically, Ireland often experiences what Britain will shortly endure; its vegetation may be in part pre-glacial, but has also been drawn from Atlantic and European sources in post-glacial times to be moulded into a distinctive form in an insular environment. For the rest, population, industry, communications, settlement and social life, the story must tell itself both in the general chapters and in the more detailed regional treatment of these pages. This book may at least demonstrate that there is abundant scope for research on the geography of Ireland; and its author hopes that those who can add to his own knowledge will do so, especially if they disagree with the conclusions reached here, for as a wise man once said, no one is convinced that he is wholly wrong except when convinced that he is wholly right.

A first necessity for the worker in geography is a sympathetic approach to people, and the author has at least found the Irish sympathetic to him and willing to give information on their farms, their families or their jobs. In many cases, the assurance that he had no connexion with any government department immediately produced a lessening of reserve and an invitation to enter the farm kitchen. There is also a debt of gratitude to the many students, clergy, teachers, shopkeepers and others who have given information on the communities among whom they live. Most of these informants have been quite oblivious of the value of their information, and few of them may read these words, but the debt remains. Much of the field work has been done with various friends, some of it in the annual field weeks of the Geographical Society of Ireland, all of which have proved interesting ventures in joint research.

On many occasions the author has been indebted to his friends on the staff of Trinity College, Dublin, and of other universities, and to many more whose assistance has been invited at various stages. A list of names is impossible in such circumstances, but three must be mentioned: Dr A. Farrington of the Royal Irish Academy, for constant help and advice, Mr J. P. Haughton, M.A., M.Sc., Lecturer in Geography in Dublin University, for so much companionship in these investigations, and Mr F. M. Synge, M.Sc., for some splendid maps. To the rest, unnamed, the author adds equally cordial thanks and the assurance that the errors are his own.

T. W. FREEMAN

Department of Geography
Trinity College, Dublin
August 1948

Preface to the Second Edition

Ten years have elapsed since the first edition of this book appeared. In that time substantial additions have been made to the knowledge of the geography of Ireland, not only in the annual issue of *Irish Geography* but in a large number of other journals. When this book was originally written from 1944 to 1948, Ireland was in a phase of adjustment to numerous changes induced by the War: since then much has been altered though emigration has remained heavy. Several chapters in this book have been virtually rewritten though in others substantial changes did not appear necessary. One purpose of writing this book initially was to bring before research workers in geography some of the richly interesting problems of Ireland and this has happened, greatly to the satisfaction of the present author. The kind reception given to the first edition by reviewers was naturally welcome: it is good also to be able to thank a number of people for interesting suggestions and corrections of errors.

T. W. FREEMAN

Department of Geography
University of Manchester
March 1959

Preface to the Third Edition

Fourteen years separate the appearance of the first and third editions of this book and the volume of material published by geographers on Ireland has increased yearly, even though many fascinating research topics are still only partially explored. The number of visitors to Ireland remains high and will certainly increase further: among them are many school and university parties on field excursions. It is no longer correct or even reasonable to speak of these islands as 'the British Isles' but rather as 'Great Britain and Ireland'. The fascination of Ireland lies in its distinction from Great Britain, its separate if related history, its population problem unique in Western Europe, its farm pattern and agriculture still in a state of evolution from the woeful fragmentation of pre-Famine times, and even in the ethos of its people. Nevertheless, structurally and climatically the relations of Ireland with Britain are so obvious that Ireland can only be understood in relation to Britain. In how many lecture rooms, one wonders, where courses are given on 'the British Isles', is Ireland relegated to a few lectures at the end of the session when the lecturer's repertoire on Britain is exhausted? And if the lecturer loses a fortnight's work through influenza, Ireland will probably be left out altogether. One can only suggest that it is worthy of more serious attention, even though it is marginally placed in Europe, a minor unit in any population study, still only partially industrial in its life and no longer the provider of a major political problem.

The main addition to this issue is a new series of population maps based on the 1961 Census, replacing those on the 1936 Census included in the first two editions. The vast labour of preparing these new maps has delayed the appearance of this edition but in much of the country the changes of twenty-five years have been so great that the delay seemed justifiable. In more than twenty years' work on Ireland the main fascination has been the population problem, solved in part through the ever-open door for workers coming to Britain. But as yet no final and unequivocal solution is in sight.

T. W. FREEMAN

Department of Geography
University of Manchester
July 1964

Preface to the Fourth Edition

It is almost exactly twenty years since the first edition went to press and in that time Ireland has changed considerably: so has this book. During this twenty years the number of professional geographers working in Ireland has multiplied several times and various geographers living elsewhere have turned their attention to Ireland. Nobody is more conscious than myself of the need for further research on the geography of Ireland, in its contemporary aspects as well as its historical aspects: nobody would be happier to see more work done by others. Life in Ireland has changed considerably but behind all the current tension lies the continuing struggle between nationalism and internationalism. It is clear that the achievement of independence in Ireland has been closely watched not only by people in Scotland and Wales, but also by nations all over the world. Not the least interesting sign of the times in Ireland is the increasing co-operation between its two political units. What does the future hold? For the Republic, a Gaelic state? For Northern Ireland, an imperialist outlook that is rooted in the nineteenth century? Or in the end, a new Ireland, having its cultural life richly cosmopolitan through the varied outlook of its people, none of whom need surrender their basic principles but all of whom must learn more tolerance and understanding for those who think differently.

Readers of this book, especially those living in Ireland, may be interested to know that the Geographical Society of Ireland has made three people Honorary members. Of these Dr Anthony Farrington has long been known for his work on the geology and geomorphology of Ireland, and for his practical interest in many societies in Dublin, not least the Geographical Society. Mr J. A. K. Grahame has given more than twenty years of devoted service to the Society and has founded a trust fund with a large donation. As the other Honorary member I can only express deep gratitude and record that work on the geography of Ireland has been a source of fascination for many years, both in the study and in the field.

T. W. FREEMAN

Department of Geography
University of Manchester
November 1968

Contents

Part I
THE COUNTRY AS A WHOLE

Part II
REGIONAL GEOGRAPHY

Maps and Diagrams

FOLDERS AT END

1. THE UPPER LAKE, KILLARNEY
The mountains are strongly glaciated and some semi-natural woodland survives in this area.

2. AN ESKER NEAR TRIM, CO. MEATH

3. FARM AND WINDSHORN TREES, CO. CLARE
The old cottages are now used as farm buildings.

4. CATTLE GRAZING IN PARKLAND, CO. MEATH

5. CATTLE FAIR, CASHEL, CO. TIPPERARY
This photograph was taken over twenty years ago. In many towns the fairs are declining, though they are not yet extinct.

6. CATTLE AUCTION, CO. MEATH
The modern auction is slowly replacing the fairs.

7. SHANNON POWER STATION, ARDNACRUSHA

This aerial view shows the power house, the head race and, in the right, the fish passage and the two 50-ft. locks for boats to ascend into the head race and the Shannon.

8. POWER STATION, BELLACORICK, CO. MAYO
Situated beside the Owenmore river, the station uses milled peat.

9. CREAMERY IN CO. CLARE

Taken over twenty years ago, this photograph shows a creamery typical of those in rural areas, though many of the farmers now bring their milk in cars (some incredibly old). Note the pile of turf (peat) by the roadside.

10. BALLYCLOUGH CO-OPERATIVE CREAMERY, MALLOW

Served by fourteen branches, this creamery produces chocolate crumb, milk powder, cheese, baby foods and butter. Other activities include the production of bacon and dairy feedstuffs, an artificial insemination plant and a cattle mart at Mallow.

11. MODERN FACTORY AT SHANNON FREE AIRPORT INDUSTRIAL ESTATE
Of Dutch origin, the factory is for the manufacture of pianos.

12. INTERIOR VIEW, GUINNESS BREWERY, DUBLIN

No attempt will be made here to explain this photograph inside the brewery claimed to be the largest in Europe, and exporting more beer (and stout is a beer) than any other in the world.

13. TANKERS UNDER CONSTRUCTION AT HARLAND AND WOLFF'S YARDS IN BELFAST

14. THE ROPEWALK AT THE BELFAST ROPEWORKS, THE LARGEST IN EUROPE

15. THATCHED COTTAGES, CO. GALWAY

The art of thatching survives in many rural areas: iron roofs perish in the warm and windy atmosphere. The young man is wearing a hand-knitted cardigan characteristic of the locality.

16. NEW HOUSING, CARLISLE DEVELOPMENT SCHEME, BELFAST
This shows an interesting combination of houses and flats.

Tables

Plates

PART I

The Country as a Whole

CHAPTER I

A General Survey

'England, that utmost corner of the west' is the description of Shakespeare in *King John*: how much more remote the island beyond, variously defined as 'the sister isle', 'Ireland the outpost', 'unhappy Ireland', or even 'John Bull's Other Island'! Mackinder shows in the first chapter of *Britain and the British Seas* that before the age of discovery the western margins of Europe formed the end of the world with the ocean as the last barrier known to man. Against the experience of centuries the records of the early Norse penetrations of America seem a mere incident; and Ireland, even more truly than England, was a western outpost, cut off from its nearest neighbours by sea. It lies, too, far from the southeast or 'continental angle' of Britain, and like the western part of the larger island is perhaps best considered as part of the Atlantic fringe of Europe, yet the Age of Discovery was not a major crisis in Irish life for most of the world's shipping passed it by. No Irish port emerged on the scene during the seventeenth and eighteenth centuries in a position comparable with ocean-minded Bristol or slave-trading Liverpool, though agricultural produce was sent out from various Irish harbours in exchange for imported manufactured goods and luxuries of various kinds. In short, neither the age of discovery nor the industrial revolution affected Ireland as closely as Great Britain, so that even now, in the last third of the twentieth century, maritime and industrial interests have not the power they possess in Britain.

Historically Ireland has a clear individuality that reflects its distinctive space relations, for it is not now and probably never could have been made merely a detached fragment of an entity known as the British Isles, or, as Professor A. G. Tansley called them, 'the British Islands'. Yet there are similarities in the physical environment, for Ireland is structurally the westward extension of Britain and climatically not infrequently Britain's anteroom, having a preview of the weather on the way to the larger island. Its natural vegetation, though floristically poorer and having certain local characteristics, is closely related to that of Britain, but as soon as the human landscape comes to the mind's eye contrasts emerge, for the farms are smaller, the towns farther apart, and industry everywhere of less significance than in Britain, if one excludes the city of Belfast and its hinterland. And there are deeper contrasts still, not least in religious allegiance, in social customs and in outlook which, though varied from one person or group to another, inevitably reflect a difference both of historical development from Britain and hardly less of the impact of current world problems and struggles on the country.

In the prehistoric period Ireland stood at the end of the world in even greater measure than Britain, and, so far as is known, no Palaeolithic immigrants ever arrived, though there are traces of Mesolithic settlements which appear to have been of small extent. In the Megalithic Age Ireland emerges more definitely on the European scene, having immigrants who must have come by sea, both from Scotland and from other parts of western Europe: the 'Atlantic sea routes', to use the term of Sir Cyril Fox, along which men travelled coastwise, making short passages across the open sea, were already important. On these routes the peninsulas of Brittany, Cornwall and southwest Wales, and the remoter havens of Spain and Portugal all figured; and there is strong evidence that Ireland, penetrated on all its coasts, had a steadily increasing population. This Megalithic phase was merely the forerunner of the Bronze Age, in which Ireland became a place of pilgrimage for its gold. To what extent settlers and traders came across the Irish sea by the several short passages, especially the fourteen-mile journey from Kintyre or the slightly longer run from Galloway, and to what extent the immigrants travelled the longer routes can only be guessed, but apparently all were used, and Ireland established a position of some consequence in the western world of the day. It lay beyond the sphere of Roman government and even of Roman invasion in the early centuries of the Christian era, though it retained contact with the remainder of the western European world, and there is evidence of emigration from Ireland across the Irish sea at this time.

An age of peculiar fascination follows that of the Roman invasion in Britain, for while England received the Anglo-Saxon invaders, Ireland remained the centre of a Celtic, western world in which Christianity acquired a strong monastic organization, and was so missionary-minded that its priests penetrated almost the whole of Britain and many parts of the continent: once more, the routes across the Irish sea and the western peninsulas of Britain, with Brittany, were used, and Irish 'saints' were known far and wide. At home, a territorial organization into small units known as *tuaths* developed, and the more ambitious rulers began to aspire towards domination of a large part or even of the whole of the country.

Norse invaders were at first ravagers of the land, but later peaceful settlers who, with prophetic acumen, chose Dublin as their chief centre, and Wexford, Waterford, Cork and Limerick as their four other main settlements. They fostered trade with ports in the west of Britain, notably with Bristol, and once again the Irish sea route had its voyagers, though the Norse people never conquered Ireland and in time were absorbed into the general population. After them came the Normans who, one century after William I came to England in 1066, crossed St George's Channel and began a penetration of the land which at first seemed likely to impose upon it the distinctive features of Norman organization. The Normans were creators of towns and framers of laws that differed considerably from the older native laws, but in the end they failed to make Ireland part of their domain, except for a small but important territory known as the 'English Pale', and in their twilight period the country was once again 'native', though it possessed more towns than ever before, had a diocesan system covering the whole of the land, and a county system that was at least embryonic and capable of expansion. The more serious days of Anglo-Irish relations still lay ahead.

Tudor and Stuart policy was to bring Ireland into closer relations with Britain by plantations or, in the case of Antrim and Down, by permitted immigrations. Once more, the northeast received a preponderance of settlers from Scotland, as during the prehistoric ages, but the immigrants elsewhere came mainly from England: the Cromwellian phase was the most crucial of all, as in it the new landlord class was established more firmly than ever before. Not far distant was the eighteenth century, the time when the landlords in their demesnes and the poor farmers or labourers outside them were the two main classes of the country: by the eighteenth century Ireland had more towns than in any previlous phase as both the Tudors and the Stuarts recognized the essential pace of the towns in any organized state. The Act of Union with Britain in 1800 seemed the logical climax of British policy towards Ireland, as from the Norman age the essential aim had been to incorporate the island into Great Britain for many reasons, including its economic value as an outlet for manufactures and as a producer of foodstuffs. But until the Belfast merchants attracted the coal and iron to build their ships, together with the raw materials for their manufactures, the impact of industrialism was slight.

The Act of Union was not, and could not be, the final solution, as the forces arraigned against it among large sections of the Irish population were too strong for it to remain. The partition between the six counties and the twenty-six gave a compromise solution in 1921 which, whatever its merits, is already deeply rooted in the life of the land. In effect, Northern Ireland is part of Britain and the twenty-six counties a republic from 1949. There are those who hope to create an all-Irish republic, Irish-speaking and having, it is said, no trace of the invasions and alien settlement of the past thousand years; they look back with admiration to centuries long past, such as that of the Celtic church, as times when Ireland had its own distinct individuality and experienced a golden age. What the future relations with Britain may be, one can only speculate; and the author will leave the reader to do that. Also of interest is the relation of Ireland and America, with whom the main contacts are neither maritime nor industrial but human, and therefore individual and personal, as there are three to four times as many people of Irish extraction on the west side of the Atlantic as at home. The westward location of Ireland has been of little significance as a source of commercial advantage: for a time it seemed that the country might become the great airport of Europe, but trans-Atlantic aeroplanes can now go direct to Britain. Even though the twenty-six counties are now a republic outside the Commonwealth, Ireland's closest human and economic relations are with Britain.

This book is divided into two parts, General and Regional: in Part I the features of the country as a whole are discussed, and in Part II there is a series of regional chapters. The subjects in Part I are structure and landforms, climate and weather, and vegetation which, taken together, are aspects of the physical environment; then follows an outline of the historical geography, distribution of population, settlement and society, agriculture, industries and communications, all of which relate to the social and economic life of Ireland. It is impossible to provide, even in a book of these dimensions, a full treatment of the historical development in each of the aspects studied, though whenever possible, some allusion is made to the years that have passed since 1845, when the Famine reversed the population trend and initiated the decline that has continued to the present. In some chapters such as population, industry and communications, a brief review of trends from the early part of the nineteenth century is given. Very clearly, the Famine years in Ireland are a crisis period of a magnitude comparable with that of the Industrial Revolution in Britain. In this book the Census year, 1891, is frequently mentioned, as by then the first great waves of emigration were spent; 1891 also saw the beginning of the classic investigations of the Congested Districts Board, which revealed the lamentable social and economic conditions of the poorest areas of the land, though happily since then social amelioration has removed most of the extreme poverty for which Ireland has long been notorious.

When this book first appeared in 1950 Ireland was far less known to visitors than now, though from the end of the war it has attracted an increasing number of tourists, mainly from Britain but also from continental Europe and from America. Thousands more are likely to visit Ireland when more accommodation is available in hotels, guest houses and cottages and also on car ferries and aeroplanes. The attractions of Ireland to tourists are unlimited. Vast stretches of sandy beach as yet sparsely occupied, fine mountain ranges for walkers, streams abounding with fish, rivers and canals for cruising, historic towns and planned demesne villages all have their appeal. To many, the attraction lies in something vague called 'atmosphere', partly explicable as the way of life and the character of the people.

A NOTE ON PLACE NAMES

Those who have visited the Republic of Ireland will be aware that each place has an English and an Irish name, and that the postmarks are in Irish only: they may also be aware that the Census form is bilingual, but that the income tax forms normally arrive in English. In the choice of place names for this book the name in general use is employed whenever possible, and the list of names given in the Census volumes has been taken as a guide in most cases, but not all. Thus Leix is preferred to Queen's County, and Offaly to King's County: similarly, Dun Laoghaire is used instead of Kingstown and Cobh rather than Queenstown. Difficulties arise in the choice between Navan and An Uaimh, Co. Meath, as the Census gives the former for the rural district and the latter for the town; again, Bagenalstown (Co. Carlow) is still in general use, though efforts are being made to popularize Muinebeag. Some familiar old names die hard, and the only concern of the author is that every place may be located: in Northern Ireland only one name, the English form, appears on the railway stations, in the Census returns, and in the various other official documents, so no difficulties arise.

CHAPTER 2

Structure and Landforms

Grenville Cole spoke of Ireland as 'the last outpost of Eurasia against the oceanic depths of the Atlantic'. Whatever its past relation with other land-masses of the North Atlantic may be, its closest physical link is with Great Britain. Ireland has two main structural lines etched out in its mountains, valleys and coasts: the older or Caledonian trend is from northeast to southwest, and the later or Armorican trend is from east to west. These, shown in figs. 1 and 2, give an essential key to the layout of the country's physical features. Ireland's mountains are less continuous than those of Scotland or Wales, and consist largely of lines of upland surrounded by lowlands. Even so, the distant vista of hills is rarely absent, except on dull days in the vast Central Lowland.

Many of the country's uplands lie near the coast, with the unfortunate result that it has been frequently compared to a saucer in elementary textbooks. Some of the uplands are clearly visible from the surrounding seas. In the east the Antrim basalt plateau overlooks Belfast lough and the North Channel, the Mournes and the Carlingford mountains rise above the fiordic lough which separates them, and the Leinster chain reaches Dublin bay; in the south and southwest the long Armorican chains stretch from the cliffed shores of Waterford to the great peninsulas that lie on either side of the Kerry rias; in West Connacht the metamorphic rocks appear both in low shores and in quartzite cliffs such as those of Achill, and in Donegal there are similar cliffs, but some low shores also. Long sea inlets lead to mountains and lowlands in the west, and the coast, with its mixture of mountain, sea, peninsulas and offshore islands, is one of particular beauty. (See geology map, Folder 1, at end of book.)

The interior of Ireland is essentially a lowland and lies mainly around 200 ft., rising gently to 400 ft. and varied by a number of low hills 600–1,000 ft. high. In this book the term 'plain' has been avoided, as it suggests a flatness which is not characteristic of the centre of Ireland: rather is there considerable topographical variety for the rockfloor is uneven, and the cover of glacial and glacifluvial materials adds to the diversity. Where the drifts are absent, as in part of the lowland near Galway bay, the scene is

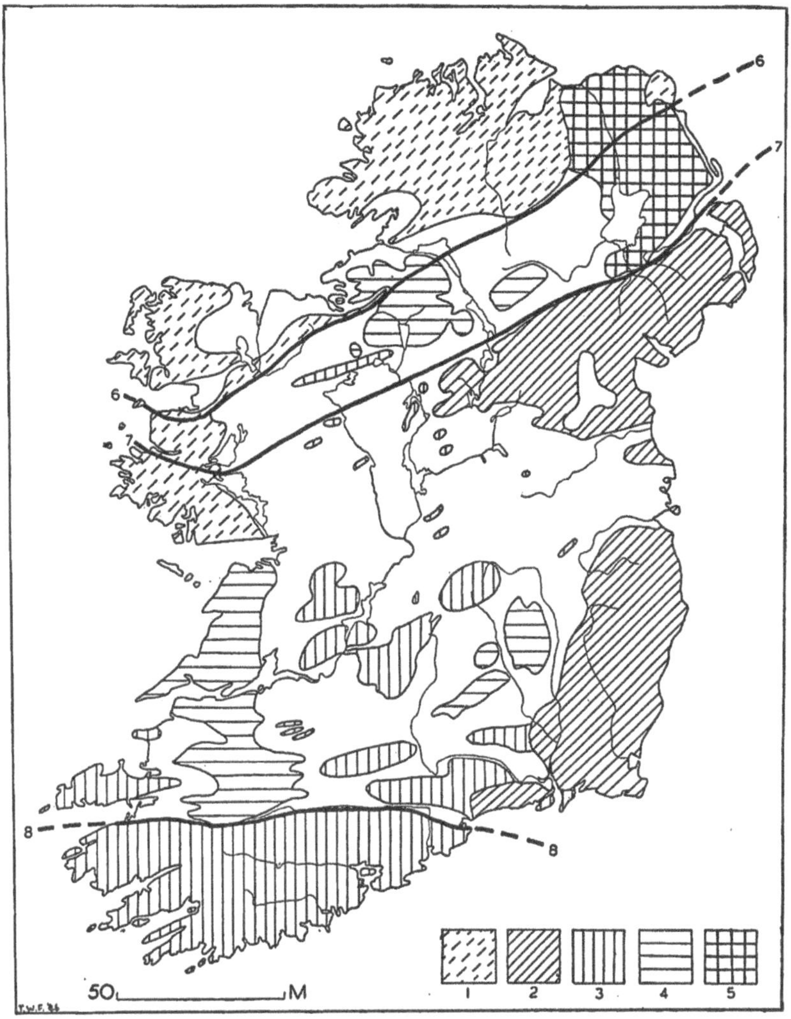

FIG. I. STRUCTURAL ELEMENTS

1, Western Caledonian masses; 2, Eastern Caledonian masses (Newry axis and Leinster chain); 3, Armorican uplands with intervening lowlands in the south; 4, Carboniferous plateaus; 5, Basaltic plateaus and lowlands; 6 and 7, continuation (6) of Highland Boundary and (7) Southern Upland boundary faults (6 *and* 7 *after* *E. B. Bailey*); 8, see text, pp. 20–1.

karstic, but the underlying rockfloor is rarely visible elsewhere, and the drifts are as much as 200 ft. deep in places. Variety is therefore the main feature of this lowland, which may be roughly delimited by a line drawn from Dundalk to Boyle, Co. Roscommon, and thence in two concave curves from Boyle to Galway and from Galway to Dublin. The Central Lowland is the heart of Ireland, and routes along the dry ground formed of drifts or eskers have been followed for thousands of years: in addition, easy passage-ways along valley and lowland corridors lead from it to every Irish shore. South of the Dublin-Galway line, for example, tongues of lowland spread round the various uplands and lead to the fertile valleys of Munster.

Around the lowland there are several types of landscape, each owing its distinctive character to its structure and history. The ranges west of the Leinster chain, though of Armorican structure, have the Caledonian trend, and it is only in the extreme south, between Co. Waterford and Kerry, that the trend as well as the structure becomes typically Armorican. Contrasting with these ranges are the plateaus of upper Carboniferous strata which survive as uplands in the 'Leinster coalfield', and in a stretch of country between Galway bay and Killarney: of little economic value or aesthetic attraction, these dreary plateaus lie off the main commercial and tourist routes. To the west, the Carboniferous limestones give place to a variety of ancient sedimentary and metamorphic rocks with many igneous intrusions, and the green fields or karsts of the Lowland are replaced by the bog-covered lowlands and towering mountains of West Connacht. In Clew bay, however, a stretch of the Central Lowland, made up of drumlin-covered limestones, reaches the coast to divide West Connacht into two parts: in the lower Moy basin there is also a drift-masked Carboniferous limestone lowland, separated from the centre of the country by the Ox mountains, a long tongue of sedimentary and metamorphic rocks with a clear Caledonian trend. The finest of the Carboniferous plateaus of Ireland, which includes considerable areas above 2,000 ft., lies between the Curlew hills and Donegal bay: in this country there are numerous escarpments of limestone or Millstone grit, some Coal Measures of quite minor economic importance and a number of corridor-valleys, of which the greatest, the drumlin-covered and lake-strewn Erne valley, opens a natural avenue from Donegal to the Ulster basin around lough Neagh.

The historic province of Ulster has a number of hill masses, but is primarily a lowland. In the northwest, consisting of the Foyle basin and Donegal, there are metamorphic rocks comparable with those of the

Scottish Highlands, and, like them, intruded by great spreads of granite. In Donegal, as in Mayo, quartzites stand up in conical peaks, such as Errigal, and the Caledonian trend appears clearly in such ranges as the Derryveagh, the Glendowan and the Urris. The Sperrins, too, are Caledonian in structure. This line is accentuated by the deep valleys that lie parallel to the mountain chains, which are penetrated by occasional transverse cuts. In the east the long tongue of Ordovician and Silurian rocks that extends through Down, Armagh, Monaghan and Cavan (the 'Newry axis') is also a Caledonian feature comparable with the Southern Uplands of Scotland or with the Leinster chain. The worn-down stump of a Caledonian range, it survives as a hilly upland, rarely more than 1,000 ft. high, except where igneous intrusions form the Slieve Croob, the Mournes, and the Tertiary ringdykes of Slieve Gullion and the Carlingford peninsula.

Analogues of the Scottish Highlands and the Southern Uplands are bounded by faults correlated with those defining Central Scotland (see Fig. 1): thus, the 'Highland Boundary' fault skirts the south side of the Sperrins, and the 'Southern Boundary' fault runs along the edge of the Holywood and Castlereagh hills. The intervening basin of sedimentation has Devonian sandstones which form an upland to the northeast of Enniskillen, lower Carboniferous beds in the Monaghan corridor and upper Carboniferous strata in the Slieve Beagh. Later sedimentary rocks, which cover a large coalfield beneath lough Neagh, include Permian, Triassic, Liassic and Cretaceous beds: of these the Triassics floor the Lagan valley and extend through the Dundonald gap to the shores of Strangford lough. Over these beds lie the basalts, which form scarps 1,000 ft. high facing lough Foyle in the west and the Lagan valley with the Antrim coast in the east: in the centre of the basalt country is the Bann valley which has, around lough Neagh, some clays of later date than the basalts which have not been precisely dated, though they are pre-glacial. The last great episode in the evolution of the Ulster landscape was the Quaternary glaciation, which left the lowland plastered with a variety of sands and gravels, and also with swarms of drumlins in parts of the Lagan and Bann valleys; the surrounding plateaus have meltwater channels and lake beds, notably in the Glens of Antrim and in the neighbourhood of Ballycastle.

Ireland's topographical variety is the result of its complex geological history, in which the Caledonian and Armorican influences emerge more obviously than the Alpine storms of Tertiary times, which were felt rather as a phase of volcanic activity and fracturing than of folding: it is not improbable that the Tertiary movements separated Ireland from Britain and left it as an island on the continental shelf of Europe. The icesheets of Quaternary times spread over a land surface that was at least broadly comparable with the Ireland of today in form and outline. The greater part of the land was covered by ice at some time, though not necessarily at the same time, and it has recently been suggested that an area extending inland from Dingle bay was never glaciated at all (see p. 27). In the glaciated mountains, the imprint is seen in overdeepened valleys, some of them moraine-strewn and partly covered with the clays of late-glacial lakes; in the lowland there are sheets of ground moraine, drumlins, eskers, kame and kettle moraines and, hardly less important, the channels of late glacial meltwaters. The present drainage therefore shows, in most of the lowland, the imperfect adjustment so far achieved during the post-glacial phase. It is, above all, the glacial influence that provides the final elements of topographical diversity in the Irish scene.

STRUCTURE AND TOPOGRAPHY

The geological evolution of Ireland is outside the scope of this work and may be studied in the works by G. A. J. Cole and T. Hallissy, J. K. Charlesworth and W. E. Nevill, listed in the bibliography (see p. 508). There can be no doubt that various peneplane levels exist in Ireland, but considerable research is needed before their extent and form can be determined: such research is made difficult by the varying thickness of the drift cover. The highest and least certain of these levels, around 2,000 ft., is said to be Devonian: it is seen on Slieve League, Donegal and on various hills in south Mayo, where the last relics of Carboniferous strata are now being removed. The second level, around 600–800 ft., was called the South Ireland peneplane by Miller, on account of its wide development in the Armorican area: Cole and Hallissy suggested that it originated by the degradation of a Cretaceous cover. Another peneplane level, more extensive than either of the above, is seen around 200 ft. in the Central Lowland and throughout the low-lying country that spreads around the various uplands. This peneplane truncates rocks of widely divergent lithological composition, and occurs alike on igneous, metamorphic and sedimentary rocks, and especially on Carboniferous limestones. In parts of the Central Lowland it rises to 400 ft. in gentle undulations, but while many traces of a levelling around this height have been seen, the recognition as a separate surface of erosion, though possible, is premature. These three levels, 2,000, 800 and 200 ft., are sufficiently prevalent to provide one key to the

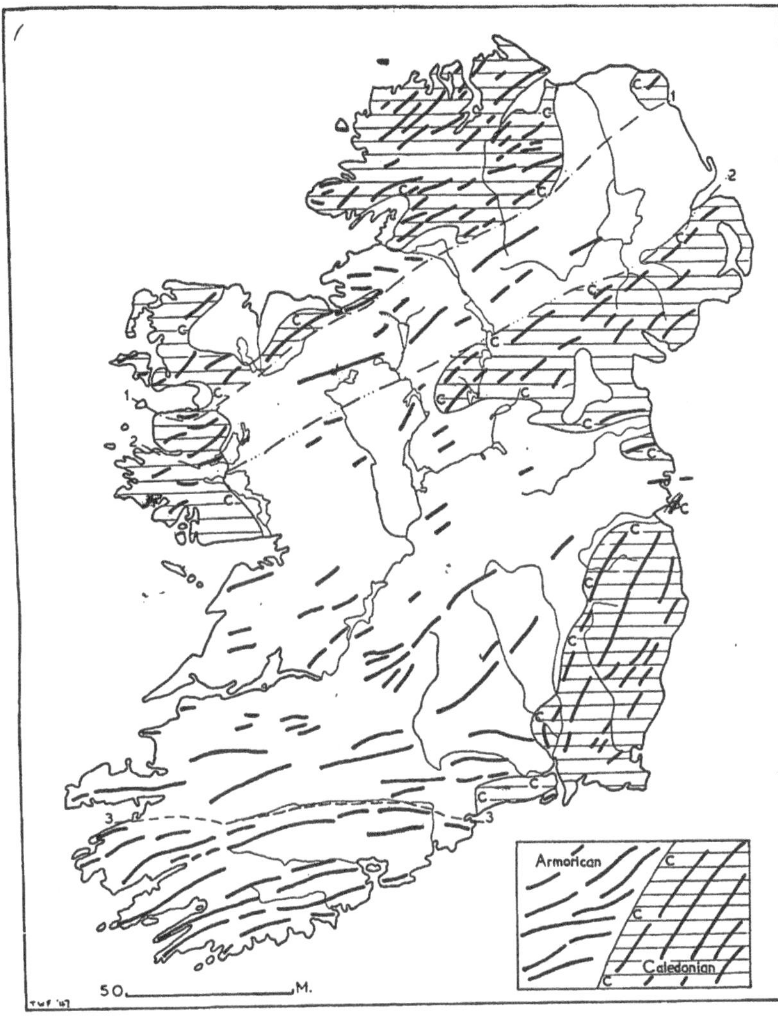

FIG. 2. TREND LINES IN CALEDONIAN AND ARMORICAN IRELAND

For 1, 2 and 3, see 6, 7, 8 in Fig. 1.

The Armorican east–west line is clearest in the south and southwest; the Caledonian line is best marked in the Leinster chain, the Newry axis country, the Ox mountains, and in the uplands of the extreme northwest.

regional geography in many parts of the land. Several of the rivers, such as the Nore, Suir, Barrow, Slaney and Boyne, flow in sluggish courses across the 200 ft. peneplane, and then incise themselves in it to reach the present sea level. Again, the Cork rivers cut through gorges in the 800 ft. and 200 ft. surfaces, and are fed by tributaries that are torrent courses in their lower reaches, and are steadily working back into the peneplanes through youthful valleys. Many of the rivers are now cutting through the Quaternary deposits, but the rockfloor has been reached in places.

The Irish landscape shows five main structural elements, each of distinct age and form (see Figs. 1 and 3); they are here indicated categorically, and then treated more fully.

1. *The Western Caledonian Masses* (northwest Ireland and West Connacht), pp. 14–17. They consist largely of pre-Cambrian, Ordovician and Silurian rocks, subjected to several phases of metamorphism by pressure and igneous intrusions. Long-continued weathering has produced extensive lowlands, but quartzites form peaks, and Caledonian mountain-building movements have produced uplands which generally have a northeast–southwest alignment.

2. *The Eastern Caledonian Masses* (the 'Newry axis' and the Leinster chain), pp. 17–19. The two masses possess a granite core intruded as the Caledonian folds arose. The Newry axis is generally hilly with a few mountainous areas such as the Mournes, where the granites were intruded in Tertiary times. The Leinster chain is a central upland surrounded by lowlands having, on the east, lines of hills marking igneous outcrops.

3. *The Armorican Fold Mountains*, pp. 19–23. Three groups of uplands may be distinguished: (*a*) the southwestern block, defined by a structural line from Dingle bay to Dungarvan; (*b*) the isolated uplands between the Leinster chain and the lower Shannon; (*c*) hills exposing Devonian inliers of the Central Lowland.

4. *The Carboniferous Plateaus*, pp. 23–5. Plateaus of upper Carboniferous strata survive in three areas: (*a*) Castlecomer–Slieve Ardagh, or the Leinster coalfield: (*b*) from Killarney to Galway bay, intersected by the Shannon; (*c*) Sligo, Leitrim, Fermanagh and Monaghan.

5. *The Tertiary Basaltic Area*, pp. 25–6. Lavaflows of basalt in the northeast form plateaus and lowland: around their scarped margins Carboniferous and Mesozoic strata outcrop and form lowlands.

1. THE WESTERN CALEDONIAN MASSES

These old masses, which are the foundation stones of Ireland, were already reduced to base level by Ordovician times, but were raised into fold mountains by Caledonian movements, which were accompanied by faulting and volcanic eruptions. They show a northeast–southwest trend, widespread faulting in the same direction or at right angles to it, and also differential weathering between quartzites, granites and schists. In West

FIG. 3. MAJOR LANDFORMS

1, Highland type, sharp peaks; 2, Major uplands; 3, Minor uplands; 4, Carboniferous plateaus; 5, Basalt plateaus; 6, Basalt lowlands; 7, Lowlands, pre-Carboniferous rocks; 8, Lowlands, Carboniferous and later rocks.

Connacht, quartzite mountains rising above a lowland around 200 ft. are the main feature of the landscape, but in Donegal the Caledonian line stands out clearly in a number of quartzite ranges with conical peaks, such as Muckish, Errigal, Aghla, Knockalla, Urris and Slieve Snacht. Slieve Tooey and Slieve League, in the southwest, have been bisected by marine attack and present precipitous scarps, rather than actual cliffs, from their summits to the sea. The granites intruded between Ardara and Mulroy bay form hills of smoother outline, such as the Blue Stack mountains, the Glendowans and the Derryveaghs: from many of the lowlands in Donegal it is possible to observe the contrast between the smoothly moulded granite hills and the conical peaks of the quartzites. In the Errigal range and the Slieve League peninsula there is some of the wildest and most typically highland scenery in Ireland, and the landscape is almost as mountainous around the Blue Stack and the Aghla ranges, the Sperrins, and also in the Fanad and Inishowen peninsulas. Much of the north-western Caledonian mass, however, consists of plateaus rather than mountains; and there are extensive coastal lowlands with occasional hills. In all probability the 800 ft. and 200 ft. surfaces are widely represented in Donegal: the presence of the 2,000 ft. surface on Slieve League has already been noted. Glacial scouring has moulded both the mountain valleys and the coastal lowlands in the west, but deposition is more apparent in the east, where the rivers are youthful streams heavily charged with detritus, which is laid down in sheltered loughs and bays.

The metamorphic mass of Connacht, which is divided into two sections by Clew bay, has an impressive offshoot in the Ox mountains. The northern part of the mass, made up of metamorphic rocks, has been worn down to hills of 800–1,200 ft., or to lowlands, except in the wonderful quartzite peaks of Achill and west Mayo. Croaghaun, at the west end of Achill island, falls precipitously from its summit of 2,192 ft. to the sea; other quartzite hills, conical and isolated, exist in the Nephin beg range, and, a little farther east, in solitary Nephin, 2,464 ft., the highest peak of all. These mountains, largely bare rock, possess a starkness of outline that is accentuated by isolation in the surrounding lowlands and sea: unlike the Errigal range, they stand alone, revealing the resistance offered by quartzite against all forms of denudation.

Ordovician and Silurian sediments are exposed between Croaghpatrick and a line south of Killary harbour. These mountains have a smoother outline than the quartzite peaks; on them rest the last remains of a Carboniferous cover and they probably possess, around 2,000 ft., relics of a peneplane. Faulting has broken the upland into fragments separated by deep valleys: two of the major faults have been correlated with the lines of fracture that limit Central Scotland. The Highland Boundary fault runs to the south of the Croaghpatrick range and along the landward face of the Ox mountains, which is a moorland around 1,500 ft. The Southern Boundary fault runs from the north shores of lough Conn westwards; here the Silurian beds fade out, and the pre-Cambrian floor of Connemara is revealed.

In west Connemara the Twelve Bens (misnamed Pins) rise above a peat- and lake-covered lowland as a splendid group of quartzite peaks more than 2,000 ft. high. The Maumturk range, also of quartzites except in the north where Silurian slates appear, has been described as 'the tangled cluster of the Twelve Bens stretched out in a curved line with their tops rubbed off'. Elsewhere lowlands prevail, though the Galway granites form country rising in gentle lines to heights of 1,000–1,200 ft. Connemara offers some of the most complex problems imaginable to the geologist, and may prove to be entirely pre-Cambrian in origin and present form, for it has no known relics of later deposits.

2. THE EASTERN CALEDONIAN MASSES

The two Caledonian masses of eastern Ireland, the Newry axis and the Leinster chain, are comparable in age and formation, but sharply different in present form. The Newry axis is mountainous only where igneous intrusions are exposed at the surface in the Carlingford peninsula, the Mourne mountains and Slieve Croob, Co. Down: elsewhere it is a lowland rising into low hills. The Leinster chain, on the other hand, stretches as a continuous upland for over seventy miles from Dublin to Co. Waterford, with numerous summits over 2,000 ft. The Newry axis was protected from erosion by a cover of Carboniferous and probably Triassic beds, but the Leinster chain has been attacked by denuding agents, probably without interruption, from Devonian times. Consequently, Silurian strata have been removed from the Leinster chain and Ordovician beds now survive only on its margins: the Newry axis, on the other hand, still retains most of its original Silurian cover.

The Newry Axis. The Newry axis extends from the coast of Co. Down through south Armagh, Monaghan and Cavan into Co. Longford, and finally disappears a few miles east of the Shannon, near the town of Longford. Granites of Devonian age appear at the surface in a belt around Slieve Croob, and form hills reaching 1,755 ft., but otherwise are hidden except at Crossdoney, near the town of Cavan: this reappearance shows that a long bar of igneous rock occupies the core of the Caledonian fold from Longford to the sea. The Ordovician and Silurian strata are, in fact, highly contorted, but the general Caledonian trend is clearly expressed in the present landscape. Drumlins, however, are so numerous that they dominate the scene: as many of them are 100 ft. high and some have a rock core, it is difficult to imagine and impossible to reconstruct the form of the rockfloor. In Cavan, Monaghan and Longford, long ridges rise to a height of 800–1,000 ft.: in Armagh and Down, there are three main topographical elements, (*a*) a low-lying surface, apparently continuous with the Central Lowland, now existing at 200 ft. or less, and well developed in the Ards peninsula; (*b*) hills around 600–1,000 ft., probably belonging to the 800 ft. surface, which is also seen on the west of the Mournes, and (*c*) the hilly country formed of intruded rocks, of both Devonian and Tertiary date. As noted above, Devonian vulcanicity is represented in Slieve Croob and the surrounding area. The Tertiary vulcanicity has left strongly marked features in two ringdykes, and in the granite knot of the Mournes. The Slieve Gullion ringdyke has a central hill of 1,894 ft. with a series of smaller hills 750–1,000 ft. high around it: in the Carlingford peninsula, immediately adjacent to the Slieve Gullion area, part of a second ringdyke appears in gabbro hills. Both these areas were also intruded by granites in Tertiary times; in the Carlingford peninsula the granites form a basin, but in the Mournes they survive as dome-like crests with pinnacles and crags above long talus slopes. The Mournes, a range of less mature outline than the Wicklows, have a clearly marked northeast–southwest trend.

The Leinster Chain. The Leinster chain is a vast anticlinal arch, with a granite core, flanked by Ordovician sediments metamorphosed into mica-schists near the actual junction; long tongues of schist are included in the granite and Lugnaquilla has a schist cap on its summit. Cambrian rocks on the east side of the chain consist of slates that have weathered into low-lying country, now drift-covered and fertile, and of quartzites forming knob-like bosses on Howth head, a line of three peaks on Bray head, the Little and the Great Sugar Loaf, the summits on the edge of the Vartry peneplane, and the Forth mountains, Co. Wexford.

This anticlinal arch is perceptible as a structural feature for over ninety miles from the coast of Waterford to the tiny island of Rockabill, off Balbriggan in the north of Co. Dublin; for seventy miles it is a well-marked upland possessing, in the Wicklow mountains, the largest area of continuously high ground in the country. Ordovician and Silurian sediments were raised into an arch by Caledonian movements and, as the arch arose, a granite magma was intruded along its core. Up to this point, its development was similar to that of the Newry axis, but henceforth its history was on quite different lines. During Devonian and Carboniferous times, the Silurian strata was stripped off, the granite exposed, and the whole mass seared by deep valleys. Some of the granite pebbles were incorporated in the Carboniferous beds. The denudation was continued from Carboniferous times onwards, and the glaciations, perhaps three in number, added their moulding influence to that of other agencies. The chain, therefore, is now an interesting mixture of mature and youthful features, for its crests, so well seen from the surrounding lowlands, are smooth and gently moulded curves where the granites are exposed, and the deep glaciated valleys are a contrasting and youthful feature. The quartzite hills maintain their distinctive form around Bray, and there are already signs that a line of hills is associated with the junction of the granites and the mica-schists: if this be so, the granites will eventually be weathered into a lowland with a fringe of mica-schist uplands.

The smooth upland surfaces of the Wicklows may be part of an ancient peneplane: they show a general accordance of summit levels. The 800 ft. level may be recognized in the Vartry and Blessington basins (see pp. 298–300), both of which are areas sloping from 1,000 to 600 ft. The 200 ft. level, clearly seen on both flanks of the mountains, truncates rocks of widely different resistance: a considerable flattening has been observed around 400–600 ft. also. But the relationship of all these possible surfaces is at present obscure, and their recognition is complicated by the difficulty of mentally removing a drift cover of varied form and thickness.

3. THE ARMORICAN FOLD MOUNTAINS

In the Carboniferous period Ireland was almost completely submerged except for a few islands such as the Leinster chain; and Carboniferous limestones still cover approximately half the surface. This phase of geological history is therefore important on this count alone, and the Armorican folding that followed is equally significant. The first indications of these movements was the emergence of Ireland above the sea level, to which we owe the presence of Coal Measures (but not of much coal) which have been largely removed by denudation. The Irish landscape still shows the lines of Armorican folding with remarkable clarity in a number of anticlines and synclines in which Carboniferous beds directly overlie Devonian sandstones, the former in the valleys, and the latter as ridges. The correspondence of the older rocks with the ridges and the younger with the valleys is remarkable, but, as Cole has pointed out, fortuitous, for had the Carboniferous rocks been quartzose conglomerates and the Devonian rocks mainly limestones, the upstanding ridges would have been synclinal and the valleys would lie along the crests of the anticlines. The parallel ridges of the south of Ireland owe their preservation, therefore, to the petrological character of the Old Red sandstone. Rocks of Silurian and Ordovician date, mainly shales and sandstones, appear as inliers in the core of the anticlinal structures; they have weathered into basins surrounded by hills of the more resistant Old Red sandstones. The Comeraghs, for example, and craggy hills of lower altitude, rise above a basin of Silurian and Ordovician rocks. Similarly, on the other side of the Suir valley, the peak of Slievenaman and a number of hills, all of Old Red sandstone, guard an upland basin that is the Silurian and Ordovician core of the anticline. Again, in the Galtees, the relation between structure and surface is equally impressive: the high mountains, culminating in Galtymore, 3,018 ft., are of Old Red sandstones, but to the west these beds have been removed to expose a basin almost entirely surrounded by the last marginal relics of the former Devonian cover. Almost all the Armorican ranges farther north have peaks of Old Red sandstone and upland basins where the older rocks are exposed.

The Central Lowland is primarily a rolling drift country with a rock floor of Carboniferous limestones, through which older rocks appear in anticlinal ridges of Old Red sandstones resting on Silurian and Ordovician beds. The adjustment to structure differs from that of the south for, as yet, the anticlines are just being revealed and marked out as hills. Though differing in present landforms, one may regard the various parts of Armorican Ireland as one structural unity: the Central Lowland may show the condition possessed by the south at an earlier phase, before its Carboniferous cover had been largely removed. Finally, it has been suggested that the underlying structure of the country between the Caledonian masses of the northeast and northwest is also Armorican in, for example, the coalfield beneath lough Neagh. Armorican folds, in short, dominate the scene in the south, are just perceptible as landform features in the centre, but are almost completely hidden in the north.

The Armorican ranges of Ireland fall into three groups as structural and landform units. They may be classified as (*a*) the southwestern block, (*b*) the isolated ranges south of the Dublin–Galway line, and (*c*) the minor hills of the Central Lowland.

(*a*) *The Southwestern Block.* This is clearly marked off as a structural feature by a more or less northward overthrust of Old Red sandstone over Carboniferous beds: this important tectonic line, which runs from

Dingle bay to Dungarvan, is continued into Pembrokeshire and probably into northern France, where it may mark the northern limit of the coal-fields. Within this area, the Old Red sandstone anticlines exist as low ridges in the east and as mountains farther west, where several peaks reach 2,500–3,000 ft., culminating in Carrauntoohill, 3,414 ft., the highest mountain in Ireland. The Carboniferous strata are varied: in the synclines forming the valleys of the Blackwater, the Bride, and other rivers around Cork and Youghal, there are limestones, but to the south there are shales hardened by pressure metamorphism into slates. These slates have been weathered into a landscape of valleys and low ridges, best developed in the extreme south of Co. Cork: they include some sandstone beds, such as the Coomhola grits, which display a sharp resistance to erosion and stand up as hilly ridges (see Fig. 65, p. 343). This southwestern block shows Armorican structure in its most interesting form: so far, none of the drainage has cut through the Old Red sandstone beds to reveal the underlying Silurian and Ordovician floor, and the most striking evidences of past cycles of erosion are seen rather in the peneplaned ridges around 200 ft., 400 ft. and 800 ft., above which level the mountains rise. The river system, evolved in pre-glacial times, has returned to its original course almost everywhere, but is now cutting through a variety of drifts and glacifluvial deposits.

(*b*) *The Isolated Ranges.* West of the Leinster chain and north of the Dingle bay–Dungarvan line, Armorican structures are etched out in a number of isolated ranges and plateaus which rise above the general level of the Carboniferous lowland. The two Carboniferous plateaus, both of them affected by Armorican folding, are discussed on pp. 23–5: the ten major hill masses are comparable in structure with the mountains of the extreme southwest, but towards the north are bent into a trend parallel with the Leinster chain. In north Kerry, there are two peninsulas that are isolated uplands approached by lowland routes: of these the Dingle peninsula has an unbroken ridge in the Slieve Mish (Baurtregaum, 2,796 ft.) and, farther west, mountainous ridges culminating in Brandon, 3,127 ft., together with some isolated peaks and considerable areas of lowland. The 'Dingle beds' of this peninsula have been ascribed to at least three periods (Ordovician to Devonian), and its structure is not clearly known. The second, and much smaller, Kerry head peninsula, on the south side of the Shannon, is an Old Red sandstone ridge some 700 ft. high; its trend is bent slightly to the south, like that of the Dingle peninsula.

Three narrow ridges of Old Red sandstone rise to more than 900 ft. in the heart of the Limerick lowland: farther east, the Ballyhoura–Galtee upland has an interior basin where the Silurian and Ordovician rocks appear, but the mountain character prevails in the Galtee range. Parallel to the Galtees is the Slievenamuck range. The association of Old Red sandstone mountains with basins of earlier rocks is also a feature of the Slievenaman range, and of the long chain of hills known in its various parts as the Kilworths, Knockmealdowns, Comeraghs and Monavullaghs. The 800 ft. level is extensively developed around these mountains, which rise to more than 2,000 ft.: the Comeraghs have wonderful corries eaten into the heart of the range with cliffs as much as 1,000 ft. high.

The bend of the Armorican folds into conformity with Caledonian trends becomes marked north of the Slievenaman–Galtee line; Cole suggests, in explanation, that the old Caledonian mass was largely eroded away in the south but still extant as crumpled hills farther north, and therefore able to influence, indeed to control the direction of the Armorican folding. In any case, the simulation of a Caledonian trend gives the countryside its characteristic stamp: roads and railways from the valleys of Munster all swing northward in their journey to Dublin. The range of hills, variously named, from Slievefelim to the Roscrea gap, the Slieve Bloom, the Arra mountains and the Slieve Aughty, all show a northeast trend; their summits vary from 1,207 ft. in the Slieve Aughty to 2,278 ft. in Keeper hill. They are for the most part irregular groups of gently swelling hills composed of Silurian strata, chiefly slates and grits, and made conspicuous in places by more or less horizontally imposed and truncated cappings of Old Red sandstone, which form such landmarks as Keeper hill, the highest summit, and the Devil's Bit mountain near Templemore. These blocks of hills owe their diversity to the superior resistance of Old Red sandstone over the Silurian beds, but they do not possess the basin and range structure of the Slievenaman, the Comeraghs and the Galtee–Ballyhoura mountains, partly because they have been influenced by powerful fault movements.

(*c*) *Armorican Hills of the Central Lowland.* The Central Lowland possesses considerable topographical diversity: its hills consist either of monadnocks that are relics of the middle and upper Carboniferous cover, of small volcanic bosses such as Croghaun, 769 ft. between Edenderry and Tullamore, or of Armorican ranges in which Devonian strata are revealed. None of these rises to an altitude of even 1,000 ft. or covers more than a few square miles, yet most of them are conspicuous features. In some of the hills the Old Red sandstones have been removed, and the underlying Silurian strata exposed; a few have igneous intrusions. The fourteen inliers of this type include the chain of hills near Kildare, Lisduff, and Ardagh (near Longford) and, on the west side of the Shannon, the Slievebawn range, which has a maximum elevation of 864 ft. The Curlew hills (818 ft.) extend as a continuous range for nearly thirty miles from lough Allen westward, but the smaller inliers of East Connacht rise only slightly above the general lowland level.

4. THE CARBONIFEROUS UPLANDS AND LOWLANDS

In all, Carboniferous outcrops occur over almost two-thirds of the country and it is not therefore surprising that many problems still await detailed geological research. But for the purpose of this general account, two major points are significant. First, the Carboniferous beds of Ireland, like those of Britain, have a clear division into a lower limestone series and an upper deltaic series, the latter represented by shales, grits and coal measures. Traditionally these upper beds were correlated with the Coal Measures, Millstone Grits and other beds recognized in Britain but it is now doubted whether this is sound. Even so, the plateaus are of upper Carboniferous beds and it remains convenient to use the terms Millstone Grits and Coal Measures. Second, although the lower Carboniferous beds are classed mainly as limestones, they include some shales and sandstones, and are of widely varying types. North of a line from Dublin to Galway they consist largely of muddy limestones, probably as far west as the Shannon, but clear water limestones appear in Co. Roscommon and in the countryside between Galway bay and the Ox mountains. Northwards again, around Sligo and Donegal bays and inland through the lowlands, the muddy limestones with shales and sandstones occur. South of this line there are bedded limestones with oolites and dolomites.

The Castlecomer–Slieve Ardagh Uplands. These form a somewhat broken ridge, of Armorican structure but Caledonian line, entirely surrounded by Carboniferous lowlands, above which they rise in sandstone escarpments that are particularly well marked on the west side of the Barrow valley. The uplands are really synclinal basins with the Coal Measures in the centre, notably in the Castlecomer area, which is drained by the river Dinin to the Nore. The Slieve Ardagh hills are cut by several gaps, one of which carried a railway line from Fethard to Thurles, but the whole of the Leinster coalfield, as it is somewhat optimistically called, forms a clear feature of the landscape from the Stradbally hills to the Rock of Cashel. (The coalfields are discussed on pp. 325–6.)

The Killarney–Galway Bay Plateaus. These plateaus, also of upper Carboniferous strata, have rounded summits around 1,000 to 1,200 ft., and a maximum elevation of 1,441 ft. near the source of the Blackwater on the Cork–Kerry border. They form long sweeping lines above the lowlands that surround them, and have an escarpment edge on the west side of the Limerick lowland and, less clearly, in central Clare. South of the Shannon the plateau is penetrated by wide valleys and consists primarily of Coal Measures with flagstones: in Clare, the Coal Measures remain only in isolated patches such as Slieve Callan, 1,282 ft. The flagstones exposed over a large part of southwest Clare form undulating plateau country which reaches the sea in the magnificent Cliffs of Moher, which are 600 ft. high. In northwest Clare, however, both the Coal Measures and the flagstones have been removed, and the underlying limestones exposed in the Burren, a karstic area consisting of an upland with ridges around 1,000 ft., wide waterless valleys reaching Galway bay and a narrow coastal lowland on the Atlantic shore. Only one small river, the Caher, reaches the sea by a drift-filled valley, but the underground waters flow through caverns and appear intermittently at the surface. A recent investigation revealed a cave three and a half miles long, the longest so far discovered in Ireland. The Burren is remarkable not only for its landforms, but also for its vegetation, which includes Alpine and Mediterranean plants in association, and shows the direct effects of wind exposure, for it is planed to ground level on the coast, but reaches woodland status in sheltered hollows (see pp. 377–80).

The Sligo–Monaghan Uplands. These Carboniferous uplands are strongly faulted and separated by deep, wide valleys with numerous lakes: some of them rise to 1,500–2,000 ft., and their sharp escarpment edges mark them off very clearly from the valleys by which they are penetrated. The escarpments are variously formed of limestones and Millstone Grits, and have weathered into long talus slopes. The most westerly outposts of the uplands are two hills near lough Arrow, Carrowkeel and Keishcorran, 1,057 ft. and 1,188 ft. respectively, and Knocknarea, 1,078 ft., overlooking Sligo bay. From these hills, uplands extend for nearly seventy miles eastward to Slieve Beagh on the fringe of the lough Neagh lowland. The area is drained by the headwaters of the Shannon, by the middle and lower Erne with its tributaries, and by various short rivers flowing to Sligo bay. The valleys have been widened and deepened by ice, and possess sheets of ground moraine with numerous drumlins, which appear as islands in the lakes, especially in the two Erne loughs. Some of the limestone uplands provide sheep pasture of good quality, but otherwise the plateaus are peat moors; they are of little economic value except in the Arigna coalfield, where workable seams exist both in the Coal Measures and in the Millstone Grits on either side of lough Allen (see p. 432). Scenically, however, this area of Carboniferous plateaus is far more attractive than either of the plateau areas discussed above, for the combined work of fracturing, subaerial erosion and glaciation has produced a landscape marked by variety rather than uniformity, in which wide valleys with lakes are surrounded by steep-sided hills, climbed by few but visible to many.

5. THE TERTIARY BASALTS OF THE NORTHEAST

Basalts poured out in Tertiary times in a series of flows, many of which are 15–20 ft. thick, still cover more than 1,500 square miles in the northeast, but were at one time much more extensive. They form gloomy plateaus over 1,000 ft. high, with scarped edges overlooking Belfast lough and lough Foyle. Later earth-movements have bent and lowered them to form the lough Neagh basin, where Tertiary clays, not as yet precisely dated, have been laid down. The basalts rest on Triassic, Liassic and Cretaceous strata, which are exposed in the escarpments. The north coast of Antrim has a remarkable series of basalt cliffs, of which the best known is the Giant's Causeway, where there is a lower layer of regular, hexagonal columns and an upper layer of thin, irregular prisms moulded into fantastic shapes called by such names as the Giant's Organ: in between these two sets of columns a fossil laterite is seen as a bright red band near the Causeway.

The northeast of Ireland owes its identity as a landform unit to these basalts, which have been faulted and warped into a basin surrounded by plateaus. The deep valleys cut into the plateau include the corridors between Belfast and Antrim, and from Larne to Ballyclare, and the cliff-walled Glens of Antrim, farther north, which possess a remoteness and individuality that marks them off from the remainder of Ulster. But in the extreme northeast, between Cushendun and Ballycastle, gnarled schists and gneisses of pre-Cambrian age are exposed; structurally continuous with the promontory of Kintyre, they belong to the same morphological province as the Scottish Highlands. The basalts mask the continuation of the Highland Boundary fault into Ireland, though it runs along the edge of the Sperrin mountains, and thence beneath the Bann valley and the Antrim plateau towards Scotland.

The margins of the basalt outcrop are marked by sharp changes in the landforms and soil almost everywhere; there could, for example, be no greater contrast than that between the bog-covered basalt and the fine-swarded chalk at Whitepark bay, a few miles east of the Giant's Causeway. Many of the excellent sheep pastures of the northeast lie on similar small outcrops of chalk. In the Lagan valley the basalt escarpment towers above the Triassic lowland, which extends in a salient to Strangford lough between low hills of Ordovician and Silurian strata. The remarkable survival of Mesozoic beds in the northeast must be ascribed partly to the protection given by a basalt cover once of greater extent than now, for elsewhere in Ireland these beds have been removed and older landscapes exposed. It is, however, the glacial phase that has influenced the agricultural life most fully as the soils of the northeast owe their variety to the mingling of many types of local and transported materials.

GLACIATION

The glacial geology of Ireland has been actively studied ever since John Ball, in 1846, suggested that the deep valleys of Mount Brandon, in the Dingle peninsula and of the Purple mountain, Killarney, were gouged out by glaciers similar to those he had seen in the Alps. Since Ball wrote, the glaciation of Ireland has fascinated many field workers, especially as many glacial forms are extremely obvious in the landscape and challenge explanation. Two main problems stand out: first, the actual history of the glaciation and its correlation with that of Britain and Europe, and second, the mapping of the glacial and glacifluvial deposits that cover most, but not the whole, of the country. The first of these problems has been treated in many papers by Charlesworth and Farrington and also in W. B. Wright's *Quaternary Ice Age*: in addition, a brave attempt at correlation has been made in *The Irish Stone Age*, by H. L. Movius. The second problem, the form and distribution of the deposits, has received some attention in the works so far mentioned, and in many more, such as Sollas's paper on the eskers of Ireland, published in 1896: unfortunately, most of the geological maps were made and the memoirs written at a time (1858–91) when the immense importance of glacial action was not understood. It is only fair to add, however, that some of the memoirs contain vivid descriptions of various glacial features.

Several workers, including Kilroe, Cole and Hallissy, drew attention some fifty years ago to the importance of soil classification in a predominantly agricultural country. Kilroe, for example, in his *Soil Geology of Ireland* (1907), produced a map on the scale of 1 : 633,600, in which he showed that most Irish soils originate from drift. Some of the older rocks, such as quartzites, certain granites and shales, weather into infertile soils, but the patches of drift laid over them support farmlands, and the cultivable area is therefore greater than the 'solid geology' map would suggest. Some of the limestone karsts in the Burren district of Clare and in the

Galway bay district are of such slight agricultural value that the drift-cover of Ireland would seem to be one of the country's most precious possessions.

Two main phases of icesheet glaciation have been recognized in Ireland, corresponding to the Saale and Weichsel glaciation of northwest Europe, the last two of the Quaternary period, though there may have been earlier icesheets. Farrington suggested that in Ireland the names Eastern General (older) and Midland General (newer) glaciations should be used, as they refer to the areas where the respective deposits are best seen. Both these icesheets left limestone-bearing drift over large parts of the Central Lowland and on the mountain sides to a varied limit that reaches at least 1,800 ft. in the north of the Leinster chain. The drifts of northeast Ireland, which are derived partly from Scottish sources, are varied and include schists both from Scotland and from Co. Antrim with chalk and basalt of local origin: in the Silurian country local materials figure prominently. The older icesheet (Eastern General) covered almost the whole of the country and part of the present seabed: the newer (Midland General) icesheet reached a line running from the mouth of the Shannon to the Leinster chain, which it surrounded and penetrated to a height of *c.* 1,250 ft. (see Fig. 4).

The Midland General icesheet left magnificent moraines, kames, kettle-holes and drumlins, and as the ice receded in an east–west direction chains of large eskers were formed in the Central Lowland. Another remarkable series of eskers, in the Mayo lowland, is correlated with ice melting from south to north.

In the southwest there were two glaciations. The earlier, the Greater Cork–Kerry glaciation, left a somewhat featureless drift that covers much of the southwest: the later ice cap, the Lesser Cork–Kerry glaciation, did not reach the coast but left extensive moraines in the Lee and Bride valleys. In the Killarney area it is apparently represented by the piedmont glaciation marked by a wonderful series of corries (cf. pp. 33–4, 364–5). Whether the whole lowland area was glaciated is uncertain but there is considerable evidence that several of the mountain ranges rose above the ice. In 1947 Farrington suggested that much of the lowland in the Dingle peninsula was never glaciated but in 1961 this theory was contested by Cuchlaine A. M. King and M. Gage, who argued that the Dingle lowland had drifts derived from an icesheet that came from the north. In the far west, at a small bay to the northeast of Clogher head, they found a drift which contained materials such as sandstone boulders of local origin with – significantly – pebbles of limestone from the north.

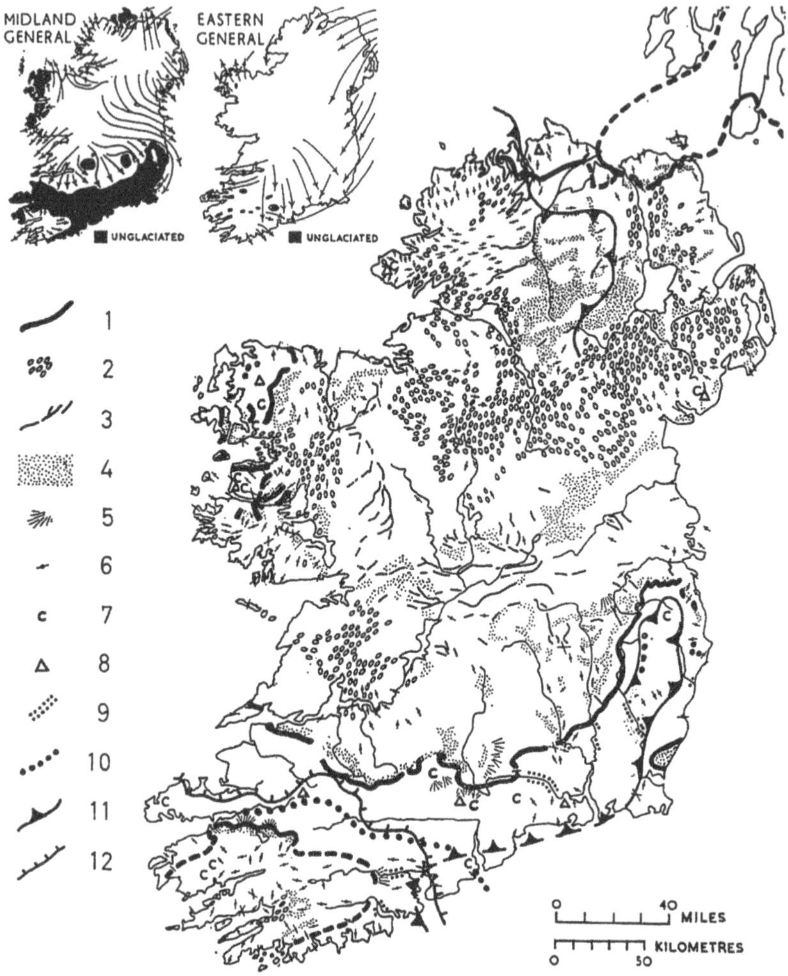

FIG. 4. IRELAND: QUATERNARY GEOLOGY

1, Terminal moraines of the last glaciation (Würm/Weichsel); 2, Drumlins; 3, Eskers; 4, Dead-ice topography (kames, kame terraces, other glacial sands and gravels, and occasionally a little boulder clay); 5, Outwash fans of sand and gravel associated with the terminal moraines of the last glaciation; 6, Striae; 7, Corries; 8, Important solifluction deposits of last glaciation age (e.g. head, fossil screes, and other slope deposits); 9, Important terraces in some river valleys; 10, Farthest extent of Cork–Kerry drift in Ireland; 11, Western extent of Antrim or Scottish erratics in Ireland; 12, Farthest extent of Galway and Midland drift in Ireland; (10, 11, 12: the limits refer to Riss/Saale glaciation).

From 'Field Studies in the British Isles'. Edited by J. A. Steers, London 1967. Reproduced by courtesy of the publisher.

Other evidence was found in the broad open valley which runs through Anascaul to Dingle and west into Ventry bay: this valley has some large, rounded and angular boulders scattered in the fields, and many more that have been collected and built into walls and gateways. On this and other evidence, the conclusion is reached that the landscape forms of Dingle and the drifts at Clogher head are the work of an icesheet of earlier date than the local corrie glaciation of the mountains, probably during the Eastern General (Munster) glaciation: this almost or completely covered the peninsula and the southern slopes of Dingle bay. Inevitably various processes have since modified the work of the earlier ice; for example, marine erosion has invaded, truncated and dismembered older glacial valleys, river valleys have become incised in older glacial floors – quite strikingly so in places – and periglacial processes have modified slopes. But by far the most interesting modification was the corrie glaciation that is seen along the Dingle peninsula from the Slieve Mish westwards to Mount Eagle, beyond Ventry, where a corrie occurs at about 1,000 ft. This is correlated by King and Gage with the much greater local glaciation, to the south of Dingle bay, which spread into the surrounding lowlands as a piedmont icesheet.

The nature and form of the deposits associated with various icesheets is of great importance through all the lowland parts of Ireland. It is not remarkable that the two words given to science by the Irish language are 'drumlin' and 'esker': Ireland has drumlin swarms of greater extent than those in Britain or America and eskers comparable with those of Finland. The drumlins are ice-moulded ridges, composed entirely of drift in many places, but possessing rock-cores in others; as few sections are available, the distinction between the two forms is not easily drawn, though in the area east of Newry and in the 'rock country' of Cavan every gradation between rock drumlins and typical drift hills may be seen. In the immediate neighbourhood of Belfast, subdued drumlins occur: they are too small to be mapped and are in effect transitional forms grading into hummocky sheets of ground moraine. The actual form of the drumlin landscapes is varied: in some places they stand up as individual hills, in others they occur in an impacted series. But the drainage is impeded by their presence everywhere, and there are numerous lakes or ponds, which are being steadily filled by silt; some have already been replaced

---

This map originally appeared in J. A. Steers (ed.), *Field Studies in the British Isles*, London 1964, p. 450, in an article on 'The Pleistocene geomorphology of Ireland' by A. Farrington and N. Stephens.

by a fen or peat bog. The extent of the drumlin belt is remarkable: it stretches from Belfast lough and the neighbourhood of Dundalk across the whole country to Donegal, Sligo and Clew bays, and there are also isolated swarms in the lower Bann valley, various lowlands in north and west Donegal, and, far to the south, around Bantry bay (see Fig. 4). As yet only parts of the drumlin swarms have been mapped, but Charlesworth has estimated that there are tens of thousands covering in all some 4,000 square miles or more, which is one-eighth of the whole area of Ireland. The smaller drumlins tend to be rounded and the larger elongated: the largest recorded, a drumlin in Donegal, was *c.* 2,150 yards long and 575 yards broad, but most are much smaller. Three groups of drumlins were investigated by Charlesworth with the following average width and breadth: Donegal, 500 by 280 yards, ratio 1·8 : 1; Bann valley, north of lough Neagh, 560 by 280 yards, ratio 2 : 1 and Ards peninsula, 420 by 245 yards, ratio 1·7 : 1. The average length and breadth is similar to that recorded in other areas, and especially in southwest Scotland, 450 by 250 yards, ratio 1·8 : 1. Drumlins are generally associated with stagnating decay of an ice-cap and their direction reflects the course of the main glacial flow during the maximum of glaciation.

Ireland's eskers have various forms. They are associated with regular recession of the ice-margin and were formed by infilling of the main sub-glacial rivers that led to the ice-margin. Their general west–east trend in the Central Lowland is apparently associated with the steady shrinkage of a sluggish ice-cap from east to west. The great kame-and-kettle morainic belt running from Kilkeel to Dundalk, Ardee and Kells apparently marks a changeover from orderly retreat to wholesale stagnation of the ice-cap (Fig. 4). The decaying ice deposited gravel in the form of kames and eskers among the drumlins, and in its final dissolution large areas of kame and kettle moraine were deposited west of lough Neagh in the counties of Tyrone and Derry and also in the southeast of county Galway. Eskers as normally developed consist of long sinuous gravel ridges at least 20–30 ft. above the surrounding countryside, some of which are marked by a line of trees or followed by a road. Many are now disappearing as they provide a useful source of gravel.

Important as drumlins and eskers are in the Irish scene, it is the many types of drift grouped as 'limestone gravel' that cover the greater part of the lowland and form the parent material of most of its soils. A primary distinction may be made between the strongly weathered but older impervious boulder clay of the Eastern General glaciation and the fresher and more gravelly drift of the Midland General glaciation. While the line

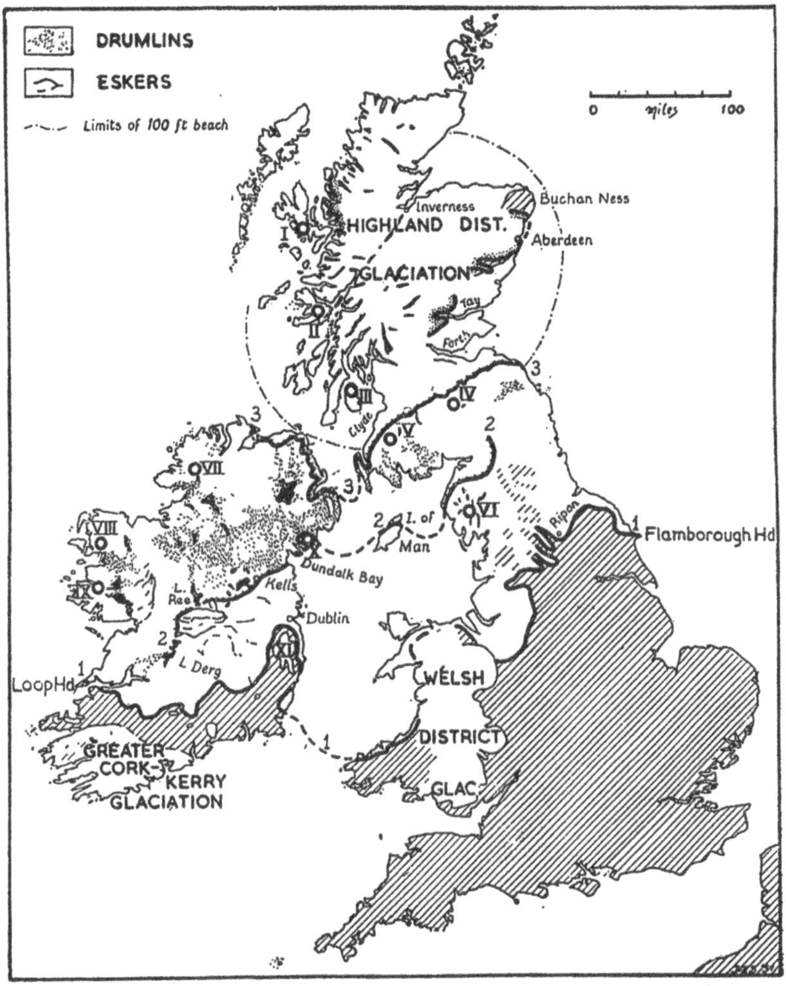

FIG. 5. RETREAT STAGES OF THE LAST ICESHEET IN THE BRITISH ISLES

This map was drawn by F. M. Synge as a contribution to the correlation of Irish and British retreat stages: the moraines are (1) the southern limit of the end-moraine of the Midland General glaciation in Ireland; (2) the Kells moraine, distinguished by F. M. Synge during 1951 and earlier field seasons, and here correlated with the morainic belt north of the Lake district: note that it forms the southern end of the great drumlin-belt of Ireland; (3) the Antrim moraine. Also shown are drumlins and eskers (see also Fig. 4, p. 28), and the main Irish centres of ice accumulation during the recession of the ice – VII Donegal, VIII Mayo, IX Galway and X Mournes.

Map originally published in Synge, F. M., 'Irish Geography', vol. 2, no. 4, 1952, p. 169

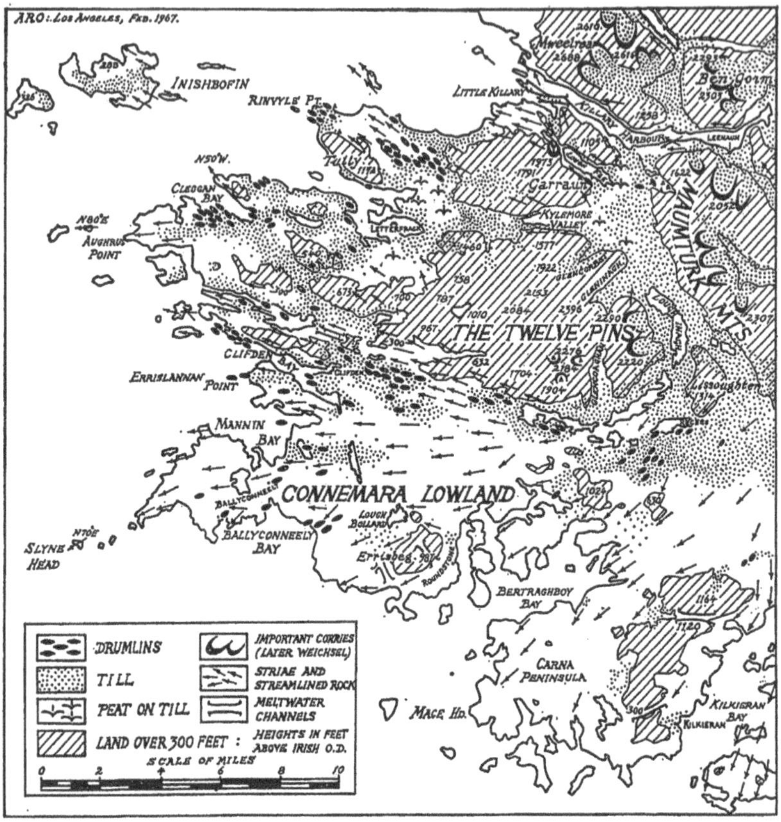

FIG. 6. DRUMLINS AND ASSOCIATED WEICHSEL PHENOMENA IN CONNEMARA

This map was prepared by Dr A. R. Orme and appeared in *Irish Geography*, vol. 5, no. 4, 1967, p. 265. It shows the effects of the Weichsel glaciation in Connemara with a strong movement of ice over the present coastline, well marked by numerous striae many of them clearly visible on rock exposed once the peat cover has been removed. Over much of Connemara there is no natural soil cover and many of the small farms have fields laboriously constructed from sand, seaweed and manure. But there are also swarms of drumlins, such as those of Clifden bay and the lowland margin beside the Twelve Bens (Pins), and in Cleggan bay. These drumlins provide some land suitable for farming and the deposits of till offer similar possibilities, as for example on Inishbofin island where the farmed area corresponds almost exactly to that shown on the map. Some of the mountains, notably the Maumturk, have fine corries thought to be of later Weichsel age.

Reproduced by courtesy of the Editor of 'Irish Geography'

of demarcation between these may be revised in time, it is at least clear that the drifts are thicker in the centre and north than in the west and south, where the cover is generally a thin rind of smoothly moulded and mature appearance. In the Cork valleys, for example, where yet another icesheet brought materials from the mountains to the west, the drift is rarely deep except in a few patches. But the newer deposits of the Midland General glaciation include the drumlins, and drifts ranging in depth from a few inches to more than 200 ft. in the Curragh, near Kildare; as a rule, the covering is thickest in the east and thins out into occasional extinction westwards, where the driftless karsts of Clare, Galway and Roscommon stand out as distinct morphological units. The absence of drift in these areas has been ascribed to strong erosive action near the centre of the icesheet, and to the removal by sub-aerial denudation of any cover that may have existed.

The mountains of Ireland show a wide variety of glacial forms which include corries and deeply trenched valleys with lateral and terminal moraines; around their margins there are evidences of meltwater channels and lakebeds that imply a drainage system controlled by the disposition of the lowland ice at various times. There were probably at least three phases of glaciation in the Wicklow mountains: the earliest, or Enniskerry mountain glaciation may possibly be an early phase of the Eastern General glaciation. The second, or Brittas mountain glaciation, has been distinguished in the Mount Leinster group, and the Comeraghs, as well as in the Wicklow mountains, where it was an icecap stretching to the coast in the east, but of unknown extent in the west. The last, or Athdown mountain glaciation, has left clearer traces than either of its predecessors; its glaciers were confined to the valleys, except on the Vartry plateau (see pp. 298–300), where a piedmont glacier was formed. On Mount Leinster and the Comeraghs, the Athdown stage was little more than a corrie glaciation.

Several other mountain masses of Ireland show clear evidence of glaciation, probably in more than one phase. In Kerry and West Cork, ice accumulated to the west of Kenmare poured northwards through the Reeks and the mountains around Killarney, and formed a series of crescentic moraines on the lowland facing Dingle bay (see Fig. 70). It was therefore a piedmont glaciation, which produced an extremely rugged landscape in the valleys and hills that were overridden: but the plucked and scoured surfaces of such areas contrast with the smooth outline of Mangerton and Stoompa, two hills that were nunataks rising out of the ice. The Reeks have the appearance of an arête, but they, too, were nunataks, and their peaks are of pre-glacial origin. Some of the corrie basins show how very deeply ice can establish itself in the heart of the mountains; everywhere, except on the summits, the gouging action of ice is apparent in this well-known area. The ice also moved southwards from the Kenmare river area and passed across the Bear peninsula to Bantry bay, finally reaching a line marked by moraines from the Shehy mountain to Skibbereen. Its southward movement was responsible for many bare rock surfaces and overdeepened valleys, but there are also considerable accumulations of drift, some of which exist as drumlins around Bantry bay. The *Memoir* of 1864 describes these as 'clearly distinguishable from the hills of rock by their smooth regular forms and their green cultivated sides, while the rocky hills are rough, and dark with heather and gorse. The drift hills are round, mostly isolated mounds of 100–150 ft. in height, and standing on a base of about half a mile in diameter, with bare rock showing in the land between them.'

The Dingle peninsula experienced a corrie glaciation which became a valley glaciation at its maximum: such hills as Brandon, Slievanea, Stradbally and Baurtregaum are corrie-worn and overlook deep valleys with lakes held up by moraines. The forms are comparable with those on the south side of Dingle bay: thus Brandon, a sharply peaked range reaching 3,127 ft., is not unlike the Reeks, though rather more thoroughly moulded by corries, and the lower hills have smooth upper surfaces above cliff-walled corries and overdeepened valleys very similar to those around Killarney. The various peninsulas from Dingle southwards possess clear indications of every type of mountain glaciation from corrie formation to valley moulding and icecap action terminating in piedmont moraines: the southwest of Ireland therefore has in its landscape the effects of glaciations comparable with those of the Alps, Scandinavia and Alaska of the present day. W. B. Wright has remarked that some of the forms, notably the moraines, possess a freshness far greater than that of even the Midland General glaciation; one may add that the scenery of this region owes its extraordinary charm largely to this last glacial phase, and especially perhaps to the large number of lakes which range from isolated tarns in corries to the deep water-filled hollows of valleys formed by ice-pressure.

Various mountain chains of the south, between the Kerry–Cork ranges and the Leinster chain, probably experienced at least two mountain glaciations: the most obvious evidences are the corries of the Galtees, the Knockmealdowns and the Comeraghs. Coomshingaun, in the Comeraghs, perhaps the finest corrie in Ireland, is cut into the heart of the mountains, and has cliffs 1,000 ft. high rising above a lake which is held up by a very large block moraine. Many of the neighbouring corries are almost equally impressive, but the range possesses a smooth upper surface comparable with that of the other mountain masses.

The mountains of Connacht and Donegal lie within the area covered by both the major icesheet glaciations of Ireland; in at least one phase, they nourished glaciations that were contributory to the main icesheet rather than independent icecaps like those of the Wicklow mountains or Kerry and West Cork. This distinction was seen as long ago as 1872 by Kinahan and Close, who showed that ice accumulated to a depth of 2,000 ft. in the Maumturk–Twelve Bens area and flowed out in every direction. Iceflows pushed down Killary harbour and moulded it into a fiord, passed south and west across Connemara to the fringing islands, north to the Moy estuary and the whole of northwest Mayo, and west through Clew bay. The iceflow that moved southeast to Galway carried its characteristic granite into the Central Lowland as far east as Mary-borough and possibly even farther. These mountains of West Connacht, therefore, were in the heart of the glaciated region: and the scouring action of ice is seen both in and around them, especially in Connemara, where many of the bare rock surfaces are covered only by the bog that has grown during post-glacial times. A later phase of glaciation is represented by corries and some of the sharply cragged valleys in the mountains.

The Donegal mountains were also a centre of ice dispersion. Although the mountain valleys were deep before the ice appeared at all, glacial overdeepening has accentuated them and made rock basins that are now occupied by lakes, of which lough Veagh is 139 ft. deep and lough Eske 101 ft. deep. In the mountain heartland of Donegal, the Caledonian line was brought out by ages of erosion, phases of faulting and the action of ice: here it is easy to see the tremendous erosive power possessed by valley glaciers that were flowing through valleys already formed, and probably in a youthful stage of the erosion cycle. On the fringes of the mountains two types of surface occur. In the extreme west, notably in the Rosses, accumulations of drift are rare; and erratics, including some of great size, are scattered across a countryside of bare, moulded rock. Elsewhere drumlins and ground moraine cover the lowlands, for example around Donegal bay and in the northern peninsulas. The western landscape in the Rosses is therefore comparable with that of Connemara and with the more remote Scottish Hebrides, but the eastern landscapes of Donegal possess affinities with the drift-covered lowlands of Ireland.

It may ultimately be possible to recognize two or more stages in the mountains of northern Ireland. The Mournes and the Carlingford mountains show evidence of more than one glaciation: in an early phase the mountains were partly, if not completely, covered with ice, but in a later phase the ice approached from the northwest, sent tongues of ice through the Mournes and spread in strength on either side of the Carlingford peninsula. The ice deepened the Carlingford lough and made it fiordic. This phase has been called the 'Carlingford readvance' by Charlesworth, who suggests that its influence may have extended from Newcastle, at the northern end of the Mournes, to Dundalk and possibly even farther into the Central Lowland. The lowlands around the mountains possess a remarkable series of moraines that were deposited during its decay, and the valleys of both the Mournes and the Carlingford mountains are thickly plastered with morainic material. Meltwater clays have added to the thickness of the deposits, which are known to be 172 ft. deep in the Silent Valley on the seaward side of the Mournes. Much of the fertile land in the Carlingford peninsula consists of drift and lake-clays, and in many places the limit of cultivation corresponds almost exactly with the spread of these deposits. The meltwater channels of this phase are also remarkable: on some of the Carlingford mountains they survive as a fossil drainage system, now completely abandoned and comparable with similar relics in the Wicklows. Finally the Mournes have a number of corries, probably formed during a late-glacial phase: the Sperrin mountains possess a similar corrie moulding and are also remarkable for their lake-clays.

The wealth of meltwater channels in Ireland is well known, but their evolution can only be explained by reconstructing each phase of glaciation on a regional basis. This is clearly impossible here, and therefore only a general statement will be made. Two points are of special importance.

(1) The meltwater channels are most obvious in and around mountain masses where lowland icesheets trapped the waters poured out from melting ice: such areas include the Leinster chain, the Carlingford mountains and many more. The Leinster chain channels indicate a drainage system for the most part parallel to the main axis of the mountains, and so at right angles to the consequent river system of the time before and after the Ice Age. The meltwater channels were formed during more than one phase, and some were links between chains of lakes imprisoned by the lowland icesheets. No general rule can be applied to more than one hundred and twenty channels, ranging from minor depressions across

fields to gashes several hundreds of feet deep and joined, in some cases, by a network of tributaries. Almost all are now occupied by small misfit streams, though they are significant features of the landscape offering useful tracks for railways and roads. The channels in and around other uplands, such as the Mournes, the Carlingford mountains, the Sperrins and others are equally striking, and the Antrim plateau has a remarkable series of deserted channels, many of which are cols from one valley to the next.

(2) The part played by meltwaters in the evolution of the lowland landscape is much more difficult to determine, but not less important. As the lowland ice decayed and the last vestiges of mountain glaciation disappeared, the present drainage began to develop. The volume of water available from even a small icesheet must have been very large, and many of the hollows existing between patches of ground moraine or drumlins in the Central Lowland were filled by shallow lakes, of which some survive to the present day, though all have been reduced in size and many extinguished altogether. Some of the lakes became fens on which peat bogs developed as the climate became damper from the Atlantic climatic period, *c.* 5500 B.C. One may still observe the fen formation around many of the lakes, and especially around the now much-reduced lough Neagh (which is, however, of pre-glacial formation). Meltwaters in the lowland also made channels, as they must have possessed considerable erosive power and were working through relatively soft materials. Some of these channels are still occupied by rivers, but others, such as those in the Collon hills, Co. Meath, now have tiny streams that are essentially misfits. While it is possible to imagine these great meltwater floods, one can only suggest at present that they played an important part in the moulding of the post-glacial river courses; it may not be too remote an hypothesis that most of the rivers in the centre and north are poor relations of the great meltwater streams. Charlesworth, in his study of the northeast, gives us a picture of a greater Lagan and a greater Bann, spreading out into lakes more numerous and vaster than those of the present, but discernible from the sands and gravels laid down on their floors.

RIVERS

The rivers of Ireland have courses of considerable variety and are attractive elements in the scenery. They cross the Central Lowland as slow-moving streams, in places surrounded by bogs and marshes, and in many cases reach their estuaries through incised valleys. Most of the rivers have senile courses on the Central Lowland, and become youthful near the sea: the Shannon, for example, having moved slowly with a drop in level of only 55 ft. through more than a hundred miles, falls more than 100 ft. in the last sixteen miles from lough Derg to the tide water level at Limerick. This fall in level has now been used for the major hydroelectric station in the Republic, and the drop of 152 ft. from lower lough Erne to the sea has been used for two dams at Cliff and Cathaleen's Falls. Most of the present rivers are cut in drifts, and have not as yet reached the underlying rockfloor; where Carboniferous limestones underlie these drifts, as, for example, in large parts of the Central Lowland, their porosity and solubility enable them to develop an underground drainage system with caves, channels and natural reservoirs. The underground water table varies from season to season, and from year to year, and reaches the surface in some of the lakes in the lowland, such as those in the course of the Shannon. The turloughs, or intermittent lakes of Clare and Galway (see pp. 391–3), are unconnected by any surface drainage, and here the underground river system reaches its fullest expression in almost driftless country. The stream density is low in many of the limestone lowlands of Ireland, but waters from underground sources reach the rivers and lakes that act as the central drainage arteries of the country.

The river systems of Ireland show at least some degree of adjustment to the structures of the country, especially in the south, where the east-west synclinal valleys between the Armorican chains are occupied by subsequent streams, such as the lower Suir, the Blackwater, the Bride and the Lee. The right-angled bends of some of these rivers led J. B. Jukes to develop his theory of superimposed drainage as long ago as 1862: the north–south reaches of the rivers were, he held, parts of an original consequent system developed on an initial Cretaceous surface since removed by Tertiary denudation. Jukes's theory was used as a basis for a modern river study of the same area by Miller in 1939, but since then a small area of Cretaceous rocks, with undeniable flints, has been formed in a valley south of Tralee near the road to Killarney. This reopens questions of the origin of Irish rivers, though the consequent line may have been north-south in Kerry and Cork, and northwest–southeast in Waterford and Wexford: as the underlying structure was revealed, however, subsequent tributaries became more important, and in some cases finally established themselves as the master-lines in the drainage systems.

Armorican Ireland has been divided into three parts (pp. 19–23 and

Fig. 1): (*a*) the southwestern block, defined by a line from Dingle bay to Dungarvan, (*b*) the country with isolated ranges and plateaus, northwards to the Dublin–Galway bay line, and (*c*) the Central Lowland proper. The subsequent lines are more fully developed in the first of these than in the second, where some of the master streams, such as Upper Suir, the Nore and the Barrow, appear to be consequents developed on the Cretaceous cover and surviving, largely in entrenched courses, to the present. The trellis pattern here is made up of major consequents with tributary subsequents: but in the southwest the subsequents are almost, though not quite, supreme. The Central Lowland with its ill-organized drainage has no clear pattern in its river system, though its numerous north–south streams, such as the Suck, the Shannon above Athlone, and the left-bank tributaries of the Boyne, may possibly be relics of Cretaceous lines. The Barrow rises at an altitude of less than 400 ft. and the Boyne, too, is a river of the lowland. The Shannon rises in the Carboniferous plateau country near Sligo bay, flows sluggishly over the western part of the lowland receiving tributaries of similarly lethargic habits, and then fills lough Derg, an ice-hollowed basin, before beginning its final spurt through rapids to the estuary.

The Caledonian southeast is a distinct morphological province with the Leinster chain as its central axis. The rivers show an interplay of consequent and subsequent elements, but only one, the Slaney, cuts through the chain: it flows through a steep-walled granite gorge in a well-defined gap between the Wicklow mountains and the Blackstairs mountains. Like its near neighbours, the Barrow and the Nore, the Slaney is an incised river for many miles above the estuary, and may be a survival from the consequent drainage of Cretaceous times; it rises in Lugnaquilla, the highest mountain in this part of Ireland, winds round the west of the chain before proceeding through it to cross the Wexford lowlands, and receives a number of left-bank tributaries running parallel to the Leinster chain which are subsequent elements in the drainage pattern. The Blackstairs mountains in the south and the Wicklows in the north both form a definite drainage divide best seen in the north, where cols lie between deep valleys drained to the east and west respectively. There are, however, strongly marked subsequent depressions such as those drained by the middle Liffey and the Vartry, on either flank of the mountains; and throughout the Wicklow–Wexford lowlands routes followed by roads and railways are drained by streams running parallel to the mountains, which are subsequent lines in the drainage. Dr A. Farrington, in careful study of the Liffey, has shown that it owes its present curious course largely to the working out in pre-glacial times of a subsequent basin (see Fig. 53, p. 299).

The rivers of the northeast show a marked adjustment to structure; the Bann, rising in the Mournes and fed by tributaries from the hilly country of the Newry axis, flows northwards in a basalt-floored depression with upstanding plateaus on either side. The Lagan rises in the Slieve Croob, runs towards lough Neagh, then bends sharply northeast and flows to Belfast lough through a depression between the basaltic plateau edge and the Ordovician and Silurian hills. The glens of Antrim and of the Mourne mountains are drained by short, swift-flowing streams; hardly less striking is the valley of the Newry river, a deeply incised and ice-worn cut in hilly country.

Easy passes lead from the Bann valley to the Foyle and Erne basins, both of which open ways to the western seaboard. The Foyle, with its major tributary, the Strule, rises in the low Devonian hills to the northeast of Enniskillen, receives numerous tributaries from the Sperrin mountains and the eastern margins of Donegal, and, having cut through the ancient metamorphic floor of the northwest, passes through Derry as a broad river to the lough. The Donegal peninsulas and mountains are drained by numerous short streams, many of which run in troughs parallel with the lines of peaks. The Erne is a river of the drumlin country from its headwater streams on the Central Lowland and in the hilly area of Monaghan to the sea in Donegal bay: its four main lakes, Gowna, Oughter, upper and lower lough Erne, all possess islanded drumlins, and it winds slowly through such country with almost imperceptible gradients, finally reaching the sea by a sharp fall at Ballyshannon. Flooding was a recurrent problem around the Erne loughs, which were connected only by tortuous channels, but the levels of the two lakes are now controlled by barrages and the river channel has been widened and deepened. Thousands of acres of land have been made productive.

The Carboniferous plateau between Sligo, Leitrim and Fermanagh is drained by the headwaters of the Erne and the Shannon, and by several short rivers which merge into estuaries around Sligo bay. The Moy, farther west, runs parallel to the Ox mountains and then swings northward, reinforced by several tributaries from the bog-covered lowlands of Mayo, into Killala bay. The Corrib, on the lowland east of Connemara, has a completely dry channel in Carboniferous limestones between lough Mask and lough Corrib, and receives many subterranean waters before it reaches the sea at Galway: its tributaries include the Clare river on the lowland and various streams from the west.

COASTS

Ireland was connected by a land bridge with Great Britain in Tertiary times, and separated from it by the Alpine movements, probably in the Miocene and Pliocene periods. The fall in sea level during the Ice Age and post-glacial times may have reached 400 ft. or more, and the present coastline shows the effects of submergence, for its river valleys are drowned and its mountainous cliffs are among the most striking in Europe. The submerged valleys of the various rivers have been distinguished by borings in estuaries such as Cork harbour, Dublin bay, or Belfast lough, and the shortening of their courses has led to heavy silting with alluvial materials that choke the tidal waterways and make dredging essential in every harbour. The little valleys on either side of the Blackwater gorge have patches of marshy alluvium due to infilling; the drumlins of Clew bay are being united one to another, partly through the accretion of mud around planted *Spartina townsendii*; and at the head of Dingle bay and other Kerry rias there are stretches of salt marsh destined to become firm land in time. Eventually the reclamation of considerable areas will be possible, and the outline of each estuary will be changed by the addition of 'intake' land. This process is accelerated by the building of sea walls and embankments, and by the dumping of waste materials as, for example, around Dublin bay, Belfast lough, the Shannon estuary and lough Foyle (see Figs. 45, 71). In these four areas the economic motive has been active, but in the remoter parts of the country it operates only to a slight degree, and therefore the reclamation process is one of natural evolution rather than conscious planning.

Wave attack is strongly felt outside the areas of sheltered water: the force of the storms is particularly acute in the west, where Atlantic spray is thrown to a height of 300 ft. The coastal precipices of the west are almost 2,000 ft. high in Slieve League, Co. Donegal, and higher in Achill island, where there are quartzites: there are smaller but still impressive drops in such places as the Cliffs of Moher, where Carboniferous flagstones reach the sea and it is possible to look down 600 ft. into the water. The attack on many of the softer materials is even sharper, and may be seen in such places as Clew bay, where the drumlins are cliffed on the west side, but surrounded by shallow waters with accumulating silts on the east. As yet, the coast has a youthful line, and its recent submergence is revealed in the many islands and peninsulas that it possesses, especially in West Connacht and Donegal. But the silting of its bays was notoriously revealed by the pier building of the Congested Districts Board; many 'harbour works' served only to accelerate the process and add to the natural reclamation in progress.

From Cape Clear to Waterford harbour there are long stretches of cliff as much as 200 ft. high, broken only by the drowned valleys and inlets of the small streams deeply set in the lowland or coastal peneplane. The bays and inlets are silting up, in some cases, such as Dungarvan and Tramore bays in the south, and Wexford harbour in the southeast, behind sand dunes. The east coast from Carnsore point to Dublin bay runs generally parallel to the strike line of the Leinster chain and has a number of wide bays due to erosion in locally softer rocks and in the drifts exposed as many miles of low cliffs. Cambrian or Ordovician beds, with intrusions, are forming a number of promontories such as Bray head (quartzite), Wicklow head, Mizen point, Arklow head and Cahore point. Some of the bays between these have sand dunes partly stabilized by planting marram grass. Elsewhere the soft drift cliffs are subject to heavy marine attack, notably between Bray head and Greystones, which stands on a small promontory of Cambrian slates and grits with quartzites: the railway line has been moved inland, and the loss of land was estimated at 4 ft. per year between 1905 and 1928, and has been similar since. In 1928, a storm removed several houses from the north beach at Greystones; the old coastal railway track is in the last stages of disintegration, and the harbour at Greystones has been almost entirely swept away. North of Bray, erosion is also severe and here, too, the railway has been rebuilt farther inland.

Dublin bay is outlined between a granite coast on the south, around Dun Laoghaire, and the Howth peninsula, of Cambrian slates and grits with quartzites, on the north. For nearly fifty miles northwards to Dundalk the coastline is a smooth stretch of low drift cliffs or sand dunes, with a wide and gently shelving sandy shore. The rockfloor is magnificently exposed on many of the beaches to the north of Howth, but the attack on the sea is more gradual here than in the area around Bray head.

The coasts between Dundalk bay and Belfast lough consist mainly of low cliffs, in which the drifts are exposed above Silurian rocks: Strangford lough has a profusion of submerged drumlins and resembles Clew bay, at the other end of the great drumlin belt which stretches across Ireland. The basaltic lavas form cliffs in Island Magee, but the plateau edge is generally separated from the sea by the narrow lowland on which the Antrim coast road is built. This lowland has several patches of raised beach, and is floored mainly by Triassic, Liassic and Cretaceous rocks. The northern coasts are of great magnificence: at Fair head, columnar dolerites form an escarpment with a talus of enormous blocks, and at Ballintoy and Benbane head, the basalts reach the coast in cliffs 300 ft. high. The columnar basalts of the Giant's Causeway are visited by thousands, but few explore the still more interesting cliff line of Benbane head, immediately to the east. Many miles of sand dunes in the bays lie between the cliffed coasts as far west as lough Foyle.

CHAPTER 3

Climate and Weather

The weather of Ireland is commonly described as 'mild, moist and changeable'; extreme cold, long frosts, heavy snows in winter and scorching days in summer are regarded as exceptions to the general routine, in which one season grades imperceptibly into the next with rain as the only common element through the year. In some years December is so mild that almost any garden will have a number of flowers in bloom and January is hardly less attractive for its warmth in sheltered places: but in 1947 the conditions from January to March were as austere as those in many parts of the English Lowland, and snow or frozen slush lay on the roads for several weeks, causing acute traffic dislocation through drifting and, later, floods. In the north, traffic ceased for some days in the early part of March, and even places on the western seaboard, such as Sligo, were cut off for several days in February. Again, late frosts and snows have been recorded in the month of May, causing partial ruin to fruit crops; every year contains some exceptional weather conditions, some of which are a considerable hazard to the farming community.

CLIMATE

Climate is in effect average weather, or a statistical assessment of conditions that have an abiding quality over a number of years, and so enable comparisons to be made between one region and another. The mean monthly temperatures for the winter months in the west are 44°–45° (December to March), and in the east, 41°–43° (November to March): this numerical difference, though slight, means that there is a more marked check to plant growth in the east than in the west. While evergreen species flourish throughout Ireland in sheltered places, they are most at home in western valleys such as those around Killarney, which have luxuriant woods where evergreens such as rhododendrons spread themselves as if native, and the Mediterranean arbutus, a low shrub in its main home, becomes a tree, as it rarely experiences any check to growth through cold or lack of rain. The mean summer temperatures in Ireland range from 60°–61° in the south to 57° on the coasts of Donegal, and the ripening of such cereals

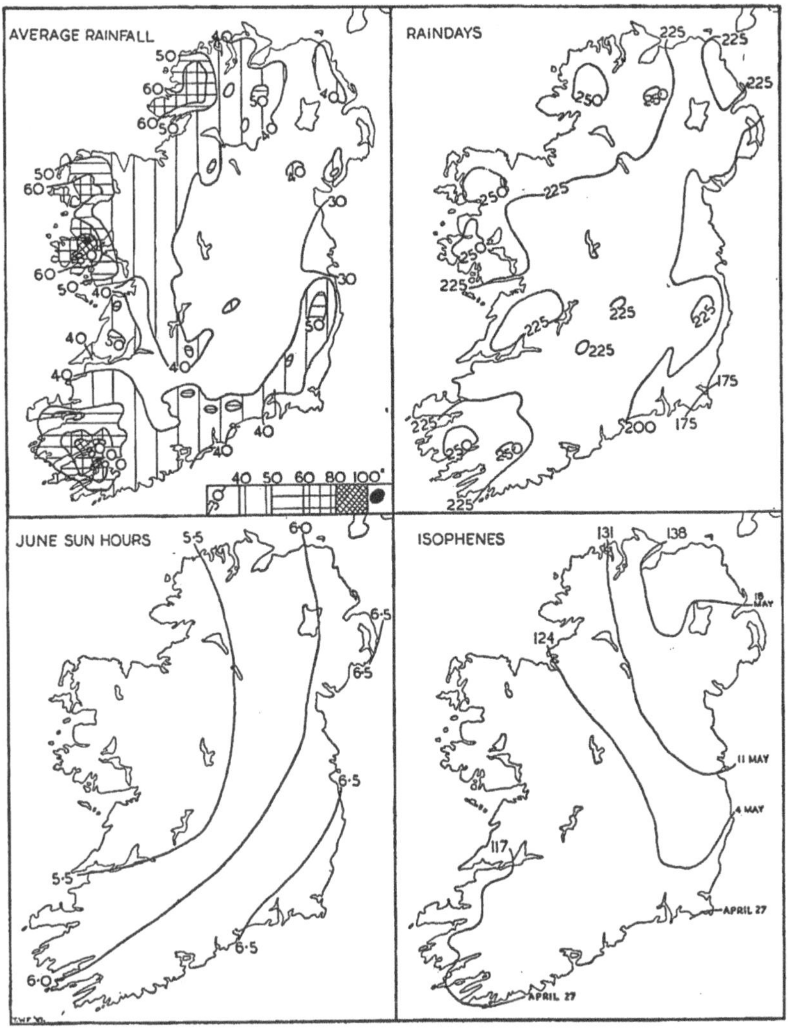

FIG. 7. CLIMATIC FACTORS; AVERAGE RAINFALL, RAINDAYS PER ANNUM, JUNE SUN HOURS AND ISOPHENES

The map of isophenes is copied from the *Land Utilisation Survey of Britain*, vol. 1, p. 442, which is based on the *Phenological Report for 1934*. The floral isophenes are for 1891–1925. The remaining maps are from *Bilham, E. G.*, '*The Climate of the British Isles*'.

NOTE. The maps in the excellent *Climatological Atlas of the British Isles*, H.M.S.O., 1952, differ slightly from those given here: but it has seemed best to leave the maps in their present form. In fact the number of stations is so small that many lines such as isotherms cannot be drawn with certainty.

as wheat and barley is hindered in most years by the frequent spells of wet weather and low proportion of sunshine. The extreme southeast, the lowlands between the Leinster chain and the sea, has an average of seven hours sun daily in June, which decreases to little more than five hours in the west (see Fig. 7): June is usually the sunniest month, after which there is a decline to the middle of winter. September (54°–58°), so crucial a month for the harvest, is warmer than May but in many years so wet that the harvest is gathered with great difficulty.

The mean annual rainfall of Ireland is, by general consent, excessive: some four-fifths of the country has from 30″–50″, whereas a similar proportion of England has 20″–40″. Only 500 square miles of Ireland has under 30″, but at the other end of the scale there are falls of 50″–100″ or more in various uplands. Relief rain figures prominently in the west from Cork to Donegal, and to some extent also in the south and east, but as Ireland's uplands are broken by lowlands, and consist rather of a number of separate ranges than of any continuous belt of mountains, the area with very heavy falls (say over 50″) is much smaller than that in Wales or Scotland. The highest falls, 90″–100″, are recorded from Kerry and West Connacht: individual stations include the Gap of Dunloe, Killarney, 94″; Aasleagh, at the east end of Killary harbour, 95″; and Delphi Lodge, in the Mweelrea mountains, 99″. Other lowland stations with very heavy falls include Garinish island, near Bantry, 70″; Kenmare, 65″; Inagh, Co. Clare, 67″ (this on the east side of Mount Callan): falls of 60″–70″ are also recorded from stations in the Sperrin mountains and the Mourne mountains. There is reason to believe that in some mountain areas, especially in the west, there are some falls greater than 100″: for example, rain-gauges placed in the high-lying valleys beside Mangerton (near Killarney) had falls of as much as 150″ in their first and second years of service (local information). As yet, the records are too short to be used as evidence, but it may well prove that the mountains of Kerry, and possibly also of West Connacht and Donegal, have rainfalls similar to those of the English Lake District.

The number of raindays in Ireland is high, and ranges from 175 in the extreme southeast to 200–250 on the west coast: the average period of rainfall at Cahersiveen is 834 hours, twice that of London. The south and southeast have the best sunshine records: the average percentage of sunshine at the normal climatological stations in all parts of Ireland, 28–30%, is more than that of the Orkneys (*c.* 25%), comparable with that of Birmingham (29%) or York (28%) though less than that of Falmouth (38%). Ireland is not so much a country of perpetual rain as one in which

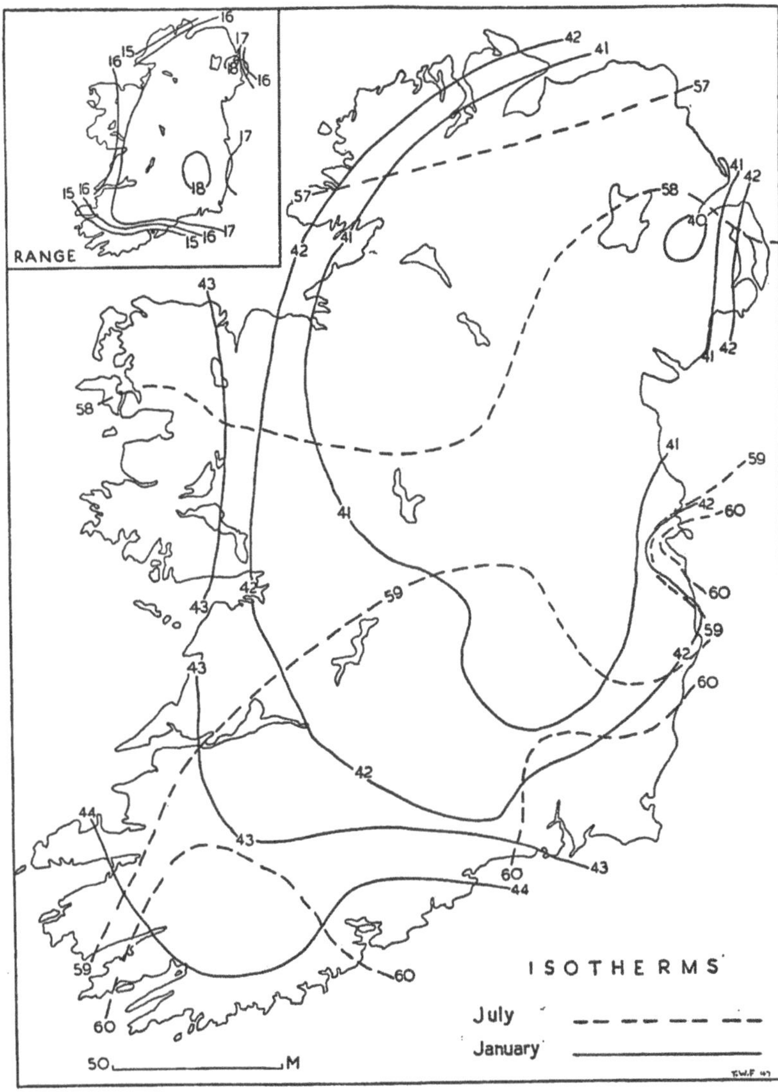

FIG. 8. JANUARY AND JULY ISOTHERMS, RANGE OF TEMPERATURE

These maps have been compiled from data in the Meteorological Office Averages of Temperature for the British Isles for periods ending 1930, and differ slightly from *Bilham, op. cit.*, pp. 151, 156–7. The lines are approximate, as the number of records is too small for certainty.

frequent showers alternate with sunny periods. There is a marked winter maximum of rainfall in the west, and a more even distribution in the east: while no station has a summer maximum, Dublin has an August maximum due to occasional thunderstorms or very heavy depression rains. One historic fall, on August 24–26, 1905, brought 2″ of rain to the whole of Ireland east of a line from Cork to Larne, more than 4″ to large parts of Down, Meath, Dublin and Wicklow, and over 5″ on some 300 square miles, with the heaviest falls in the extreme north of the Leinster chain and on the lowlands of Meath. Almost all the rain fell on August 25, and was associated with a northeasterly wind, which apparently caused the southwesterly air to rise and lose its moisture rapidly. This storm, comparable with the famous East Anglian rains of August 25–26, 1912, caused widespread damage, especially around Bray, Co. Wicklow, where the water flooded the lower part of the town to a depth of 4 ft.; furniture and barrels of beer floated out to sea. Three bridges were swept away, a weir was destroyed, many trees were uprooted, roads were undermined by the waters, and the level of the river bed was permanently lowered. Dr H. R. Mill, commenting on this heavy fall, said that it recalls Symons' famous dictum that 'a fall of 4″ in one day may occur any place in the British Isles'.

The variability of rainfall from year to year is less in Ireland than in Great Britain: the mean percentage deviation from the normal ranges from 6% in the northwest to 12% in the southeast. Fig. 9 shows the annual falls at Queen's University, Belfast, and the yearly average for Ireland computed by the staff of the British Rainfall organization. At Belfast the variability averages 9% and inspection of the diagram will show that over two-thirds of the seventy years had 90–117% of the average recorded fall. The average for Ireland shows a narrower range, of 7–8%: out of seventy-five years, four-fifths had from 94 to 110% of the calculated mean. In each case the tendency for unusually wet or unusually dry years to occur is apparent: the extreme outer range for Belfast is 33% more or less than the average, and for the whole country, 25%.

The rainfall is sufficiently variable to offer major problems for agriculture, generally from excess rather than deficiency, and not least in the harvest periods which in some years are delayed through September into October. Potatoes, oats and grass are better able to withstand these hazards than wheat or barley, which are normally restricted to well-drained soils in the eastern counties. One of the essential controls in the layout of a farmer's fields is natural drainage, for sloping lands dispose of

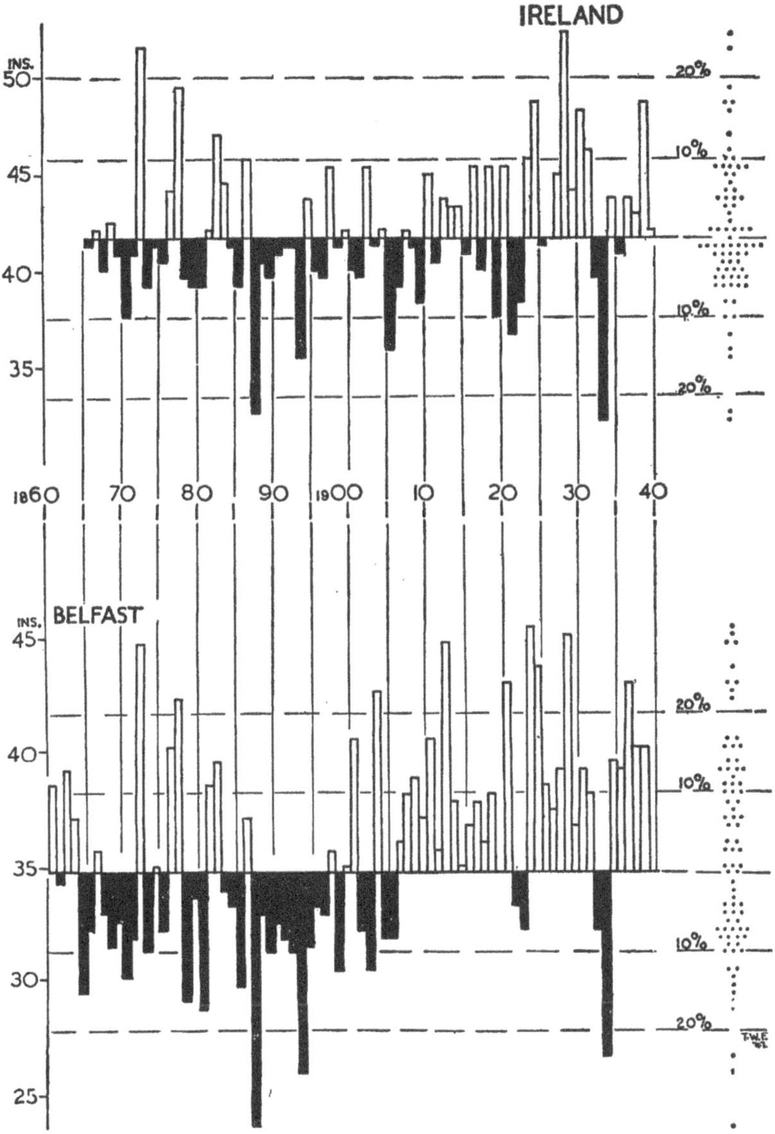

FIG. 9. RAINFALL: UPPER GRAPH, AVERAGE FOR IRELAND; LOWER, QUEEN'S UNIVERSITY, BELFAST

The figures are taken from '*British Rainfall*' and from *Salter, M. de C. S.*, '*The Rainfall of the British Isles*', 1921, p. 226. There is a tendency for a series of dry years to be followed by a series of wet years; few years have the exact average, though most are within 15% less or more. On the right-hand side the number of years having each percentage of the average fall is noted.

too-frequent rains capable of making level ground water-logged: a microclimatological study of the land would show that a patch of hummocky ground moraine, a drumlin, or a hillslope of gentle gradient may provide an excellent site for arable fields while the low ground between them is indifferent pasture or even peat bog. Similar contrasts are seen in the natural or semi-natural vegetation, for woodlands and dry heaths are found in areas having quick, natural drainage: under the present climatic conditions bog represents the climax vegetation of most of the uplands and many of the ill-drained parts of the lowlands, particularly along the western seaboard, where most of the farmland has been won from bog and, if neglected, speedily returns to its primitive state. The luxuriant woodlands of sheltered valleys in the west, of which those around Killarney are best known, show a prodigality of growth that makes them almost useless for economic purposes: mosses, lichens, ferns and other epiphytes cover the tree trunks in profusion, rhododendrons and even bamboos spread through the woods like weeds, and growth proceeds through the mild, moist days of winter almost unchecked by cold, for the decay and renewal are simultaneous. In the east, however, the growth of woodlands is much less hindered by the spread of parasitic vegetation and there is an abundance of magnificent trees, many of them in parklands: many of the older forests of the Commission are in the east but in recent years numerous sites have been acquired in the west.

Westerly winds predominate in Ireland, though naturally a climate experiencing the passage of so many depressions has winds from every quarter at some time during the year: at Donaghadee, Blacksod point and Dublin, nearly two-thirds of the winds are from some direction between south and northwest. Cahersiveen has 14% of calms, a higher proportion than that for any other major station in the British Isles, but its winds are generally similar to those of other Irish stations. Winds of gale force are most frequent in the northwest, where they average 40 per year, against 36 in the southwest and slightly less in the Irish Sea. Most of the gales are from a westerly direction: they are so severe a menace to shipping that many inhabitants of islands and remote coastal districts in the west profit from periodic salvage hauls. The light railways in Donegal, Clare and Kerry (see p. 242) took precautions against gales, and anemometers were installed at Quilty, Co. Clare, and Dunfanaghy Road, Co. Donegal, to give warning of danger. On the West Clare railway, a warning was sent to all stations when the wind reached a velocity of 60 m.p.h.; all services ceased when the wind reached 80 m.p.h. and were not resumed until half an hour had elapsed with no wind of 60 m.p.h. or over. At

Quilty, a gust of 112 m.p.h. was recorded on January 17, 1922, and there are also records of 109 m.p.h. from Dunfanaghy Road, on January 28, 1928, with several winds of more than 90 m.p.h. on other occasions. In 1961, gusts of 114 m.p.h. were noted (p. 53).

The absence of trees on the exposed western coasts of Ireland is usually regarded as a direct result of the exposure to westerly winds, chiefly of moderate strength, but including some gales. Where the country is level or gently sloping, bog is generally developed, but on the steeper slopes and other areas having good natural drainage, heaths occur. In some areas near the coasts a halophytic vegetation, which owes its presence to the effect of salt spray, has been recorded at exposed spots between Kilkee and Loop head, Co. Clare, and at Kinclassagh head, Co. Donegal. On the Clare coast the stronger westerly winds carry flecks of spray in such quantities that maritime or halophytic species are dominant up to a height of 100 ft., and frequent up to 300 ft., where they disappear and a heather moorland is established. At Kinclassagh head, halophytes compete with alpines and various other plants characteristic of the adjacent rocky or peaty areas to make a curious local association up to an altitude of 100 ft., the 'spray level': even on a fine day large flecks of spray are driven inland by west winds, and the air is moist and salty. Some form of *Plantago* sward has also been recorded by Dr R. L. Praeger on the south slopes of the Great Blasket island, Co. Kerry, on Clare island, Co. Mayo (to 50–100 ft.), on Inishkeen and Inishturk, and at the western end of Achill island (to 300–400 ft.). The author has observed similar vegetation on Inishbofin island, Co. Galway, at 200 ft. and on Horn head, Co. Donegal, at 500 ft.

Individual woods and trees show wind-moulding, for example on lough Mask, Co. Galway, where the trees on islands are about 30 ft. high on the east but much smaller on the west. On the islands of lake Muckno, Co. Monaghan, the same shearing is to be seen, but not in any one direction, as the winds appear to vary from one part of the lake to another. Further, at some points on the east coast, individual trees are bent westwards, implying the killing of buds and shoots by strong east winds: more frequently, however, trees are bent eastwards. Under the present climatic regime the natural tree limit, conditioned both by landforms and by the weather, is almost at sea level along the western coasts, around 600–800 ft. in the western mountains, but much higher in the drier east, probably 1,000–1,500 ft. Even in the west, a slight degree of shelter will allow a shrub vegetation to develop: the Burren of Co. Clare, a district of limestone pavements and tabular hills, has a low-growing ground flora on the windswept coastlands and exposed slopes, but in a few slight depressions hazel scrub is found, and there are woods in the valleys away from the coast (see also pp. 378–80).

The climates of Ireland are more favourable to grassland than to arable farming, and crops are won with a hard struggle in many years: it is perhaps not remarkable that the Irish speak of 'saving the harvest'. The scheme of stock farming is based on grass, for in the south and west it is possible to leave cattle in the fields for almost the whole of the year, and in the east and north natural pastures are available for as much as ten months. The replenishment of the herds is conditioned largely by the relative times when spring follows winter: in the 1891–1925 map of floral isophenes, it will be seen that twelve selected plants bloom around April 27 in the southwest, and three to four weeks later in the northeast. This may be correlated with the classic work of J. Hooper, who showed that calves were born some weeks earlier in the southwest than those in the Central Lowland, which in turn were available before those from Donegal, Cavan and Monaghan. It is in such ways that climatic factors influence an agricultural population.

WEATHER

'Mild, moist and changeable' is a suitable text for a sermon on weather, but such conditions are not invariable as those who experienced the weather from January to March 1947 will be aware. No systematic treatment of this cold spell is available for Ireland, but its main features are well known: sheep were buried in snowdrifts; roads and railways were blocked for some days in places; snow lay for weeks in most parts of Ireland, and was still seen in the Wicklow mountains in the middle of May. But such conditions are exceptional though it has been stated that only the four months June to September are immune from snow on the lowlands of Ireland and Great Britain. Data from northern Ireland shows that on the average snow lies for 5–15 days over much of the country below 500 ft. and that it is seen even more rarely in eastern county Down, in the lower Bann valley, the Roe valley, the lough Foyle lowlands and the Strabane basin. In the Republic, similar conditions prevail, and the figure of 18 days for the exposed Phoenix Park station at Dublin close to the mountains hardly invalidates the general statement that 5–15 days may be expected on lowland sites, with only 3 at Roche's Point, Cork and 5 at Cahersiveen.

Floods due to heavy rainfalls are naturally more frequent than those caused by snow melting, though the Shannon experienced both in the autumn and winter of 1946–7. In September 1946, the harvest season was so wet that a national appeal was made for volunteers to help to bring in the crops, many of which were standing in flooded fields. The Shannon overflowed and covered many of the 'callows' or low-lying pastures through which it flows in its middle course above lough Derg, and the volume of water was great enough to impose a severe test on the various drainage works along the river. In March 1947, floods comparable with those of the previous autumn were caused by the melting of snow combined with heavy rains. To add to the difficulties of the country, the peat supplies stored in the open air, as they must be owing to their bulk, were saturated by rain in the autumn, then covered by heavy winter snows which on melting impregnated them with water a second time, and finally drenched with rain.

Periods of deficient rainfall also create difficulties in the Shannon basin, as the Ardnacrusha power station depends on the inflow of water from lough Derg, which acts as a reservoir for the whole of the catchment basin: the level of the Shannon lakes is partly affected by the level of the underground water supplies, which are drawn from Carboniferous limestones. Records of the average intake exist from 1894, and an average was struck as a basis of expectation for the flow feeding the tailrace (see Fig. 72, p. 376): between 1932 and 1944, only one year, 1938, gave more than the computed average, and in four, 1932, 1933, 1934 and 1941, the deficiency was more than 20%. In recent years, the effects of weather on Irish electrical production have been complicated, as a bad year for river flow may be a good one for harvesting peat. In 1959 heavy summer and autumn rains increased the output of the hydro stations but interfered with the peat harvest, and the relatively dry winter gave a low water level when peat supplies were short. In 1960 conditions were more favourable: though the dry summer gave reduced water supplies the peat harvest was good, and during the wet winter the rivers were full and peat plentiful. In 1961, a very wet year, the electricity generated at the hydro stations was 30–40% above the average, but on September 16 the hurricane 'Debbie' hit Ireland, ruining all the exposed vegetation so that near most western coasts all the fuchsias were killed off. Gusts of as much as 114 miles per hour were recorded and some five hundred wood poles and over 5,000 conductors were ruined by falling trees. The electricity supplies in the south and west were interrupted by salt deposits on the insulator surfaces. Obviously, averages have only a limited usefulness under weather conditions as varied as those of Ireland, but a year with a high river flow is helpful economically. Both in 1965 and 1966, the water flow was heavy, reaching almost a record amount in the former year. But in some years dry periods occur: for example in 1965 there was a water shortage in February and there were similar low-water periods in June to September 1963 and from December 1963 to February 1964.

There are few records of first and last frosts at any Irish station, but Roche's Point and Cahersiveen have never recorded temperatures below 32° in the months May to September, and Malin head has a similar immunity from June to September. Inland stations are far less favourably placed: at Armagh, near an important fruit-growing area, only June to August are definitely frostless, and at Markree, Co. Sligo, frost has been recorded in every month except July. Similarly, the Dublin (Phoenix Park) station, in an exposed position, has recorded temperatures below 32° in every month except June, July and August. An interesting seventeen-year record at Roundwood, Co. Wicklow, altitude 700 ft., from 1920 to 1936, showed an average of only thirty frosts per annum, the first occurring between September 28 and December 6, and the last between March 7 and May 19. Coastal areas are generally less liable to frost than areas inland. In Northern Ireland, two coastal stations, Castlerock in Co. Derry and Kilkeel in Co. Down have no frosts from June to September and an average of one or a fraction of one in May and October but there are later spring and earlier autumn frosts inland, notably in some of the depressions surrounded by drumlins and other hills where cold air may slide downhill and displace the warmer air on clear nights.

The examples of weather drawn from various years show the importance of the unexpected element in the cycle of the seasons; yet a wide range of weather is natural enough under a climatic regime in which winds are associated with cold northeasterly or warm southwesterly conditions; on many occasions families of depressions cross Ireland, so that the country has a preview of Britain's weather. Severe cold from the westward spread of the continental conditions in winter, and the genial warmth of anticyclones from southerly sources in summer, reach Ireland less frequently than the east or south of England. The climatic affinity of Ireland is with the west rather than the east of Great Britain: Kerry resembles the southwest of Wales, Donegal, such parts of Scotland as Skye or Kintyre, and East Ulster resembles the southwest of Scotland. The east of Ireland and the greater part of the Central Lowland is perhaps most comparable with the drier parts of North Wales or with Cheshire, for it has a moderate, rather than a heavy, rainfall and a seasonal range of temperature much lower than that of areas in the English Midlands or the east coast.

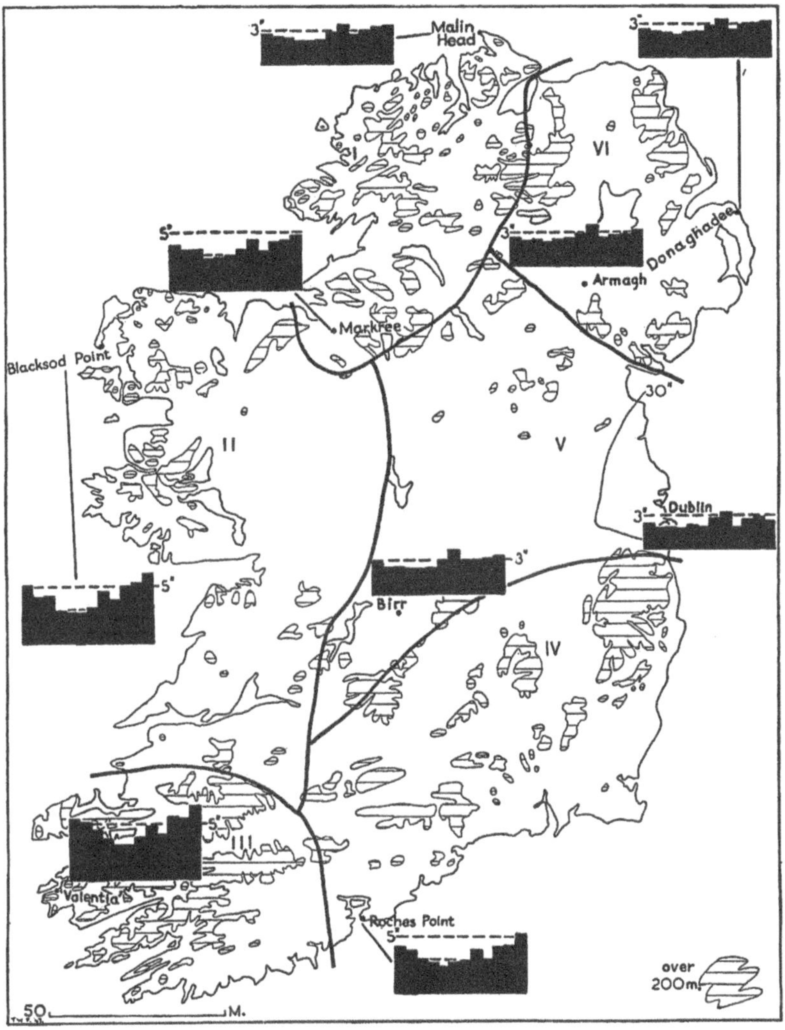

FIG. 10. CLIMATIC REGIONS, WITH RAINFALL AT SELECTED STATIONS

The six regions are described in the text; the areas over 200 m. are shown as they must inevitably be regarded as separate climatic enclaves, though there is little definite information of the ways in which their weather and climate differs from that of the lowlands.

CLIMATIC REGIONS

Ireland, though small in area, is sufficiently large to show a variation in its climates that is of considerable botanical and agricultural significance; its uplands, though almost devoid of weather records, are so different from the lowlands that they have been separately indicated in Fig. 10. Of the six types distinguished here, the first three might be called 'Extreme Atlantic', to use Professor A. G. Tansley's term, and the remaining three might be termed 'Irish lowland type', as in them the maritime influences, though strong, have not the dominance seen on the Atlantic coasts.

1. *The Northwest.* The rainfall is from 40″ to 70″, except in the north of Inishowen (Malin head, 32″) and a few sheltered lowlands: the maximum fall is in winter. The mean temperatures for the coolest month, January or February, are around 42° near the coasts and slightly less inland (e.g., Markree 41° from December to February), and the summer means are around 57°–58°. Gales are frequent in winter, and persistent westerly winds inhibit tree growth along the coast. The combination of high rainfall and mild winters allows luxuriant plant growth in sheltered places, but bogs are developed where drainage is poor; oats and potatoes are the crops most suited to the climatic conditions.

2. *Connacht and Clare.* In general character similar to region 1, this area has slightly milder winters (42°–43°), very slightly warmer summers (August, 58°–59°), and even more abundant rain, with falls of around 100″ in the heart of the mountains and 40″–60″ per annum on all the lowlands except those near the Shannon and a strip of inland Clare. Strong winds limit tree growth on the coasts and peat bogs are developed on the ill-drained lowlands: Mediterranean species are locally abundant. Agriculture is pastoral, with sheep on dry limestones, cattle elsewhere, and the crops are mainly oats and potatoes.

3. *The Southwest.* The rainfalls are 60″–100″ in the mountains of Kerry and West Cork, little less in many of the mountain valleys, and 50″–60″ on the adjacent lowlands – e.g. Cahersiveen 56″, Killarney town 55″ – though some coastal stations have less than 50″: there is a pronounced winter maximum. Excessive rain is combined with a winter warmth (December to March means, 44°–45°) that invites comparison with Mediterranean resorts, but there the comparison ceases, for the southwest has rain on three days out of four, and an average of little more than one hour's sun per day in December and January. The summers are warmer than in regions 1 and 2: sheltered places show a particularly luxuriant

vegetation and Mediterranean species, evergreens of all kinds, and sub-tropical exotics, thrive. Bogs are developed on areas of poor drainage, and the farming is mainly pastoral, with oats and potatoes.

4. *The Southeast.* The rainfall in the lowlands is 30″–40″ in the east and centre, but over 40″ in most of Co. Cork and Co. Waterford (e.g. Roche's point, at the entrance to Cork harbour, 42″): the uplands have falls of 40″ upwards. The winters are mild, with means of 44°–45° along the south coasts, and around 42°–43° on the east coasts, where sheltered places show a luxuriance of vegetation not unlike that of southwest Ireland. The summer months are comparatively warm, with July means of 59°–61°, and the sunshine records show that the lowlands between the Leinster chain and the sea are more favoured than any other part of Ireland, with a June average of seven hours' sun per day. The mountains are bog-covered, though in most places the bog is being heavily eroded under the present climatic conditions: the lowlands have bogs only in particularly ill-drained patches, and there is no blanket bog similar to that of the west (see pp. 65–6). This climatic region has some excellent arable land on sloping ground, producing rich crops of oats, barley, potatoes, turnips, sugar-beet and, in certain situations, wheat: its pastures are equally good and the climate is far more favourable for farming than that of the western regions. The tree limit in sheltered valleys is probably around 1,000–1,500 ft., and forestry plantations are reclothing bare slopes with trees up to the natural limit. In many places around the Leinster chain, the upper limit of settlement and farming is around 1,000 ft., itself an expression of the favoured climatic position of the region.

5. *The Central Lowland.* This, the area of the lightest rainfall in Ireland, includes two areas with less than 30″ per annum, of which the larger extends from Dublin northwards to Clogher head (near Dundalk) and inland for some twenty miles, and the second covers a few square miles around Banagher on the Shannon: the latter is not well attested. The stations outside these areas show mean annual falls of 30″–40″. The mean winter temperatures are around 40°–42°, varying to some extent according to exposure: the Dublin records are of particular interest here, for the Trinity College station in the city, at 12 ft. above sea level, has January means of 42·9°, while those from the Phoenix Park, at 155 ft., are 2° lower. The extraordinary mildness of the low-lying parts of Dublin and its suburbs is reflected in the early flowering of shrubs and trees, which are sometimes as much as a fortnight in advance of those in the outer suburbs standing only 100 ft. higher. Exotics of sub-tropical origin flourish along the coasts near Dublin almost as richly as in Cork or Kerry.

The agriculture of the Central Lowland is pastoral in emphasis, for both the heavy and impermeable boulder clay and the limestone drifts give rich grasslands, but excellent crops are also grown, especially in well-drained situations: these crops include wheat, barley, oats, sugar beet, beans, peas, potatoes, turnips and mangels. The present utilization is not necessarily the most intensive possible under the climatic conditions, as there is very little market gardening or fruit growing, though some small pastoral farms near Dublin have been transformed into orchards and fruit farms with conspicuous success.

6. *The Northeast.* Although the northeast is in many ways the most intensively farmed area of Ireland, its climate is cooler throughout the year than that of the remainder of the country. In winter the mean monthly temperatures are around 40° – e.g. Lisburn, *c.* 40° from December to February, Armagh, 40·5° in the same three months – but conditions on the coasts are warmer, and there are averages of 42°–43° at Donaghadee. The mean temperatures in the summer months are around 58°–59°, probably less on the north coast: the maps of floral isophenes show that the springs come approximately three weeks later than in the southwest. The rainfall is 30″–40″ on the lowlands, but higher in the Mourne mountains (*c.* 40″–60″, significantly less than in Donegal). Grassland predominates on most farms, with a substantial arable acreage given to oats and potatoes, but normally very little wheat or barley: the region is renowned for its excellent potato crops, including seeds, and for its cattle pastures rather than for the bread and brewing cereals, which are more at home farther south. Bogs and fens occur in depressions and around the many lakes in the drumlin belt: the natural tree limit is difficult to assess, but probably not less than 1,000 ft. in sheltered valleys, and the limit of farming is little lower than 1,000 ft. in such places as the western side of the Mourne mountains.

CHAPTER 4

Vegetation and Soils

Two-thirds of Ireland is improved agricultural land, and most of the remaining third is rough grazing, woods or forestry plantations. Of the areas that are uninhabited one-third to one-half is peat bog, though the lack of a general Land Use survey in the Republic makes this statement speculative. In the Northern Ireland Land Use survey, bogs were separately shown and the 1″ maps clearly presented the distribution of peat in such low-lands as the Bann valley. How much peat has been cut no-one can say, but it is the traditional fuel of the country, and there are districts in the west of Ireland where its removal, at least from the most accessible areas, is already causing problems. During and since the 1939–45 war, peat has been commercially exploited by Bord na Mona (the Turf Development Board, established in 1935). The bogs are of varied types, and a summary account of them is given on pp. 63–6. One of the major problems is the use of the land, of widely varying fertility, left behind once the bogs are cleared. In the mid-nineteenth century, many writers regarded such areas as potential farmlands to be colonized by the landless, and hoped to settle the country's population problem by this form of reclamation.

Other areas within the 'rough pasture' category include some limestone karsts in the area around Galway bay and in Roscommon, of which the Burren, Co. Clare, is discussed on pp. 377–9. Nowhere could the effect of wind exposure on tree growth be seen better, for shrubs reach up through the cracks of the limestone pavement to ground level, and not an inch above it unless there is some shelter. Similarly, hazel or thorn scrub will survive under the shadow of a wall giving protection from the west winds. West Clare and the Aran islands, having no tree growth, have a rich development of gentians and *Dryas octopetala*, intermingled with familiar wild flowers in coastal situations, but away from exposed positions shrubs and trees appear once more, notably hazel and – rarely – ash. Even though many workers have been attracted to these karstic landscapes, many problems still remain. There are, for example, patches of heather on limestones, rooted in a soil which may be residual after solution of the lime content, or may be the last relics of some glacial deposit in an area where the main visible trace of glaciation is numerous erratics.

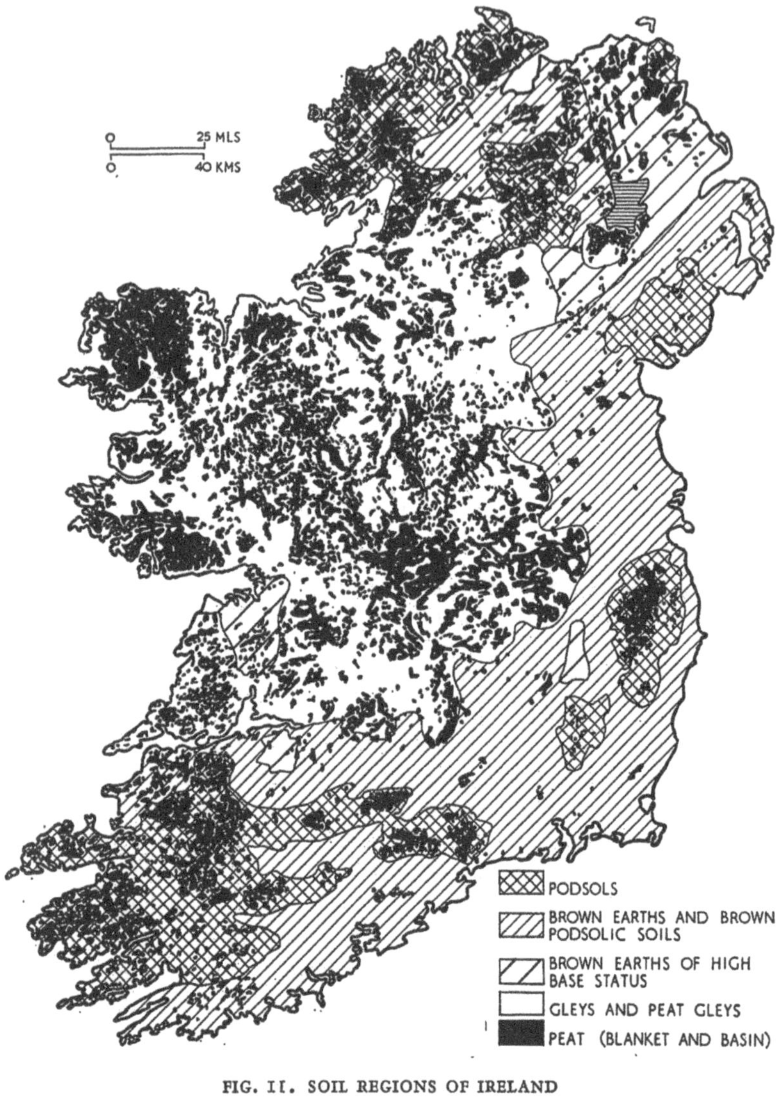

FIG. 11. SOIL REGIONS OF IRELAND

This map appeared in an article by L. J. Symons and J. G. Cruickshank, published in J. A. Steers (ed.), *Field Studies in the British Isles*, London 1964. The complex interpenetration of peat bogs with farmland in the Central Lowland is well shown, and in the mountains, particularly in the north and west, there is an extensive cover of blanket bog. The characteristics of the soils distinguished are discussed in the text.

From 'Field Studies in the British Isles'. Edited by J. A. Steers, London 1967. Reproduced by courtesy of the publisher.

Many other types of lowland rough pastures exist, and visitors to the west may be surprised to observe the extent to which rough boggy ground is used as cattle and sheep pasture. On Inishbofin island, Co. Galway, almost all the rough pasture, or commonage, is of this type and it has been said locally that 'every year the feed of twelve sheep is burned in the grates' by using the 'scraws' or sods of thin peat that cover the rock. It is abundantly clear that the area of exposed rock has been vastly increased by the removal of the thinner sods, so easily gathered. One could not assume that there is much sign of a 'natural landscape' in much of the west, though the uncultivated karsts of Clare are perhaps of this category. Heather, bilberry and grass moors exist, and two examples, both from the drier east, are given on pp. 291 and 300. But large areas of the mountains are covered with blanket bog, much of it under active erosion at the present time.

The history of Irish woods is an economic rather than a political study. Much of the Irish landscape has the appearance of a well-wooded countryside, but that is due primarily to the abundance of single trees in the hedgerows or in the pastures, or to the planting of windbreaks beside farms. Little natural (or semi-natural) woodland survives, except for some patches of oak-birch forests on the steep slopes of mountain valleys in the Wicklows and other hills, and the famous Killarney woods, which have been carefully studied by A. G. Tansley and others (see p. 69). Mention should be made also of the surviving woods of islands in lakes and of the woods in the west, which are sheared from ground level on the windward side to normal height on the leeward side. Many of the most interesting woods have been planted by the owners of demesnes, who in some cases also made many of the roads within their territory into avenues: some of these demesnes have been taken over by the Forestry Commission, either as a whole or in part, and the Killarney woods are now within a national park. It may well be that much of the Central Lowland would naturally be covered with oak wood in some form but the land has been agriculturally far too necessary for such woods to survive.

SOILS

Soils develop in Ireland under conditions of high rainfall, with precipitation to evaporation normally in a ratio of 1 to 2–3, or even more. The abundant moisture may percolate through the soil, carrying downwards the soluble material, or it may be stagnant on or near the surface until removed by surface run-off or artificial drainage. If the parent material is permeable, podsols or brown earth will develop, but if it is a clay or silt a gley will occur. Fertilization and liming may alter the character of a soil, for example by changing a podsol into what is generally termed an 'artificial brown earth'.

Podsols normally have a surface horizon (A1) of a peaty organic nature, overlying a strongly leached horizon (A2) which may be grey-white in colour, commonly referred to as bleached. The B-horizon has an iron pan, or accumulation of materials from the A-horizon. Such soils are common in coarse-textured soils, such as the sands and gravels of the eskers, kames and moraines that occur so plentifully in the Central Lowland, and also in such areas as Ballycastle, which has a wonderful series of glacial-lake deltas. There are also many sandy areas in the glacial drifts of Ireland where similar soils develop.

Brown earths normally develop on loams derived from granite drifts, as for example in Down, Armagh, Wicklow and Wexford or mica-schists as in Tyrone and Derry, and also from some Devonian (Old Red) sandstones. Equally such soils may be derived from basaltic drift, especially from the lateritic layer that is so notably visible at the Giant's Causeway and exists inland. There are similar soils on limestone drifts. The brown earths owe their productivity to good tillage with the help of lime and other fertilizers. Many of the upper parts of drumlins in the east have brown earths though the lower slopes may be waterlogged.

Gley soils have a high water table and occur on clays and silts. The heavy precipitation of most of Ireland favours the development of such soils, which for example replace brown earths on drumlins as one moves westwards: they have, for example, been noted around Dromore, Co. Tyrone. And such soils are widespread. Much of the Central Lowland and the drumlin–drift belt from Donegal bay to the Shannon estuary has gley soils, which give many lush but poor-quality pastures that need closely-spaced field drains with the continued eradication of rushes and sedges.

Peat soils are of varied depth and occur widely in all the Irish lowlands where drainage is impeded and fen developed under drier and warmer climatic conditions (cf. p. 64). They are found, for example, between drumlins and spreads of ground moraine and may provide good soils for potatoes, oats and grass given adequate drainage. In mountainous areas, and in lowlands with a high rainfall (cf. p. 65) peat may lie on a soil that is heavily leached with a pronounced hard pan: indeed it has been estimated that podsolization is widespread from a height of *c.* 450 feet. Where the peat cover is thin, it is easily cleared, leaving a soil that may be a heavy clay in the lowlands of little use except for pasture. In many western areas peat lies as blanket bog (pp. 65–6) on the soilless rock, which remains barren when the peat cover is removed.

VEGETATION TYPES

Four types are discussed here; bogs, fens, woods and sand dunes, with which areas of salt marsh are commonly associated. The first three of these all have now considerable economic significance. The bogs are sources of fuel for farmers, for processing into a welcome substitute for coal, even for power-stations; the woods are a valued home source of timber; and the fens, given drainage, may be turned into excellent pastures, as, for example, beside the river Bann in Northern Ireland. Such areas, known as 'callows' beside the Shannon, have traditionally been subject to occasional flooding but are valued as natural cattle pastures in the drier periods. Salt marshes may form useful natural pastures, in time reclaimable by embankment in places: similarly there are some excellent grasslands on salt and freshwater marsh behind areas of sand dunes or shingle ridges. So far no detailed work has been done on any Irish area of dunes, though they are known to be highly unstable on the west coast, and in some places likely to invade agricultural land (cf. p. 74).

BOGS

Few lowland bogs, if any, are now in their natural state for the population have been cutting the edges of all the more accessible bogs as fuel supplies for centuries. It has been estimated that half the total area has been removed, and while this may be true, it does not necessarily mean that half the supply of peat has gone, for many deposits of considerable thickness still remain. It was noted above that their peat deposits, in some cases of only a few inches, are easily cleared, and some of the smaller, shallow bogs have been completely cut away. In the Republic an annual estimate is given of the amount cut by farmers for their own use and for sale: during the war of 1939–45 this exceeded 4,000,000 tons in some years but since then it has gradually diminished so that by 1965 it was less than 1,500,000 tons. In addition Bord na Mona cuts some 2,000,000–3,000,000 tons of peat each year, mostly for electricity generation. Most of the farmers' peat comes from the bogs of the west but Bord na Mona is exploiting twenty-six bogs of which twenty are within the Central Lowland. Through the combined efforts of private and public enterprise, it is said, the bogs of Ireland will be largely if not entirely removed within the next sixty to a hundred years. Apart from a limited use as a fertilizer, peat is at present used entirely as a fuel and it may well be that within sixty years no-one will be interested in peat as a fuel any more.

Peat is still the major fuel of the country, because coal is expensive, wood scarce and substitute fuels little used. In many parts of the west, exhaustion of all the nearer supplies may be a real problem, and some of the islands, notably the Aran group, are now entirely without home supplies, so peat is shipped from the mainland. Obviously interest centres in the soil left after the clearing of bogs: in much of the west there is bare rock, a sodden marsh or even a lake. But it is possible that there may be some reclaimable land in the Central Lowland, either for forestry or even for agriculture. There are drained patches of cleared bogs in some places which carry crops of potatoes or other crops, and pastures. In some places trees are grown on cleared areas, and already there are several plantations of conifers in the Central Lowland. Successful planting has also covered many square miles of upland bog with trees, notably in Northern Ireland.

Irish bogs have been divided by Tansley into two types, raised and blanket, of which the former occur primarily on lowlands having a rainfall of less than 50″. According to Bord na Mona material,* the main area of raised bogs lies to the east of the Shannon, though some are found to the west of the river. The second type, blanket bog, is found in the west and also on some mountains farther east, notably the Wicklows. Bogs of the blanket type are in areas with an average rainfall of at least 40″, and commonly very much more: they occur on lowlands to the west of the line from Clonakilty to the Fergus estuary, to Galway bay, Sligo bay and finally from Donegal to Londonderry – more simply, they are characteristic of West Cork, Kerry, Clare, West Connacht and Donegal. Within this area there are lowland bogs of vast extent, such as those of Kerry, of Connemara to the west of Galway city, and of north-east Mayo (the largest continuous expanse of more than three hundred square miles), and of west Donegal. These bogs have always seemed to be a limitless reserve and are now used for power stations at Gweedore, Co. Donegal; Bellacorick, Co. Mayo; Screeb, Co. Galway; Milltown Malbay, Co. Clare and Cahersiveen, Co. Kerry.

Raised Bogs. Raised bogs are built up on fen and marsh in areas of poor drainage. Some occur in small basins and cover only a few score areas, but others cover twenty square miles or more and some of the best examples are on level ground forming watersheds. At present there is a

* British Association, 'A view of Dublin', 1957, in article by H. M. S. Miller, p. 127.

bog-free belt extending inland for some twenty miles from the east coast but it may be that within this area any bogs have long since been cleared. Where ground moraine provides reasonably good conditions of natural drainage, the land is farmed, but in ill-drained hollows and flat water-partings, fens developed after the glacial period, and as the climate degenerated bogs grew above them. It is, however, by no means certain that the removal of the bogs will reveal a potentially rich fen under existing Irish climatic conditions, though some valleys and hollows in the east have fens, mainly drained for cattle pastures.

Seen from a railway line or a road, many of the bogs of the Central Lowland have slightly convex surfaces, which in one measured case proved to be 1 in 100, or 3°. Estimates of the thickness of bogs are varied, though the raised bogs are thought to be at least 20 ft. thick and in places much more. In the natural state there is no surface drainage and the waters collect up in a marginal ditch or 'lagg', but in practice this is rare as the bogs have been cut for fuel round the edges and many artificial channels have been made for drainage. As the surface dries, heather and its associates will appear, possibly with a few birch trees and pines. Seen in section, the bog may be sufficiently dry at the surface to support such vegetation, but there may be brown and sodden decaying vegetable matter beneath, in some cases down to the water table. Many of the rivers retain the brown colour of peat for many miles from their sources in the raised bogs of the lowlands or the blanket bogs of the mountains.

Blanket Bogs. So named because they cover the ground like a blanket, these bogs are apparently restricted to areas having an average rainfall of at least 40″ and generally of more than 50″. Characteristic of the west, they also occur on various mountains elsewhere, notably in the Wicklows. Spread indiscriminately over every type of surface, they have been observed on slopes of as much as 15°: in thickness they vary from a thin rind or 'scraw' on the rocks to a continuous bed 10 to 20 feet deep or more. Such areas are well seen in parts of Kerry and Connemara where the roads are left intact, supported by the remaining uncut peat. In north-west Mayo the bog is so extensive that the road from Crossmolina to Belmullet runs for over twenty miles through a landscape virtually uninhabited except for a few farmsteads on small areas of alluvial land beside streams. In many places the bogs are broken by numerous lakes and lakelets, varying in size from small pools to considerable areas of water, in some cases enclosed by rocky rims with heath or scrub. Bogs are broken also by outcrops of rock, many of them *roches montonnées*, covered with heather and gorse, or in a few exceptional and well-sheltered places with dwarf woods of oak (*Quercus sessiliflora*) and birch (*Betula pubescens*) or scrub of hazel, hawthorn, willow, rowan and holly.

The blanket bogs have been formed during the past two and a half millennia (the sub-Atlantic period), and the pine and birch forests of sub-Boreal times are exposed by cutting. In their present form, they show more variety of vegetation than one might expect: in places there are spreads of cotton-grass, of heather and its associates, or tufts of *Molinia*, and patches of bog-myrtle, *Myrica gale*. These bogs have a colourful aspect at times; for example, when the heather is in flower the bog-myrtle has its deep-brown fruits or the cotton-grass turns red in autumn. In the mountains springs give flushes of bright green moss and grass made the more vivid by the sombre vegetation around them. At present there is every sign that the bogs on the mountains are being rapidly eroded, partly through cutting for fuel which accelerates natural runoff and allows gullies to develop. It may be that sheep on the mountains so reduce the vegetative cover that gullies develop, and many moorlands consist of peat haggs separated by ditches several yards wide and several feet deep, obviously in process of rapid expansion.

FENS

No comprehensive survey of fens has been carried out in Ireland but it may well be that fens are more widespread than is generally known. It is probable that large areas beside rivers, notably the callows of the Shannon which form such good cattle pastures in summer, and possibly also the similar pastures beside the Barrow, Nore and Suir, are natural fens. Beside the Erne with its two large loughs there were until recently some 11,000 acres regularly flooded and 18,000 acres waterlogged, all of this of limited agricultural use and of virtually no use at all in wet seasons. Much of this is apparently fen rather than peat bog, and the new drainage scheme stabilizes the level of the loughs and provides a good drainage channel between them instead of the tortuous water courses that existed previously. Flooding has in consequence been greatly reduced but the maintenance of field drains will be necessary for the scheme to remain successful. The lower lake is a glacial rock basin and one purpose of the scheme is to maintain the water level at the Cliff hydro-electric station (see p. 471).

Ireland's largest single area of fen is around lough Neagh, notably on the south side. Drainage schemes aimed at the control of the river Bann and lough Neagh may enable the area given to market gardening and soft fruit production to be further expanded. It may be that much of the agricultural settlement in this area is comparatively recent for the Halls, writing shortly before the Famine, spoke of the use of 'turf so black and hard that it gladdens the housewife's heart' and of the 'squatters . . . who have cultivated patches of this arid common into productive land'. These fens have a neutral or very slightly acid soil (pH 6–6·5) as the rivers draining into the lake are plentifully supplied with basic materials derived from the basalts through which they flow. The area is considerably moister than the English Fenlands, with an annual rainfall of *c.* 32″ compared with 20″–25″, and the 'ramparts' or surviving ridges of bog left for access are built up of yellowish-brown peat, composed of decayed *Sphagnum*, *Eriophorum* or *Calluna*, which is acid in reaction. Apparently clearing of the surface layers has exposed the fen, which in its natural state may be flooded to a depth of 4–5 ft. in winter and early spring. Careful botanical work was done on this area by the late Dr J. M. White, but the flooding which caused an interesting zonation of plant species is now limited by the new sluices at Toome, at the north end of lough Neagh.

Fens may occur also in small patches between the drumlins, and around the numerous small lakes or ponds that develop in such areas of impeded drainage. It may be too, that removal of some of the low-lying peat will reveal fens: Dr White showed the author patches of fen in the Blessington lake-basin, now submerged (see p. 298 and Fig. 53). Many of the lakes in the drumlin-drift belt are surrounded by considerable areas of marsh and possibly fen: in one case investigated by a party of the Geographical Society of Ireland – lake Muckno, Castleblayney, Co. Monaghan – the area of water had been reduced from 3 to 1½ square miles (see Fig. 82, p. 442). The old lake floor, it was observed, was partly used for peat-working, but with some difficulty as its surface consists of a pulp-like black mud which must be pounded into shape and then carefully dried before it can be used.

WOODLANDS

The greater part of lowland Ireland appears to be well wooded, and there is a marked absence of trees only in some western districts on or near the coast, and in the major peat bogs of the Central Lowland. The appearance is deceptive for Ireland is the least forested country in Europe (apart from Iceland), and only 3% of its area is covered with woodlands. There are numerous shelter-belts, single trees in hedges or fields, copses or woods in private parks, but few traces of natural woodland are to be seen. Such woodlands occur mainly on hillsides never cleared for grazing and cultivation, or on inaccessible islands in lakes. Though some of these woods have been invaded by sub-spontaneous pines from plantations and occasionally by other trees (see pp. 69–74), there are a number of natural oak woods (*Quercus sessiliflora*), along with birch and ash in places. The oak woods are highly varied, and a number of trees form a lower stratum or exist as co-dominants with oak to form a mixed woodland or scrubland. Such trees include holly, hazel and rowan and more rarely hawthorn, aspen, yew and, around Killarney, arbutus. Willows and alders have their normal dominance in damp places and some remarkable hazel scrubs are developed on limestone crags in the west, notably in the Burren of Clare. Relics of natural woodland may be seen in some of the Wicklow valleys, chiefly on slopes too steep for pastures, and also in the Killarney area. But as the Forestry Commission has extended its planting, many of these old woods have been submerged though some still occur in remote areas.

No reliable estimate of the tree limit under the present climatic conditions can be made. There is evidence that in the sub-Boreal period forests of pine and birch covered many hills and lowlands, where peat bog is now the climax vegetation. It is worthy of note that the tree limit is at sea level along the exposed western coast, but it rises sharply to 600–800 ft. in Kerry, and in the east of Ireland it may be as high as 1,500 ft. in some sheltered valleys. The splendour of individual trees in gardens, parks and hedgerows suggests a former forest vegetation of some luxuriance over lowlands now cultivated, though many of the finest trees in Ireland today are introduced species, such as beech and sycamore and various conifers, such as pines, spruces, firs and larches. Among the conifers only the juniper and the yew are native, but the introduced species are widespread in plantations, and the Scots pine at least is able to sow itself and compete with birch on many dried peat bogs and hills. The planting now in process, which is mainly of coniferous trees, will remove the spacious and open aspect of many mountain valleys.

Natural and Semi-Natural Woodlands. The term 'semi-natural' is perhaps safer than 'natural' in any discussion of existing Irish woodlands, as few of them have been immune from timber gathering, if not actual clearing, sheep grazing and even cattle pasturing. In modern times some of them have been invaded by sub-spontaneous pines and others have been 'beautified' by the planting of exotics. A few woods have managed to escape some or all of these influences, and it is reasonable to hope that remnants of natural woodland will survive, though some lie within areas now controlled by the Forestry Commission.

The famous woods of Killarney have an evergreen quality which distinguishes them from the deciduous woods of Britain and continental Europe. It is therefore proposed to consider them as a special Atlantic type of woodland most fully developed around Killarney and also seen near but not on the western seaboard of Ireland. The climatic influence of mild winters (January mean temperature 43°, with mean minimum 37° at Killarney), the infrequency of frost and the heavy rainfall spread through the year with constant high humidity, are largely responsible for the development of a forest type of some richness, or more accurately of constant growth and constant decay.

The Killarney Woods (*Atlantic Type*). These woods* may be regarded as natural over wide areas, as there has been only local planting of exotics, though many are seen in the natural park near Killarney. Like most Irish woods, they have *Quercus sessiliflora* as the dominant tree with a subordinate layer of *Ilex aquifolium*. These oak-holly woods are the most widespread type, but some patches of pure holly wood have been observed. The arbutus, generally found in the margins of the woods, may grow to 30 ft. in the Killarney area. The woods of Glengarriff, on Bantry bay, are like those of Killarney in general aspect, as oaks are dominant with holly in the second layer and some arbutus on the margins. The curious spread of *Rhododendron ponticum*, apparently at the expense of holly, has been observed in the woods of Glengarriff. Similar woods occur around Kenmare and the arbutus has also been recorded as a native tree around lough Currane, near Waterville. To the north, the woods of lough Gill, Co. Sligo, are very similar to those of Killarney; oak is dominant, with birch, hazel and holly as a second layer and arbutus on the margins.

Evergreens are frequent in the Killarney woods and many introduced species appear sub-spontaneously along with the native holly and ivy. Among them are the *Prunus lusitanica* from Portugal and the Atlantic islands, *Rhododendron ponticum* from the Caucasus and the Balkans, *Laurus nobilis* and *Viburnum tinus*. The native evergreen yew is widespread and forms some pure woods, especially on patches of bare or thinly covered limestones, and its present distribution may be due partly to sub-spontaneous sowing. In the woods of Killarney there is a rich growth of epiphytic vegetation on fallen logs and even tree trunks; boulders are generally covered with a matted vegetation. The plant cycle on boulders is first liverworts, followed by heavy mats of tall mosses, filmy ferns

* A full analysis of the Killarney woods is given in Tansley, A. G., *The British Islands and their Vegetation*, 1939, pp. 327–40.

(*Hymenophyllum* spp.), but the mats of moss are frequently washed off by rains and the cycle begins again. A similar development is seen on the trees and logs, which are colonized by comparable, but not necessarily identical, species, including the filmy fern, herb robert (*Geranium robertianum*), *Polypodium vulgare*, *Ilex*, *Saxifraga spathularis* and *Rhododendron ponticum*. The general impression of these natural and semi-natural woods is that their very richness makes them practically useless: in them one species preys upon another and there is little good timber available, for beauty of scene and fascination of detail are combined with perpetual decay.

Other Irish Oak Woods. The remaining oak woods of Ireland are confined to two main types of environment, hillsides and lake islands, and although the surviving patches cover only a small area, the evidence available suggests the former existence of oak woods of some variety. A wood near lough Gill, Co. Sligo, has hazel and ash as co-dominants with oak covering moss-grown limestones with a rich ground flora of anemone, goldilocks (*Ranunculus auricomus*), wood sorrel, bluebells and primroses; the abrupt change to metamorphic rocks is seen in the sudden appearance of bilberry and the disappearance of nearly all the flowers except wood sorrel and bluebells. The ash and the hazel also disappear, and are replaced by birch and holly. The woods on islands in lough Erne have oak dominant, with ash, birch, alder, aspen, holly, rowan, crab-apple, hazel, hawthorn, and bushes such as the spindle-tree, buckthorn, guelder-rose, blackthorn and willows, underneath which there are several species of fern and also primroses, wood anemones and bluebells. These dense highly mixed woodlands are on islands of drift-covered limestones and sandstones. The oak woods near Pontoon, at the southern end of lough Conn, Mayo, have *Quercus sessiliflora* dominant, with birch (*Betula pubescens*), rowans, holly, ash and hazel. Ivy and honeysuckle climb trees and cover rocks; and the rich ground vegetation includes bracken, ferns, bilberry and a wide variety of flowering plants. Some patches of natural or semi-natural woodland in this neighbourhood are rich in alders. Many of these woodlands are of the rich western Irish type, which in all probability was once widespread in areas away from strong winds. But wind-cut woodlands also occur: some of the rocky islets in the many small lakes of Connemara have planed scrub with rowan, juniper, willow, holly, ivy and, less frequently, birch, yew and oak. The windward side of such islands may be devoid of trees altogether and have only heather, *Ericas*, gorse (*Ulex gallii*) and bracken.

In eastern Ireland any form of natural or semi-natural woodland is extremely rare except in some of the valleys in the Wicklow mountains, especially those drained by the Avoca and its tributaries. These are oak woods (*Quercus sessiliflora*) with hazel and holly as a second stratum. The ground flora is fairly limited and one wood in Glenmalure has only wood sorrel and wood sage, but in most cases there are also wood anemones, bluebells, foxgloves, wood rushes, bracken and various ferns, according to local shade conditions. Nearly all these woods have now been taken over by the Forestry Commission and some are already replaced by plantations.

Planted Woodlands. The history of the Irish woodlands has been carefully investigated by A. C. Forbes, who shows that the decline in the natural forest cover was a process of slow attrition by cutting, burning and grazing from Neolithic times onwards. Although the maps and records of Norman and post-Norman times are virtually useless as historical evidence, there appear to have been a few Royal forests similar to those in England. However, the name 'forest' does not survive in any part of Ireland as it does, for instance, in Sherwood Forest or the New Forest, and the forest laws were hardly applied except perhaps in the Glendalough area which covered part of the Wicklow mountains. Fallow deer were brought to Glencree, north Wicklow, in 1224 from Chester, and until recently the herd in the Powerscourt demesne included some of them. The Wicklow upland may well have been a largely uninhabited and partly forested area in the thirteenth century, and there is an interesting record of clearing of woodland at Glendalough in 1229.

Ireland was very sparsely wooded by the seventeenth century, and this has been attributed to the heavy export of timber in Tudor times. In Queen Elizabeth's reign the destruction of all woods in which the Irish might hide was advised, but commercial considerations won the day, and speculators bought up any existing lots of timber for export trade. The chief trade in Irish timber was for pipe staves and charcoal burning, but in all probability its general use at home was more destructive than its cutting for export abroad.

The need for afforestation was recognized in various ordinances of the sixteenth and seventeenth centuries, but it was not until the eighteenth century that the planting of introduced and native trees became widespread. The yew is probably the oldest planted tree, partly on account of its religious significance. The *Topographia Hibernica* for the fourteenth century states that 'yew is commonly planted in cemeteries for ornament', and some old yews, singly or in avenues and walks, may have been planted around monasteries. There are scattered records of planting in private parks during the sixteenth and seventeenth centuries, among which the most interesting is that 'firs' (Scots pine?) were planted in many parts of Ireland, and also that Sir Walter Raleigh planted cherry trees, elms, walnuts, sycamores and spruces at his estate in Co. Waterford.

Trees planted before the end of the seventeenth century included Scots pine, stone pine (*Pinus pinea*), English and Dutch elms, the Spanish chestnut, lime, walnut, hornbeam, the evergreen oak, plane, mulberry and *Robinia* (locust tree or false acacia). In the eighteenth century many demesnes and deer parks were planted with belts and clumps of trees or single trees so placed as to beautify grasslands. In most cases landscape effect was the prime consideration, but some useful timber reserves were made where plantations were laid down. The many demesnes in lowland Ireland and in such areas as eastern Wicklow give the countryside its present wooded appearance. The trees planted were chiefly beech, oak, elm, chestnut and lime, with a mixture of conifers, pine, spruce, silver fir, and in some cases the Lebanon cedar. The European larch was introduced about 1750, and all the trees so far mentioned are excellently developed in Ireland.

Some records of planting around Dublin are of special interest. The development and planting of the Phoenix Park began in 1740, and the two botanic gardens in the city, at Glasnevin (the National Botanic gardens) and Ballsbridge (Trinity College gardens), were opened before the end of the eighteenth century. The magnificent trees in these parks and gardens include many exotic species such as palms, eucalypti and evergreens which may also be seen in sheltered areas throughout Ireland. The Royal Dublin Society gave premiums for tree planting between 1741 and 1807, but the scheme was not very successful as, according to Arthur Young, too little attention was given to the fencing of plantations, and the peasantry were in such constant need of timber that the many woods were spoiled. Arthur Young also commented on the youth of many of the plantations that he saw on his travels, and held the opinion that they could be multiplied almost indefinitely.

The nineteenth century saw the development of planting for timber production, but the area wooded showed no great increase during the century. The planted area in 1791 was estimated to be 105,000 acres, but this figure apparently excludes a great deal of natural woodland that was subsequently cleared and replanted. The pressure on the land during the pre-Famine years drove the level of cultivation higher up the valley sides and at the most some 4,800 acres were planted annually. This was merely an extension of demesne planting on rather more advanced lines.

Some of the demesne owners brought in Scottish nurserymen between 1825 and 1850, and plantations were laid out in which American conifers from the Pacific states were used. The most important introductions were the Douglas fir (*Pseudotsuga taxifolia*) and the Sitka spruce (*Picea sitchensis*), but a large number of conifers was also tried. Various famine relief schemes took the form of timber planting, and by the end of the century the area of woodland was over 300,000 acres. The Japanese larch was not introduced until the end of the century, but its reddish colour has already made it familiar in the forest landscape.

The Irish Forestry Commission was founded in 1908, when a detailed inquiry into the possibilities of forestry in Ireland was carried out. This Commission reached the conclusion that about 250,000 acres of suitable land would be required. By 1967 over 60,000 acres in Northern Ireland and 450,000 acres in the Republic had been planted. Enthusiasts have argued that several million acres could be forested. The Congested Districts Board included forestry as one of the remedial measures for the poor regions under their control, but their one experiment was a dismal failure. Knockboy Forest, near Roundstone, Connemara, was planted on a drained bog with imported trees, then with native trees and finally with trees raised from local seed. The Atlantic winds killed the young trees or gave them a stunted inclination to the northeast. £10,000 was wasted on this experiment, and the only result was the provision of all-too-abundant cover for woodcock and hares.

Private planting has virtually ceased in Ireland and the policy of both governments is not only to set out large plantations, but also to develop some woods of a few hundred acres. Most of the land acquired is not forested, but 'bare land', including portions of old demesnes not suitable for agricultural development, and land above the present limit of cultivation, or heaths and dried bogs. Conifers are the main trees planted, but both in the Republic and in Northern Ireland the intermingling of belts of trees differing in colour adds to the scenic attraction of the new woods. In Northern Ireland a state forest park has been established at Tollymore, near Newcastle, Co. Down, on the edge of the Mourne mountains. Situated in a tourist area, it has attracted a large number of visitors. In the Republic the first state forest park was opened at Gougane Barra, Co. Cork in 1967, and others are to be developed at Glendalough, Co. Wicklow, a much-visited area from Dublin, and – more remotely placed – lough Key, Co. Roscommon, in a fascinating scene of lakes and hills.

In both parts of Ireland the forestry authorities have taken over various demesne woodlands, of which many in Northern Ireland were greatly depleted during the war. Northern Ireland is estimated to have some 30,000 acres of private woodlands, probably increasing by 400 acres a year, so that in time, perhaps fifty years, there will be 50,000 acres. But even with the 150,000 acres of state plantations, the 200,000 acres will only cover 6% of the area, for in Northern Ireland, as in the Republic, forests occur almost entirely on land unsuited to crops and pasture.

SAND DUNES AND SALT MARSHES

Little attention has been given to the sand dunes of Ireland though they are plentifully developed along all the coasts. The main problem on the west coast appears to be their stability under high winds, with acute danger to neighbouring areas during gales. During a field work visit to Dunfanaghy in the spring of 1957, a Manchester party studied the dunes joining Horn head to the mainland and found that substantial changes had occurred since the 6″ map was published in 1904. The sand, moving northeastwards, had covered a hillside to its summit at 300 ft., and even swept down the other side into a valley. A large lake had been formed in the middle of the dune belt, where in 1904 only a blowout was recorded on the map. Part of the instability was said locally to be due to the heavy cutting of marram grass, especially in the early part of the 1939–45 war when fuel was scarce. On the west side of Horn head an area of some 150 acres had been covered with sand since 1904, partly as a thin cover with a rough growth of grass, and partly in dunes still in active movement. In this case the dunes do not function as a shield for a salt marsh or a lagoon, though there are many such cases in the west, for example along the coast to the west of Horn head. Each year the Land Commission spends a few hundred pounds on measures to fix sand dunes in the west.

Post-glacial sand and shingle deposits are extensively developed in Northern Ireland at the head of Dundrum bay, at a few places in the Ards peninsula, at the mouth of the Bann and in the Magilligan area at the entrance to lough Foyle. Magilligan point consists of a number of sand and shingle beaches which in places merge into sand dunes, liable to move as they are incompletely fixed by marram grass and other binding agents. Sand also covers large areas of the post-glacial raised beach that is developed around the coast of Northern Ireland, in some patches in the form of dunes: much of this sand area is used for golf courses (including some of considerable fame), rifle ranges and caravan sites.

There are also many fine stretches of dune coast on the east of Ireland.

At Newcastle, Co. Down, the dunes have been subject to severe erosion partly through path formation but few dune areas have as many visitors though problems of sand instability exist elsewhere. These dunes have been partly fixed by planting marram grass and providing groynes to check drifting on the beach. Farther south there are some splendid stretches of dunes, notably on the coasts of Wicklow and Wexford. The dunes at Brittas, Co. Wicklow, are changing constantly in form but without any perceptible expansion or contraction of total area. Similarly there are many stretches of dunes on the south and west coasts of Ireland, some of which are subjected to occasional heavy wind erosion.

Salt and freshwater marshes are also well seen, for example between Greystones and Wicklow, where shingle ridges hold up a belt of marshes and lagoons almost a mile wide. There are also stretches of salt marshes in many of the shallow estuaries and inlets the country possesses. A recent survey of salt marshes in Co. Dublin showed that the smaller ones in the estuaries were being eroded, but that the North Bull island was apparently stable. This island extends from the Bull wall, built from 1820 to 1823, north towards Howth head. Apparently a between-tide bank existed as early as 1685, but the present island only developed after the wall was built. Approximately three miles long and half a mile wide at its maximum, it consists of a fine beach backed by a line of young dunes, and salt marsh on the landward side. This marsh is apparently not now growing.

Many salt marshes exist behind dunes and shingle ridges but they have not received detailed investigation as yet. The most notable example of reclamation of salt marshes and alluvial slob lands covers several square miles on the east side of lough Foyle in C. Derry. The embankment, constructed and strengthened from the nineteenth century, and the associated pumping works, were made with the help of the Irish Society from the nineteenth century, and the railway line to Derry from Coleraine runs along the embankment. Many of the soils are of great fertility but some areas have been taken over for airfields and industrial sites, notably at Eglinton, near Derry city. On the west side of the Foyle estuary the seawall has been breached and several hundred acres of agricultural land lost, but given proper pumping and a guarding embankment the deep silty loam soils are well suited to arable and grassland farming.

CHAPTER 5

Historical Geography

Ireland, like England and France, is a country with a tradition of settlement and agriculture extending over several thousand years, and therefore, in human terms, an old land and not new, like many parts of America, Australia, or even the Soviet Union. Its fields have existed for generations, even for centuries, though the farmers change the boundaries from year to year, according to their needs and enterprise, so that the tide of agricultural settlement and enterprise has advanced at times and receded at others: a multitude of small changes have moulded so old a landscape into its present form. In all the changes something abides, such as the tendency for farming units to be small in many parts of the country and, perhaps even more important, for tools and implements used by the farming population to have a form distinct from those used elsewhere. E. E. Evans, in *Irish Heritage*,[1*] has shown that this feature of the country's life is as clear and definite in the immediate neighbourhood of the northern towns as in the remoter west. Ireland has a considerable part of its population living at a simple level that might be called 'peasant' (a term not loved by many people in the country), preserving many traditional practices and implements that are of undeniable antiquity: and much of the non-material culture is being investigated by the Folklore Commission, which is well able to sift the chaff of fantasy from the wheat of authenticity.

The preservation of highways is less frequent in Ireland than in England, where so many prehistoric and Roman roads are still recognizable as ridgeways on chalk downlands and other hills, and not a few of the main roads preserve the straight courses of the Roman engineers. Many roads in Ireland have been lost through changes in the distribution of population in a country of disturbed political history, combined with the late and somewhat sparse development of towns. The multiplicity of roads in Ireland is such (see p. 232) that any lane will be abandoned and forgotten once it has served its usefulness, though some fords on rivers have a long tradition of use: the lower Bann fords, for example, are excellent sources of prehistoric materials. Visible antiquity of settlement is a less obvious feature of Ireland than of England, for the village with an old church and

* For references, see pp. 515–18.

ancient houses is a rare phenomenon: many of the older churches are now abandoned and roofless, and almost all the humbler houses have a restricted life, and in time are replaced by others. All over Ireland there are ruins of stone houses, the best of which are used for animals, the worst merely a pile of stones: even less durable are the mud-walled cottages of which, in some cases, no trace remains within a few years. In general, while it is not difficult to detect the imprint of the eighteenth and nineteenth centuries in the landscape, it is only with a more sweeping imagination that earlier centuries emerge to view: few of the towns have even a single Tudor house, though several have streets with a number of stately Georgian houses and public buildings.

The changing landscape of Ireland is a palimpsest in which the earlier efforts of man are less obvious than the later ones, though traditions show a remarkable degree of survival in the daily lives of the inhabitants. Much of the political tension has centred around the problem of poverty and lack of opportunity, and one essential feature of the historical geography is the fight for land between the native population and various immigrants differing both in outlook and, since the Reformation, in religion. The life of the poorest people of the land still involves unceasing labour for a very small return, times of stringency when food resources are low through bad harvests or poor catches of fish, and even the near threat of starvation on occasion. The reports of the Congested Districts Board from 1891 onwards give a picture of the very poor not unlike that provided by many travellers of the nineteenth century, and the Gaeltacht Commission of 1925,[2] quoting a Royal Commission of 1909, spoke firmly of various western communities as 'to a large extent the wrecks of past racial, religious, agrarian and social storms in Ireland', who, 'in a bad year, are saved from extreme privation only by relief measures, and so constitute a serious financial danger to the nation, though they are a most valuable potential asset'.

In this present age, the thought of Ireland as a country of emigration comes most easily to the mind, but through the centuries immigration has been not less important, and the position of the country in relation to the rest of the world, or its 'space relations', are therefore worthy of consideration. The ideas of Sir Cyril Fox in the *Personality of Britain* are suggestive in this connexion:[3] he divides Great Britain into two parts – the 'Highland zone', which comprises Scotland, Cumbria, Northumberland and Durham, the Pennines, Wales and its borders, and the 'Lowland zone', which includes the southeast of England and the Midlands. Obviously, the 'Highland zone' stands nearest to Ireland, which is its natural if now insular continuation: Ireland is not, however, primarily a highland country, but rather one in which uplands and lowlands interpenetrate. Nor is that all, for an important environmental influence in the life of Ireland through the ages has been the contrast within its lowlands between the bogs and the dry ground made up of morainic material, glacifluvial deposits such as eskers and outwash gravels and, in a few restricted areas, actual rock outcrops.

A further idea of Sir Cyril Fox, that in the Highland zone types of culture and social organization arrive late if at all, but have the quality of continuity, is more fruitful in relation to Ireland. It is clear that many prehistoric ways of life lingered on into historic times in Ireland, and that the Norse, Anglo-Normans, and later immigrants, had only a very partial control of the island they entered as traders or conquerors; again, the Romans and the Anglo-Saxons, who left so vital and abiding an imprint on the English Lowland, never penetrated Ireland at all. Although Ireland has qualities in common with the western or Atlantic side of Britain, it is yet sharply distinct from it: even the northeast, though part of the United Kingdom, has a social complexion quite different from that of Scotland, and a rural landscape of small farms that is peculiarly its own.

In the remainder of this chapter, the position of Ireland is considered period by period, but only in broad outline, though some tentative suggestions are made on the regional geography of past ages, for the aim of historical geography must be the elucidation of the life of men in relation to the environment at definite periods of time. Much is transitory, but some things abide, unshaken by revolution, famine, invasion, political reorientation, industrialization and other changes of a cataclysmic nature (see Fig. 12).

PREHISTORIC TIMES[4]

In late-glacial times Ireland lay beyond the limits of the inhabited world, though Scotland was already occupied by wandering groups of hunters; only in post-glacial times was Ireland penetrated, and the first settlers, with a culture of Mesolithic affinities, probably came in the Boreal climatic period around 6000 B.C. The time of gradual transition from Arctic conditions to the warmer Boreal climatic phase is one of great importance in the moulding of the Irish landscape, as the rivers, fed by melting snow and ice, were larger than at present and eroded wide channels, and many of the glacial deposits were re-sorted into outwash sands and gravels. In this phase the birch, the pine and the willow began

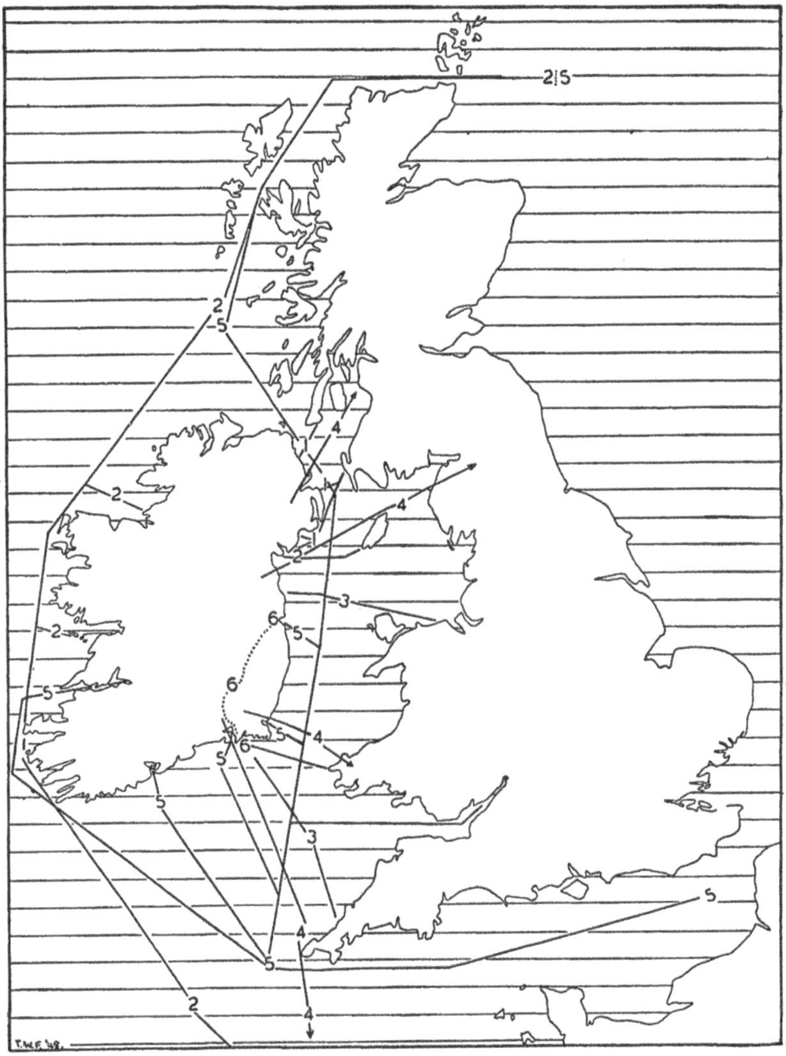

FIG. 12. SPACE RELATIONS OF IRELAND

The main lines of overseas movement affecting Ireland are shown diagrammatically: 1, Mesolithic; 2, Neolithic; 3, Bronze Age; 4, Age of the Celtic Church; 5, Norse; 6, Norman.

to form forests as the climate ameliorated from tundra to the more 'continental' conditions of Boreal times, in which the summers were probably similar to those of the present but the winters colder and the precipitation less; during this phase, around 6800–5600 B.C., hazel and pine became dominant, with elm and oak in the later stages as the climate improved towards its optimum around 4000 B.C. The land emergence of Boreal times probably extended to the present 20-fathom line, but it is impossible to say whether the land-connexion with Britain was complete. The sites chosen by the first settlers, and not now submerged or lost, include many open beaches, especially those rich in flint derived from the chalk of northeast Antrim and spread by ice along the present eastern coastlines. Down and Antrim have the main traces of Mesolithic occupation and consequently the name *Larnian* has been proposed for the culture associated with northeast Ireland, southwest Scotland and northwest England. Traces of occupation of Mesolithic affinities also occur along various rivers, notably the Bann, and on various sandhills, but many of these are of much later date than the Boreal phase and some communities, regarded as epi-Mesolithic, remained at a primitive level, living mainly by fishing, collecting and possibly fowling until the early part of the Christian era. The whole story of the Mesolithic Age and its long survival as a primitive cultural relic into later ages cannot be told until more excavations reveal its full extent: so far, the northeast stands out as an area of special importance.

Around 5600 B.C. the Atlantic climatic period, which covers three millennia, began: in its earlier stages conditions became warmer and moister, so that a climatic optimum was reached somewhere between 4000 and 3000 B.C., after which cooler conditions returned once more. The sea rose above its present level and forests were thicker than ever before, with pine prominent at first and a mixed oak forest, with elm, birch and alder as significant elements, appearing later. The climate, at once warm and moist, favoured peat accumulation in the many hollows left between the various glacial and glacifluvial deposits, many of which were fenlands under the earlier and somewhat drier climatic regime. The Atlantic period was followed by the cold, dry sub-Boreal phase, in which peat accumulation was arrested (2500–750 B.C.), and trees, including pine, grew on the drying bogs, in some of which various relics of the Neolithic and Bronze Ages have been discovered beneath a further layer of peat accumulated during the sub-Atlantic phase, in which wet oceanic conditions prevailed once more. The epi-Mesolithic culture, as noted above, survived through all these phases, but the arrival of Neolithic settlers some time around 2000 B.C. (probably earlier) is of vastly greater importance, as it marks the time when Ireland was opened to immigrants from several parts of Europe and assumed its significant position on the Atlantic seaways of Europe.

Ireland has a remarkable variety of megalithic monuments, many of which were built long after the first Megalithic settlers arrived and even, it is suggested, in the late Bronze Age. Exact mapping has shown that the horned cairns found in the north are comparable with those around Solway Firth, in the Clyde area and also in the Isle of Man. Farther south chambered cairns are found, notably in the Boyne valley, and at Baltinglass hill, on the west side of the Wicklow mountains, which are comparable with those of Caithness. Various derived forms occur, such as the 'lobster claw' type, which is related to the horned cairns of the north, and there are also many simple dolmens, apparently late and degenerate forms. It is clear that the Megalithic people came along the Atlantic sea route, possibly both direct from Brittany and even Spain, and also from Great Britain and the Isle of Man, and the actual routes of entry probably included various estuaries and bays around the coast, and not least Sligo bay and Carlingford lough, both of which have numerous megaliths in the vicinity. Galway bay and the Aran islands should also be noticed, as they too lie in an area with a remarkable number of megaliths.

Stone circles and standing stones are also plentiful in Ireland, but generally of uncertain age, though an excavation of particular interest at lough Gur, Co. Limerick, has led to the dating of a circle at 1700 B.C.: a beaker was found on the site. So few beakers have been found in Ireland that this phase of the Early Bronze Age was formerly thought to be of little significance, but it may eventually prove that the beaker people, racially a cross between Nordic and Alpine elements, settled in at least a few places. The first megalith-builders and the beaker people herald the Bronze Age, when Ireland became the focus of a widespread trade in gold ornaments. Although the country is not rich in minerals and its copper and gold mines have long since been abandoned, the copper lodes of West Cork and Co. Waterford, and possibly also those on the eastern side of the Leinster chain at Avoca, were exploited in the Bronze Age, and smelted with tin (imported from Spain or Cornwall) to make bronze. The source of the gold was probably the same auriferous tract of Co. Wicklow, the only Irish locality in which gold has been found in economic quantities in modern times, though as yet few Bronze Age relics have been found in this area. While estimates of the extent of occupation can only be speculative, the Bronze Age saw the emergence of Ireland on the

European scene, still closely linked with the Atlantic sea routes, preserving its megalithic traditions, and, without question, having an increasing population.

In the later phases of the Bronze Age, around 750 B.C., the sub-Atlantic climatic phase, a time of damp weather conditions, replaced the drier sub-Boreal phase, and peat accumulation became important once again not only in the west of Ireland but also in the drier centre and east, where many hollows have been steadily filled up with peat, in places of a fenny nature, to this day. Many crannogs originally planned as lake dwellings are now completely surrounded or covered by bog, which has buried numerous other prehistoric relics such as implements, and even more substantial evidences of occupation such as roads built of logs in the Bog of Allen and elsewhere. One example of a road in the Bog of Allen was ascribed tentatively to the Bronze Age, as a bronze spearhead was associated with it. Iron probably came to Ireland around the beginning of the Christian era, though relics of the Iron Age are singularly few: at this time the population included people whose equipment was still characteristic of the epi-Mesolithic, Megalithic and Bronze Ages, and Ireland lay beyond the easy reach of the invading hordes of Iron Age people who spread from continental Europe to the English Lowland.

The Roman legions never landed in Ireland, though Agricola, in A.D. 81, contemplated an invasion with a force of only one legion and auxiliaries, certainly an underestimate of the force he was likely to need.[5] Before the end of the second century, and still more during the third and fourth, the menace of raids from Ireland impelled the Romans to fortify parts of the Welsh coast, much as they fortified sections of the east coast against the Saxons or kept a guard on the Roman wall against incursions from Scotland. By the beginning of the fifth century, people from Ireland, the Scotii, were established in some strength both in north and south Wales, in Cornwall and in Argyll, but there is little trace of these settlements in the next two centuries, and it is not certain whether the people were expelled or absorbed. All the evidence suggests that the population of Ireland must have increased steadily during the prehistoric ages, and it is not improbable that considerable numbers of refugees, escaping from the Romans and later from the Saxons, went to Ireland and that a number of slaves, some of whom were Christians, were taken across the Irish sea.

THE AGE OF THE CELTIC CHURCH (AGE OF THE SAINTS)

At no time is the distinction between the western section of the British Isles and the eastern clearer than this, the traditional golden age of Irish history according to many of its people.[6] The eastern lowland of Britain received the Anglo-Saxon hordes, but the west was Celtic and Christian and from the Irish Church missionaries penetrated all parts of Europe. The church in Ireland received no shock comparable with that of the heathen Anglo-Saxon invasion of the English lowland, and developed distinct characteristics. The contrast between the Celtic and continental churches centred around the date of Easter and the monastic system. Although the Council of Arles, 314, and the Council of Nicea, 325, had given rulings on the date of Easter, the Celtic church kept the older date, but the controversy became acute only during the sixth and seventh centuries, when contact with the continental churches became general. Similarly the Irish monastic system was a product of isolation; although it is possible that St Patrick, whose arrival is generally dated 432, was sent to establish the diocesan system, every early Irish Church was monastic. A few of the Irish churches date back to the fifth century and many more were established during the following centuries. Some of the 'saints' founded several churches to form a *paruchia* which might be scattered over the whole country, though many were associated with the *tuaths* or smaller political units of the times (on these see pp. 84–5). Again, many of the churches were in remote places, including rocky islands off the coasts of Ireland and Great Britain, and before the end of this missionary phase the churches of Irish monks were established in all parts of Ireland, in Great Britain, and in many parts of the continent: for example, the work of evangelization carried the Irish into southwest Germany, where the people were still pagan during the seventh and eighth centuries, and through the Frankish domains from the seventh century onwards.

There is an apocryphal element in some of the stories of Irish saints, and not least of St Brendan, who was associated with Clonfert and lived during the sixth century.[7] St Brendan visited a number of luxuriant islands, long celebrated as the Fortunate Isles or Isles of the Blest, sojourned for a time on one of them, and, it is said, reached a coast whence he penetrated the interior by a forty-day journey, halting on the banks of a great river in a land of fruit and gems. He saw volcanic action, thought he was viewing the infernal regions, and observed, according to one account, Judas in perpetual torment. The narratives point to the possibility that St Brendan reached Madeira and neighbouring islands, but the wanderings of the saint were sufficiently extensive for him to view icebergs as well as tropical islands set in warm seas.

In time the association of the Celtic church with the older date of Easter made their missionaries less welcome, and the clash between Celtic and orthodox views came to a head at the Synod of Whitby, 664; but the conflict between the missions of the Celtic Church and those of St Augustine, spreading from Iona and Canterbury respectively, failed to deter the missionaries from their enterprise both in Britain and on the continent. The Venerable Bede (*c.* 672–735), himself working in a diocese where Celtic Christianity had been powerful, wrote of the constant contact between Ireland and Britain. Some of the wide influence was achieved by the successors of St Columba at Iona, with which Durrow, Kells and Derry were associated; at one stage missionaries from Iona were the dominant influence in Northern Ireland, Scotland and Northumbria.

The learning and art of the monastic churches has been studied by many workers, who have shown the singular perfection of Celtic forms, notably on the crosses still preserved beside ancient churches. Kenney writes of the secular literature as the 'mental revelation of an early Medieval people, practically unmodified by ecclesiastical control, in which primitive, pre-Christian and pre-Roman elements are mingled'.[8] During the ninth century Irish influence in European scholastic circles reached its height; many of the exiles were teachers of grammar or scribes, and the surviving manuscripts include the book of Kells, in which the gospels are copied with an endless wealth of exquisite decoration. How far the monastic church influenced the life of the people remains obscure, though some of the monasteries were surrounded by small communities of people, and in time some of the abbots became lay lords, whose family held the office from one generation to another, with the monks as farm labourers. The relation between the ecclesiastical and secular authorities is also obscure, though there are indications that the local ruling families commonly held control of the monasteries, at least in the later phases.

During this phase, and for some centuries before and after, Ireland was politically organized into a number of *tuaths*, which were in effect small states,[9] probably about eighty to one hundred in number, which tended to form associated groups under the leadership of one particular state in the group: Irish life was not tribal, for the *tuaths* were in effect political divisions administering laws through their own judicial and executive organs. Citizenship was given to those holding property in land, to men of learning including clergy, and to those versed in the liberal crafts such as housebuilders, chariot-makers, the builders of ships and boats, harpers but no other musicians, and to many more, but not to tenants, craftsmen, strangers and those defying the laws. The laws were those of Ireland, but their administration was through the local assembly over which a king presided: he was also the commander of forces in war and a judge in the public courts. A source of weakness was the elective system by which a new king was chosen from any son, grandson or great-grandson of a former king, or in other words descendants of one great-grandfather: not unnaturally this tended to split up the royal family into several rival factions. There were no standing armies except possibly during the third and fourth centuries A.D., and the long resistance of the *tuaths* to the Norse and the Anglo-Normans was sustained by civilian levies. In social organization the *tuath* was vastly different from the city-state of the Mediterranean, for MacNeill says, 'the Irish state remained a rural city, a city of the fields. Its chief occupation was agriculture, and all its magnates were agriculturists.'

The combination of *tuaths* into groups probably formed the basis of the provinces: thus *Dal Cuinn*, the division of Conn beyond the Shannon, was the foundation of Connacht, and one of its families held the high kingship of Ireland from *c*. 404 to 1002.[10] This kingship was centred at Tara, Co. Meath, on land wrested from *Laigin* (Leinster) and possibly *Ulidia* (Ulster) to make the kingdom of *Mide* (Meath), which consisted of the eastern part of the Central Lowland, in what is now the counties of Meath, Westmeath and Longford. It is commonly said that many of the early roads centre on Tara and wind across the eskers of the Central Lowland, but exact confirmation of their courses is generally lacking, and the word 'esker' does not, as commonly supposed, itself mean 'road', but merely a ridge or elevation.

In many ways the age of the Celtic church presents a picture of a civilization of distinctive quality developed in an insular environment which, by its remoteness, was able to preserve ancient traditions and to build something with a quality of greatness. Sir Cyril Fox's point about the continuity of tradition in the Atlantic world is apposite here: never was the conservatism of Ireland more thoroughly seen than at the Synod of Whitby in 664. Even so, the civilization did not keep itself to itself, for never were the Irish more eager to spread their religion and learning. The shock of external contact came later. Irish monks remained on the continent of Europe at several famous monasteries, including Ratisbon, Wurzburg, Nuremberg, Constance, Vienna and many more, but eventually they were replaced by Germans or others: at Ratisbon, the last Irish monks were ejected in 1515 on the ground that the monastery was founded by the Scotii who must, it was argued, have come from Scotland and not from Ireland.

THE NORSE INVASION AND SETTLEMENT

The Norse peoples, coming both from Denmark and Norway, arrived off the Irish coasts around 795, and for two centuries they plundered monasteries and spread ruin through large parts of the land; sailing up the rivers in their long-boats, they established themselves at times in suitable bases such as estuaries and inland waterways, among which lough Neagh was important.[11] Several of the monasteries were sacked by the Norsemen on numerous occasions: thus, there are records that Armagh was sacked at least nine times, Clonmacnoise ten, Kells five, Glendalough four, Lismore six and Kildare sixteen: the Round Towers in Ireland were probably built to defend the churches, to store treasures, to watch for invaders, and, in times of peace, to serve as belfries. The aims of the Norse peoples were partly to acquire loot, but their attacks, though savage, were intermittent, and the Church was by no means submerged either at home or abroad, for many Irishmen still went to continental Europe during this phase, though they were teachers of grammar and scribes rather than ascetic monks: the saint had become the scholar. Ireland was hindered in its resistance to the Norse peoples by the lack of any unified control, for the High Kingship existed more as a symbol than as an effective controlling force, but the Norse peoples settled down as traders and Christians rather than as pirates and pagans, and in time were absorbed into the general population, as elsewhere in Europe. Although the tales of their ravages come from every part of Ireland, their final settlement was limited to small kingdoms around Dublin, Wexford, Waterford, Cork and Limerick.

The Norse kingdoms of the ninth and tenth centuries include the four major ports in central and southern Ireland, Dublin, Waterford, Cork and Limerick. All these ports were trading centres beside a river, on sites not unlike that of Copenhagen, which originated as a wharf beside a sheltered creek: in Dublin ships used the Dodder estuary at Ringsend and a riverside wharf on the Liffey near the centre of the city: at Cork, a creek between two islands in the Lee (see pp. 339–41); at Waterford, the confluence of the St John's river and the Suir (see p. 315), and at Limerick, a sheltered haven guarded by the islands on which the older part of the city is built (see pp. 385–7). The Kingdom of Dublin, which appears on a map as a forerunner of the Pale to which the Normans were reduced in their less prosperous phases, controlled the vital crossing of the Liffey in the tidal reaches and for some distance inland (probably as far as Leixlip, the 'salmon leap'), and included two minor river ports, Wicklow and Arklow, and various other havens rarely visited by modern shipping but well suited to the navigation of a thousand years ago.

Only a small part of Ireland was under the direct control of the Norsemen in the early years of the eleventh century, and the remainder was divided into 'kingdoms' which show an interesting relation to the ground-plan of the country. Thus, the Kingdom of Ulidia foreshadows the east part of Ulster, and the Kingdom of the North is its westward continuation through varied mountain and lowland country to the western seaboard. Connacht was the land to the west of the Shannon, though Clare was linked with Munster by crossings at two ancient ecclesiastical centres, Limerick and Killaloe. Tributary to Connacht in the eleventh century was the Brefni kingdom, which straddled the west part of the great drumlin belt and the plateaus south of the Erne valley. The Meath kingdom, one of the most interesting and persistent in Irish history, covers the Central Lowland from the Shannon to the sea; Meath is not now recognized as one of the provinces, though the name still survives in the dioceses and in two counties. The Leinster kingdom is based on the valleys and surrounding lowlands between various mountain ranges, and its major rivers are the Slaney, which alone pierces the Leinster chain, the Barrow and the Nore: with the last of these the kingdom of Ossory is associated. Finally, Munster consists of the east–west ranges and lowlands that extend from the neighbourhood of Waterford into the magnificent peninsulas and bays of Kerry, together with the less attractive and generally low-lying country around the long dreary estuary of the Shannon, with Clare.

The impact of the Norse people on Ireland is difficult to assess, and much of the tradition centres around the plundering, of which it has been written, 'From the fury of the Northmen, Good Lord deliver us.' But in later years they laid the foundations of towns destined to be important for centuries. The 'Ostmen' of Dublin founded Christ Church Cathedral in Dublin, on a ridge beside the Liffey in 1040, and their appointment of a bishop consecrated at Canterbury helped to orient the Church towards the rest of Europe. Some of the monasteries pillaged by the Norse never revived, but others, such as Clonmacnoise, maintained themselves and their monastic tradition: the craftmanship achieved a new perfection during the Norse period both in stone work, for example the High

Crosses of Tuam and Monasterboice, and in gold and bronze working, seen in the Cross of Cong. Architecture was similarly successful, and Romanesque tradition was linked with Irish design to produce such buildings as Cormac's Chapel, on the Rock of Cashel. There was, in fact, the same continuity of tradition that marked previous ages, combined with the first clear signs of town growth: there was also the political instability which made possible the Norman invasion from Wales in 1170, at the invitation of Dermot, King of Leinster.

THE NORMAN 'CONQUEST'

At the time of the Norman invasion the *tuaths* were still in existence and life in Ireland was essentially rural, for towns were unknown to almost all the population except in the vicinity of the Norse centres. The Irish used most of their land as rough pastures, and had improved very little and cultivated still less, though they had crops of wheat, barley, rye and flax: possessing several arts, they found idleness more attractive than work as a general rule. In the Church an effort was made to divide the country into dioceses, but contemporary accounts say that the clergy still preserved their monastic traditions rather than a more pastoral and organizing outlook, and were in fact given rather to the life of contemplation than to the improvement of their flock.[12]

The Normans failed to make Ireland into a single kingdom, though from the beginning of their invasion they recognized the strategic importance of Dublin, which submitted to Strongbow within a few weeks of his arrival in the late summer of 1170. Dublin was given its first Charter in the winter of 1171–2, under which the city was to be inhabited by men from Bristol, with whom there were already considerable commercial contacts: the early records suggest that Dublin had inhabitants drawn from several other towns as well, who settled on the south side of the river within the old walls, while the survivors of the Norse population moved across the river to a suburb known as Villa Ostmanorum, or Ostmanstown, a name still surviving as Oxmantown. Various strategic points were fortified by motte-and-bailey castles and by more permanent stone structures, and the Normans established themselves firmly in the old Norse centres, Waterford, Wexford, Cork and Limerick, and in various other places including New Ross, Kilkenny, Carlow, Arklow, and many more that are still important market towns. Having these bases, the Normans hoped to conquer Ireland through the great Anglo-Irish nobles, but in time many of their lordships became virtually Irish, and the English kings were never able to secure the measure of unity and control that they established in the English Lowland.

During the thirteenth century, Anglo-Norman power spread through the country leaving, however, many areas unsubdued or owing merely nominal allegiance to the Crown. The area most effectively controlled was Leinster, though even here the Wicklow mountains remained unconquered and the country now known as Leix and Offaly preserved its independence. Munster had lordships based on the valleys and the lowlands of the south around Cork, and the Decies covered what is now Waterford: a kingdom of Limerick extended into what is now Kerry. The de Burgh family fought for Connacht from 1201 with varying fortunes and established the town and castle of Galway against the attacks of the 'men of Connacht'; this town, the 'key to all Connacht', was walled from 1270 onwards and became a lone outpost of English influence, differing in language and culture from the surrounding country but able, through its strong strategic position, to maintain itself as an important town through every stage of historical development.

The Anglo-Norman lordships were strongly held until the early fourteenth century, but the Gaelic chiefs, aided in their designs by the quarrels of Norman families, were gradually able to regain parts of their former territory. There was no effective anglicization of the country, nor was there any centralized force to subdue rebellion and, even more important, the Normans were adopting Irish customs. Neither the English language nor the English law was known in the greater part of Ireland. The Statutes of Kilkenny, 1366, endeavoured to prevent further absorption of the English by the natives, and the preamble states the 'now many English, forsaking the English language, fashion, mode of riding, law and usages, live and govern themselves according to the manners, fashion and language of the Irish enemies, and have also made divers marriages and alliances with the Irish enemies aforesaid': it was therefore enacted that the English should not adopt Irish customs, should speak English only, marry wives of their own nationality, and that, in the part of the country controlled by the English, no Irishman could become a beneficed clergyman. The 'land of peace' envisaged in these laws was the counties and liberties of Louth, Meath, Trim, Dublin, Kildare, Kilkenny, Carlow, Wexford, Waterford and Tipperary: here alone, it then seemed, was there any hope of maintaining the English ruling-class as a strongly entrenched force. Although the laws remained in force for more than two hundred years and were re-enacted on several occasions they failed to establish a social order in which the conqueror and conquered remained distinct one from

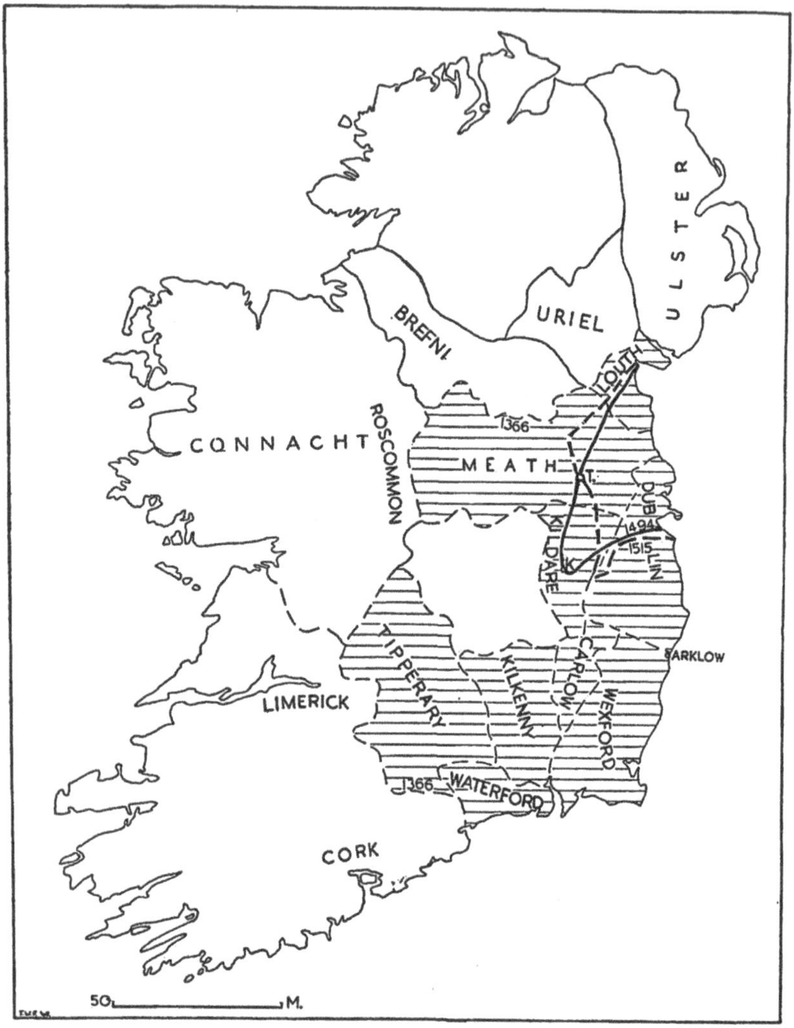

FIG. 13. THE 'LAND OF PEACE' AND THE ENGLISH PALE

In 1366 the Statutes of Kilkenny aimed at establishing the 'land of peace' in nine counties. With Meath is included the liberty of Trim and the modern county of Westmeath. The boundaries shown here are only approximate; the town of Arklow was allied to Kildare in the fourteenth century, and no county of Wicklow existed. At no time was the 'land of peace' effectively held throughout the area shown.

Thanks are due to Dr Jocelyn Otway-Ruthven for assistance in the compilation of this map and also of Fig. 14 another, for many of the English were absorbed to become 'more Irish than the Irish themselves'. Not all, however, for Anglo-Irish families still remain a not inconsiderable element in the 'land of peace' and their apartness from the general population has been made sharper by a divergence in religion from the sixteenth century: it is in the area mentioned in the Kilkenny Statutes of 1366 that demesnes or parks are most numerous: here, too, the farm units are larger than in the land either to the north or to the west, and substantial Protestant communities are still to be found (see Fig. 13).

The final refuge of the Anglo-Normans was the English Pale, a strip of country which was based on Dublin and the coast to the north and south, with inland boundaries varying considerably from one time to another: in times of weakness it was hardly more than the modern county of Dublin, but in times of strength it reached far into the lowlands to compass part, if never all, of the 'land of peace'.[13] At the beginning of the fifteenth century the English lordship had almost ceased to exist, and Ireland was under the control of a Gaelic and Norman aristocracy, with only a few eastern counties as a relic of the 'land of peace'. The Kilkenny Statutes were hardly observed at all, and even the English ecclesiastics were reported to be in full flight, especially from the counties of Wexford, Waterford, Kilkenny and Louth, the very areas where the spread of English customs and culture seemed most probable. In 1435 it was said that the king's writ hardly extended more than thirty miles from Dublin, and that beyond this enclave the power lay with 'enemies and rebels': some thirty years later the Pale was defined, apparently for the first time, as the counties of Louth, Meath, Dublin and Kildare, but in fact it was considerably smaller. Even the Pale was not secure, for many Irish ways prevailed there too, and raiders entered it to steal cattle. In 1494, after the Tudors had been in London for nearly ten years, it was suggested that the Pale should be protected from insurgents by a double ditch, to run from Dublin to Kildare, and thence northward to Trim and on through Meath to Dundalk (see Fig. 13): in 1515 it was reported that 'the folk daily subject to the King's laws included half the country of Louth, half of Kildare, half of Dublin and half Meath', of which the last meant the modern county of Meath with Westmeath excluded.

Although the Normans failed as conquerors, the Ireland of 1500 was in many ways different from that of 1170; towns[14] were the great gift of the Normans to Ireland; the first of them were the old Norse or Ostman centres which became royal towns with mayors during the thirteenth century. The establishment of Dublin as a capital city was of particular importance: made an ecclesiastical city by Ostmen, having a Norman charter and a mixed population, a port trading especially with Bristol, situated within sight of the hills but at the seaward end of lowland routes from all over Ireland, it became the great political, economic and social centre of Ireland, but always more cosmopolitan than the country as a whole, as London is in England or Paris in France. The other towns were less securely entrenched in the life of the country, and of those shown in Fig. 14 a number have perished or failed to grow, partly because the ancient custom of selling goods at rural fairs allowed the 'enemy', according to an Act of 1431, to 'take great customs and profits to the depression of the boroughs and the trading towns'. Some of the Norman towns, however, have had a continuously successful career since they were founded: thus the west coast towns, such as Galway, Limerick, Tralee and Dingle, had a strong maritime trade with France and Spain, from which wines and cloth were imported in exchange for wool and hides. Fig. 14 shows that towns were most numerous in the southeast of Ireland, and a comparison with Fig. 23, p. 135, will show that this is still the case, though in the twentieth century Ulster has a number of towns depending on industry, of which some were established in Tudor and Stuart times. There is considerable evidence that in time the Irish conquered the towns by infiltration, and it is reported that Irish was heard more frequently than English in the streets of Galway during the reign of James I. The names of families prominent in Galway include Lynch, Kirwan, Nolan, Fallon and Tully in Galway around 1350, and the effort to keep the Irish and English apart failed in the end, so that no meaning was left in by-laws stating for example that 'no O' or Mac (shall) strut or swagger through the streets of Galway'.

THE TUDOR AND STUART PLANTATIONS[15]

In 1500 Ireland's veneer of urban settlement, though important in the country's trade, was an insufficient basis for the political and commercial integration that the Tudor monarchs had in mind, yet a county organization covered most of the country and the dioceses were well-established units comparable, if not exactly identical, with those existing today. The impact of the Anglo-Norman phase of occupation was seen most fully in some of the towns and around some of the strongholds of the nobility in Leinster and Munster, and least clearly in Ulster and Connacht. From Tudor times onward the effort to conquer Ireland was made more firmly than before, either by the conciliation and anglicization which some

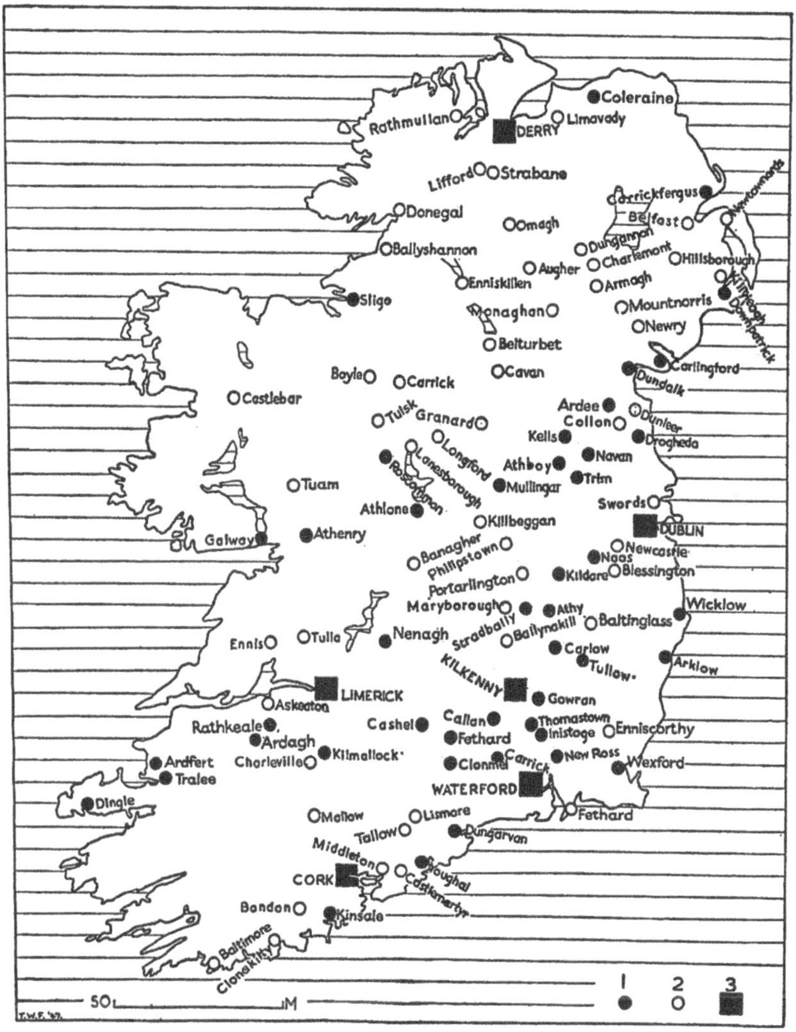

FIG. 14. TOWNS IN IRELAND

1, Norman foundations; 2, Tudor and Stuart foundations; 3, Cities. For sources, see note 12, p. 517. There is a list of southwestern towns in *Orpen, G. H., 'Ireland under the Normans'*, vol. 4, p. 49; see also *Moody, T. W., 'The Irish Parliament under Elizabeth and James: A General Survey'*, *Proc. R. Irish Acad.*, vol. 45 C6, 1939, pp. 41–81.

Most of the towns shown had charters, but there were many more medieval towns not included, of which a large number have little importance today. Some shown here are now mere villages, but most of the major towns are included; the towns of Ulster are mainly associated with the plantations.

thought possible or by war. Henry VIII hoped, at least before the effects of the Reformation were seen, to win the country by a natural process of pacification in which the chiefs were to give their allegiance to the Crown in return for titles to the land and English law. At first the towns were loyal to the Tudors, as their citizens expected greater trade, security from the predatory attacks of the Irish, and the maintenance of the privileges granted in their Charters at the beginning of the sixteenth century: at this time Dublin and several other towns were said to be 'adorned with fine buildings, prosperous, secure in their medieval gild organization, and in every way comparable with the major towns of England'.

The Plantations brought new elements into the population of Ireland even more surely than the Anglo-Norman settlement, for one of the essential aims was to make parts of Ireland 'for ever English' (or Scots in some cases), as civilized nuclei of a country ruled by the Crown. Hardly less important was the establishment of many new towns of which some acquired Charters, especially under the Stuarts, and the rehabilitation of many old centres. In some ways the plantation was feudal in conception and designed to attract people to settle down as farmers rather than to wander over large areas with their cattle. The only Irish villages were clusters of houses around mansions or churches, and the houses were made of rods or wattles plastered over with clay and covered with straw, in effect circular huts supported on poles, with no window, but a hole in the roof to let out the smoke from a fire on the floor. Some of the historic sites were neglected and even Armagh was reported in 1586 to be a 'small village near the church and friaries' which were 'for the most part broken or neglected'.

PLANTATIONS IN LEINSTER AND MUNSTER

Leix and *Offaly*, immune from direct Norman influence, were among the first parts of Ireland to be planted, and their now-abandoned names, Queen's and King's county respectively, commemorate the making of the two shires under Philip and Mary in 1556 on lines similar to those of their neighbours, Meath, Kildare, Kilkenny and others. Parishes were formed and the main centres were to be 'for ever' Maryborough (Leix) and Philipstown (Offaly). But in 1575 it was said that 'now both counties are much spoiled and wasted, by the race and offspring of the old native inhabitors which grow great and increase in number; and the English decay, both in force and wealth, and are not of ability to answer their rents and services, but let their land to Irish tenants'. The main danger came from the O'Moores of Leix and the O'Connors of Offaly, who were banished to the neighbourhood of Tarbet and various other parts of Munster and Connacht in 1608: some fourteen years later, the plantation was said to be successful. As a whole, the Leix and Offaly plantation showed features typical of others, among them the fear that the whole purpose would be defeated by the settlement of the Irish on land given to the English, the transportation of refractory elements to the west, and the founding of new towns.

The MUNSTER plantation began around 1586, and large parts of the province were divided into seignories, or lots, of 12,000, 8,000, 6,000 and 4,000 acres, to be held by the 'undertakers', together with various common lands. The various attractions laid before the settlers included descriptions of the fertile soil 'apt for wheat, rye and barley', the abundance of good timber, deposits of ironstone (ore) and lead, delicacies such as oysters, mussels and cockles, all kinds of edible birds including wild swan, cranes, pheasants, partridges, quails and woodcock, and a cost of housekeeping less than one-quarter that of England. The whole account, which was written in 1590 by Robert Payne, an undertaker in Co. Cork who hailed from Nottinghamshire, reads like the prospectus of any new company exploiting virgin resources, but other accounts suggest that the new settlers were not always fortunate, and also that Ireland became a bolt-hole for undesirables. In 1588, one of the Pembroke family said that 'our pretence in the enterprise of plantation was to establish in those parts piety, justice, inhabitation and civility, with comfort and good example to the parts adjacent . . . our drift now is, being possessed of land, to extort, make the state of things turbulent, and live by prey and by pay'. Many of the undertakers would not let land to settlers for a period of more than twenty-one years or three lives, and found it profitable to let land to Irish tenants who were generally willing to pay higher rents, but the 'better kind', it was said, really tried to plant their territory with Englishmen, who were drawn in many cases from Dorset, Somerset, Devon, Lancashire and Cheshire.

Two men of renown, Sir Walter Raleigh and Edmund Spenser, were among the undertakers: the former was conspicuously successful and the latter not, though he wrote very attractively of the countryside in which he lived for some years. Raleigh acquired various lands in Co. Waterford, and subsequently added 40,000 acres in Cork, Tipperary and Waterford, including the town of Youghal, already a port, and the salmon fisheries on the river Blackwater. In 1604 he sold those lands to Sir Richard Boyle, 'the first and great earl of Cork', who rebuilt various towns, and founded the frontier towns of Bandon, Clonakilty, Enniskean and Castletown to hold the settled country against the Irish of West Carberry (Co. Cork) and Kerry. The Earl employed some hundreds of workmen and 'created several corporations of Protestant English traders'. Spenser had a much smaller estate of some 3,000 acres in Co. Cork and lived in an old castle at Kilcolman, three miles from Doneraile. In 1598 he complained that the settlers had been attacked by the Irish, and that they had adopted the ways of their persecutors to become predatory in their turn: his Memorandum to the Queen said, 'out of the ashes of desolation and wasteness of this your wretched Realm of Ireland, vouchsafe most mighty Empress our dread Sovereign, to receive the voices of a few most unhappy ghosts... which lie buried in the bottom of oblivion, far from the light of your gracious sunshine....' Gloriana, it seems, was too far away, and Spenser, after his house was burned, retired first to Cork and later to London.

At the beginning of the seventeenth century, the most advanced parts of Ireland were Leinster and Munster: the Pale had some timbered houses in the countryside and a number of towns, and the remaining areas of these two provinces had several walled towns and cities, a number of minor towns and a considerable export trade, especially with Spain. An account of 1611 shows that the major ports were Dublin, 'to which all the kingdom resorts'; Waterford, the next in importance, but with only one-third the trade of Dublin; Cork, Drogheda and Galway, three towns of approximately equal commercial importance; New Ross, 'a poor ruined town out of trade': Dungarvan, 'a very poor fisher town', Youghal and Kinsale. Much of the export and import trade was conducted through London, Chester and Bristol: the exports included hides and skins, timber, tallow, fish, flax, leather, wool, linen yarn and frieze (homespuns), all of which may be grouped as raw materials or the surplus of domestic industry. The imports included wine (always prominent), iron, salt, spices and a wide variety of manufactured articles including pewter, textiles (notably silk), groceries and ammunition. There was spinning and weaving in the towns and some domestic production of yarns and cloths.

ULSTER UNDER THE TUDORS AND STUARTS

Ulster in the sixteenth century had very few towns, though a number of Anglo-Norman castles and some ancient ecclesiastical buildings survived, most of them in ruins. In the reign of Elizabeth the Earl of Essex was given the government of Ulster, but he reported that the difficulties of colonizing 'this waste country' were almost insurmountable, owing to the opposition of the natives and the failure of the 'adventurers' who, 'having not forgotten the delicacies of England and wanting resolute minds to endure travail', were eager to return home, though some English had settled in Ulster during the reign of Elizabeth as during the Anglo-Norman phase. A survey of 1586 suggests that the whole province was in a primitive state with, however, great agricultural potentialities: in Antrim and Down there had been many immigrants from the western islands of Scotland who had settled along the coasts, encouraged by the O'Neills. In the effort to encourage the immigration of lowland Scots and English, the British Government gave the MacDonnells a large part of Antrim, defined as the *tuaths* between the Bush and the Bann, to which they added the Glens of Antrim and the Route early in the seventeenth century. In Co. Down there was also a tradition of Scottish immigration, and in 1606 the Montgomerys were established in the county: on their arrival the country was so desolate that 'thirty cabins could not be found, nor any stone walls, but ruined roofless buildings, a few vaults at Grey Abbey, and a stump of an old castle at Newtownards'. The Montgomerys brought over smiths, masons and carpenters to Donaghadee and Newtownards, who built houses of stone and humbler cottages and booths with sods and saplings of ash, alder and birch, using rushes for thatch and bushes for wattles. In summer, Donaghadee was only three hours from Port Patrick, and boats crossed two or three times a week bringing more immigrants: within a year the colonists, aided by good harvests in the years 1606 and 1607, were selling grain in Scotland and Newtownards had become an important centre trading with Belfast, then a small but growing town.

The 1586 account speaks of Co. Louth, now included in Leinster, as part of the English Pale, possessing corporate towns at Drogheda, Dundalk, Ardee and Carlingford, but liable to frequent incursions from the interior. Co. Monaghan, which was not included in the Plantation except for some local grants, had 'no buildings, only a few defaced monasteries'. According to an account dating from the middle of the seventeenth century, this county was 'the most poor, barbarous, and despicable in the kingdom', except for a few areas planted by the Earl of Essex and by Lord Blayney, who founded the town of Castleblayney in 1641.

The plans for plantations under the Stuarts included the granting of lots of 1,000, 1,500 or 2,000 English acres, together with woods and bogs, to undertakers who were to be either English or 'Inland Scottish' (that is, not those from the Western Isles) and servitors, military men who also let out land to tenants. The counties of Down, Antrim and Monaghan were not included in the plantation; of these the first two had a considerable number of immigrants, whereas Monaghan was little influenced by the Scots or English. The counties to be planted were Fermanagh, Tyrone, Donegal, Derry, Armagh and Cavan. The Derry plantation, for example, which began in 1610, was preceded by an appeal to the corporation of London to provide settlers for the enterprise: like several other plantations, it brought numerous immigrants, some of them English, but more of them Scots, and also settled a considerable Irish population on the land. The countryside was described as well-watered (a fact which no one would question), with plenty of fuel available both from the woods and the peat bogs, fit for all kinds of husbandry and for the breeding of horses and cattle, suited to 'English sheep', with an abundance of wild life, including deer, sheep, rabbits, martens (probably the pine marten, still occasionally found in Ireland), and squirrels. Flax, madder, hops, woad and even hemp could be grown. Fish, especially herrings and eels, abounded, and the rivers were liberally stocked: pearls could be found in the oysters of the Foyle estuary. In all these favourable circumstances, the area to be planted could supply London with beef, pork, fish, rye, barley and beans; the hemp and flax could be made into yarn, cloth and sails for the shipping, and the local stone and wood could meet the needs of the building trade. Moreover, the favourable position of Derry's excellent harbour guaranteed trade with Spain and the 'straits' (of Dover) and Newfoundland. Optimism marks the documents inviting the settlers, caution the recommendation that those who came to view the promised land should 'be led by the best ways' and lodged 'in their travels where they may, if possible, receive English entertainment in Englishmen's houses'.

There was no immediate displacement of the native population in the six counties escheated in 1610, for the aim, similar to that of earlier plantations, was to make islands of British territory as oases of 'civility and religion' in the midst of Irish barbarism and superstition.[16] Some of the land was given to English and Scottish undertakers who settled British colonists on their estates, or to servitors or 'deserving' native grantees who were free to have natives as tenants, or for the support of the Church, chiefly to maintain bishoprics: on these lands the natives were to remain. In all, the undertakers, servitors and native grantees numbered only a few hundreds, and the plantation did not remove the natives from the lands, partly because Ulster was then sparsely populated. The natives who remained were, however, reduced from the status of proprietors to that of 'tenants-at-will', though they were driven off the more fertile land only gradually, and that more by economic forces than by state policy. While there was some transference of population, expulsion was limited to the 'swordsmen' or armed followers of the Irish lords who were in theory sent to Munster and Connacht but in fact went mainly into the various remote fastnesses within the Ulster province. From the British point of view the plantations in Ulster, as in Leinster and Connacht, were only partially successful, as they failed to subdue Ireland: the more difficult years of Anglo-Irish relations still lay in the future, which held in store both the Cromwellian settlement and the Penal laws.

Cromwell confiscated more than half the whole area of Ireland for his soldiers and settlers, the Adventurers, who in many cases sold their debentures to speculators and returned to England, like many more before them. It was in this phase that transfer to Connacht and other parts of the west became more serious: it was in this seventeenth century that the last traces of the old Irish *tuaths* disappeared: it was now that a new landlord class was installed in Ireland, and the power of the old Catholic landlords was finally broken.

According to Petty,[17] the 'profitable area' of Ireland in 1670 was 18,750 square miles, about three-fifths of the whole island, of which the new Cromwellian settlers had 7,000 square miles, the Catholics 5,500 square miles, and the older Protestant settlers the rest: the estimated population was 1,300,000 (but probably more), of whom 800,000 were Catholics and the rest Protestants, chiefly Presbyterians and Anglicans. Petty's estimates show that the land was more thoroughly occupied than at any previous period, though his 'profitable area' probably consists rather of land regarded as potential farmland than land actually exploited. Having undergone an agrarian revolution in which more land was settled than ever before, Ireland was ready for a commercial expansion, prevented partly by the Navigation Acts which, in 1663 and 1670, forbade direct Irish trade with the colonies and canalized her overseas commerce through England. The cattle trade with England ceased in 1667 and not long afterwards, in 1691, the Penal laws against the Catholics made the relations between the Church of Ireland and the remainder of the population still worse. The aim was twofold: to maintain the dependence of Ireland on England and to establish the immigrant aristocracy which had been planted in the country and enriched with the greater part of its landed wealth.

ARTHUR YOUNG'S IRELAND[18]

Arthur Young's travels and residence in Ireland covered the years 1776–1779 and one merit of his work was his ability to see Irish life through the eyes of people at various social levels. He was not blind to the poverty of large sections of the population and suggested that the gentry might usefully employ more men on their too-abundant acres and allocate holdings to various families; he had actual experience of land settlement on the Kingsborough estates at Mitchelstown, Co. Cork.

The period of Young's sojourn in Ireland antedates the French Revolution, which spread radical political ideas and created new trading conditions throughout Europe; it also precedes the era of the Corn Laws, 1784–1846, which induced Irish farmers to till increasingly large areas of land to supply wheat to Britain, while themselves living on a diet in which the potato figured prominently. Young saw the beginnings of an arable expansion, which he regarded as in some ways against the best interests of the country, though he advocated with emphasis and maddening reiteration the cultivation of turnips for the winter feeding of livestock. The export trade in 1779 was chiefly of agricultural raw materials, and was growing steadily in volume under the stimulus of increasing demand in Britain. Industry was still primarily domestic, though there were a number of small-scale factories in various towns supplying the home market; of the industries, linen and wool figured in the export trade. The first signs of the industrial age were seen in the various canal projects and the improvement of the inland waterways, the expansion and reconstruction of various towns and the mounting interest in the supposedly considerable coal deposits of the country: there was also an increasing population, though emigration, it was said, drew off a number of the more adventurous elements. Young saw Ireland at a time when hope seemed reasonable and there was little to suggest the Famine conditions experienced seventy years later. True, the social and economic advance of the Roman Catholic majority, estimated by Young to be about four-fifths of the entire population, was impeded by the application of the Penal laws, though many recognized the futility of any continued attempt to convert, or alternatively to subjugate, so large and unconquerable a section of the population. Young's picture of Ireland is therefore at once gloomy and optimistic; depressing conditions are painted faithfully and no class of the population escapes censure. The government is blamed severely for its 'interference', commercial interests in both Ireland and England were shortsighted and selfish, but probably a reasonable mind would prevail among many in the end, with new towns occupied by workers in factories, more overseas trade for the ports, not to mention the widespread expansion of turnip fields, better crop rotations, improved cattle breeding and other ameliorations. But like Sir

Robert Kane, who wrote hopefully in 1844 and dismally in 1884, Young died a disillusioned man.

In the late eighteenth century landscape demesnes or parklands figured prominently and in many cases covered at least two square miles, and often considerably more, much of which was in an unkempt condition and of little use except as pastures for the cattle and horses kept by the owner. The natural wealth of the grazing area, which included Limerick, Clare (possibly the east only), Waterford and Meath, was apparent, but the rich lands in Limerick showed fences, drains, and buildings in ruins, and fields full of weeds, as waste as the mountains of Kerry. In the arable areas such as Louth, Kildare, Kilkenny, Leix and Offaly the industry of the population was more in evidence than in the pastoral areas, but generally the farms were too small and the tenants too poor to make the best use of the land; having no capital, they grew oats and wheat for sale, used teams of weak horses, manured the land inadequately, and possessed very few implements. Many of the houses were wretched hovels, consisting normally of one windowless room with walls of kneaded mud and straw, and the door was the only outlet for smoke. The thatch was straw, potato stalks, heather, or sods of turf: although the worst houses were in Connacht, there were several as bad in Limerick and elsewhere. Young appreciated the vital fact that the lowest economic and social level in Ireland was represented not only in the poorer western districts but throughout the country. He also reported that the farms of the north were even poorer than elsewhere and rarely consisted of more than 10 acres and generally of much less, and the main interests of the occupiers lay in the spinning and weaving of flax rather than the cultivation of the land.

The absence of forests was so marked that 'Baltic fir' was imported for building purposes: 'the greater part of the country exhibits a naked, bleak, dreary view for want of wood, which has been destroyed for a century past with the most thoughtless prodigality, and is still cut and wasted.' In all probability, most of the timber was cut for fuel, either domestic or charcoal, though some had been cut for profit. Very little semi-natural woodland has survived in Ireland (see pp. 67–71), and most of the present forests, apart from those of the Forestry Commission, have been planted by the landed gentry. Young notes the great success of many recent plantations, which had such trees as larch, beech, Lombardy poplars and oaks, and also comments on the numerous ash trees planted in hedgerows. These, however, were never cut but allowed to grow for thirty or forty years, by which time they were full of gaps. Ash trees are still a frequent feature in Irish hedgerows.

Many of the gentry built new houses during this period, and it is fortunate that so many of the stately Georgian houses have survived not only in the countryside but also in the towns, where they housed the middle and upper classes of the population. On the other hand, this age was strangely indifferent to the needs of the poor, who lived in squalor both in town and countryside, their thatched hovels grouped shapelessly together in or near the town or spread planlessly through the countryside. The very desolation and poverty of the countryside encouraged Young to hope for change. The short leases, twenty-one years to Catholics and three lives to others (a form of lease still surviving, though generally with the addition of a certain number of years), gave the tenants little abiding interest in their land, and in the event of eviction they became 'squatters' who built a house in a day or two, planted potatoes, and used any available pasture for the animals they brought with them. At the bottom of the farming scale the casual labourer appeared; unprotected by the Poor Law, the wandering families settled in any convenient place, for example under a dry bank where, 'with a few sticks, furze or fern, they made up a hovel much worse than an English pigsty, support themselves by work, begging and stealing: if the neighbourhood wants hands, or takes no notice of them, the hovel grows into a cabin'. Young goes on to comment that around Mitchelstown he had often seen a family installed one morning where none was seen the day before, their rough shelter occupied by a man, a woman, six or eight children and a pig.

Some of the labourers employed on a seasonal full-time basis by the gentry were paid primarily in land rather than in money; on this land they grew potatoes and kept a cow, so that their main diet was potatoes and milk rather than bread, though in the north more oatmeal was consumed. The very poor had to hire a patch of land on which to grow potatoes on the 'conacre' system, but those with even one acre could grow sufficient to feed a family of eight for a year. Times of stringency there must have been, but Young found the people 'as athletic in their form, as robust and as capable of enduring labour as any on earth'. He adds, however, that this capacity was carefully reserved for occasions when the worker wished it to be used. In some cases groups of farmers rented a mountainside and, working together on a co-operative basis, gradually made considerable improvements. Each man had a potato patch for his family, kept a couple of cows, a few sheep, pigs and poultry, and acquired a small cash profit to pay the rent from butter, wool, oats and the sale of a few young cattle and lambs. If possible, the farmers let out some land to the landless on 'conacre' for growing potatoes and oats, or took in cattle for part of the year. Apparently the farmed area was expanding in both the mountainous areas and the lowlands, and Young held the opinion that it would be possible to provide holdings for a substantially increased population.

The main produce of the farming was pastoral in the late eighteenth century, and there were substantial imports of barley, wheat, oats and flour. Already the arable acreage was expanding in many of the richer counties, chiefly those of Leinster, rather than in the poorer counties where subsistence farming prevailed. Young views this critically, asking, 'Ought you to turn some of the finest pastures in the world into the most execrable tillage that is to be found on the face of the globe?' The corn crops were full of grass and weeds, the harvest a tedious struggle with wet autumns, and manuring was little practised, though some lime or limestone gravel and even shell marls (reported from the Shannon and Wexford harbour) were applied to clay soils. Cattle and sheep were being improved by breeding, partly through the efforts of various landlords who introduced English stock. In 1837[19] it was said that when a farmer in Co. Limerick ploughed up extra land, his neighbours regarded him as a man of sinking fortunes, falling towards the subsistence level from the status of a grazier, and it is clear that pastoral farming is deeply entrenched in many parts of Ireland by custom and respect as well as by factors of climate and soil, such as heavy rainfalls and the wide distribution of heavy clays that nurture rich grasslands but are hard to work for crops.

The industries of Ireland were few during the latter half of the eighteenth century, and the country's imports consisted mainly of manufactured goods. Linen cloth and yarn were the only significant industrial export, and the people of the north had already acquired a skill and tradition that made their wares famous, though as yet the industry was almost entirely domestic. It was the cotton industry that brought factories into East Ulster and, in time, drew the flax workers from their rural homes into drab towns; Young thought that the people would be much happier in towns than on the land. Their farms were used merely to provide a supply of potatoes and oats, a little pasture and the flax needed for the spinning wheel; 'all the crops are nothing but filth', and 'agriculture is extirpated, annihilated'. Some people believed that flax should be grown on land already exhausted by several crops of oats, whereas in fact it needs a rich soil plentifully manured; as the linen trade expanded during the century, more and more flax-seed, undressed flax and undressed hemp was imported. Ulster alone showed signs of industrialization on any large scale in the late eighteenth century; Ulster alone has achieved it since.

Woollen manufactures were declining in the eighteenth century, partly because Britain protected her own production against Irish imports and thereby restricted the Irish producers to the home market. No heavy industries were found in Ireland, and no one foresaw the place that Belfast was to occupy in shipbuilding and allied industries a century later, yet it seemed that the coal resources were considerable and that their exploitation awaited only the provision of capital. No Geological Memoirs existed to show that most of the fields are highly faulted, difficult to work, and in any case contain only a limited amount of coal compared to those of Great Britain. Charcoal was the main industrial fuel of the day, though the woods were already few and far between, and the Cumberland coalfield had a considerable export trade with Ireland. The iron founders bought ore from Lancashire, or pig iron from Bristol, Chepstow, Workington or Carron (Glasgow): five of the foundries were in Dublin, two at Belfast and one each at Enniscorthy, Mountrath and Newry. Much of the manufacturing enterprise of Ireland was in the hands of traders who kept workrooms rather than factories and sold goods of high quality and fine craftmanship to the gentry; there is, for example, a tradition of handmade boot and shoe making in Dublin, and the wealthier homes of Ireland generally contain some old Irish silver, furniture and glass of superb quality and design. A few of the factories now operating in Ireland trace their origin back to the eighteenth century.

Roads and canals, with river waterways, were being constructed and improved during the late eighteenth century. 'Presentment' highways were made (see pp. 232–3) with public money, which was also available for the canals. One of these, the Newry Canal, was designed largely to carry coal from the neighbourhood of Coalisland, west of lough Neagh, to the sea. Most of the canals have wharves, warehouses, hotels for travellers, tollhouses and bridges of neat design, and, near Dublin, rows of elms planted to provide the city with waterpipes. Young was scathing about the route followed by the Grand Canal from Dublin, as it struck through the heart of the Central Lowland instead of proceeding to the coalfields that he supposed to be so rich: there would, he said, with undue pessimism, be nothing to carry but turf, and 'who at Dublin would burn that with Whitehaven coals at their present price?' The public works of the day also included harbours, piers, bridges, churches and schools, of which some have proved useful and many not: those familiar with the Irish countryside will know some harbour that never holds water, and probably never did, or some church for which no congregation ever existed.

Social conditions in eighteenth-century Ireland were marred by the continuance of the Penal laws against the Catholics, which prevented the vast majority of the population from rising above a humble level of society either in the countryside or in the towns, though there were some rich Catholic merchants and shopkeepers. Although these laws were abrogated or allowed to fall into disuse towards the end of the century, the social divisions remained acute and hampered the effective rise of a substantial middle class in both town and countryside. The landlords, some of whom were absentees in England or residents in Dublin, commonly employed middlemen to arrange the letting of land. Not all the landlords were pernicious in their habits, but their power was excessive; 'even in this polished age,' comments Young, 'there are instances of a severe carriage towards the poor'. The sons of the gentry rarely engaged in trade or any other occupation and many of the country houses had a number of relations who were 'idlers . . . whose best employment was to follow a hare or fox'. A more estimable element was the farmer of a couple of hundred acres, in effect the minor country gentleman, who possessed some capital and provided a certain amount of employment. Much of the trade was in the hands of the Protestants, and among the traders there were many whose power of exploitation was considerable, especially in the countryside. The Quakers were 'the only wealthy traders in Ireland', according to Young, and while this is probably an exaggeration, the Quakers are still a numerically small group of respected citizens, strongly represented in the commercial life of the country. For the rich in the towns, especially Dublin, life was a gay round with many diversions such as theatres, concerts, dinners and overcrowded parties, but for the poor it was a hard struggle in pestilential slums. One section of the population had too much money and the other too little, yet capital was so scarce that industrial activity was insignificant.

The population was apparently increasing sharply;[20] Young believed it to be less than 3,000,000, though there are good reasons for believing that it was at least 4,000,000; even so, it was doubled within the period 1780–1845. Many factors favoured an increase, among them the ease with which a family could be fed with milk and potatoes, the pathetic simplicity of the housing, the acceptance of poverty as a fact of life by the majority of the community and the universality of marriage. This last feature is in sharp contrast to conditions today, when Ireland has a greater proportion of bachelors and spinsters than any other country from which reliable data can be drawn, but the widespread celibacy has arisen gradually during the past hundred years. Young thought that children were 'not burthensome', and that the happiness and ease of a family increased in proportion to their numbers. Even the servants in the great houses were married and had a family somewhere in the rear. Emigration was already a feature of Irish life, especially among the Presbyterians of the north who went to America. Apparently content with little, the Irish increased and multiplied so that Irish workers were available for the growing towns both inside the country and Britain. There was very little immigration except among the professional classes, for labour was cheap and Ireland was therefore a country 'to which no workman in the world ever yet emigrated'.

THE NINETEENTH CENTURY AND AFTER

During the first forty-five years of the nineteenth century Ireland's population continued to increase though the country did not share in the industrialization of Great Britain at this time, but continued to be an exporter of meat, butter, grain, meal and cattle, while supporting its own population on potatoes with milk, butter and oatmeal, but only rarely meat or bacon. Fishing was practised only to a limited extent. Dietetically the population was not under-nourished, monotonous as the diet may have been, but there was a drastic struggle for land and probably a continuing decline in fertility. It is probable that many new areas were reclaimed, for example in the west, where the opening of roads to Belmullet and Clifden were beneficial to trade and encouraged the reclamation of new land on the coasts, on hillsides, or on patches of river alluvium. During the first half of the nineteenth century it was widely believed that the rural population of Ireland would be provided with land from reclaimed bog or on mountain sides: such views were probably based partly on contemporary experience. And there were other grounds for hope. Domestic industry prevailed throughout the country, especially the weaving of linen and woollen cloths, and it was hoped that as canals supplemented roads, trade would increase and the mineral resources provide a basis for further industrialization. Later, much the same hope was held for the railways and few foresaw the dangers of famine even though there were years of acute distress in parts of the country, notably the west, and poverty in town and countryside was recognized to be rampant. Any man who had forecast a famine on the scale that it came in the late eighteen-forties would probably have been regarded as the most senseless pessimist.

Four crucial dates stand out in this period of social history – (*a*) 1845, when the first catastrophic failure of the potato brought widespread distress, wholesale emigration of probably a million people within five years, the death of many thousands, variously estimated at between 500,000 and a million, and the rapid decline of population from its peak, which was at least 8,300,000 and possibly almost 8,500,000: (*b*) 1869, when the Church of Ireland was disestablished and the transfer of land to tenants as purchasers was begun with the glebe, a relatively small measure which was the forerunner of vastly important legislation: (*c*) 1891, when the Congested Districts Board began its survey of the poorest areas of the land, and money was found for many schemes of social significance, including the creation of new farms, the improvement of stock, the encouragement of industries and of fishing, the building of piers, and the construction of new roads and light railways: (*d*) 1921, which divided Ireland into two political units, Northern Ireland (so often mis-named Ulster, of which historic province it is only part), and the Irish Free State, or Saorstat Eireann, which in 1937 became Eire and in 1949 the Republic.

Since the first edition of this book appeared, an excellent historical survey of the Famine period[21] has been published; the seven authors all make it clear that the Famine was so catastrophic an event that the country could not 'return to normal' but must follow new lines of development. Not the least significant of these was the vast emigration overseas, especially to the United States, and the lesser emigration to Britain, which with the high famine death rate reduced the population almost to half by the end of the century. The question remains whether Ireland was overpopulated before the Famine. That some change in its land distribution, some expansion of trade, more schemes of social welfare were needed, no reasonable man could doubt. In the state of its economic development in 1845, Ireland was definitely overpopulated, but there are those who maintain that a population of eight millions – or more – could be supported given the right economic arrangements. At all times regret is voiced at the extent of Irish emigration but the outward movement still continues in spite of commissions of inquiry in the Republic and the widespread advertisement of Northern Ireland's facilities for new industries in British papers. All Irish problems have one common root – population and resources, therefore the geographical aspects are discussed in the remainder of this book in the regional chapters of Part II and under five major headings in the rest of Part I.

1. *Population.* Ireland has for over fifty years had virtual stability of total population, though the birth rate is high and offset by emigration. There has been a marked movement to towns, primarily to Dublin and Belfast (Chapter 6).

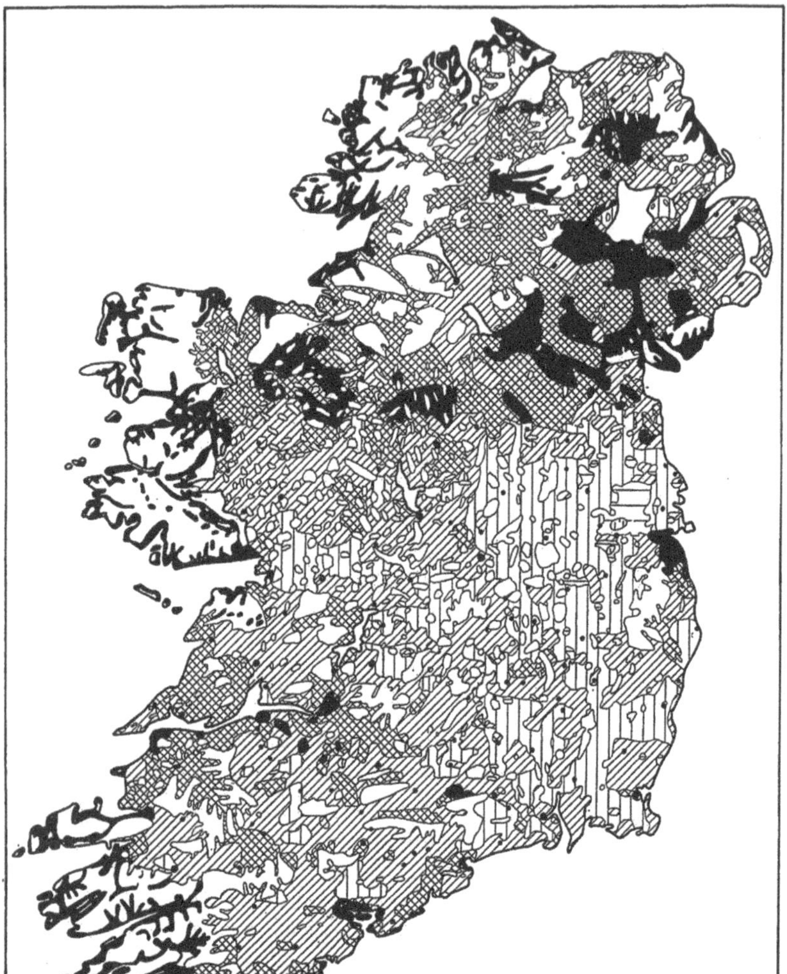

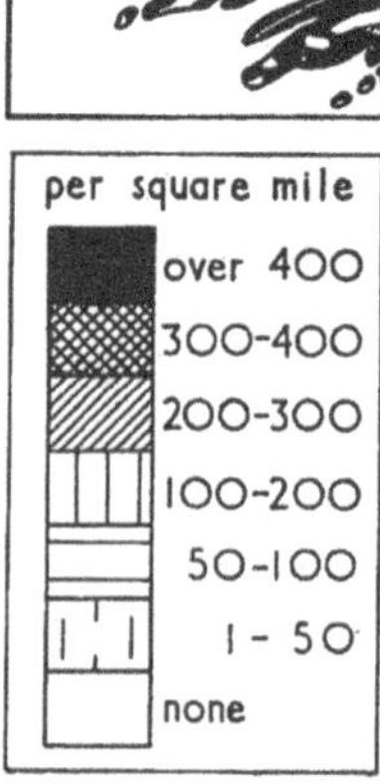

FIG. 15. POPULATION DENSITY, 1841

Much of rural Ireland had at least 200 persons to the square mile at this time and substantial areas over 400. The lowest densities were in the eastern 'grazing districts' of Meath and Kildare. Comparison with Fig. 21, p. 131 and Fig. 23, p. 133 is revealing.

Originally published in 'Pre-Famine Ireland', Manchester, 1957, p. 18.

2. *Social Factors.* The division of Ireland in 1921 has not removed for ever all causes of internal strife and dissension. By 1921 the problem of settlement on the land was largely solved and post-Treaty legislation merely developed the schemes already marked out in principle and practice. The rift between Catholic and Protestant remained strong, and the Protestant community declined numerically in the Republic from 1921, primarily by emigration. Encouragement of the Irish language by the Dublin government, actuated by the wish to develop national consciousness, caused a further rift with Stormont. One social factor of special interest is the lack of villages in many areas though market towns are strongly entrenched in the life of the countryside. Whatever the historic pattern of settlement may be, it is clear that the number of hamlets, or 'rundale villages' practising a form of communal farming, has declined by the re-allocation of land into separate and distinct holdings. Many of the villages are associated with demesnes and some are places established as towns which failed to develop, but a large number appear to be of recent growth, in some cases around churches, parish halls, an inn, a couple of shops beside a crossroads. Monthly fairs may add to the strength of a village if they are well supported and it is felt by many people that more and better villages are needed as social centres. Such concern is due in part to the wish to stem emigration, which is certainly due in part to the dullness of country life and the natural desire for more entertainment (Chapter 7).

3. *Agriculture and the Land.* Since the Famine the area under crops has steadily declined, except during the wars of 1914–18 and 1939–45. Since 1921, agricultural policy has diverged considerably in the two political units of Ireland (Chapter 8).

4. *Industries.* Already in the eighteenth century it seemed likely that parts of the Ulster province were likely to become the major industrial area of Ireland, a forecast abundantly justified in the nineteenth century through the steady expansion of Belfast and its satellites and the concentration of cotton factories and, later, the linen trade combined with shipbuilding in what is now Northern Ireland. Elsewhere industry existed mainly to supply home demands and never figured prominently, except for a few articles, in the export trade. In general, the direct impact of the Industrial Revolution on Irish life was far less severe than in Great Britain or Germany, but also far less remunerative. The modern trends in the Republic are interesting as they show a planning of industry with electricity as the main source of power, in which the social desirability of reviving many country towns has been recognized, though

Dublin has grown more rapidly than any other town (see Chapter 9).

5. *Communications.* Canals seemed the acme of modernity in the late eighteenth century, but few then constructed have not outlived their usefulness. Railways were added in the mid-nineteenth century, but not until the twentieth century did they reach the poorer parts of the western seaboard, long hindered economically by isolation: many of the transport gaps had been filled by the 'bians', horse-drawn cars named after Charles Bianconi, an Italian who opened the first routes around Clonmel in 1815. The main trunk roads were already in existence by 1800, and the pattern of by-roads became steadily closer, partly through 'presentment' construction, partly through various relief schemes and finally by the work of the County Councils (see Chapter 10).

SUMMARY

Ireland belongs to the Atlantic world of Europe. Useful as Sir Cyril Fox's concept of Highland and Lowland Britain has been, its modification to include an Atlantic sector, covering the western coastlands of Britain, the Isle of Man and Ireland, means more, for it gives recognition to the physical relations between Ireland and the west of Britain, and to the use of the seaways in prehistoric and early historic times. Up to the Norman period, these ways were vital routes; but from the Norman period onwards the main movement to Ireland came directly from Britain by the short sea crossings. The age of the Celtic church saw Ireland deeply Catholic yet distinct in its ecclesiastical organization from Rome, divided between many rulers yet a potential nation-state. To the Norse immigrants Ireland owes the foundation of some of its major ports, but its culture and Christian allegiance remained unsubmerged. As Ireland was drawn closer to Britain in its life, the diocesan system of the Church prevailed over the monastic organization of the early Christian phase, the country was divided into shires and towns were established for the first time, beginning with Dublin: it was not, however, until Stuart times that towns were found throughout Ireland. The Tudor and Stuart plantations, with the efforts at religious and political conquest they implied, brought Ireland into closer contact with England than at any previous phase: from this time the land question, the economic status of the country as a whole and of its various social groups, became crucial. Out of fear there arose the Penal laws of 1691, which tried to subjugate the Catholic majority for ever in the interests of a Protestant ascendancy, with the result that a sharp dichotomy of wealth and poverty arose and much of the land was underfarmed. Meanwhile, the signs of Ulster's coming industrial greatness were apparent, but the linen industry was still domestic. The nineteenth century saw the life of Ireland altered almost beyond recognition through industrialization, land redistribution, new means of communication, population increase to the summit of 1845 followed by mass emigration, and the increased strength of the towns: all these changes were perhaps less obvious to the outer world than the political tensions that raged so constantly. The ultimate solution, acceptable to some but not to others, came in the twentieth century, and gave the country two political entities which, it is already clear, differ so radically in policy and outlook on many social and economic questions that it is hard to see how they may eventually be united.

THE DIOCESES OF IRELAND[22]

The division of Great Britain and Ireland into diocesan units has received little attention in geographical literature, though Mackinder[23] gives a map of the pre-Reformation dioceses in England. In Ireland the territorial organization of the church shows a division of the country which antedates the county organization by several centuries, and the boundaries shown in Fig. 16 are those of the Church of Ireland dioceses which have, in many cases, been altered comparatively little since medieval times, though there are records of other dioceses that have existed at various dates: among these are the dioceses of Mayo and Annaghdown, both of which were absorbed by the archdiocese of Tuam in the sixteenth century. Many of the dioceses are of monastic origin, and remarkable ruins in remote places recall the Christian worship of more than a thousand years ago: Glendalough diocese, for example, already had a long tradition of scholarship and missionary enterprise when it was united to Dublin in 1216, but it has now no Cathedral church.

Episcopal organization is normally but not inevitably associated with sees in towns, but at the time of its growth in Ireland there were few towns that could establish their right to be the sees of territorial bishops. By the end of the eleventh century the Norse people in Dublin, Wexford, Waterford and Limerick were Christians, and their churches had bishops who apparently ministered only to the towns themselves and to their immediate hinterlands. There are still dioceses, comparatively small in extent, based on Dublin and Waterford, though the Limerick diocese was apparently large from its first beginnings and has since been reduced

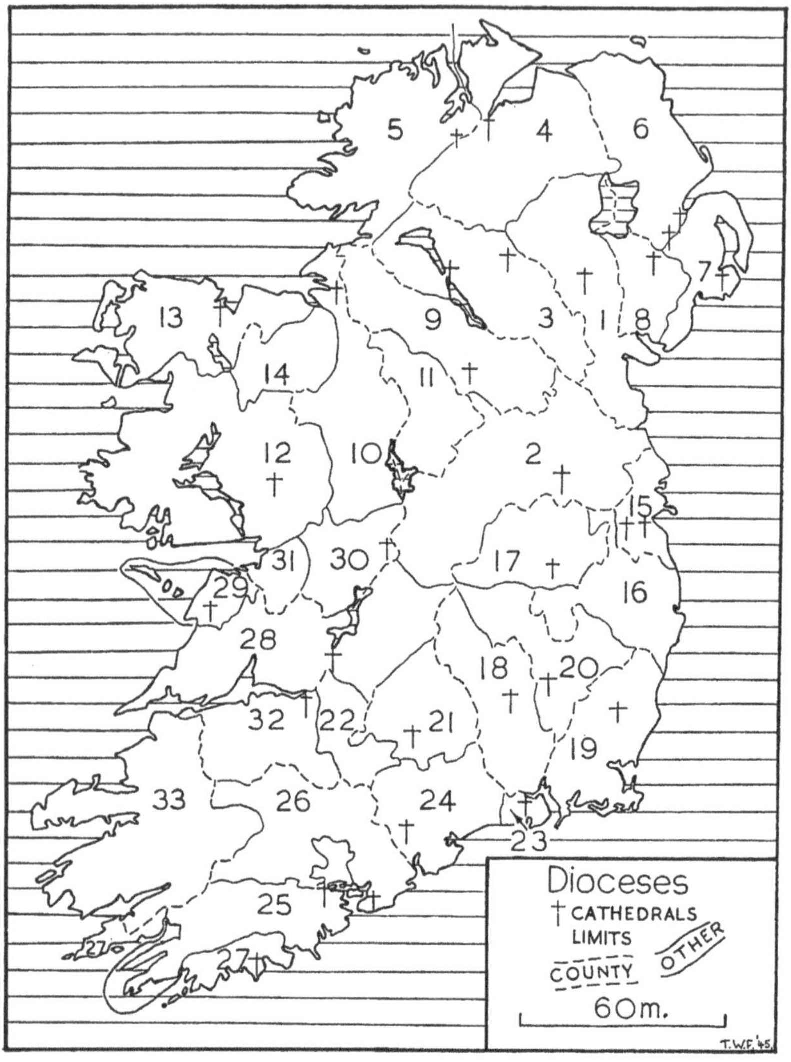

FIG. 16. DIOCESES IN THE CHURCH OF IRELAND *(see opposite)*

PROVINCE OF ARMAGH

Diocese	*Cathedral*
1. Armagh, archdiocese	Armagh
2. Meath	Trim
3. Clogher	Cathedrals at Clogher and Enniskillen
{ 4. Derry	Londonderry
{ 5. Raphoe. United to Derry, 1834	Raphoe
6. Connor. United to Down, 1450: separated from Dromore and Down, 1945	Lisburn
{ 7. Down	Downpatrick
{ 8. Dromore. United to Down and Connor, 1842	Dromore
N.B. There is also a cathedral in Belfast	
{ 9. Kilmore	Kilmore
{ 10. Elphin. United to Kilmore, 1854	Sligo
{ 11. Ardagh. United to Kilmore, 1692–1751; to Tuam, 1751–1839; to Kilmore from 1839	No cathedral
{ 12. Tuam, archdiocese to 1839	Tuam
{ 13. Killala. United to Tuam, 1834	Killala
{ 14. Achonry, to Killala, 1622	Achonry

PROVINCE OF DUBLIN

{ 15. Dublin, archdiocese	Dublin (Christ Church)
St Patrick's Cathedral, Dublin, is a national cathedral	
{ 16. Glendalough. United to Dublin, 1216	Christchurch, Dublin
{ 17. Kildare. United to Dublin, 1846	Kildare
{ 18. Ossory	Kilkenny
{ 19. Ferns. United to Ossory, 1835	Ferns
{ 20. Leighlin. United to Ferns, 1597	Leighlin
{ 21. Cashel, archdiocese to 1839	Cashel
{ 22. Emly. United to Cashel, 1562	No cathedral
{ 23. Waterford. United to Cashel, 1832	Waterford
{ 24. Lismore. United to Waterford, *c.* 1358	Lismore
{ 25. Cork	Cork
{ 26. Cloyne. United to Cork, 1835	Cloyne
{ 27. Ross. United to Cork, 1617	Rosscarbery
{ 28. Killaloe	Killaloe
{ 29. Kilfenora. United to Tuam, 1661–1742; to Clonfert, 1742–52; to Killaloe, 1752	Kilfenora
{ 30. Clonfert. United to Killaloe, 1836	Clonfert
{ 31. Kilmacduagh. United to Clonfert, 1627–1836; to Killaloe, 1836	No cathedral
{ 32. Limerick	Limerick
{ 33. Ardfert and Aghadoe. United to Limerick, 1661	No cathedral

Ardfert cathedral, in ruins, is a national monument. In 1954, Trim parish church became the Cathedral of the Meath diocese; St John's, Sligo, has replaced Elphin Cathedral, now demolished.

Other dioceses absorbed include Clonmacnoise, to Meath *c.* 1560; Mayo, to Tuam in 1573; Annaghdown to Tuam in the mid-sixteenth century; and a diocese of Ardmore, to Lismore in the late twelfth century.

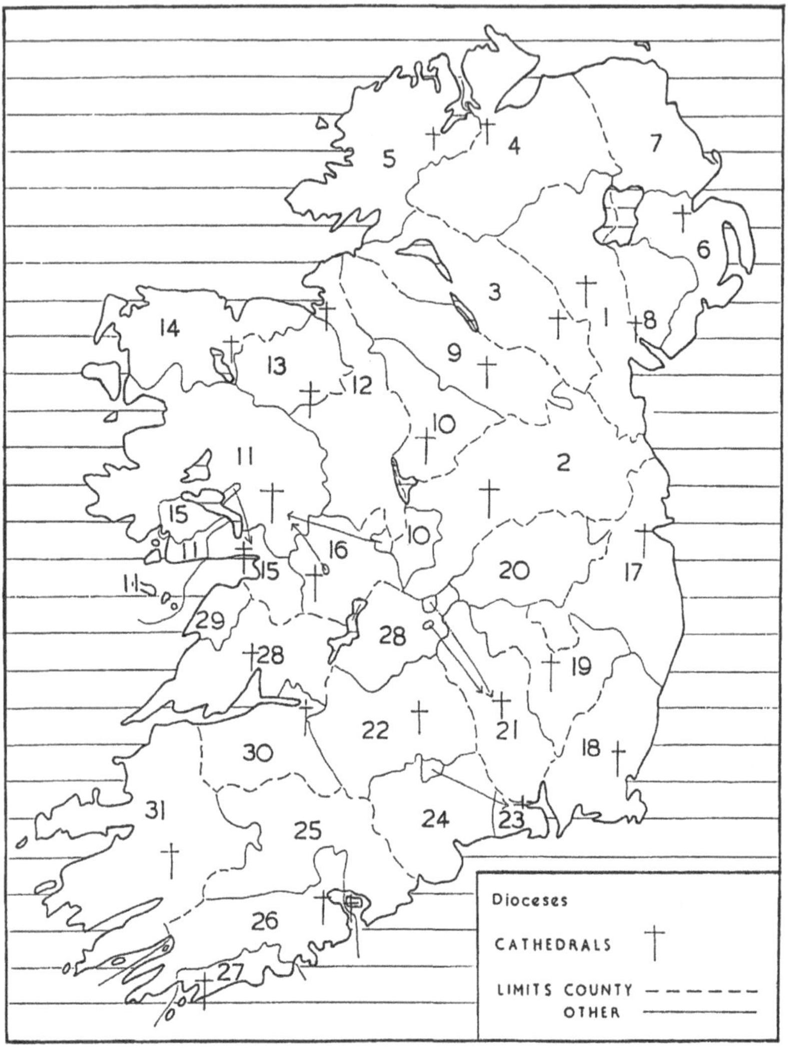

FIG. 17. ROMAN CATHOLIC DIOCESES (*see opposite*)

PROVINCE OF ARMAGH

Diocese	*Cathedral*
1. Armagh, archdiocese	Armagh
2. Meath	Mullingar
3. Clogher	Monaghan
4. Derry	Londonderry
5. Raphoe	Letterkenny
{ 6. Down	Belfast
{ 7. Connor	Belfast
8. Dromore	Newry
9. Kilmore	Cavan
10. Ardagh and Clonmacnoise	Longford

PROVINCE OF TUAM

11. Tuam, archdiocese	Tuam
12. Elphin	Sligo
13. Achonry	Ballaghaderreen
14. Killala	Ballina
15. Galway and Kilmacduagh (and Kilfenora in the province of Cashel)	Galway
16. Clonfert	Loughrea

PROVINCE OF DUBLIN

17. Dublin, archdiocese	Dublin
18. Ferns	Enniscorthy
{19. Kildare	Carlow
{20. Leighlin	Carlow
21. Ossory	Kilkenny

PROVINCE OF CASHEL

22. Cashel, archdiocese	Thurles
{23. Waterford	Waterford
{24. Lismore	Waterford
25. Cloyne	Cobh
26. Cork	Cork
27. Ross	Skibbereen
28. Killaloe	Ennis
29. Kilfenora (with Galway)	Galway
30. Limerick	Limerick
31. Kerry (Ardfert)	Killarney

See the *Catholic Encyclopedia*, New York, 1910, vol. 8, p. 114, and the annual Irish Catholic Directory and Almanack, Dublin. Comparison of Figs. 16 and 17 will reveal both the interesting similarities and differences, of which the location of cathedrals should be noted.

rather than extended: the Wexford diocese has disappeared, though there is still a diocese of Cork. These foundations provided only for a very small part of Ireland, and outside them the church was still primarily monastic in the early twelfth century, though there was already some form of diocesan organization in Meath, probably around several centres such as Trim, Kells, Dunshaughlin, Clonard, and possibly also Clonmacnoise. In addition, there were archiepiscopal sees at Armagh and Cashel, of which the former has remained supreme to this day. In 1110, at the Synod of Rathbreasal, an attempt was made to divide the country into twenty-four dioceses, twelve in the north and twelve in the south: in 1152, at the Synod of Kells, four archbishoprics were established, at Armagh, Dublin, Tuam and Cashel. The choice of Dublin was a recognition of its position as the most important town in the country, and its burgesses had already claimed metropolitan status in the civil sense in 1074. Dublin alone had been in some sense submissive to Canterbury, but it acknowledged the primacy of Armagh in 1152.

The dioceses are equal in number to the counties but their boundaries correspond only to a limited extent, though they show some of the country's major regional divides, such as the Shannon above lough Derg, the river barriers of the Nore, Suir and Barrow, and of the Bann valley in the north. Some later subdivisions of the country are at least suggested by the diocesan units: the diocese of Ardfert and Aghadoe bears a strong resemblance to the county of Kerry and the Dublin diocese is closely similar to the county of Dublin. But there are also dioceses that spread amorphously across the Central Lowland, where the Shannon alone forms any effective regional divide; Meath diocese, for example, spreads from the Irish sea to the Shannon and Elphin extends from the Shannon to Donegal bay. The diocese of Killaloe has its centre at an ancient crossing place of the Shannon and includes most of Co. Clare with part of the Central Lowland. The dry limestone country of north Clare, with the Aran islands, forms the small Kilfenora diocese, remote and thinly populated, but rich in early Christian ruins. Perhaps the most curious of all is the Ross diocese, which had a bishop in the sixteenth century and consists of two separate parts of southwest Cork.

Many dioceses were at first associated with the *tuaths*: one of these, Raphoe, corresponds fairly closely with modern Donegal, except for the Inishowen peninsula, and was probably the land of the Cinel Connail. Several of the dioceses have changed their boundaries; the Derry diocese is based on the territories of a people around the Carntogher mountains, who pushed their enemies eastwards to the Bann from the Roe valley.

The Clogher diocese, now partly in the Republic and partly in Northern Ireland, was at first very small and perhaps merely titular, but during the twelfth century the present Monaghan and Louth counties were added and the see was placed in the village of Louth. At a later stage the Louth area was added to the Armagh diocese, and the Clogher diocese spread westward almost to Donegal bay. The remotely placed cathedral of Clogher occupies the site of a monastery whose traditions date back to the end of the fifth century, and some forty years ago the parish church of Enniskillen was raised to Cathedral status in the Church of Ireland. Armagh diocese extends from the Boyne to the northern limit of lough Neagh, but was much larger in its earlier form: the dioceses of Down, Connor and Dromore now have their main population centres in the Lagan valley, but their original cathedrals were in remoter places, Downpatrick, Connor and Dromore (see Fig. 16).

CHAPTER 6

Population

Alone among the countries of Europe, Ireland has had a declining population from 1845 almost to the present time, though from 1961 to 1966 there was an increase of 2·3% in the population of the Republic. Apart from a small increase from 1946 to 1951, this is unique in the post-famine period and suggests that industrial policies are at last stemming the emigrant tide, at least partially, for the Republic had an outward emigration of 212,000 people from 1956 to 1961 but only 80,600 from 1961 to 1966. In Northern Ireland the official emigrant figures only cover persons going beyond Britain, but as the natural increase is about 10 per 1,000 annually, and the increase of population from 1951 to 1961 was *c.* 54,000, from 1,371,000 to 1,425,000, there was an outward movement of approximately 92,200 almost entirely to Great Britain. In both parts of Ireland the unemployment rate is normally at least 5% and at times considerably higher; in both parts of Ireland the farming population is declining and the towns – or at least some of them – are growing. But emigration still has its natural appeal. In this chapter a short historical survey of population is followed by a discussion of its present distribution, with particular reference to the current growth of towns, and notably of the larger towns.

POPULATION TRENDS FROM *c.* 1700

No one knows what the population was before 1821, when the first complete Census gave a figure of 6,802,000. Various workers have assessed the earlier evidence, notably K. H. Connell, who quotes figures ranging from 2,791,000 in 1712, to 3,584,000 in 1772, though this may well be an underestimate. In 1672 Sir William Petty had calculated that there were some 200,000 families and a total population of 1,100,000 but in 1687 he revised his estimate to 1,300,000. If these figures were accurate, the population doubled within twenty-five years to 1712, which is scarcely credible. In 1788 a Revenue commissioner, Mr G. P. Bushe, gave a figure of 3,750,000 and expressed the opinion that many people did not want to be recorded at all by the tax-collectors, the magistrates or the clergy. Even if, as some workers have apparently thought, all the early figures are underestimates, one thing is clear – there was a heavy increase from some time about 1770 to 1845: this may have indeed been a continuation of a trend already well established. It is tempting to regard Arthur Young's comments in 1778 on rising rents, more trade, large families, universal marriage, land reclamation, and the expansion of farmed land, as indicative of an increasing population. Two returns based, like others before them, on the number of hearths, gave 4,207,000 in 1791 and 5,395,000 in 1805: if these are compared with the 6,802,000 of 1821, there was an average increase of 17% in each decade.

Heavy increase is not a unique demographic phenomenon. In Ireland it was a rural increase, for in 1841 only one-fifth of the population lived in towns and villages with more than 200 inhabitants. The 1831 figure, 7,767,000, is generally regarded as far too high, because the enumerators were paid according to the number of persons they returned, and what could be easier than to add a few pages of Murphys and O'Neills? It is improbable, for example, that Connacht added 21% to its numbers within ten years. But the 1821 figure was reasonably accurate and the 1841 figure, of 8,175,000, even more reliable. This twenty years had an increase of 1% per annum, which could exist with a high birth rate and a high death rate, making allowance for the emigration of the times. In 1841, it was estimated, the loss by overseas emigration during the previous ten years was 403,459, of whom more than half went from Irish ports; there were 419,256 Irish-born people in Britain, of whom possibly one-quarter had been added since 1831. In all, half a million people left Ireland during this decade, and it has been estimated that the total emigration to the United States and Canada between 1780 and 1845 was *c.* 1,140,000, and some 600,000 went to Great Britain in the same period.

Long before the Famine, emigration was a well-established feature of Irish life and one could argue that as it has apparently existed for two hundred years, it is futile to suppose that it can cease now. Emigration does not make Ireland unique: what makes her demographically distinct from other countries is its fall of population since 1845, to half. Emigration has affected every class of society, and in many cases has been financed by relations already settled abroad. The emigrants, having come largely from rural areas, have settled primarily in towns as jobs were most easily found: in 1841, one-fifth of the Irish in Britain were in Liverpool or Manchester, in each of which they numbered more than one-tenth of the population. But there were also the 'harvesters' who went to Scotland in considerable numbers, of whom many remained permanently. In Ireland there is no clear correlation between areas of poverty and heavy emigration, for counties like Meath lost as heavily, even more heavily, than western counties such as Kerry or Donegal. Everything depends on local circumstances: in some western districts many men go away for a time as contractors' labourers but retain their homes with their families on the original holding. In other – and more numerous – cases a definite break is made, especially by farm labourers and relatives of farmers. Within six years, from 1961 to 1967, the number

TABLE I

Population of Ireland, 1841–1961

	Total	% Decrease	% Increase
1841	8,175,124	—	—
1851	6,552,392	19·5	—
1861	5,798,564	11·5	—
1871	5,412,377	6·5	—
1881	5,174,836	4·6	—
1891	4,704,750	8·8	—
1901	4,458,775	5·1	—
1911	4,390,219	1·6	—
1926 (15 years)	4,228,553	3·4	—
1936–7	4,204,476	0·5	—
1946–7	4,289,275	—	2·0
1951	4,329,587	—	1·0
1961	4,243,803	1·9	—

of males on the land dropped by 70,000 to 310,000 in the Republic. Clearly this is the continuation of a movement by which Ireland has lost over six million people since 1845.

In 1845, the population cannot have been less than 8,300,000, and probably was 8,500,000, so the recorded decrease from 1841 to 1851 must be an understatement. Even so, the evidence is dramatic enough. Every county in Connacht lost at least 27%, and the average loss for the Munster counties was 22%, ranging from 16% in Waterford and 19% in Kerry to 24% in Cork and Tipperary, and 26% in Clare. In Leinster the decreases ranged from 16% in Louth to 29% in Longford, but Co. Wexford lost only 11%. All the Ulster counties lost heavily, with an aggregate of 16%, but Donegal's decrease, perhaps surprisingly, was only 14%. The decade 1851–61 saw a decline of 12% over the whole country, unevenly distributed with losses of more than 20% from some of the richer counties such as Kilkenny (−21%), Meath (−21%), Wexford (−20%), Tipperary (−25%) and Clare (−22%), but of less than 10% from Kildare (−5%) and several of the poorer counties, including Donegal and the whole province of Connacht except for Galway. Between 1861 and 1881 the rate of loss diminished and the decline of 7% in the first ten years was spread fairly evenly through the country, but in the second decade, with a net loss of 5%, Connacht's population decreased by only 3% and Kerry even recorded an increase of 2%.

Nobody knows how many emigrants left during the Famine period but it has been estimated that in 1847 alone more than 250,000 landed at Liverpool as part of a great human tide received not only at ports but even at remote beaches along the Welsh coast. Not all of these stayed in Britain but many of them settled in the growing manufacturing towns. In 1851 there were 962,000 Irish-born people in the United States, 727,000 in Britain and 227,000 in Canada; by 1861 there were 1,611,000 Irish-born in U.S.A., 806,000 in Britain and 286,000 in Canada. Within Ireland itself, only a small fraction of the rural unemployed could be absorbed in towns such as the growing Belfast, which alone showed a rate of increase comparable with that in English manufacturing centres: the decline of home weaving and spinning, a gradual process lasting from 1825 to 1865 (see p. 228) left thousands of domestic workers free for factory employment, far more, in fact, than the mills could absorb. In some years bad harvests acted as a stimulus to emigration, for example in the winter of 1879–80, and there were similar disasters during the decade from 1881, a time of heavy outward movement which lowered the population by 9%. Schemes of assisted migration were financed by the charitable; one was that of Mr Tuke, who organized the movement of 10,000 people, known as the Tuke emigrants, who left Galway in the years 1882–5 to settle in St Paul. But difficulties remained: in 1885 the potato crop was deficient in west Mayo, Galway and islands off the coast, and cattle prices were low, and in 1890 the potato crop gave only one-quarter to one-half of its normal yield along the entire western seaboard from Donegal to Kerry.

At no time, before the Famine or after, has emigration been regarded as the only possible solution of the population problem. In 1891, the Congested Districts Board was given the responsibility of caring for those areas having a rateable valuation of less than £1 10*s.* per head, which covered the areas shown in Fig. 18. The reports of the inspectors showed that poverty was of varied incidence: in Donegal for example, the gap between income from the farms and the income needed was met by seasonal labouring in Britain, combined with domestic industry. Seasonal labourers also went in large numbers from Mayo and Galway but not from Kerry, where the men and women relied rather on intermittent work nearer home or on fishing. In effect, the Congested Districts were

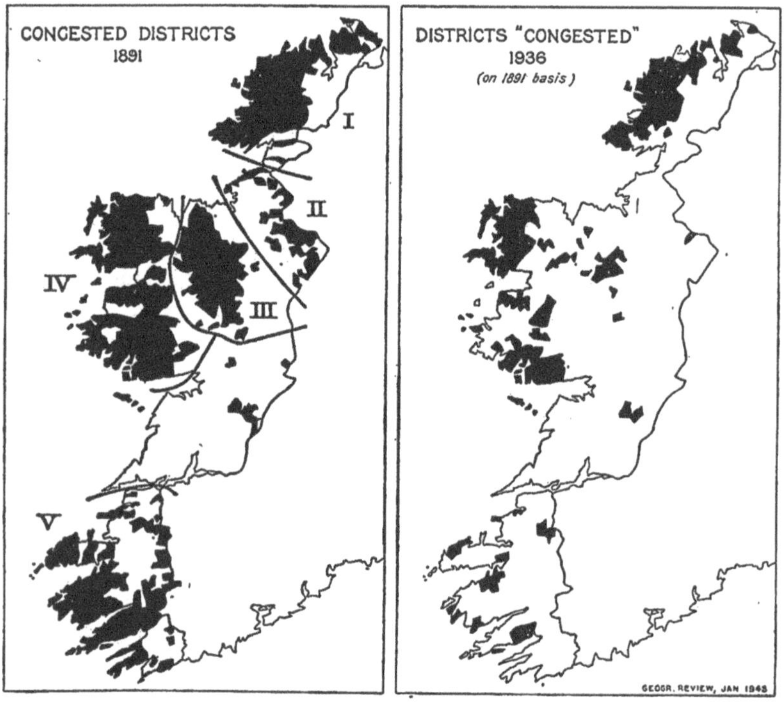

FIG. 18. THE CONGESTED AREAS, 1891 AND 1936

The Congested Districts of 1891 were the poorest in the land, with an average valuation of less than £1 10s. per head: by 1936 the decline in population has resulted in a substantial decrease of the congested area. By 1966 very few areas were 'congested'.

First published in 'Geogr. Rev.', vol. 23, 1943, p. 5

those areas where the population had an inadequate amount of land to maintain life except at a very low standard: at first, hopes centred around the enlargement of holdings, the encouragement of domestic and small factory industry or the improvement of fishing, but while all these have materialized to some extent, emigration has been the major palliative: the population of the Congested Districts, as defined in 1891, has since declined by 51% from 541,000 to 264,000 in 1961. The few areas that would rank as congested on the 1891 basis are in Donegal, Mayo and Galway, including the Aran islands; there are none now in Sligo, Leitrim or Roscommon, Kerry and West Cork (using the 1966 figures). In Fig. 19 it will be noted that the rate of population decline differs widely from one area to another, for some areas have lost as much as two-thirds of their population since 1891, and in others the loss has been relatively slight.

There is no obvious correlation between poverty and emigration rate. Only the most detailed local work can supply the answer to the question raised here. In parts of Donegal, for example, the rate of outward movement is comparatively low because the farm population have a variety of resources, including seasonal labouring in Britain or on various local public works, employment in peat cutting, tomato growing in the Gaeltacht from Falcarragh westwards, domestic industry, a limited amount of factory employment, government subsidies to Irish speakers, remittances from emigrants, and unemployment insurance for part of the year. Field work in the west of Ireland always raises one fundamental economic question, 'What does it cost to maintain this population?' If one finds a family consisting of two or three single or widowed people living on a few acres in comparative comfort, enjoying various minor luxuries yet having almost nothing to sell from their land except one beast a year and a small quantity of eggs, obviously there is some extra-mural source of income. Rarely is it fishing: a glance at the dole queue or the tactfully masked question about postal orders may supply the answer. The answer is often indirectly given in the casual mention of a daughter who is a cook at Blackpool, a sister who is doing well in Chicago, or a son with a good job in Birmingham. All this emigration has been of benefit to the west of Ireland and so too has the work of the Congested Districts Board and the Land Commission, discussed on pp. 177–82. Without past, present and presumably future emigration, the prospect for the west of Ireland would be grim indeed.

But emigration is not, and cannot be, a phenomenon seen only in the parts of Ireland least well-endowed with natural resources. In the Republic, the natural increase was 9·2 per 1,000 per annum from 1956 to 1961, the highest since 1871 when the first reliable figures appeared. Within the same five years, there was a decrease of 2·8%, giving a net emigration rate of 14·8 per 1,000 per annum, the highest recorded since 1891. In the five years from 1961 to 1966 the population increased by 2·3% and the emigration rate was 5·7 per 1,000 (the natural average increase per

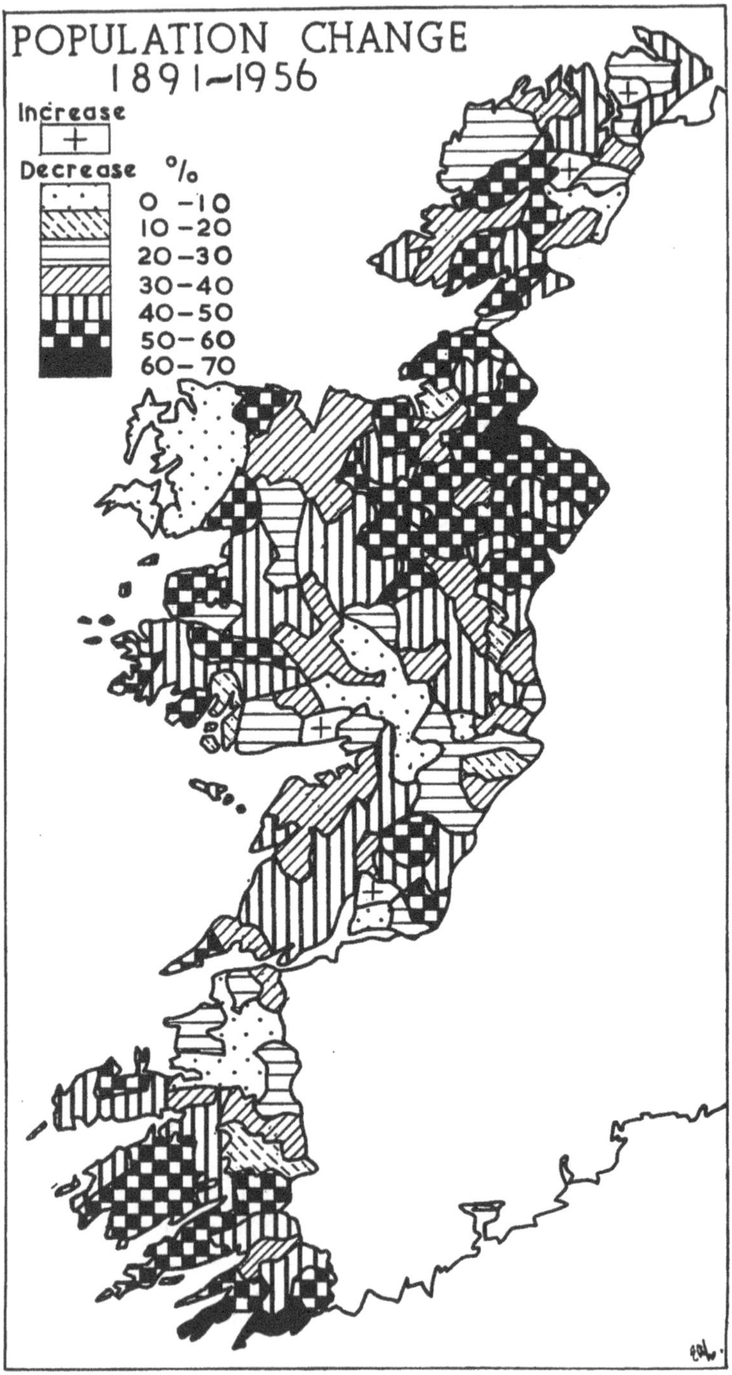
POPULATION CHANGE
1891–1956
Increase
+
Decrease %
0 –10
10 –20
20 –30
30 –40
40 –50
50 –60
60 – 70

1,000 of 10·3 less 4·6, the average actual increase annually, or 80,600 people). Although there was a decline in the number of emigrants within these five years, there was still heavy outward movement from the Connacht and Ulster counties of the Republic. The five Connacht counties had an annual average loss of 13·6 per 1,000, varying from 10·6 in Galway to 19·1 in Leitrim, the highest of any county: these same five counties had a mean annual rate of 18·3 per 1,000 in 1956 to 1961 varying from 16·2 in Galway to 22·7 in Leitrim. In the three Ulster counties the rates for 1956 to 1961 were Cavan 21·3, Donegal 17·9 and Monaghan 26·5 and the losses for 1961 to 1966, though lower, were still considerable, at Cavan 13·6, Donegal 15·0 and Monaghan 12·9. Elsewhere the losses were generally slighter though from 1961 to 1966 there was an emigration movement of at least 10 per 1,000 from six Leinster counties (Carlow, Kilkenny, Leix, Longford, Offaly and Westmeath) and from two counties of Munster (Kerry and Tipperary South Riding): these counties have no large towns as effective magnets for rural migrants. From 1961 to 1966, Dublin (including the county and Dun Laoghaire) became once more an area of net immigration, of *c.* 18,000 people. During the previous ten years from 1951, the Dublin area (as defined above) had a net emigration rate averaging *c.* 10 per 1,000 annually. To limit emigration, not by preventing people from going elsewhere should they wish to do so but by providing employment and amenities that will make home seem more desirable, has been the aim of every Irish politician for a long time: at present there are signs that a change is beginning.

All this shows that emigration is not something that belongs to an historic past but a living present. From 1871 to 1951, the net emigration was 3,300,000, of whom one-fifth went from Northern Ireland: from 1891 to 1951, when the heavy wave of outward movement was spent, the loss was nearly 1,900,000, and even from 1901 to 1951, almost 1,500,000 people left. Only to a small extent have Irish towns absorbed rural migrants except in Belfast and Dublin (the latter until the late 1940s) though Tables III, IV and V show that since 1891 the towns of Ireland have grown considerably in population. Table IV shows that the main increase

---

FIG. 19. POPULATION CHANGES IN THE CONGESTED WEST, 1891–1956

The changes are remarkably varied, but the decreases have been heavy in Co. Leitrim, in southwest Cork and in parts of Kerry. There is no obvious correlation between rate of decline and poverty and some coastal areas have remarkably small declines. This map covers the areas included in the wider definition of Congested Districts worked out in 1909.

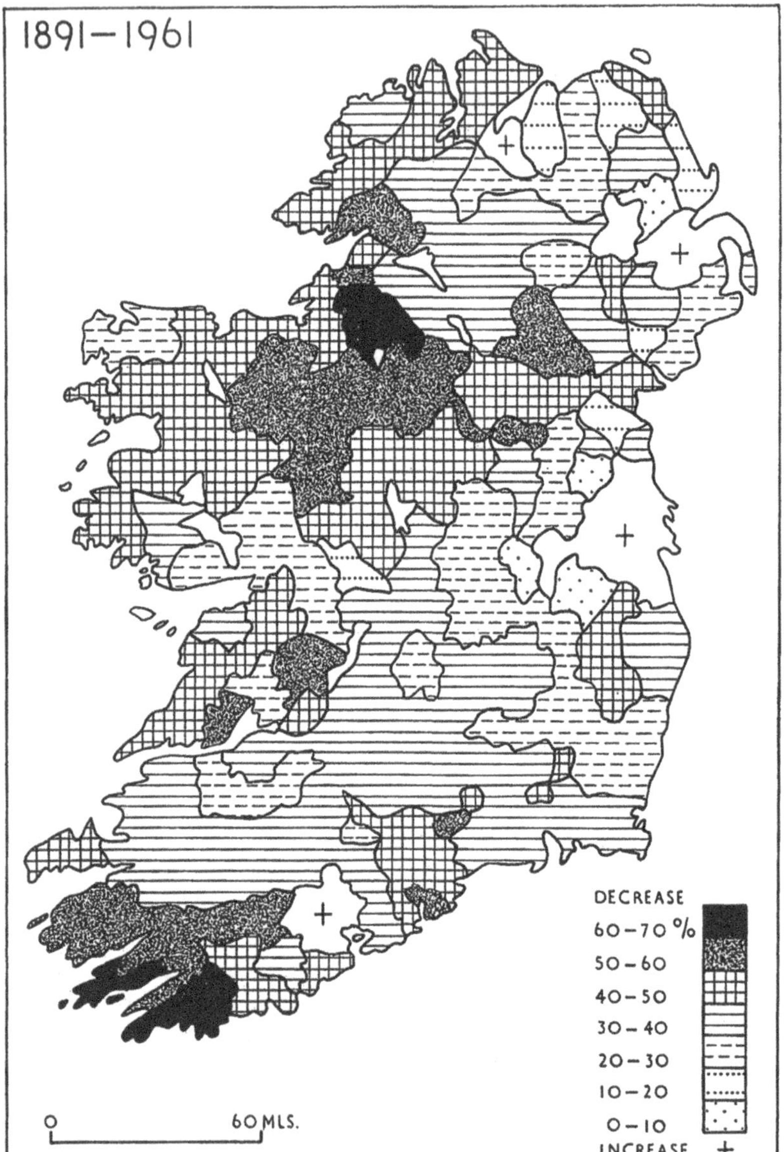

FIG. 20. INCREASES AND DECREASES, 1891–1961

In some parts of the country the losses are remarkably heavy, but increases are found only around the cities, though in recent years they have also been recorded in rural areas of Dublin and Meath counties.

has been in the larger towns, conspicuously through the growth of Dublin: with Dun Laoghaire and County Dublin its population has increased by over 250,000 since 1921 and growth still continues. During the 1950s towns of all sizes showed signs of growth, and even the villages increased in size. If one includes as 'towns' all places having more than 500 people, in 1961 the Republic had less than half its population in rural areas for the first time in its history. In Northern Ireland, on the other hand, the towns have grown markedly in population for more than a century, and the figures shown in Table V, from 1891 to 1961, merely indicate the continuance of a trend apparent long before 1891. Except for Belfast, the census authorities in Northern Ireland, unlike their colleagues in the Republic, have not indicated the extent to which town populations have spilled over into adjacent rural districts so the virtual stability of the rural population shown in Table V is probably more apparent than real. Until 1937, the Northern Ireland population was virtually stationary, but since then it has increased slightly. And emigration persists: from 1951 to 1961, it was 92,200, following an outward movement of 67,000 from 1937 to 1951, though this figure includes forces and merchant navy deaths overseas: from 1961 to 1966 the loss was approximately 38,000. Within the shorter period 1936–46 the Republic lost 187,000 people by emigration, followed by 120,000 more within the next five years, 197,000 from 1951 to 1956 and 212,000 from 1956 to 1961, that is, over 700,000 in twenty-five years.

Although the rate of emigration from the Republic and Northern Ireland has declined during the nineteen sixties, it is unlikely that emigration will cease though both governments would wish it to do so. At present the Belfast government is seeking new industries with the zeal of the development areas in Great Britain, and in the Republic a similar policy is followed, necessarily with the home market primarily in view, though with hopes of some export trade. But new enterprises of this type can only take up a part of the labour already available among the unemployed or – in the rural areas – the underemployed also. The key to the problem lies in the failure to provide, in town and country alike, enough new jobs to equal a high natural increase of population. The rural population diminishes annually as small farms are united and mechanical methods reduce the demand for labour. The analysis of the farm population given on pp. 190–2 shows that both labourers and farmers' relations must have contributed substantially to the outward movement. Crisply put, Ireland is a reservoir of labour on which Britain can draw at will – and America or the British Dominions. This view is unpopular within

Ireland: unfortunately it is abundantly confirmed by a study of the outward movement and its destination.

Only at one time was there an inward balance of movement, of 7,000 to Northern Ireland and 8,000 to the Republic, from 1931 to 1935. This was not due to the economic attraction of Ireland at the time – far from it – but to the depression elsewhere. Once the economic tide turned, in 1935, emigration began again, but primarily to Britain, whereas up to 1914 and even after the 1914–18 war many of the emigrants went to America. For ten years from 1940 an analysis of migrants was made, on which the Table below is based. In 1941–3 there was a considerable

TABLE II

Migrants from the Republic in thousands, 1940–50

N.B. Only those going to employment are included in these figures (see text)

Year	Grand total	Total males	Unskilled workers	Agri-culture	Total females	Domestic	Nursing
1940	26·0	17·1	5·9	5·4	8·9	5·3	1·6
1941	35·2	31·9	21·0	1·8	3·3	1·3	0·8
1942	51·8	37·3	23·8	4·8	14·5	6·0	2·2
1943	48·3	29·3	18·3	3·6	19·0	9·1	2·8
1944	13·6	7·7	4·3	1·4	5·9	2·8	1·1
1945	23·8	13·2	6·2	3·1	10·6	4·7	3·5
1946	30·0	10·8	5·9	2·4	19·2	12·1	3·9
1947	31·2	12·5	5·6	4·6	18·7	13·2	2·5
1948	40·1	21·7	11·6	5·4	18·4	12·4	0·9
1949	25·5	12·5	6·1	3·5	13·0	8·3	0·4
1950	17·4	8·5	3·4	2·8	8·8	5·2	0·3

movement of men, chiefly unskilled labourers, many of them engaged in aerodrome building. In 1944 outward movement declined sharply, as for some months before the invasion of Normandy traffic to England almost ceased, but in 1945 there was again heavy emigration, especially of men. But in 1946 women predominated, though later on the number of men and women was approximately equal. No such exact analysis exists for later years.

At all times Irish emigrants have settled mainly in towns. In 1841, four-fifths of all the Irish-born in England, and nearly three-fifths of those in Scotland, were in towns of at least 10,000 people, and many of the remainder were in other industrial areas. In 1961, 72% of the 951,000 Irish born people living in Great Britain were in towns of 50,000 or more. Of the Irish-born, 847,500 were in England, where they numbered 2% of the population but only 23,000 were in Wales, 0·9% of the population. Within the county of London (since extinguished) the Irish-born were 5·1% of the population, and they were prominent in several of the central boroughs such as Chelsea (5·1%), Fulham (6·0%), Hammersmith (10·7%), Hampstead (7·8%), Holborn (7·1%), Islington (7·3%), Kensington (7·5%), Paddington (10·6%), St Marylebone (5·5%), St Pancras (8·7%) and Westminster (7·3%). The whole of London County had 162,100 people of Irish birth. With northern England, it was a magnet for the Irish immigrant as long ago as 1841, but by 1961 the number of new migrants, Irish-born, was only 15,700 in Liverpool (2·1%) and 29,300 in Manchester (4·4%) compared with 59,000 in Birmingham (5·3%) and 19,400 in Coventry (6·4%). Though Scotland still attracts Irish migrants, approximately half of them from Northern Ireland, the number of first generation settlers in 1961 was only 80,500 (1·6%), of whom 31,100 (2·9%) were in Glasgow with another 21,500 in the neighbouring counties of Dunbarton, Lanark and Renfrew. Obviously the Irish, now as in the past, have settled where suitable employment may be found: the numbers of males and females of Irish birth living in England are virtually equal. Fortunately prejudice against Irish settlers appears to have diminished recently partly because many of those having such feelings in Britain have become racialists. Who will they hate in ten years' time?

DISTRIBUTION OF POPULATION

Within the past hundred years, the balance of country and town population has been materially altered in Ireland. In 1841 only one-fifth of the population lived in towns and villages, that is 1,655,100 out of 8,175,000, but in 1961 three-fifths of the population, 2,490,571 out of 4,244,803, were in towns and villages. These figures are analysed in Tables III–V; all the figures given for 1891 and later refer to all the people living within any nucleated settlement noted in the Census reports. In 1841 these were defined as groups of twenty houses or more, a definition which still survives in the Republic. In Northern Ireland, however, separate figures are given only for places having at least fifty houses, so exact comparisons cannot be given. But it is surely significant that in 1961 there were

1,519,000 people in the Republic living outside all places recognized by either census authority as nucleated settlements. True, one must remember that in 1841 there were over six and a half millions in rural areas, so the past hundred-odd years has seen the reduction of the country population by at least half in most areas, in some by two-thirds or more. The current decline of the rural population, by no means finished over the greater part of the country, is an adjustment to rural economic circumstances still continuing after nearly four generations of outward movement.

Approximately three-quarters of Ireland is occupied farmland, having a density of population which ranges from less than 50 to more than 400 to the square mile. This disparity must be stressed as one reviewer of the first edition spoke of the 'even lowland distribution of population'. It is by no means even, nor was it at any time. Fig. 21, for 1936, shows that over most of the Central Lowland, south of a line from Dundalk to Galway, the density was 50–100 per square mile, but that to the north of this lowland, the lowlands had 100–200 per square mile. Why there should be this contrast is unexplained, but it occurred also in 1841 when rural densities were far higher. But a clue is given by Fig. 32, p. 184, which shows that north of this same Dundalk–Galway line the farms are, on the average, smaller, and clearly all the adjustments of the past century through emigration have failed to give Ireland uniformity of farm size. Why should they? A square mile having four large farms, of 150 acres or so, might have four households with twenty persons, ten labourers of whom perhaps four would be married with fifteen dependants, giving 45 people within a square mile. A square mile having twenty farms of 30 acres or slightly more, will probably have 100 or more people, of whom the vast majority will be in the farm households. In between, vast lowland areas have farms averaging 40–50 acres: fifteen such farms may support a population of 70–80. Areas with such a density now had 200 or even 300 to the square mile in the years before the Famine.

The last surviving relics of congestion survive in parts of the west, where densities of as much as 400 to the square mile are recorded; in northwest Donegal, for example, such densities cover a considerable area from Dunglow northward. Other areas of the Atlantic fringe are almost as closely settled, for example around Belmullet, and on some of the islands off Donegal, on Inishbofin, Co. Galway, or on Achill. From Galway all round the coast to Clew bay, and in the extreme west of Mayo, the population lives in a narrow coastal belt, not generally more than a mile wide, where fields have been made with sand, seaweed and

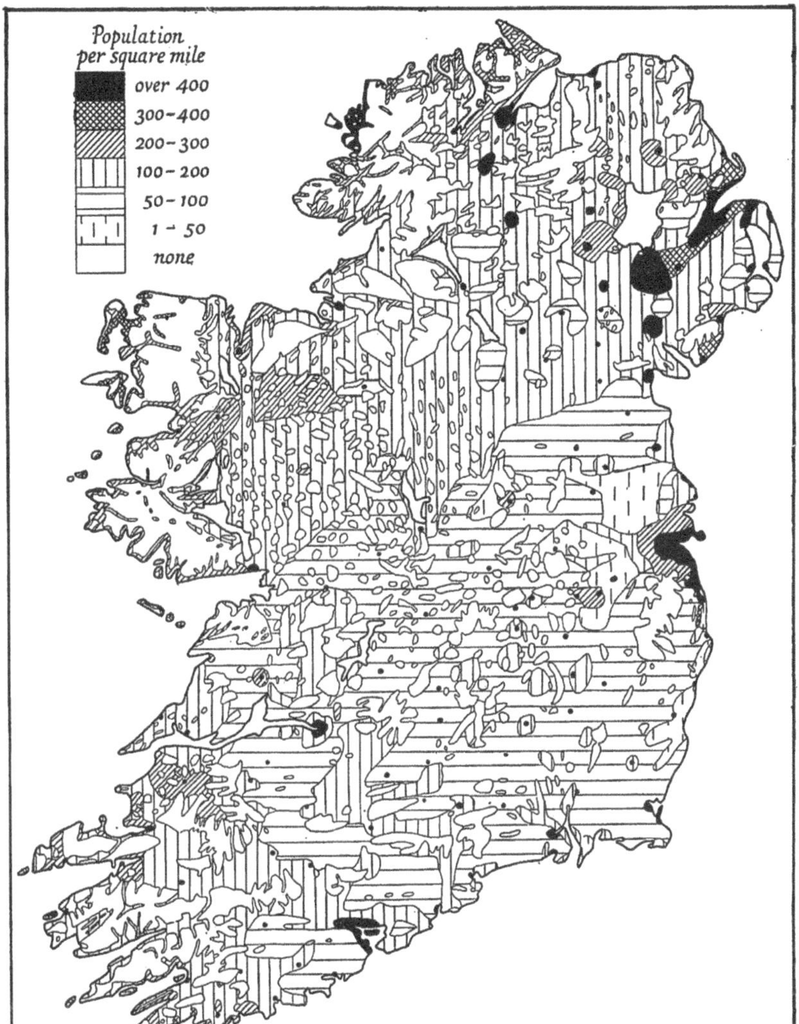

FIG. 21. DENSITY OF POPULATION, 1936

In the rural areas the density per square mile range from under 50 to more than 400, and was generally higher in the north and west than in the south and east.

First published in 'Geogr. J.', vol. 110, 1947, p. 52

the local peaty soil: this belt and a few favourable inland patches of land are closely settled with farms that generally have only a few acres of improved land, but vast expanses of rough grazing. In southwest Clare the farms are relatively small, averaging 40 acres, and densities are generally well above 100 per square mile, but north Clare, facing Galway bay, now has farms of 80–90 acres and a density of approximately 50 to 70 to the square mile. Before the Famine, it had tiny farms and a density six times as great. No area in Ireland shows more clearly the vast renovation of an agricultural landscape made possible by emigration. Kerry and West Cork, though included within the Congested Districts, never had farms as small as those of West Connacht or Donegal, but the loss of population has been remarkably heavy, partly through the decline of fishing.

Fig. 22 refers to conditions in 1961, and is a generalization of the dot maps given in Chapters 12–22. Comparison of these maps with those in the earlier editions of this book will show that in many areas the population has declined markedly since 1936: the general effect is shown if Figs. 21 and 22 are compared. Several areas show particularly heavy losses. These include three of the western counties with which the Congested Districts Board was closely associated – Donegal −20%, Mayo −24% and Kerry −17%: also included are Clare −17%, Sligo −20% and Roscommon −24%, parts of which were formerly congested. But the remarkably high declines in Cavan −26%, Monaghan −23%, and Longford −20%, are associated not with areas of marked poverty but rather with those divided mainly into small farms that, as noted on page 189, are no longer regarded as providing an adequate basis for the upkeep of a family. To a great extent the present emigration is hardly a flight for mere subsistence but rather an indication of an entirely natural desire for a higher standard of living. In Northern Ireland the decreases of the past twenty-five years (actually 1937–61) have not been as heavy as in the Republic, but there were losses of 14% in the rural districts of Lisnaskea, Co. Fermanagh, of 13% in Kilkeel, Co. Down, and of 11% in Lurgan, Co. Armagh, Castlederg and Clogher, both in Co. Tyrone. But some areas such as northwest Donegal, or northwest Mayo, which might be expected to show heavy declines, have not done so and there is no obvious correlation between emigration and limited natural resources.

Even though the loss of workers from the land in the Republic of Ireland has been heavy, agriculture remains a major source of employment in the country as a whole: at the 1961 Census the 336,000 males working on the land represented 43% of all the gainfully employed males in the country and the 41,700 females employed on farms were nearly

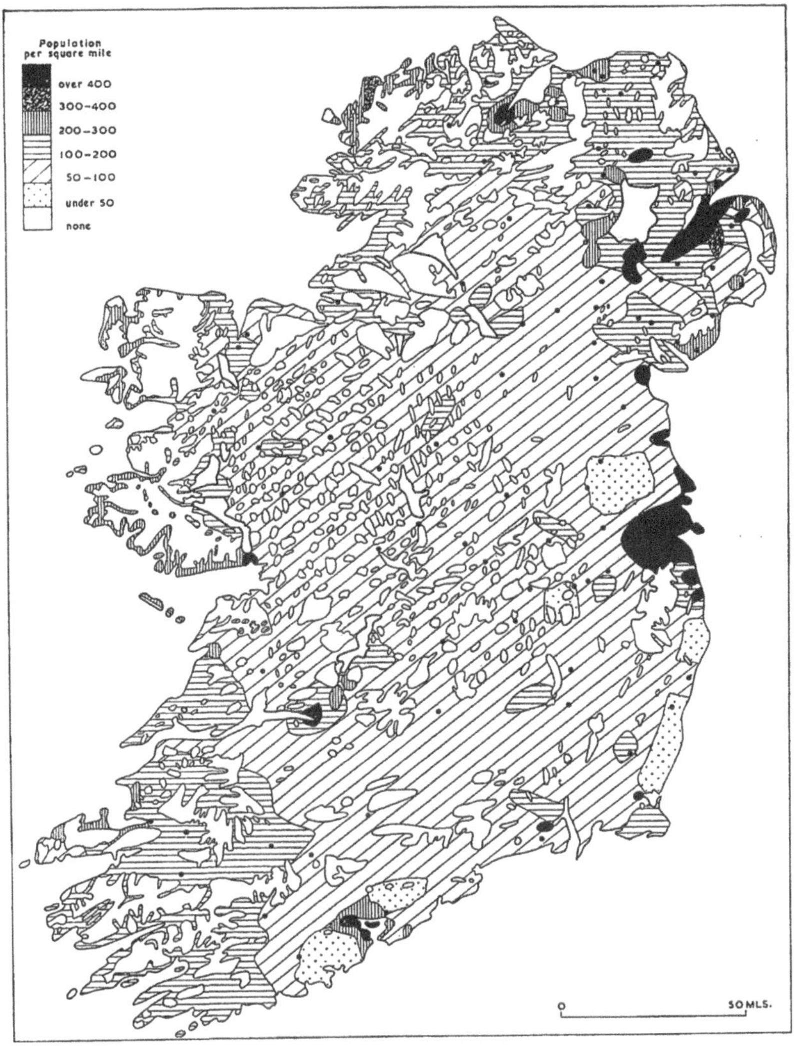

FIG. 22. DENSITY OF POPULATION, 1961

This map is based on the dot maps of Chapters 12–22.

15% of all women money-earners. But the number of farm workers has declined sharply since 1936 when 596,400 people were recorded as employed on the land (cf. p. 192). The decline has been continuous and apparently is still in progress for in 1966 it was estimated that the number engaged in farm work had declined to *c.* 321,000, of whom 24,000 were on a part-time basis. There are many labourers in rural areas who work intermittently on the land, but reference to page 192 will show that their numbers have declined to less than half from 1936 to 1961. In Northern Ireland the trends have been comparable. Although farming provides employment for some 13 per cent of the insured population, during the past ten years there has been a reduction of approximately 20,000 farmers and about 10,000 regular male workers, one quarter of the labour force in 1956. There are now some 89,300 workers on the land in Northern Ireland, of whom 38,600 are owners of holdings, 16,400 wives of owners and 34,300 are family or hired workers, male or female. The current decline in rural population is a continuation of a movement seen from the Famine period and obviously likely to continue for some time. It is associated especially with the combination of small holdings into larger farms, with the mechanization of agriculture and with the reluctance of men to become farm labourers.

So far attention has been given mainly to the rural population, which is now settled primarily in isolated farmhouses or roadside cottages. When the first Ordnance maps were produced, from 1829 to 1842, parts of Ireland had most of the agricultural population in rundale villages, around which the arable land was split up into numerous patches so that – in theory – everyone had equal shares. A large part of the work of the Land Commission has been to break up these villages to give everyone a compact holding, but this invariably involves a reduction of population, a new dispersed settlement pattern and the break-up of the old hamlets or villages. Relics of such settlements can be seen on the ground as well as on old maps but it is by no means certain that they ever prevailed over the whole of Ireland. Villages as generally understood probably date back to the Anglo-Norman period in the form of settlements around castles and churches. It is unlikely that the Norse settlers established more than a few villages, if any at all, though they contributed to the growth of Dublin and other ports. The lack of Roman town foundations, combined with the monastic tradition of the Celtic church rather than a diocesan organization, both affected Ireland permanently, for no part of the country has its rural population clustered in villages around parish churches.

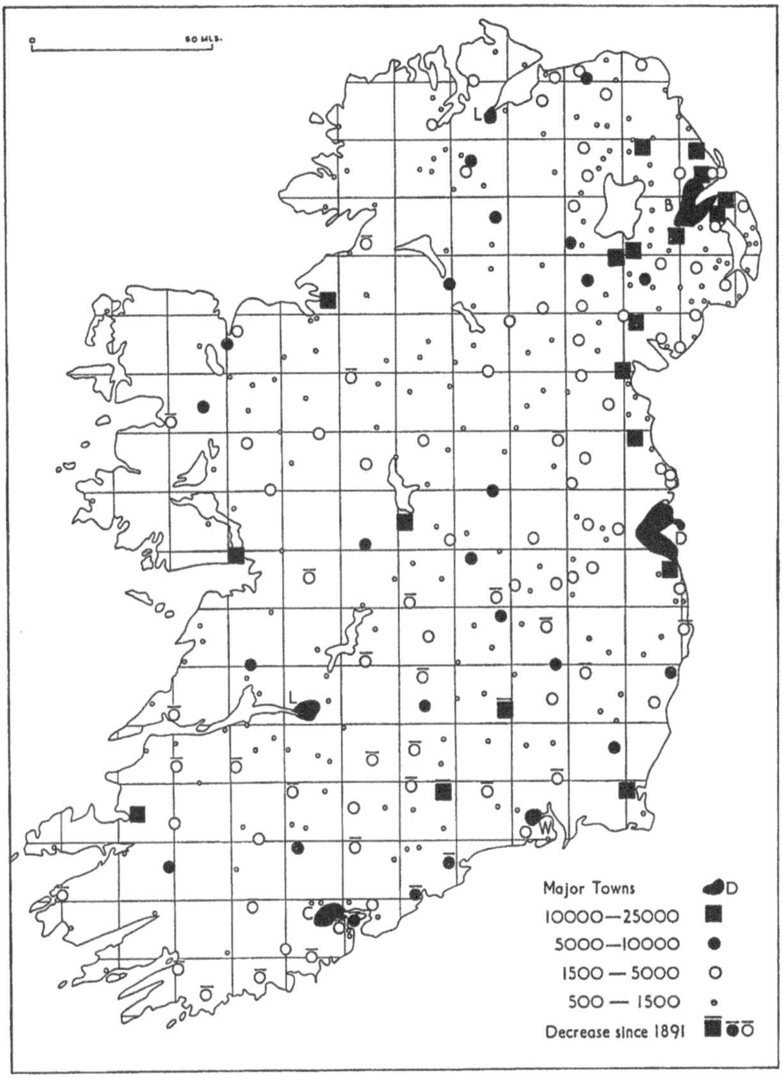

FIG. 23. THE TOWNS OF IRELAND, 1961

Leinster, the east of Ulster and the east of Munster have more towns than the remaining area of Ireland. Many of the towns were established in Anglo-Norman, Tudor and Stuart times, though not all those shown on Fig. 14, p. 93, are still important. Attention is drawn to the large number of very small towns with less than 1,500 people.

Each square covers 250 square miles

Even so, many villages and towns have been established. In the Anglo-Norman period, many charters were given for the holding of markets and fairs, though some of the latter were in places deep in the countryside, where roads met but few if any houses existed. All the invaders from the Anglo-Norman phase gave charters to numerous places, as all recognized that towns were necessary both to the administration and commerce of the country. Equally, many towns and villages arose beside demesnes, and from the eighteenth century for example, some stately streets were built by the landlords, with market halls and churches: these places attracted the weekly markets and fairs at longer intervals. In time, after the Famine, most of the weekly markets ceased, but the fairs became firmly established on a monthly basis. Gradually the country towns evolved, but in no country is it harder to separate towns from villages by any statistical limit. Table III, adapted from the Republic's 1966 Census, shows that there are 102 places with more than 1,500 people, but over eight hundred others called 'towns and villages'. Northern Ireland (1961) has only sixteen places with more than 5,000 people, twenty-four with 1,500–5,000, sixty with 500–1,500, and 101 with fewer people but at least fifty houses. Exact comparison between the two divisions of Ireland is impossible owing to statistical differences and the existence in Northern Ireland of industrial villages, of which there are few in the Republic, yet throughout Ireland many places are towns in function even if they have only a few hundred inhabitants.

A glance at Fig. 23 will show that over much of the country these small places are the only trading centres of significance: in north and west Donegal, for example, the two rural districts of Dunfanaghy and Glenties cover 608 square miles with a population of 28,300 (1966), of whom 6,100 (20%) are in towns and villages, of which the largest are Killybegs (1,100), Dungloe (900), Ardara (540) and Glenties (740) and on the north Dunfanaghy (360) and Falcarragh (470). These are the 'towns' with banks, shops, fairs, employment exchanges, cinema shows, dance halls, bus services, doctors, a medical dispensary, solicitors and even small industries. Not every place has all these facilities: for example, Dunfanaghy's fairs no longer exist but Falcarragh has excellent fairs, a few bright shops and a new bank. And a new secondary (grammar) school and a Gaelic theatre stand outside any village on the Atlantic road. Faced with the problem of brightening rural life, many people favour a policy of encouraging village growth, by adding schools, new halls, or small industries and, when possible, placing labourers' new cottages beside existing settlements, where services such as electricity and piped water may be available.

TABLE III

Town and Country Population in the Republic, 1961 and 1966

	Pop. within boundary	Pop. suburbs or environs	Total in 1966	Per cent of total pop.	Total in 1961	Per cent changes 1961–6
DUBLIN CONURBATION						
Dublin C.B.	568,772	81,381	650,153		595,288	+9·1
Dun Laoghaire M.B.	51,772	33,042	84,814		68,101	+24·1
Bray U.D.	12,699	969	13,668		12,615	+7·9
	633,243	115,392	748,635	26·0	676,004	+12·1
CORK AREA						
Cork C.B.	122,146	3,137	125,283		115,689	+8·2
Cohh U.D.	5,613	1,113	6,726		6,142	+9·1
Passage West town	2,686	85	2,771		2,665	+4·1
	130,445	4,335	134,780	4·6	124,496	+6·8
Limerick C.B.	55,912	2,170	58,082	2·0	51,732	+7·5
Waterford C.B.	29,842	0	29,842	1·0	28,216	+5·4
Other towns over 10,000 (9)	129,392	9,107	138,499	4·8	132,723	+3·9
Towns 5,000 to 10,000 (17)	97,301	12,802	110,103	3·8	100,169	+9·9
Towns 3,000 to 5,000 (21)	77,528	7,944	85,472	3·0	84,139	+1·2
Towns 1,500 to 3,000 (46)	98,258	2,280	100,538	3·7	101,668	−0·9
ALL TOWNS OVER 1,500 (102)	1,251,921	154,030	1,405,951	48·9	1,299,147	+8·0
Places of 500 1,500 (167)	138,906	624	139,530	4·8	131,122	+6·0
Villages 200–500 (255)	79,450	0	79,450	2·7	78,709	+1·0
Other villages (388)	50,075	0	50,075	1·7	51,784	−2·7
ALL TOWNS AND VILLAGES	1,520,352	154,654	1,675,006	58·1	1,560,762	+6·1
COUNTRY POPULATION			1,208,996	41·9	1,257,579	−3·8
TOTAL	1,520,352	154,654	2,884,002	100	2,818,341	+2·3

Even so, villages appear and disappear in consecutive censuses in the most puzzling way, due to the demolition of old houses or the addition of new ones: the Dingle peninsula, for example, had the well-established neat town of the same name, with six villages in 1891, seven in 1901, eight in 1911, nine in 1926 and 1936, eight in 1951 but six in 1961 and 1966.

Over most of the country, at least three-quarters of the population live outside towns and villages. The majority of the larger towns lie within a belt stretching from East Ulster through Leinster into Munster: Leinster has a tradition of town life dating back to Norman times and a farming population with considerable buying power supporting strong market centres of which a number have industries. Most of Munster has as much as one-third of its population in towns, except in Kerry, though many of them have declined since 1891, in some cases through the loss of garrisons after 1921, and more recently through the failure to attract industries. All country towns want industries: by no means all have found them. In Connacht there are a few towns of considerable vitality, such as Galway, Sligo and even Ballina, but most of the remainder are small market centres. As there are considerable economic differences between the two political parts of Ireland, their towns will be considered separately.

Table III is based on the admirable material given in the 1966 Census which includes for any town that has spread beyond its bounds the population of the outer area. Dublin, with Dun Laoghaire and Bray, now has a higher percentage of the Republic's population than ever before, and to some extent this shows the tendency to expand marked in many capital cities. Cork with its environs has shown signs of population growth since 1961, following a period of stability for ten years. Limerick, the largest town on the West coast, has shown signs of growth in recent years but has hardly experienced the vast expansion forecast when the Shannon airport was constructed. Waterford has increased in size only slightly. These four cities, with Wexford, share the distinction of a Norse settlement: all are ports. So too are the next four towns in size, Galway* (26,300), Dundalk (21,700), Drogheda (17,900) and Sligo (13,400). Dundalk and Drogheda are both vigorous commercial and industrial towns: Galway and Sligo are minor regional capitals of the west. Bray, included with the Dublin conurbation, is partly a residential and holiday centre and partly a market town on a most attractive site with the Leinster chain behind it. Kilkenny (12,000), Clonmel (11,500) and Athlone (11,000), are all major inland towns with a variety of industries: Wexford (12,700) is a strong trading and industrial town, with little shipping now,

* 1966 figures, including 'suburbs or environs'.

FIG. 24. TULLAMORE, CO. OFFALY, 1947

1, Substantial houses; 2, Smaller houses and cottages; 3, Outbuilding; 4, Shops and licensed premises; 5, Industrial premises; 6, Public buildings.

This town is more thoroughly industrialized than most provincial centres, and a distillery is prominent. The nucleus is compact, but housing is spreading in ribbons along the various roads. Population in 1946, 5,897; in 1966, 6,874.

Based on the Ordnance Survey, 25" to 1 mile, 1912, by permission of the Minister for Finance of Ireland

Drawn by F. M. Synge and first published in 'Irish Geography', vol. 1, No. 5, 1948, p. 145

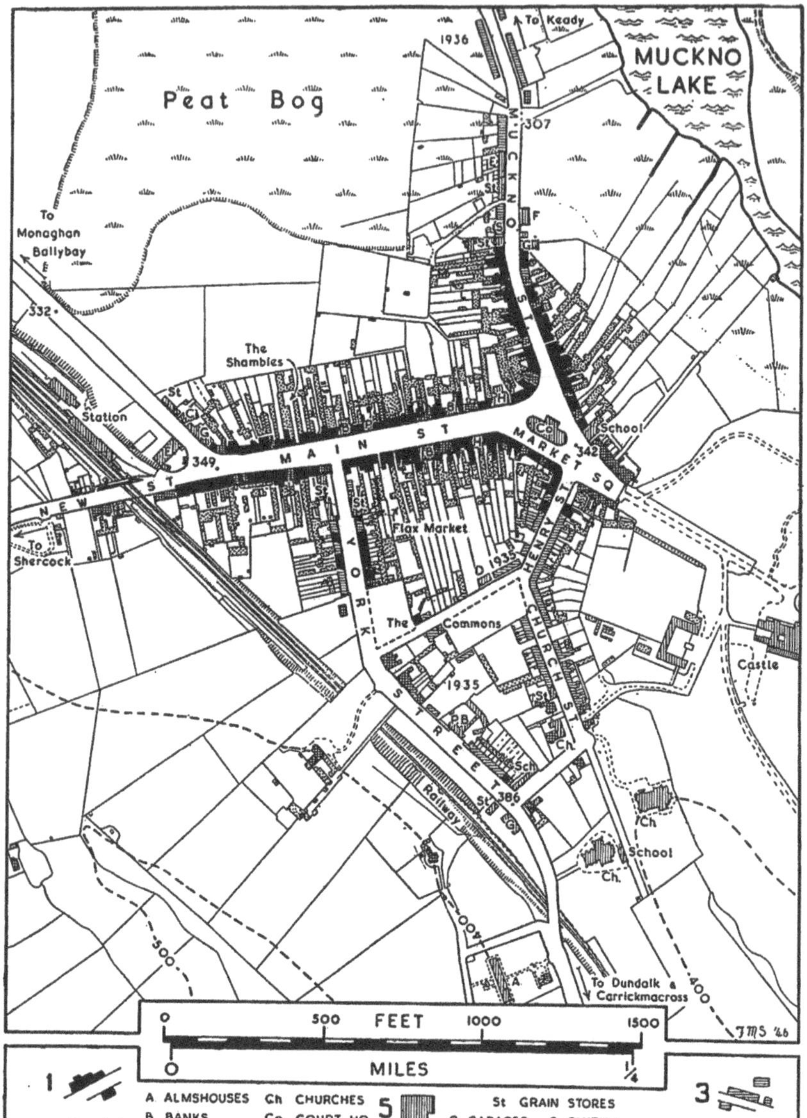

FIG. 25. CASTLEBLAYNEY, CO. MONAGHAN, 1946

1, Shops, including licensed premises; 2, Substantial houses; 3, Smaller houses and cottages, with dates of recent additions; 4, Outbuildings; 5, Public buildings.

Like many small towns, Castleblayney has one wide main street and several minor streets, but since this map was made the railway has been closed. In 1946 the population was 1,765 but in 1966 2,083.

Based on the Ordnance Survey, 25″ to 1 mile, 1909–1910, by permission of the Minister for Finance of Ireland

Drawn by F. M. Synge and first published in 'Irish Geography', vol. 1, No. 4, 1947, p. 121

and Tralee (12,500) is by far the largest town of Kerry, with a notable bacon industry and a vigorous local trade.

The seventeen towns with 5,000–10,000 are the more successful of the smaller market towns: some such as Carlow, Tullamore (Offaly), Ennis (Clare), Maryborough (Leix), Castlebar (Mayo), Monaghan and Mullingar (Westmeath) are also county towns. The degree of industrialization varies but Tullamore, shown in Fig. 24, has during the past hundred years added to itself a variety of enterprises including distilling, milling, bacon-curing and, notably, a large worsted yarn factory in the grounds of the old prison. Not all these towns have been industrially as successful, though a number of them, if not all, are strong trading centres with a good range of shops and well-patronized fairs. The smaller market towns, of 1,500–5,000 people, are the most usual type: Fig. 25 shows one such, Castleblayney, Co. Monaghan, a demesne town built chiefly towards the end of the eighteenth century and now serving as a market where the farmers can sell grain and potatoes to merchants, shop, and visit the monthly fairs. In this small town of 2,100 people, there is a wide range of town facilities. Mention should perhaps be made of a few places that draw part of their living from visitors, such as Greystones, with Delgany (3,950) in Co. Wicklow, Tramore (3,200) in Co. Waterford, or, in the west, Buncrana (3,100) or the smaller Bundoran. The last lives largely on tourists, but of the others Greystones and district is a magnet to the retired, Tramore is partly residential and Buncrana has been industrialized in such a way that it could serve as an awful example for planners – a demesne town, similar to Castleblayney, has had factories and houses added in every direction.

Northern Ireland as a political unit has a more even balance of town and rural population than the Republic. Table V shows that only 4000 of the population lives in rural areas or in small towns and villages of 1,500 people or less. But this is an overestimate of the essentially rural population, which is probably not more than one-third of the whole. Nevertheless any account of Northern Ireland which speaks of it as primarily industrial bears a closer relation to trade figures than to the landscape, for apart from the concentration of industry, commerce and housing around Belfast lough and in the Lagan valley, there is no other urban area of considerable size. Londonderry, focused on its fine hilly site above the first crossing place of the Foyle, is comparable in size only with Limerick, with its bridge over the great Shannon river. Through the increase in its towns, Northern Ireland has maintained its total population to a greater extent than the Republic. In 1841 the population of Northern Ireland was 1,649,000 but by 1891 it had declined to 1,236,000: since then, however, there has been a steady rise to 1,485,000 in 1966, which was an increase of almost 60,000 (+4·2%) from 1961. In 1951 it was noted that during the previous forty years the population of Northern Ireland had increased by 9·6%, compared with 7·0% for Scotland and 7·3% for Wales and that within the same period the Republic's population had declined by 5%.

Belfast and Londonderry are both of Stuart foundation, and the rise of Belfast during the nineteenth century gave Ireland its one major industrial city. Cotton-spinning was introduced in the late eighteenth century, but its main growth came with the success of the linen trade, the improvement of its river-port, and the development of its great shipbuilding and engineering industries. At all times a market town, it has become the one great commercial and administrative centre of Northern Ireland, now having with its outer suburbs a population of approximately half a million. Londonderry (55,700), sufficiently far from Belfast to be a regional capital, has since 1922 had part of its natural hinterland, in Donegal, cut off by a customs barrier. The main growth of Londonderry has been associated with its famous (though declining) shirt industry from the eighteen-fifties, but it preserves in character as a market centre and port, now with a widening range of chemical and engineering industries.

Of the remaining larger towns, only Newry (12,200) and Coleraine (13,600) are outside the major industrial area based on Belfast. Newry was a prosperous port and market centre in the late eighteenth and early nineteenth century, but like many of the smaller ports it has for seventy years been virtually stationary in population. Coleraine, at the first Bann crossing, is a vigorous market and industrial centre. Six other towns of this class – Larne (17,300), Ballymena (16,000), Lisburn (21,500), Lurgan (20,700), Newtownards (13,000) and Portadown (20,700) – are all within a radius of thirty-five miles from Belfast and have grown since the linen industry left the homes of the farmers in the mid-nineteenth century for the factories. Bangor (26,900) has been included as part of the Belfast conurbation, as it is part of the attractive residential and holiday belt of settlement which extends along the shores of the lough. The smaller towns include six of 5,000–10,000 and twenty-four of 1,500–5,000. Most of these are the ordinary market towns of the countryside though many have a linen mill. Holywood (8,000) is part of the Belfast conurbation, and Carrickfergus (11,000), an earlier settlement than Belfast round its castle, is now virtually joined to the city. Armagh (11,000) is the great cathedral centre, Banbridge (6,800) a linen centre once famous for its bleach greens. Enniskillen (7,200), Omagh (9,900), Dungannon (7,300) and Strabane (8,800) are all market and route towns with some industry, but though the proportion of towns in Northern Ireland is greater than in the Republic the market towns of the countryside are the main type.

SUMMARY

During the period from 1891 to 1966, certain major trends have appeared.

1. The growth of towns has been greater in Northern Ireland than in the Republic.

2. The increases in the towns of the Republic are more marked in the larger than in the smaller centres.

3. The rural population has declined by more than two-fifths in the Republic since 1891. An exact comparison with Northern Ireland is difficult as the rural figures include a considerable amount of urban overspill, but up to the 1937 Census the loss in rural areas differed little in the Republic and in Northern Ireland.

4. Emigration has been approximately equal to the natural increase from almost all the classes of towns in the Republic except Dublin, with the result that many towns have fluctuated little during the past seventy years, and even the past century. Dublin has at times absorbed some of the rural emigrants, but its fortunes have fluctuated from one period to another.

5. The increasing proportion of the people in the towns does not indicate any considerable immigration from the countryside, but shows rather that the towns have not shared in the general decline of population. In the life of Ireland as a whole they are now more significant than ever before, and the rural–urban balance is now not unlike that of France.

In the following Tables, which are intended as a commentary on the text, the recent trends of population are analysed in the case of large towns (over 25,000), medium-sized (over 10,000 and 5,000–10,000), and small towns (1,500–5,000). The still smaller 'towns' and villages are included in the rural population which, as explained above, has been declining markedly for the past century. The Republic's population for 1966 is analysed in Table III, page 137, so the 1961 figures are used in Table IV. For Northern Ireland, the 1961 figures are used as only provisional returns exist for 1966: the population figures of the towns noted on page 142 are from the 1966 returns.

TABLE IV

Population, 1891–1961: The Republic

	1891	% Total pop.	1936	% Total pop.	% Change 1891–1936	1951	% Total pop.	1961	% Total pop.	% Change 1936–61
Dublin, Limerick, Cork, Waterford	518,063	14·9	699,040	23·6	+36·2	846,424	28·6	880,448	31·3	+26·1
Other towns over 10,000	81,142	2·3	92,806	3·1	+13·3	101,840	3·4	132,723	4·7	+43·1
5,000–10,000	89,091	2·6	90,191	3·0	+ 1·2	97,281	3·3	100,169	3·6	+10·8
1,500–5,000	183,292	5·3	175,702	5·9	− 4·0	170,799	5·8	185,807	6·6	+ 5·9
Rural areas	2,597,106	74·9	1,910,681	64·4	−26·5	1,744,249	58·9	1,519,194	53·8	−20·5
TOTAL	3,468,694	100·0	2,968,242	100·0	−14·4	2,960,593	100·0	2,818,341	100·0	− 4·7

Note:

(1) For 1891 and 1935, the Dublin 'conurbation' was taken to include the city, Dun Laoghaire, Bray and Howth: so defined, its population was 362,000 in 1891 and 523,000 in 1936. By 1951 the suburbs of Dublin and Dun Laoghaire had extended far beyond the administrative boundaries: the Census Report gives the estimate that these outer Dublin suburbs had 53,805 people, and those of Dun Laoghaire 10,565. Including these figures the total population of the conurbation was 646,535. For the same areas in 1966, see Table III, page 137.

(2) The Cork conurbation includes the southeast and southwest suburbs as defined in the Censuses from 1936 onwards, with Cobh and Passage West. Including these areas, the population was *c.* 92,500 in 1891, 107,000 in 1936 and 120,000 in 1951. The table on page 137 shows that Cork and district is now increasing in population.

(3) In 1951, of the 1,744,249 people living in 'rural areas', 204,659 (6·9% of the whole population) lived in small towns and villages, i.e. places having more than 20 houses. If these are excluded, the 1,539,590 people remaining (51·8%), were those in dispersed habitations or hamlets. In 1961, 261,615 people were in small towns and villages, 9·2% of the population, with 1,257,579 people (44·7%) in dispersed habitations and hamlets. By 1966 the number living in hamlets and dispersed habitations had diminished to 1,208,996 (41·9%).

TABLE V

Population, 1891–1961: Northern Ireland

	1891	% Total pop.	1937	% Total pop.	% Change 1891–1937	1951	% Total pop.	1961	% Total pop.	% Change 1951–61
Belfast conurbation, Londonderry	312,891	25·4	563,233	43·9	+79	604,366	44·0	642,501	45·1	+ 6·3
Other towns over 10,000	56,696	4·6	77,246	6·1	+36	96,228	7·0	105,011	7·4	+ 8·1
5,000–10,000	28,849	2·3	40,456	3·2	+40	40,759	3·0	45,924	3·2	+12·7
1,500–5,000	32,419	2·6	51,527	4·1	+58	51,530	3·8	58,169	4·1	+12·9
Rural areas	805,201	65·1	547,283	42·7	−32	578,038	42·2	573,857	40·2	− 0·6
TOTAL	1,236,056	100·0	1,279,745	100·0	+ 4	1,370,921	100·0	1,425,462	100·0	+ 4·0

Note:

The Belfast conurbation is here taken to include the city, Holywood, Bangor, Carrickfergus and Lisburn for 1891 and 1937: so defined its population was 280,000 in 1891 and 477,000 in 1937. For 1951 and 1961 it also includes the new urban district of Newtownabbey and parts of the rural districts of Castlereagh, Hillsborough and Lisburn, which with the city had a population of 502,700 in 1951 and 528,900 in 1961. The figure of 642,501 also includes the neighbouring towns of Holywood, Bangor, Carrickfergus and Larne, as well as Londonderry, where the population increased from 50,092 in 1951 to 53,744 in 1961.

There are seven towns of over 10,000, six of 5,000–10,000 and eighteen of 1,500 to 5,000 included in the above table. Not included in the last group are six towns, Greenisland, Randalstown, Killyleagh, Rathfriland, Maghera and Sion Mills, that had over 1,500 people in 1961 but not in 1951. The Northern Ireland Census lists places having fifty houses or more: in 1951 these had 73,323 inhabitants and in 1961, 81,385 inhabitants. These are here included, with the six towns noted above, among the rural population. Note that the definition of a village differs from that used in the Republic.

CHAPTER 7

Aspects of Social Geography

Many travellers to Ireland have left accounts of the country as they saw it and several nineteenth-century writers, not to mention Arthur Young and others in earlier times, have given notable contributions to the literature. Most of them were realistic in approach though not invariably so: Thackeray, for example, had deep religious prejudices and his contemporaries, Mr and Mrs Hall, occasionally lose themselves in a welter of legends and fairy tales. All the travellers agree that improvements could be made, and the Irish have been liberally supplied with good advice. 'Grow turnips'; 'reclaim the bogs'; 'plant acorns'; 'work harder'; 'keep your hotels cleaner'; the chorus goes on. Nevertheless, most of the travellers showed a sympathetic approach and, like Henry VIII, apparently thought that the real need was to raise the standard of living. Steadily Ireland is entering the modern affluent society, though not too rapidly. Electricity is now provided almost everywhere and shops in the remotest West are will stocked with appliances. But there is still a gentleness of pace in much of rural Ireland that leaves an abiding impression on delighted tourists.

Political life in Ireland has never been dull. The division of Ireland into two units has proved at least a partial solution of vast problems, but the cleavage between unionists and republicans remains. Thousands of Irishmen are determined to keep the six counties as part of the United Kingdom: thousands are determined to make a thirty-two county republic. Sufficient time has elapsed since 1922 to make the two divisions into separate economic units, but although this is a vital factor in the situation, social tensions rest also on religious differences, on educational policy, especially in relation to the Irish language, and on the attitude to Great Britain and the Commonwealth. Many past problems have been solved but not all. The land has been transferred to the people who farm it as owners rather than as tenants. Religious persecution belongs to a past now distant, but remembered too often. Rather is the current problem differences of policy that may make eventual union difficult to achieve so this chapter, in dealing first with settlement types, and then with religion and language, also considers the political problem as it now appears.

COUNTRY AND TOWN POPULATIONS

As noted briefly on pp. 109, 134, the present distribution of farmhouses dates back only for a few generations and the general effect of the agricultural re-settlement has been to break up the hamlets which still existed in the early nineteenth century, notably in the extreme north of Co. Antrim and Co. Derry, in the greater part of Donegal, in a wide belt from the Moy valley to Galway bay, with an extension eastward to the Shannon at lough Ree, in the Dingle peninsula, in the countryside around the Nore and Barrow rivers, and in various other areas. Those listed were taken from the original 6″ to 1 mile maps of 1830–42, and may show the surviving patches of a pattern of nucleated settlement which at one time covered the whole country, or a large part of it. But the immigrant landlord aristocracies, from Anglo-Norman times onward, have all influenced the settlement pattern and at present, as shown on p. 181, the last hamlets are disappearing under land re-settlement schemes. Dispersion has such variety that no single controlling influence now exists, and the farmed landscape is dotted with houses spread in what appears to be a completely haphazard fashion over the cultivated land, though most farms have reasonably good access to the roads. In many areas the farms are placed beside the roads and in the west, where the holdings are very small, the houses are so close together that the country road may resemble a village street. In parts of the Central Lowland, where cultivation is restricted to patches of ground moraine surrounded by bogs, the houses in many cases line the roads. The houses of new farms made in some land settlement schemes on former demesnes are also situated mainly along the roads, and many of the new cottages are similarly placed.

The housing standards of the rural population have been vastly improved by various government schemes, and the poor cottages, thatched and attractive in appearance but inadequate as dwellings, have in countless cases been abandoned in favour of new cottages with pebble-dash walls, manufactured tiles or imported slates. These new cottages are perhaps aesthetically alien to the landscape, yet they are already part of it. They are occupied either by farmers of a few acres or by labourers: each cottage has a patch of ground for the cultivation of vegetables. Some of the old cottages are used as cowsheds or stores, but unfortunately many of them are left to moulder into decay with the result that some villages appear to have suffered from war devastation on a substantial scale. In many parts of Ireland, both in the countryside and the towns, a demolition squad charged with the job of removing all derelict and useless remains of buildings would be a great social asset. Each county normally has its own cottage design: in Co. Dublin and Co. Wicklow, the cottages have white walls, but in Co. Meath the walls are salmon pink and the roofing materials also are varied, though red tiles are frequent; in Co. Kilkenny there are also asbestos slates. It is sometimes possible for the traveller along the roads to mark his passage from one county to the next by the change in cottage design. The author has never forgotten his first journey by car across central Ireland which left the impression of a far higher standard of rural housing than reading had led him to expect.

On the whole the farmhouses correspond in size fairly closely to that of the farms around them. The medium-sized holdings, of 30–100 acres, generally have two-storey houses and farm buildings grouped around a yard. While every part of Ireland has farms of large, medium and small size, the medium-size farm, with its two-storey house, dominates most of the Central Lowland and almost all the south from Wexford to Kerry. The holders of small farms generally live in one-storey cottages, some of which cover a sufficient ground space to give the family several rooms. Co. Meath and Co. Dublin have some of the largest farms in Ireland, around houses correspondingly substantial; here, too, cottages are numerous, for a considerable number of labourers are in regular employment. The Irish rural scene still has a wide range of standards, both in housing and in economic organization, though the extremes of wealth and poverty characteristic of the eighteenth century are no longer general: far from being inhabited by a standardized peasantry, the country shows contrasts similar to those between the farmers of the rich country around Aberdeen and the crofters of the Western Isles.

Two-fifths of the population of the Republic live outside all towns and villages and the dispersion makes difficult the carrying of electricity and piped water to farms. But rural electrification is now well advanced and many of the new cottages are placed beside villages so that services may be provided conveniently. Even so, water is a problem on many farms, and daily journeys of some length may be necessary for drinking water. Of the 302,000 farm dwellings in the Republic only one-tenth have a fixed bath or a hot water tap and less than one-quarter have a piped water supply. And two-thirds of all these dwellings have neither flush lavatories, dry or chemical closets. In county Leitrim one woman spoke as if paradise would dawn if only the village could have one 'fountain' (pump). But if elementary amenities are lacking, the tradition of giving neighbourly help remains strong. (No drains, but good neighbours.) At busy times on the farms, especially on the humbler farms, labour will be shared and tools lent for use in mowing, spring sowing, shearing, harvesting, hedging or turf cutting. But emigration has removed so many people from the land that in many countrysides there is an acute shortage of labour, for many of the farmers' relations naturally prefer a steady job overseas to the dependent status and intermittent employment they experience at home. Many of the poorer farmers having only a few acres of land supplement their living by working on the larger farms or as general labourers in the district.

Traditionally, much rural social life centres around particular houses, around lane ends which are meeting grounds, some of which have a boarded floor for Irish dancing. Meetings in farm kitchens for conversation or 'a bit of a dance' happily survive and in some parts of the country there are local games such as skittles, played on the country roads or in special alleys. Handball alleys, rather like squash courts without the roof and the back wall, puzzle some visitors: these are seen in villages and even beside roads in the countryside. Much good work has been done by people who organize social and athletic gatherings in the countryside for the general complaint is of dullness, and the only resort the drinking shop. Almost all the fairs are now held in towns and villages, for few of the remoter gatherings have survived. A fair-day galvanizes many a small town or village into life and attracts farmers from a wide area with their cattle, sheep and pigs, which line the streets and lanes from an early hour. The fair is, however, surrounded with abuses as a system of buying and selling for the prices are varied and unpredictable, and depend very much on the persistence of the buyer and seller. An inquiry in Northern Ireland showed that returns to the farmer were liable to vary from fair to fair, fairer prices and auction marts, where they have been established, give to both buyer and seller. In many cases farmers will bring cattle into the fair from a radius of ten miles, driving them along the roads; they have already lost some value as a result of their long journey and sale is almost imperative, as few owners will be willing to take them back. But the fair is a social function attended by scores of people merely to see their friends.

To draw a line between towns and villages in Ireland is virtually impossible. There are some villages which have obviously been built largely by the owners of the demesnes beside them, but many have apparently arisen in a haphazard way beside a crossroads or a church. At all times some are in decline and others adding to their trade: if a village loses its fairs, its shops will suffer for much of the trade is done on fair-days. Markets for produce have virtually ceased to exist, for most of the selling of produce other than stock is handled by grocers who buy eggs, potatoes, grain, home-churned (country) butter, and even flax and feathers. Some of the co-operative creameries have become mercantile organizations on a wider scale by dealing in eggs and other produce, and by selling ironmongery and general groceries. Even in villages, individuals of enterprise may achieve a quite extraordinary range of commercial power. One such, in a place hard to rank as a town, was described on his billhead as an 'auctioneer and valuer, family grocer, corn and seed merchant, corn buyer'; in addition he kept an hotel, had a farm immediately behind it, and sold agricultural machinery. As a rule several miles separate any town or village from the next, and the resultant pattern of nucleated settlement is a mere skeleton of that in many parts of the English Lowland or the Paris Basin. Places with a few hundred people in Ireland are generally towns in function, for their monthly fairs and their shops may serve an area of fifty to one hundred square miles.

Nevertheless there are many towns in the Irish countryside which have an importance out of all proportion to their modest size. Like towns elsewhere, they have four main functions: they provide commercial contacts between a rural community and the outside world; they are centres of banking, education and professional services; they manufacture part of the agricultural output, and they have factories which manufacture goods for the home market or for the export market, most notably in the case of the linen industry. The first two functions are common to all towns, but the third and fourth imply a more advanced stage of economic organization in which agriculture has passed from primary production into processing and industry has entered deeply into the life of the countryside. Although some are attractive and lively places, others are capable of considerable improvement and not a few still have wretched housing, though new estates have been added recently. Schools of the secondary type, banks, special services such as those of lawyers, auctioneers and estate agents, and a range of public transport are generally available: in them, too, there is generally at least one cinema, a dance hall, the public library service, various clubs and a remarkably high number of licensed premises, though observation suggests that these are diminishing. In the local idiom, towns are known as 'good' and 'poor'. A 'good' town is one which possesses a well-established trade with the countryside, one or more hotels, a few workshops or even a factory providing employment, and some amenities. It may be small or large: thus Graiguenamanagh, with 1,200 people, and Clonmel with 11,600, are both regarded as 'good'. Some towns have been successful and others have not and their fortunes have been conditioned largely by the enterprise of the people living in them. There are generally a few prosperous traders whose shops, in the smallest towns, cater more obviously for the sowing of fields and the care of cattle than for the furnishing of homes and the stocking of larders. In the larger provincial towns, however, the shops are frequently of a surprisingly high standard and carry enormous stocks, as their trade covers a wide radius.

The proportion of town dwellers to rural population is highest in the richer parts of the country: in most parts of Leinster and Munster two-fifths of the whole community is in towns and villages as against one-quarter in Connacht and the three counties of the Republic within Ulster. Northern Ireland has, through its industrialization, a higher proportion of the population in towns and villages everywhere. But as recently as 1936 Co. Leitrim and Co. Roscommon, with a total area of 1,500 square miles, had only 11% of their population in towns and villages, though by 1966 this proportion had increased to 20%. In much of Ireland there is a degree of dependence on agriculture which recalls eighteenth-century conditions. Many of the Irish towns have decreased very considerably during the past hundred years, and the efforts to revive them by industrialization based on electric supply have arrested their decline rather than added to their population in any considerable degree. As most of the new industries depend on imported raw materials, the natural advantages possessed by the ports are obvious, and in the absence of government planning the Republic's new industries would probably be placed mainly, indeed almost entirely, around Dublin, or else in the three other regional capitals, Cork, Limerick and Waterford, all of which have a long-standing industrial tradition. The Georgian façade in each of these four towns shows that they are not industrial creations of yesterday, but the well-established ports, commercial centres, cathedral cities and manufacturing towns of a country depending primarily on the land. Unfortunately they all possess severe housing problems which are due to two main causes: first, the decay of some of the Georgian property into one-room tenements, particularly in Dublin, and second, the heritage of wretched cottages built from the eighteenth century onwards to house an increasing number of families. In Belfast there are large areas of small working-class cottages similar to those of English towns. Both the major cities have slum-clearance schemes similar to those of British cities.

DISTRIBUTION OF RELIGIOUS GROUPS

Ireland is the only part of the British Isles to list the religious allegiance of its population at each Census, and all save an insignificant number of people give the information required. The religious count was first made in 1861, since when the relative proportion of Catholics to Protestants has been three to one. In 1871 and 1881, the proportion of Roman Catholics, at 76·7%, was actually higher than in 1936, with 75·5%. And within the six counties of Northern Ireland the number of Roman Catholics fell from 534,000 (39%) in 1871 to 428,000 (33·4%) in 1937, but rose to 498,000 (34·9%) by 1961. In 1961 the Republic had 145,000 Protestants, 5·1% of the population, compared with 171,000 in 1946, when the proportion, at 5·7%, was 12% less than in 1936, and a little more than half the proportion of 10·8% in 1871. Until the 1914–18 war, over one-tenth of the Republic's population was Protestant: in 1961, less than one-twentieth belonged to the Church of Ireland (3·7%), the Presbyterian Church (0·7%) or the Methodist Church (0·2%). Clearly there is a difference between the two parts of Ireland, for in the Six Counties the Protestants have remained approximately two-thirds of the population and in the Republic they are declining.

Mere mapping of the distribution of the religious groups proves very little, but Figs. 26–28 reflect some interesting social, historical and political factors. Although the number of professed atheists in Ireland is small, the number of people in Northern Ireland who are 'of no stated denomination' rose from 5,865 in 1951 to 28,418 in 1961. The category of 'other religions' included 69,362 people (5·1%) in 1951 and 97,717 (6·9%) in 1961: among them were Jews, a small Society of Friends group, Plymouth Brethren, Baptists, Congregationalists, Unitarians and some reformed Presbyterians. All these, with the Methodist community, must be regarded as in general sympathy with the major Protestant groups, the Church of Ireland and the Presbyterians. In general it is a fair assumption that the Protestant community bears a close resemblance to the Unionist party, though this is by no means invariable. But it is nevertheless true that ever since 1921 Northern Ireland has been governed by the Unionist party and so far Labour candidates have had few successes. In the Republic there are a few constituencies where 'the Protestant Vote' is regarded as a factor in elections. There have been many exceptions to these generalizations, and various Nationalist leaders and Gaelic enthusiasts have been Protestants; but the argument that certain Northern Ireland counties

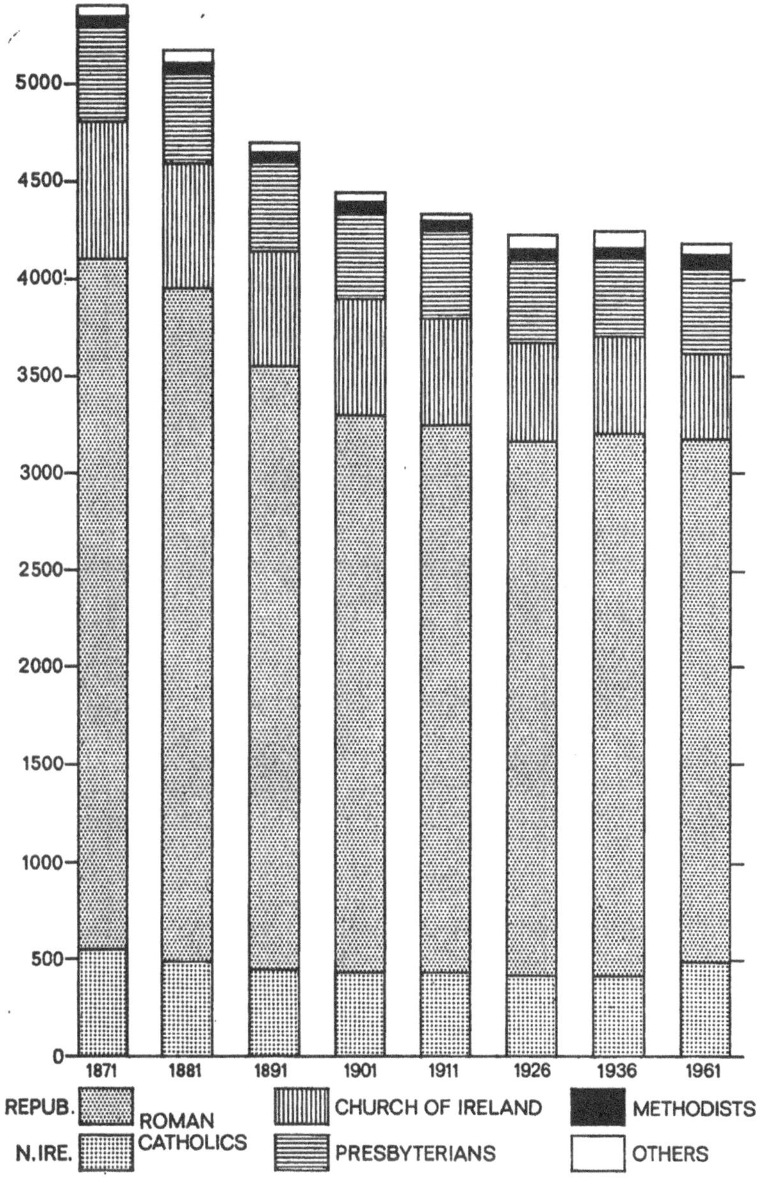

FIG. 26. RELIGIOUS ALLEGIANCE IN IRELAND, 1871–1961

This graph includes the whole population of Ireland, and the number of people belonging to each communion is shown: for the Roman Catholics, the number in the Republic and Northern Ireland is separately distinguished.

Based on the Census Reports

have Catholic majorities is commonly used as one for their union with the Republic.

Everywhere the population is mixed. The figures given for 1961 both in Northern Ireland and in the Republic only cover towns and rural districts, and the last detailed figures for the six counties are for 1911. These have been mapped in Fig. 27 which shows that no clear line can be drawn between the main religious groups. County figures mean little

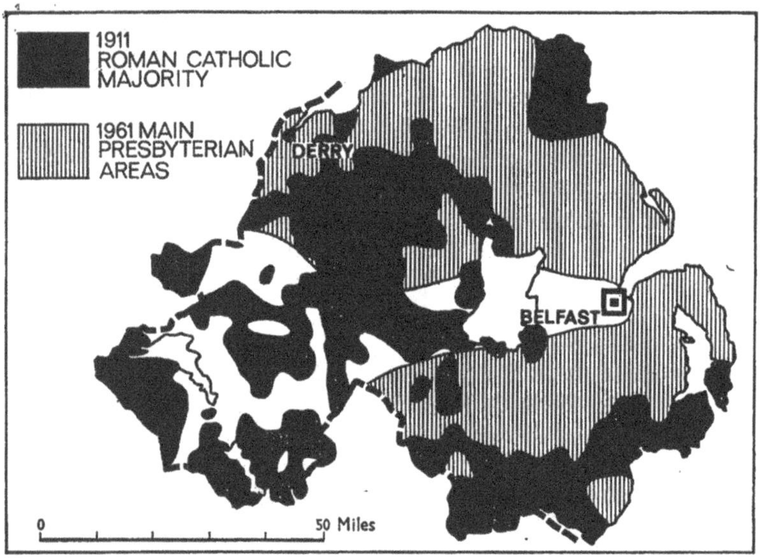

FIG. 27. ROMAN CATHOLIC MAJORITIES IN NORTHERN IRELAND, 1911: MAIN PRESBYTERIAN AND CHURCH OF IRELAND AREAS

Since 1911, the returns have been available only for rural districts (see Fig. 28). This map shows the mixture of Catholic and Protestant west of the Bann, and also the substantial Roman Catholic communities of south Down, south Armagh and north-east Antrim. It also shows the main Presbyterian and Church of Ireland areas.

where there are such mixed communities: in Londonderry city, for example, 67% of the population is Roman Catholic but the rural areas around it have Protestant majorities, which are also found in the north and east of Co. Derry. All the other country west of the Bann, that is the south of Co. Derry, with the counties of Fermanagh and Tyrone, is of mixed allegiance, and there is a wide belt with a Roman Catholic majority extending through the south of Co. Armagh and Co. Down, including Armagh city (58% Catholic) and Newry (84%). The extreme northeast, including the small town of Ballycastle and the northern glens of Antrim, has a Catholic majority said to be composed partially of people descended from Catholic Scots who settled here before the major plantations and sponsored immigrations began.

There is a general statement in a recent Ulster Yearbook that the six counties have a population composed almost equally of Irish, Scottish and English elements. Whether this has sound historical justification or not, there is no doubt that immigrants came into Ulster in considerable numbers during Tudor and Stuart times and began the improvement of the north, then so desolate. It is at least interesting to note that the Presbyterians predominate among the Protestants in Co. Down and Co. Antrim through to Co. Derry – and across the border in the east of Co. Donegal, as well as in Co. Monaghan. The Church of Ireland adherents are more numerous than the Presbyterians in the Lagan valley, in Co. Armagh (except for the extreme south) and in Fermanagh and Tyrone. It would be an over-simplification to relate this distribution entirely to ancestry but it is at least one element in a complex situation: for example, there has been some mixture of populations through industrial growth and many of the towns differ in their relative allegiance from the countrysides around them.

Although many people regard themselves as allied to some Church organization and yet never attend its services, one feature of the country's life is the very large attendance at church everywhere. Apart from its unassessable religious significance, the gatherings on Sundays and various holy days at least helps to preserve the solidarity of many communities in the rural areas. In the cities, the number of worshippers at most churches is so large that such gatherings lack the direct social importance of Sunday morning services in rural areas, except through the demonstration of a common disposition towards the spiritual life. Minority groups, such as the Protestant bodies in the Republic, are particularly conscious of their solidarity, and the Sunday congregations in the rural areas may include almost the whole community. This frequently gives visitors a fallacious impression of the numerical strength of the minorities, though their social cohesion extends deeply into the web of daily life through marriage, sports clubs and organizations of all kinds, education and general outlook. It is perhaps not too optimistic a comment that many ancient barriers are being broken down to some extent, especially in the Republic, where the disparity in numbers makes the political power of the minority quite negligible under the present regime.

In Northern Ireland the proportion of Roman Catholics in rural districts ranges (1961) from 8% in north Co. Down to 73% in the south of Co. Armagh and across the border, from 69% (1961) in part of Co. Monaghan and 68% in the Stranorlar R.D., Co. Donegal, to more than 99% over much of Ireland west of the Shannon. Although every town of more than 1,500 people (except for Spanglehill, Co. Cork) had at least one Protestant in 1946, at the 1936 Census, when fuller details were available, at least one-fifth of the Republic had an exclusively Catholic population: about half of the district electoral divisions in Clare and Galway, and about one-third of those in Mayo and Kerry had no Protestants at all. Lack of similar figures prevents exact comparison, but Fig. 28 shows that over much of the country the Protestants numbered less than 1% of the population in 1961. In a country town or a wide rural area, a congregation may consist of only one or two families, of whom it has been written 'How proudly the five maiden ladies hymn their defiance'. Many churches have been closed, or are kept open only for summer visitors: the author recalls one fine church which was kept open for three people until three maiden sisters from Belfast took the old rectory and doubled the congregation. The key to the situation lies in emigration and a table given in the 1946 Census report* showed that from 1936–1946 the emigration rate among Protestants aged 10–34 was 50% higher than among Roman Catholics of the same ages.

Except in the small areas noted in Co. Donegal and Co. Monaghan as having relatively strong Presbyterian communities, Church of Ireland people are the main Protestant group, but there are also small Methodist societies scattered through the country, many of which trace their existence back to the days when John Wesley made his tours through Ireland. All these bodies minister assiduously to their people, however small their numbers: the Methodists, for example, maintain several hundred preaching places, many of which are supplied on weekdays only, and also hold services in the open air at various fairs. In fact, the Methodists are the only Protestant body who have added to their numbers in the last seventy years: in Northern Ireland they have increased from 27,000 in 1871 to 72,000 in 1961 (5% of the population), though in the Republic they have declined from 19,000 in 1891 to 6,700 in 1961 (0·2%). The Presbyterians, on the other hand, declined steadily in both parts of Ireland from 498,000 in 1871, to 419,000 in 1936–7 (391,000 in Northern Ireland), though there was an increase to 413,000 in Northern Ireland in 1961, and the Republic had 19,000 in 1961. The Church of Ireland adherents now number approximately 11% of the whole population, compared with 12% in 1926, 13% from 1891 to 1911, and 12·3% in 1871 and 1881, but they have

* Census of Population 1946, Second Interim Report, 1950, p. 3.

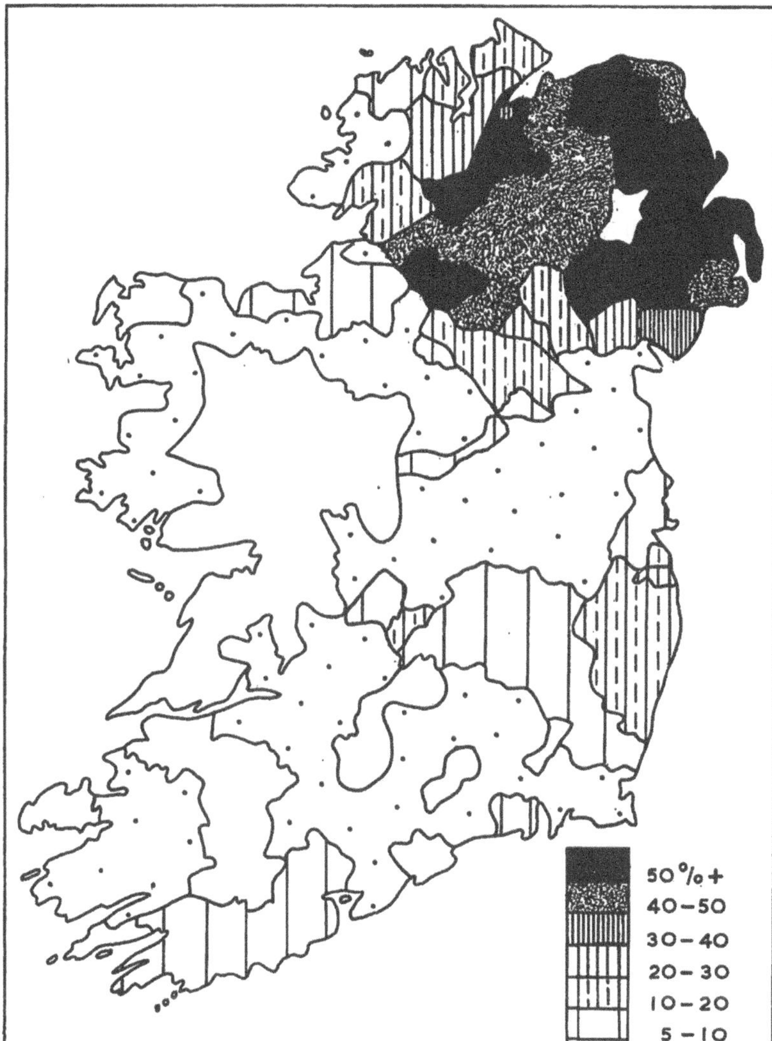

FIG. 28. PERCENTAGE OF PROTESTANTS IN EACH RURAL DISTRICT, 1961

There is a clear Protestant majority in parts of Ulster, substantial minorities in most of the remaining rural districts of Ulster, and a significant representation in Dublin, Co. Wicklow, West Cork and a few other areas. Over approximately half the country, the population is almost exclusively Roman Catholic.

increased from 329,000 (1871) to 345,000 (1961) in Northern Ireland and have declined in the Republic from 339,000 in 1871 to 104,000 in 1961. Until 1911, they formed 8% of the population in the twenty-six counties, but now under 4%, so that their once significant representation over the greater part of the country is a thing of the past, and they have steadily become a Church with its main strength in Northern Ireland and a few parts of the Republic.

A religious Census inevitably reveals a large number of divisions, but the Irish population is fairly clearly divided between Roman Catholics and the three major English and Scottish forms of Protestant worship. In 1870 the Church of Ireland was disestablished, and no religious communion is established in either division of Ireland today, though the Eire Constitution of 1937 stressed the 'special position' of the Roman Catholic Church in its territory, while at the same time recognizing the Church of Ireland and other bodies, and maintaining the general principle of liberty of conscience in such matters. It now remains to consider the three main groups separately, with some indication of its general social position in the community as a whole.

THE ROMAN CATHOLIC CHURCH

Dominant over the whole of the Republic, and including 94·9% of the population in 1961 compared with 89·2% in 1871, the Roman Catholics are alone maintaining their numerical support among the population. There were 113,000 fewer in 1961 than in 1946 as this period was a time of heavy emigration, but this affected the Protestant figures at least as much, if not more than those of the Roman Catholic majority. In 1961, the Roman Catholics were in a strong demographic position, as only 32% of their people were over 45 years of age, compared with 46% of the Protestants; and the ratio of children per 100 women aged between 15 and 44 was 269 among the Roman Catholics, and 245 among the Protestants. In 1946 the number of children per 100 married women aged 15–44 was 287 for the Roman Catholic and 221 for the Protestant population: since then the disparity has lessened considerably. There is now no substantial source of replenishment for the Protestant community, which before 1914 received members from the army, the police, or commercial and professional workers. In recent times a few professional workers have come to Ireland with some immigrants attracted by jobs in the new industries. In 1961 the Irish Census included an enquiry on the date when people born in England and Wales settled in Ireland, which showed that in most years approximately 1,000–2,000 came though some 2,750 came in 1960. There has never at any time been any considerable opportunity for English settlers in Ireland apart from a few special posts in industry, commerce and the professions.

Many social services in Ireland are provided by the Church and its monastic communities. Education in primary and secondary schools is normally under the partial control of the Church, though technical education is a direct state service. Numerous boarding schools are run by monks and nuns, especially those in country towns for farmers' sons and daughters. Hospitals in some cases are staffed partly by nuns, and in some counties the County Homes also have nuns in charge. Various religious communities provide help for people in special need through infirmity or delinquency; all this helps to explain that there are over 22,000 people in the direct service of the Church, including over 13,000 nuns, lay sisters and postulants, nearly 6,000 clergy and monks, and more than 3,000 theological students (in 1961), many of whom will ultimately leave Ireland for service among the Irish in dispersion or in the widespread foreign missionary organizations. These workers account for 30% of the professional class in the Republic. Naturally they are a strong social group, able to exercise a great deal of leadership, especially in those areas which are almost exclusively Catholic.

In Northern Ireland, the Roman Catholics commonly regard themselves as in a less fortunate position than in the Republic. Difficulties arise not so much from official provisions for education and other social services, but through employment problems, not to mention a fatal attachment on all sides to 'old unhappy far-off things and battles long ago'. The problem of the Partition of the country remains unsolved, though the present boundaries were sanctioned by legislation of 1925 in Westminster and Dublin: in 1949, when the Republic was proclaimed in Dublin, the Government of Ireland Act passed at Westminster ruled that 'in no event will Northern Ireland or any part thereof cease to be part of His Majesty's dominions and of the United Kingdom without the consent of the Parliament of Northern Ireland'. This means that boundary revision, on a basis of political sympathy, has been outlawed both by the Acts of 1925 and 1949; there is little point in attempting to draw new boundaries. Assuming that the Roman Catholic population of Northern Ireland corresponds fairly closely to the Nationalist party (on which see p. 171), no new boundary could be drawn without making a 'minority problem'. In any case, the Unionist policy of 'not an inch' appears to be inviolable for the present.

Some careful work has been given recently to Belfast by Dr Emrys Jones, who has used the 231 enumeration districts of the city to map the distribution of Roman Catholics: this gives a valuable refinement of the ward figures available in the 1951 Census. The proportion of Roman Catholics in the city declined from 34% in 1871 to 23% in 1926, though the numbers grew from 41,000 to 96,000, with further increases to 104,000 (24%) in 1937, and 114,000 (27%) in 1961. Of this 114,000, over 57,000 are in four wards, Dock, Smithfield, Falls and St Anne's, which have a total population of 81,650. Two wards have over 90% Roman Catholics, Falls with 91% (31,680 out of 34,860) and Smithfield 90·2% (8,031 out of 8,903). The main concentration is along the Falls road, leading from a point near the city centre southwestwards to the city boundary, and covering the entire socio-economic scale from unskilled labouring and manual work near the city centre, to the professional classes in Andersonstown, near the city boundary. In some parts of this belt, the concentration is even more marked than the figures would suggest. Dr Jones found that some enumeration districts were populated entirely by Roman Catholics. Similarly, there are other concentrations within the city, including several close to the centre such as those along the York road, which is flanked by mean streets overshadowed by factories and warehouses. Similarly the small Roman Catholic population in Ballymacarrett was found to be concentrated in a small area of artisan housing. At all times the Antrim side of the river has had the vast majority of the Roman Catholic population and this remains characteristic even though slum clearance has resulted in considerable changes through the building of housing estates on the edge of the city. In 1961, the Census figures showed that 18% of the population was Roman Catholic in the outlying areas built up in Co. Antrim, but only 7·4% of the comparable areas in Co. Down.

Religious and social segregation is also a feature of Londonderry. It is commonly said that if one stands on the walls near the Church of Ireland cathedral and looks northward to the hillside suburbs, the intervening valley of working-class homes is the main Roman Catholic quarter, with the modern cathedral at its summit. True, much-needed slum clearance has given the city peripheral housing estates mixed in population. Drawing of ward boundaries is an interesting exercise, especially as the city council has a Unionist majority: the North and South wards each have six councillors and two aldermen, and Waterside three councillors and one alderman. Readers must draw their own inferences from these figures, taken from the 1961 Census:

TABLE VI

Religious Distribution, wards of Londonderry 1961

Wards	Total pop.	Roman Catholic	%	Presby-terian	%	Church of Ireland	%	Others	%
North	12,479	5,005	40·2	3,426	27·4	3,097	24·8	951	7·5
South	29,579	26,609	90·0	1,076	3·6	1,148	3·9	746	2·5
Water-side	11,704	4,459	38·1	3,364	28·6	2,871	24·6	1,010	8·7
Total	53,762	36,073	67·2	7,866	14·5	7,116	13·3	2,707*	5·0

* Of these, 1,035 were recorded as Methodists.

Taken from Census of Population 1961, County and County Borough of Londonderry, Belfast, H.M.S.O., 1964, p. 35.

THE PRESBYTERIAN COMMUNITY

The Presbyterians in Ireland are traditionally associated with Scottish settlement, and they form an absolute majority of the whole population in the whole of Co. Antrim (except in the Lagan valley and the Glens of Antrim) and in the northeast of Co. Down. They form a majority of the Protestants around Belfast and in the city, in central Down around Banbridge, and in the predominantly Roman Catholic parts of south Down and south Armagh. They are also more numerous than other Protestants in the whole of Co. Londonderry and an adjacent area in north Tyrone. But several towns in these areas have Church of Ireland majorities, and in almost every case the Presbyterian dominance is more complete in the rural areas. However, the Presbyterians are more numerous than the Church of Ireland people in northeast Donegal and in the rural districts of Monaghan and Castleblayney, Co. Monaghan, all of which have Roman Catholic majorities. They have shown a slight relative decline in Northern Ireland with a numerical drop from 436,000 in 1871 (32·1%) to 391,000 (30·5%) in 1937, but 413,000 in 1961 (29·0%), whereas in the Republic they numbered 62,000 in 1871 (1·5%) but only 19,000 in 1961 (0·7%). Their main strength lies in part of Northern Ireland, and they become statistically insignificant even in Co. Fermanagh, but yet have at least one representative in over four-fifths of the hundred-odd towns and in most rural districts of the Republic, where their members include many people of Ulster extraction and some descendants of Scottish gardeners brought over to work on estates.

THE CHURCH OF IRELAND

The Church of Ireland possesses a diocesan and parochial organization that covers the whole country, but almost threequarters of its people live in Northern Ireland, compared with only one-half in 1871. In 1951 there were 346,000 Church of Ireland people in Northern Ireland (24%) compared with 329,000 (24%) in 1871, but in 1961 only 104,000 (3·7%) in the Republic, compared with 339,000 (8%) in 1871. The marked decline since 1921 has been due to various social changes including the withdrawal of people from garrison towns and of the Royal Irish Constabulary, but the more recent large decreases must be ascribed largely to emigration. Even so, the Church of Ireland may justly claim to be more deeply rooted in the Republic than the Presbyterian communion, partly because it includes among its members many more people who are not recent immigrants. But its members are very few in many parts of the country, and while they are represented in almost all the towns and every rural district, in some places their numbers are hardly one in a thousand. Within the Republic there are three areas with a comparatively strong Protestant community, primarily Church of Ireland:

1. Dublin city and county, Co. Wicklow and the north (but not the south) of Co. Wexford, with scattered communities of some numerical strength in a belt extending across the Central Lowland to the Shannon, but not beyond it. Within the past twenty years however the proportion of Protestants has declined heavily both in rural areas and in the country towns.
2. Eastern and southern Donegal, Cavan and Monaghan – i.e. part of the old province of Ulster, with some local extensions into Sligo and Leitrim.
3. In south and southwest Cork. In the extreme southwest Schull R.D. 16% were Protestants in 1961, much the same percentage as in 1946 though there had been a heavy decline of population.

1. *Dublin and its hinterland.* The Protestants in Dublin city and county numbered 68,000, 11·5% of the population in 1936, but only 61,000, or 9·7% in 1946; those returned as Church of Ireland declined from 50,500 in 1936 to 38,100 in 1961. In 1936, the main concentration of Protestants was in the favoured south suburbs and those along the coast to Dun Laoghaire and Dalkey on the south, and from Clontarf to Howth and Malahide on the north. In 1946, Dun Laoghaire and Malahide had 18·5% and 19·4% respectively of Protestants, and in the Rathdown I rural

district the percentage was similar. In 1961 the Protestants numbered 16% of the population of Dun Laoghaire and 12·5% in Co. Wicklow: though fewer, they are still strongly represented among the professional and commercial classes. Equally they are relatively numerous among the retired people living in and around Dublin, including the north of county Wicklow where the Rathdown II rural district, which includes Greystones, had over one-fifth of its population listed as Protestants in 1961. The Dublin Protestants include people at every financial level, though their survival is perhaps made easier by comparative affluence. There are also Presbyterians and Methodists, and small groups of Baptists, Plymouth Brethren and others, including members of the Society of Friends.

The main strength of the Protestant community in Wicklow and North Wexford, away from the outer suburban fringe of Dublin, lies among the holders of large farms. There is a story, at least partially true, of a Church of Ireland rector and a Catholic parish priest who had respectively one-fifth and four-fifths of the population under their care, but the Protestants had as much land as the Catholics. In 1936, one-quarter of the farmers of Wicklow were Protestants, and they had three-fifths of the 224 farms of more than 200 acres, and over one-third of the 611 farms of 100–200 acres. There is every reason to believe that this situation is much the same now: the rural districts of Baltinglass I, Shillelagh and Rathdown in Co. Wicklow, and Gorey in Co. Wexford, all had 10·8–18·0% of Protestants in 1961, whereas the towns of Wicklow, Arklow and Gorey had 4·7–8·9%. In Co. Wexford also the Protestants have a considerable proportion of the large farms, but they number only 1·5% of the population in the town of Wexford and only 3·4% in the neighbouring rural districts.

Fig. 28 shows that there are Protestant communities of some strength in a wide belt of country extending through the Central Lowland to the Shannon. Until 1946 many areas within this belt had over 5%, and even more than 10% of Protestants, but since then their members have declined considerably, though they are still well represented in many rural areas, especially among the farmers having over 100 acres. They are less numerous in the towns. The Church of Ireland has the main allegiance of the Protestants though there are some small groups of Presbyterians and some interesting Methodist communities which trace back their origins to the visits of John Wesley, who travelled both on horseback and even on the canal boats to various places in Ireland.

2. *Donegal, Cavan and Monaghan.* The significant Protestant minority in these three Ulster counties forms a social cross-section of the population

to a much greater extent than elsewhere and includes people at all income-levels. Twelve rural districts with Protestant minorities of 14–37% had in 1936 a total population of 187,000, of whom 147,000 were Roman Catholics and 40,000 Protestants. The latter included nearly 20,000 Church of Ireland people and nearly 17,000 Presbyterians. By 1961, the population in these same twelve rural districts had declined to 136,000, of whom 21,700 were Protestants, including 10,800 Church of Ireland people and 9,800 Presbyterians: the Protestant minority ranged from 12% to 30%. In the eleven small towns that serve this area there were 25,400 people of whom 2,050 (8%) were Protestants including 1,300 (5%) Church of Ireland. These three Ulster counties show something of the religious mixture characteristic of the rest of the province, but their Protestant minorities are declining absolutely and relatively; in 1961 they included only 13% of the population compared with 17% in 1936, 23% in 1926 and 26% in 1911. As a whole they form well-integrated groups, strong here and weak there, but generally more significant among the rural population than in the towns. In 1936, Protestants numbered 17–23% of the farmers in the three counties, they were proportionally more numerous among the holders of large farms: for example, in Co. Monaghan they had about one-tenth of the farms of 1–30 acres, nearly two-fifths of the medium-sized farms of 30–100 acres, and more than half the few large farms.

3. *South and southwest Cork.* The Cork group of Protestants is spread through the southern part of the county from Cork city westwards: they include 2·3% of the population of Cork city and 5–10% of most of the rural districts near the south coast, with 16% in Schull rural district. Five-sixths of them belong to the Church of Ireland, though there are small Methodist and Presbyterian communities in the area. Outside Cork they include a few relics of the landed gentry, but are mainly farmers: in 1936, 6·4% of all the farms in Co. Cork were in Protestant hands, though they formed only 5% of the population. In 1961, they were only 4·5%, and there has been a marked decline since 1911, when they formed 9% of the population. This is partly due to the withdrawal of the British military, whose departure decimated the Church of Ireland congregations in some towns and reduced the Methodists and Presbyterians almost to extinction. However, the Protestants still form fairly stable communities in many rural parts of south Cork, and include families whose religious allegiance goes back for many generations. Near Cork they hold medium-sized and large farms, but farther west they are represented in a wider range of Irish society. As one journeys westward through Skibbereen to Schull

and on to the extreme west of the Schull peninsula, there are not only parish churches of the Church of Ireland, generally in villages, but some bare and plain Methodist chapels much like those to be seen in many parts of the English countryside. Some are closed now but in several of them small congregations still assemble week by week.

SUMMARY

Virtually every novelist who has dealt at all successfully with Ireland has realized that differences of religious allegiance enter closely into the fabric of rural and urban life. Devoted rectors of the Church of Ireland see their congregations diminishing year by year, chiefly through the emigration of young people: in the end churches will be closed (or prudently sold to the Roman Catholics) or kept open for the congregations of the visitors so welcome in the summer months. Far to the west in the Dingle peninsula there is a church, closed for much of the year, of grim bastion-like appearance but brightly painted in white and green inside and offering a welcome to those who come during the holiday period. Even though the number of Protestants is so few in much of Ireland, as Fig. 28 shows, they have maintained a position of dignity and even of influence in Irish life. Their distinction is maintained partly by education and the family ethos, for the home remains a strongly unifying force in Ireland. And they are well represented among the intelligentsia, the professional classes, the holders of substantial farms and prosperous tradespeople. Only in Northern Ireland is the present numerical strength of the Roman Catholics and Protestants likely to remain fairly stable, for in the Republic emigration has drawn away so many Protestants and others have, not unnaturally, been absorbed into the general Roman Catholic community by marriage.

THE IRISH LANGUAGE

Efforts to encourage the use of the Irish language in day-to-day contacts rest on the assumption that a separate language is a mark of distinct nationality. One could argue that many nations exist without possessing a language of their own, or that the Finns have two languages and the Swiss four, yet in both cases the national consciousness is strong and indeed highly respected. The association of English with oppression in many minds has led to a policy of teaching children Irish and making a knowledge of the language essential in certain professions, notably the law and school teaching. Article 8 of the Constitution (1937) states that:

1. The Irish language as the national language is the first official language.
2. The English language is recognized as a second official language.
3. Provision may, however, be made by law for the exclusive use of either of the said languages for any one or more official purposes, either throughout the State or in any part thereof.

From its foundation in 1893 the Gaelic League has worked to encourage not only the language and literature, but also Irish dancing and distinctive games such as hurling and Gaelic football. The most obvious impact on the visitor is the use of Irish on railway stations and street signs: laws, Census forms, income tax literature, are printed in both languages. Less obvious, but far more enduring, is the fine work of scholars both in language and literature. The opinion is often given that one-half of the population were Irish-speaking at the beginning of the nineteenth century, but by 1851 when for the first time the language question was asked in the Census, the number had fallen to one-quarter (1,500,000). By 1911, the proportion had been reduced to one-eighth (580,000), most of whom (554,000) were within what is now the Republic. In 1925 the Gaeltacht Commission was appointed to inquire into the condition of the areas in the west of Ireland, where the language survived among a substantial proportion of the population. The areas with which the Commission dealt were similar to those studied by the Congested Districts Board some thirty years earlier, and admitted to be the poorest in the land. The Commission reported that all the alleviations already achieved had not removed poverty, but felt that the maintenance of a substantial Gaeltacht population was essential, as it was an Irish-speaking area from which the language might expand eastwards. The Commission distinguished two types of Gaeltacht: (*a*) the *Fíor-Ghaeltacht*, where at least 80% of the population speak Irish, and (*b*) the *Breac-Ghaeltacht*, in which 25–80% speak Irish. The maps shown in Fig. 29 show the distribution of the language in 1926 and 1936: the map for 1926 approximates to the more elaborate maps given in the Commission's report as the result of a special enumeration.

Both these areas are considered separately below. In 1946 it was reported that 589,000 people over three years old, 21·2% of the whole population excluding infants, were able to speak Irish, compared with 667,000, 23·7% in 1936. This means that within ten years there was a decline in Irish-speakers of 78,000, or 11·7%, though the general population decrease was only 13,000, or 0·4%. The loss of Irish-speakers in the Gaeltacht was 19·0%; the *Fíor-Ghaeltacht* decrease was 14·8%, and the *Breac-Ghaeltacht* decrease 23·7%, but in the non-Gaeltacht the loss was only 7·6%. In 1956 the Gaeltacht was redefined to cover areas in which

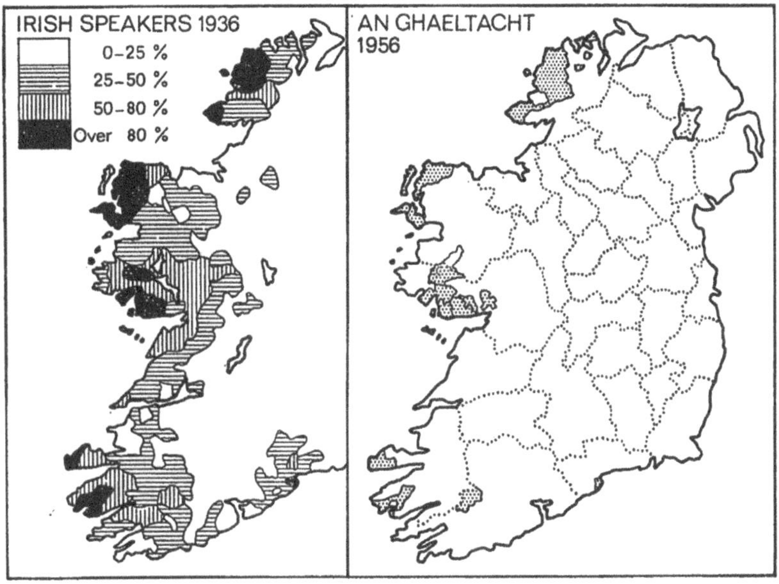

FIG. 29. IRISH SPEAKERS, 1936, AND THE 1956 GAELTACHT

The *Fíor-Ghaeltacht*, or area with more than 80% of its population Irish-speaking, is shown in solid black and the remaining shaded areas are the *Breac-Ghaeltacht*. The original delimitation of these areas was achieved by a special enumeration in 1925: see Report of the Gaeltacht Commission, Dublin, 1926. Note that from 1956 the distinction between *Fíor* and *Breac-Ghaeltachta* was no longer made.

Based on the Census Reports of the Republic

the vast majority of the population spoke Irish, with a general average of 87%. These areas correspond broadly to the *Fíor-Ghaeltacht* in the counties of Donegal, Mayo, Galway and Kerry, with small areas in Cork and Waterford: the population of the new Gaeltacht was 85,700 in 1956 and 78,500 in 1961.

THE 'FÍOR-GHAELTACHT'

Predominantly rural in character, this area includes only one town, Dingle (1,400), Co. Kerry, though there are various market villages such as Dungloe (900), Co. Donegal. The Aranislands, at the entrance to Galway bay, various islands off the Donegal coast, the extreme northwest of Co. Donegal and Connemara, west of Galway town, are areas to which people go in considerable numbers to learn Irish. In general the *Fíor-Ghaeltacht* is an area of isolated farms but they are so close together in some places, such as the west of Donegal around the Bloody Foreland, that they form an almost continuous line of settlement along the roads. But in most of the *Fíor-Ghaeltacht*, isolation is the salient characteristic of life, for vast bogs separate small groups of farms located near rivers, along roads, or on the coast. This is especially true in some parts of Mayo and Kerry, though all possible transitions between the two extremes will be found.

The population of the *Fíor-Ghaeltacht* declined from 168,000 in 1926, 156,000 in 1936 and 122,000 in 1961, of whom 77%, 83% and 72% respectively were Irish-speakers. The proportion of children aged 3–9 able to speak Irish increased from 57 to 70% between 1926 and 1936, which may be due partly to the encouragement given in the schools and partly to the fact that a substantial proportion of the children learn the language at home. On the other hand, 82% of those aged 30–39 spoke Irish in 1926, and 77% of the same group in 1936, and the proportion of those over 40 years of age speaking Irish also showed a slight decline in the same ten years. The loss through disuse between the school ages and those of young adults has been carefully estimated by the statisticians of the Census office, who have suggested that the 'lapse rate' between 1926 and 1936 was only 2% in Galway, 7% in Mayo, and 10% in Donegal, figures which may seem serious enough but yet are much smaller than those for the *Breac-Ghaeltacht* or the non-Ghaeltacht areas. It is clear that the most favourable chances of building up an Irish-speaking population exist in the *Fíor-Ghaeltacht*: this population will, however, be bilingual. Between 1936 and 1946 the population of the *Fíor-Ghaeltacht* over three years of age declined from 148,200 to 137,200 (−7%) but the number of Irish-speakers fell by 15% from 123,100 to 104,900. There was a further decrease to 83,600, 35% of the population, due to population decline, by 1961.

THE 'BREAC-GHAELTACHT'

Originally defined as the area having 25–79% of its population able to speak Irish, this area, continuous with the *Fíor-Ghaeltacht*, covers part of the counties of Donegal, Mayo, Galway, Clare and Kerry, with West Cork, and also a large area near the south coast between Cork and

Waterford, where one district around Ardmore has a sufficient proportion of Irish speakers to be regarded as a *Fíor-Ghaeltacht*. Many types of landscape and grades of productivity are represented in the *Breac-Ghaeltacht*. In Donegal, it corresponds largely to the former Congested Districts and does not include the richer lands of the east towards the Foyle. In Mayo it also covers a large part of the Congested Districts, but in Galway it stretches eastwards into the richer sheep-rearing lands around Tuam. In Clare, from Galway bay to the Shannon, the *Breac-Ghaeltacht*, though not a rich area, is not one of marked poverty. The part of Kerry included here consists mainly of the old Congested Districts, but the areas of Co. Cork and Co. Waterford are in much richer regions.

The population of the *Breac-Ghaeltacht* declined from 311,000 in 1926 to 296,000 in 1936 and 244,000 in 1961, and the proportion of Irish-speakers was respectively 40%, 35% and 41%. Closer investigation shows that the language lost ground among the people over 30 years of age between 1926 and 1936: 31% of those aged 30–39 could speak Irish in 1926, but only 25% of the same group ten years later, and similarly, 55% of those over 40 could speak Irish in 1926 against 42% in 1936. Among young children of 3–9 years the proportion of Irish-speakers rose from 17% to 29% in the same period, and of those aged 10–14 there was a rise from 45% to 67%, obviously the result of school education. The 'lapse rate' through disuse between the school ages and young adults was high, 15% in Kerry, 28% in Clare, 28% in Cork and 32% in Waterford (excluding the cities of Cork and Waterford). From 1936 to 1946, excluding children under three, the population of the *Breac-Ghaeltacht* declined by 6·3% but the Irish-speakers declined by 23·6% and numbered only 34% of the total. By 1961, however, it had increased to 35% though the number of Irish-speakers over three years of age had declined from 89,000 to 81,000.

SUMMARY

In 1961 it was reported that 716,400 people, 27% of the Republic's population over three years old, were Irish-speakers. This was the highest percentage ever recorded since figures were first available in 1851, when the number of Irish-speakers was 24·5%. The redefinition of the Gaeltacht in 1956, as noted on p. 167, covered an area smaller than that of the original *Fíor-Ghaeltacht* but bore a general resemblance to it. In Co. Donegal it included a population of 28,900 in 1956, reduced to 26,400 by 1961, located in the northwest and west: in Co. Mayo, with 18,200 in 1956 and 16,300 in 1961, the Irish speakers were in the extreme west, including Achill island and the Belmullet area, and in Co. Galway, with 24,600 in 1956 and 22,900 in 1961 they were in Connemara west of the city of Galway. Co. Kerry has Gaeltacht areas in the west of the Dingle and Bear peninsulas, with 9,400 people in 1956 and 8,700 in 1961. In Co. Cork the main Irish-speaking area is in the Macroom rural district and in Co. Waterford there are some 900 in the Ardmore area of the southwest. Within these areas there are various schemes to retain the population, such as tomato growing from central depots in Connemara and northwest Donegal and the encouragement of new industries or those, such as tweed weaving, based on traditional crafts.

THE DIVISION OF IRELAND

That Ireland possesses a geographical unity no one can doubt: it would appear to be designed by nature to be one political unit. Unfortunately such an ideal appears to be remote, as many events in recent history have served to draw Northern Ireland and the Republic farther apart than ever. There does not appear to be any probability of revising the Border itself in any near future: as the chance of doing so was missed in 1925 with the consent of both sides, the disputed areas of that time have been welded into Northern Ireland's economic system. Dr J. H. Andrews has recently discussed the proposed new boundaries which were shown on a map published in the *Morning Post* of 7 November 1925: they indicate an extension of Northern Ireland into the east of county Donegal, with small gains in South Donegal and Belleek. The Free State was apparently to be given a belt of land in south Armagh, but not the towns of Newry and Keady. An area of Co. Fermanagh within the market area of Clones (where the border runs past the end of the main street) and – less certainly – an area including Rosslea in southwest Fermanagh were to be added to the Free State. But, as Dr Andrews shows, there was controversy on the accuracy of the newspaper report (Fig. 30).

The fundamental guarantee of Northern Ireland lies not in British arms but in the votes of two-thirds of its population. The six counties share fully in the modern social benefits of the Welfare State, with higher rates of benefit than those possible in the Republic. In its modern form Northern Ireland depends for its prosperity on a full share of international markets, especially necessary for its major industries, linen, shipbuilding and aircraft. Some industries, such as shirt making in Londonderry, have been depressed in recent years but there, as in other areas with declining industries, hope lies in attracting growing enterprises such as chemicals and electrical trades, not behind a tariff wall but with free access to the British and overseas markets. In other words, the economic interests of the six counties lie with Britain as firmly as those of the six counties of the north of England mentioned earlier. Much of the policy of the Republic was initiated in the nineteen-thirties with the idea of economic independence, the protection of home industries and – above all – the achievement of a deeper sense of nationality.

And nationality is a key word here. Vital as the economic argument may be – no doubt must be, in Northern Ireland – it is hard to imagine that the economic argument would ever be crucial in the Republic. At all times, most of the Republic's exports have been sent to Britain, obviously a good market, yet the search for other markets constantly continues, with little result. It is only with reluctance that many people in the twenty-six counties recognize the inevitability of relationship with Britain, especially in view of the remarkably heavy emigration movement of recent years, most of it to Britain. Far more sensible has been the outlook on the tourist trade, so vital at present as an invisible export: much has been done to popularize Irish holiday facilities in Britain, as well as in America. But during the 1939–45 war, the British publicists who, in excellent weekly journals, pointed out the economic advantages of participation on the Allied side, made no effective appeal to Irish opinion. The anti-British feeling, let it be faced, was too strong for any such appeal to succeed. Nor was it realistic to expect Ireland to hand back to Britain ports which had been returned to her control in 1938. In all the circumstances, two courses were open to the twenty-six counties—full membership of the Commonwealth with war and possible economic advantages, or neutrality and a relationship with the Commonwealth that might be broken at any time.

In 1949 the break came, and the *feu de joie* sounded on O'Connell Bridge in Dublin in the early hours of Easter Monday. At that time it was believed by many people in Ireland that partition would end comparatively soon but in the 1960s the prospect still seems as distant as ever – perhaps even more remote. Some thought that the United States would compel the reunion of Ireland, others that Britain, giving concessions all over the world to nationalist sentiment, could not withhold it from Ireland. It was argued that a majority of the population of Ireland as a whole favoured union, which should therefore be granted, and that the Stormont government, far from representing Ulster, in fact represented only part of the six counties, having a Protestant majority and an industrial life that is somewhat alien to Ireland as a whole. It is undeniable that the main numerical support of the northern parliament is concentrated within little more than half its territory.

There is much to be deplored in the present situation, and it would be unjust to fail to recognize the strong feelings of the Nationalist party in Northern Ireland. But the Northern Ireland case for retention of the union with Britain is strong. On the basis of fundamental principles, the argument is that a sentiment of Irish nationality need not involve independence of – still less hatred of – Britain: the wish is to maintain unity with Britain, by keeping a valued link with the Crown, and with the United States. Secondly, the majority view in the north rejects, or at least fears the clerical influence that is alleged to exist in the government from Dublin: along with this, the Belfast government offers 'freedom from censorship' of publications. In trade, Great Britain is at once a supplier of raw materials and a consumer of finished products, having all its markets freely open to the industrialists of Northern Ireland. Forward then, with Britain, says the Ulster unionist: no truck with Irish, 'a dead language . . . another example of how Eire puts the past above the present' . . . 'in Eire your reading is chosen for you' (not in the author's case, incidentally) . . . the Ulsterman has the right 'to be a pilgrim'. A shade smug, perhaps?

Still there are deep rifts in Irish life, and great problems for both governments, some of which are indicated in the following chapters. In this chapter attention has been drawn to the increasing numerical dominance of the Roman Catholic church in Ireland, and the sharp division between the Protestant majority and the strong minority in Northern Ireland: it has been noted also that the policy of encouraging the Irish language has had a very limited success and makes no appeal in Belfast. Recent history has done little to bring the Belfast and Dublin governments closer together. On a broader historical basis, one can recognize the continuing effects of the sponsored immigrations of plantations which made Ulster different from the other provinces. (Incidentally one regrets that the six counties are officially referred to as Ulster, instead of Northern Ireland.) Any sign of co-operation is welcomed, and the two governments have recently combined for power development and for railway control. Under the lough Erne scheme to supply electricity, some power is available for Northern Ireland, and the control of the river's level has been amicably settled by the two governments (see p. 471). The impending bankruptcy of the Great Northern Railway was prevented by a scheme of joint nationalization so that this vital railway was kept open: this railway is now run by Coras Iompair Eireann and the Ulster Transport Authority. These are small but useful matters. The divergence rests on the two differing conceptions of nationality and any immediate change is improbable, though the attitudes of political leaders are becoming more friendly. Had Britain entered the Common Market, the Republic of Ireland would almost certainly have done so too, and the economic changes might have led naturally to closer co-operation in other fields. A hopeful sign from 1965 has been the interchange of visits between the premiers in Belfast and Dublin.

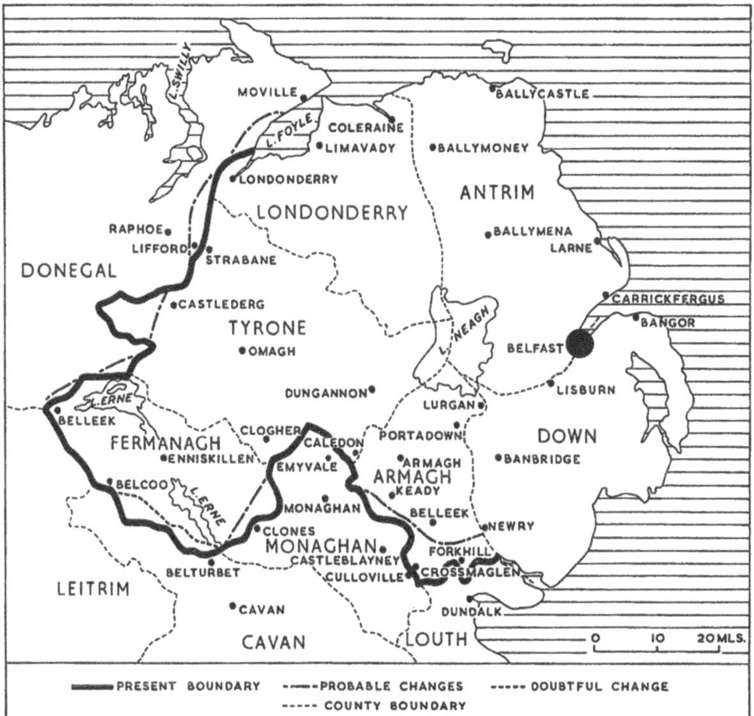

FIG 30. POSSIBLE BOUNDARY CHANGES OF THE NINETEEN-TWENTIES
Redrawn from a map in Andrews, J. H., 'The "Morning Post" line', 'Irish Geogr.', vol. 4, no. 2, 1960, pp. 99–106

CHAPTER 8

Agriculture

In spite of the heavy emigration from rural Ireland, agricultural occupations still support one-third of the employed population in the Republic and one-eighth of those in the six counties. Agricultural produce is the major export item in the Republic and a useful contribution to the export trade of Northern Ireland. Apart from the world need for farmers to provide all they can, the prosperity of the greater part of Ireland in area rests on agriculture which is the main support of its trading towns and – perhaps to an inadequate extent – a supplier of industrial raw materials. Although much has been done to improve farming, still further improvements are desirable, in fact necessary. Of all past political issues none was more vital than the struggle for possession of the land, partly solved by legislation from 1870 onwards. But 'land hunger' still exists and the effort to provide holdings of reasonable size still occupies the Irish Land Commission in Dublin.

The fragmentation of the land in Ireland into minute farms was extreme before the Famine, despite the efforts of some landlords to make larger holdings: in 1841, there were 685,000, of which only 7% were of more than 30 acres. The hopes of the time rested in the reclamation of new land on mountain sides or from peat bogs, and it was generally believed that thousands of people could be given new farms within the country. Even though there are still hopes of reclaiming bogs and adding fields on hillsides, the total area cultivated has been virtually stationary since 1870, though there was some expansion of the cultivated and improved area from the late eighteen-forties to 1870. One writer of the Famine period, Jonathan Pim, expressed his belief that a great deal of land could be reclaimed on mountain sides, and gave as an example the Forth barony, in the south of Co. Wexford, an area which had received a good deal of publicity for its high standard of farming. He suggested that many of the country's debt-ridden or 'encumbered' estates should be sold to tenant farmers and quoted a Scottish witness before the Devon Commission, who advocated a division of the land into farms of 30–100 acres, among which smaller holdings should be 'dovetailed'. These small farms, of 4, 8 or 10 acres, could be occupied by men who were labourers, tradesmen or even small shopkeepers, working the land with their families. They could reasonably be expected to pay a higher rent than the full-time farmers, and could supply butter, cheese, eggs, poultry and pigs for sale. That some change was needed, everyone agreed, though Pim was in advance of his time in proposing the removal of certain landlords, or at least those landlords eager to sell. A fundamental problem was control of the land, as at the humblest level of Irish life land was a guarantee against starvation. Some of the social aspects of life on the land are discussed on pp. 147–51, and the distribution of population in rural areas is treated on pp. 129–34. There are also some comments on the historical aspects of farming in Chapter 5.

It was not foreseen in the Famine period that the ultimate solution would be to eliminate the landlords altogether, and even Mr Pim passed by that possibility. Nor was it foreseen that Ireland's rural problems would still be acute over one hundred years later, when the opportunity of migrating from farms would be eagerly seized not by the farmers themselves but by their families. And although the number of farmers is now little more than half what it was before the Famine there are still, as Table VIII shows, a large proportion of farms of 30 acres or less, some of which provide only a very modest living for their holders. Therefore, one must regard the adjustment of population and land, begun in Famine times, as a process by no means complete.

LAND TENURE

At the end of the eighteenth century many landlords were eager to increase the number of small farms, as by a law of 1793 each holder of lands valued at more than £2 was given a vote, so the landlords made as many holdings as possible to increase their political power and income. The demand for crops remained good until 1815, but depression followed and many of the tenants were evicted, especially after 1829, when a law was passed raising the voting qualification to £10 per annum. The evicted, martyrs in the eyes of their fellows, emigrated to America, Britain or the growing town of Belfast, or squatted on a patch of waste land, or else swelled the already formidable army of casual labourers and beggars. Many of the landlords were becoming more and more extravagant and overspent their resources, with the result that the number of 'encumbered estates' increased: in short, the landlord and tenant relationships of the eighteenth century were obsolete, but few people realized this at the time and much of the contemporary literature consisted of advice to landlords to reform

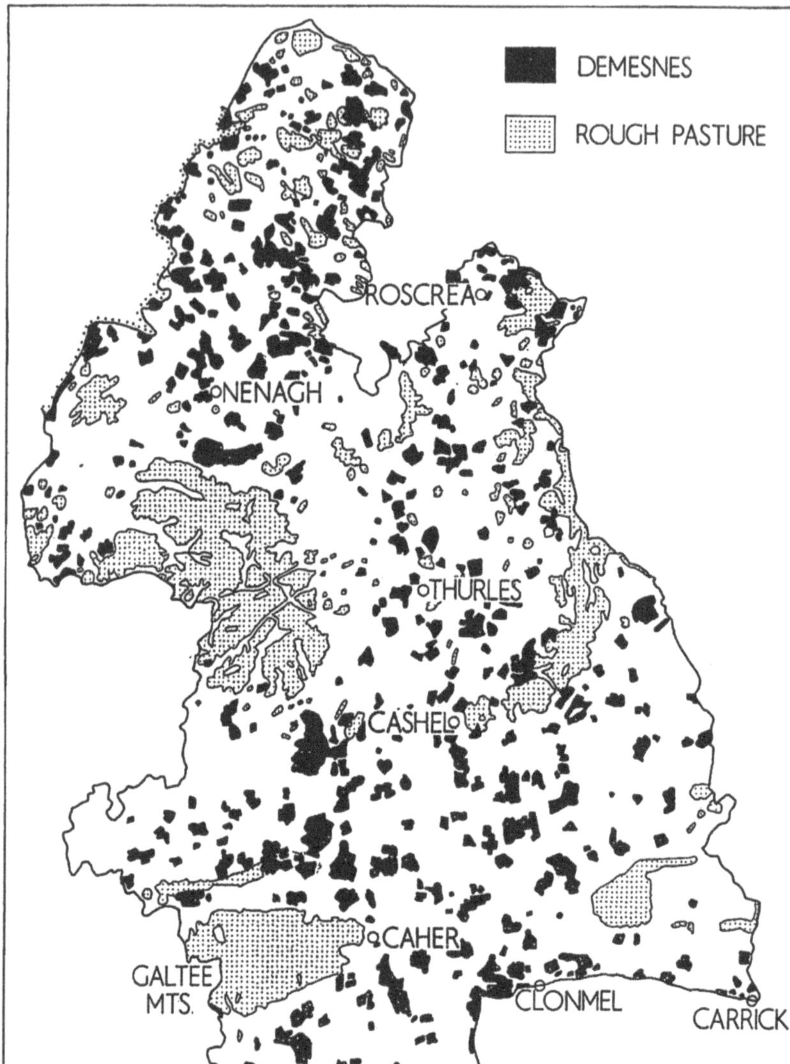

FIG. 31. DEMESNES AND UNIMPROVED LAND, CO. TIPPERARY, *c.* 1841

This county was notorious for agrarian unrest: the demesnes were relatively numerous and many were of considerable size. But over most of Ireland before the Famine, the position was much the same. This map was prepared from the 6″ to 1 mile maps of Co. Tipperary, which was surveyed from 1839 to 1841.

First published in 'Pre-Famine Ireland', Manchester, 1957, p. 215 their ways. Some temporary difficulties were removed by employing people on public works, a feature of Irish life from the late eighteenth century, and in the Famine years these included various schemes of land reclamation, drainage, the building of river embankments and, most needed of all, farm premises. But evictions still continued, as many farmers either would not or could not pay the rent. And the amount of land in demesnes was considerable (Fig. 31).

When the first Land Act was passed in 1870, the land was owned by some 19,000 proprietors, of whom one-fifth possessed 80% of the cultivated area. There were five classes of tenant under the landlords – leaseholders, middlemen, annual tenants, cottiers and labourers. The leaseholders, in all numbering 135,000, held the land in perpetuity, or under an agreement defined by the duration of two or more lives, and were mainly Protestants, as the Penal Laws abrogated in 1827 had forbidden long leases to Roman Catholics; the descendants of these leaseholders survive as farmers of substantial holdings. The middlemen were the agents of the landlords and let off land to the remaining three classes. The annual tenants, some 525,000, formed the typical 'small farmer' class settled mainly on lands valued at less than £15 per year, and were liable to eviction on six months' notice. The cottiers lived in poor cottages generally located on someone else's land, and had very little land of their own, though they normally hired a patch of 'conacre' (pp. 189–90) to grow a crop of potatoes or pasture their stock.* Least fortunate of all were the labourers, who had no land at all, though they too would commonly hire a patch of conacre: incidentally, the distinction between the labourer and the small farmer has always been very difficult to draw in Ireland, as one class grades into the other. In many parts of the country land was so severely subdivided that some of the holdings were in a large number of microscopic patches, separated one from another in a 'rundale' pattern of great complexity.

The Land Act of 1870 did little to transfer land to the farmers, but it prepared the way for what was called 'dual ownership', by which the landlord was made to pay the tenant some compensation for improvements and for disturbance or a change of tenure. In 1881 Gladstone introduced a bill which established throughout Ireland virtually what was known as the Ulster custom, or tenant right. The slogan of the times was the 'Three F's', or the right of the tenant to fixity of tenure (so long as he paid his rent and observed his covenants), the free sale of his interest on his holding (a feature of tenant right) and a fair rent. This last was to be

* Conacre land is hired by auction, for eleven months (crops) or seven months (pasture)

fixed by the new Land Commission established by this 1881 legislation. These first two Land Acts, however, transferred very little land to the farmers, though the Land Commission was empowered to loan money to the holders in 1881, and similar provisions had been made in 1870. A more successful early effort was the transference of the glebe lands on the disestablishment of the Church of Ireland in 1869: three-quarters of the holders became owners by paying one-quarter of the purchase price and the remainder on a thirty-two-year mortgage. The failure of the Acts was due to the lack of capital among the farmers, who had to find one-third (1870) or one-quarter (1881) of the purchase price: only 1,600 holdings were transferred, averaging 60 acres in 1870 and 42 acres in 1881. No widespread distribution was made until the acts of 1885 and 1891 provided all the purchase money as loans and made the annuities less than the rents. Under the 1885 Act, 16,000 tenants bought their holdings within the twenty-six counties and the 1891 Act gave tenant-ownership to 42,000 more: in the six counties, 20,800 farmers bought their holdings, which averaged 25 acres each, under the Acts of 1870–96. The Wyndham Act of 1903 was even more successful, as it made 195,000 tenants into purchasers in the Republic, and 57,400 in Northern Ireland (average holding 21 acres), and the Birrell Act of 1909 added 66,500 more, including 5,500 in Northern Ireland. Special provision was made in 1907 for evicted tenants, who were settled on land bought by compulsory purchase, a new principle in the legislation.

Land legislation had steadily brought the smaller farms of Ireland within its range: the Acts of 1885 and 1891 dealt with farms averaging 32 acres. But in 1891 the problem of the uneconomic holdings was attacked for the first time and the Congested Districts Board was established to deal with the poorest areas of the land. These are shown on Fig. 18, and were defined by the district electoral divisions having a rateable valuation of less than £1 10*s.* a head. Many ways of improving conditions were tried, including the re-allocation of land into compact holdings, the building of roads and light railways, the breeding of better stock, the encouragement of fishing and the foundation of industries. Not all of these were new and not all have been equally successful. It soon became clear that so much land was needed for resettlement schemes that estates must be found outside the original congested districts, so in 1909 the special legislation was extended to cover the whole of counties Donegal, Sligo, Leitrim, Roscommon, Mayo, Galway and Kerry, with the west of Co. Cork and the greater part of Co. Clare. The purchase of estates had been facilitated by the 1903 Land Act, which provided that the state should subsidize the prices given to landlords, especially in cases where a complete estate was acquired. The Estates Commissioners were set up to implement this measure. Powers of compulsory purchase were given in 1909, both to the Land Commission and the Congested Districts Board, which was now able to move tenants from densely settled areas in the west to estates many miles away, and even in time to the east of Ireland.

After 1922, land was allocated to occupiers in both parts of Ireland. In Northern Ireland, under an Act of 1925, 38,500 purchasers acquired 805,500 acres, an average of 21 acres: roughly four-fifths of the land is now owned by occupiers. In the new Free State the position was far more complex, for all the poorest areas of Ireland were within its territory and the work begun by the Congested Districts Board was by no means complete. The first new legislation, in 1923, transferred the land settlement work of the Board to the Land Commission and delegated its other activities, such as agricultural improvements or road building, to appropriate government departments. The Board had spent £3,000,000 on improvements, and retired full of honour and respect. The essential problem is the large number of 'uneconomic holdings' that are scattered all over the country especially in the west. Some estimates suggest that at least half, or even three-quarters of all the holdings in the country are uneconomic, but the solution does not necessarily lie in the addition of more land: in 1962 the Minister for Lands, not unfairly, asked 'all landholders to ensure that their holdings are at the point of maximum productivity before setting their eyes on extra acres'. Before 1922, 316,000 holdings, covering 11,000,000 acres, had been bought, and 750,000 acres of untenanted land distributed among 35,000 allottees to enlarge existing holdings, or create new ones, of which 600,000 acres were distributed in the congested counties. But the new Dublin government still had to transfer 110,000 holdings, of 3,000,000 acres, to their tenants and find for the relief of congestion some 1,500,000 to 2,000,000 acres.

Modern legislation begins with the Land Act of 1923, subsequently modified by others without any general change in policy, though in 1933 the annuities were cut to one-half. The main provision of 1923 was an extension of the powers of compulsory purchase, to complete the creation of a tenant proprietary, and also to help to eliminate the 'uneconomic holdings', especially the great concentration of such holdings in the west, by the compulsory acquisition of suitable lands for their enlargement. Provision was made also to prevent the subdivision of holdings by their owners into uneconomic farms. The land vested in the Commission includes most of the tenanted land, all untenanted land situated in a congested county and any untenanted land elsewhere that the Commission might require. This last provision has made it possible to move 1,000 families from the west to new holdings some distance away from their original homes, a re-arrangement discussed below. This work is now the main task of the Land Commission, whose first work, rent fixing (for which it was founded), is no longer necessary: since 1923 its second task, the settlement of 113,000 tenants as owners of 3,100,000 acres, has been virtually completed. Carrying on earlier traditions, the Land Commission has spent £18,000,000 on improvement works, two-fifths of it on buildings. To summarize the work of the Land Acts in the Republic, they have dealt with 459,000 holdings, covering 15,600,000 acres (nearly nine-tenths of the country). The main present work of the Land Commission is to regroup scattered plots into compact and workable farming units. In a recent year 480 holdings, covering 11,200 acres, were rearranged and most were enlarged in the process. Since 1950 over 7,000 holdings have been rearranged and almost 1,200 migrants have been settled on new farms.

Ruinous subdivision of land has long been characteristic of the west and occurs where a rundale village is surrounded by strips of land re-allocated annually to the cottagers. Many of these hamlets had ten or more cabins, sharing a townland, and numerous examples are shown on the 6″ maps of the first edition of the Ordnance Survey. The origin of this system is obscure, but in 1845 it was estimated that some 2,000,000 acres were held either in common or in joint tenancy. Various accounts survive from pre-Famine times of rundale villages and estates, such as that of Lord George Hill in the Gweedore district of Co. Donegal (1837): in some cases a form of seasonal movement, or transhumance, was practised so that the rough pasture, or 'commonage', could be more thoroughly used. There is evidence too that subdivision of land was a consequence of the increase of the rural population from 1770 to 1845, even though some landlords tried to prevent it, but it is also said that the landlords favoured the increase of population as every cottager was a potential payer of rent. To make new farms in such areas has always involved the reduction of the population, normally as emigrants or seekers of work off the land, and the evicted have at all times been regarded with sympathy. Many 'strip farms' in the west are due to re-settlement schemes by landlords, by the Congested Districts Board, or its modern successors.

Various examples of modern re-arrangement have been given by the Land Commission. Rarely can the rundale areas be transformed without either taking over extra land in the vicinity, or moving people to other areas, or both: in one not untypical case, nineteen holdings shared 120 acres of improved land in 265 plots (one holder had eighteen separate plots) and 130 acres of commonage. Compact holdings were made for thirteen tenants on 290 acres with access to 389 acres of commonage. In another case, Kilvine, Co. Mayo, 1,588 acres were divided into one holding of 270 acres, 240 acres of bog and 1,078 acres occupied by 108 tenants in 486 plots. The houses were very old, thatched and generally unsound, and thirty-five of them, with outbuildings, were in a cluster with a radius of ninety yards. The Land Commission acquired the 270-acre farm and let it out in parcels of 7–15 acres: instead of 108 holdings there are now 84, and instead of 486 plots, only 120. New dwellings and outbuildings are provided for fifty families and the new farms average 20 acres, with a poor law valuation of £10.

In the Gaeltacht Commission report the hope was expressed that large numbers of Irish-speaking families could be sent to the Midlands, partly to encourage the use of the national language. In fact the emigration has been on a small scale, and included 799 families by March 31, 1962, who surrendered 24,860 acres and received in exchange 27,039 acres. Far more important is the use for new holdings, forests, school grounds, playing fields or other purposes of some 625,000 acres within the Gaeltacht areas of Waterford, Galway, Mayo, Donegal and Kerry. All this cost £4,802,000 and another £40,000 was spent on reclaiming waste land or fixing sand dunes. The largest scheme of group migration was in Co. Meath, in and around Gibstown, where 122 families (772 persons) were settled from 1935 to 1940 on holdings with an average of 24 acres, each of which had a new dwelling house and outbuildings and a turbary plot. Each immigrant family received seed, stock and equipment, had its land partly ploughed and was welcomed with a week's provisions and a year's peat. The land, 5,233 acres, was previously held by seven owners, and was in grass except for 1,000 acres let in conacre: some 2,304 acres was given to tenants already there. In 1952, 111 of the original colonist families were still there and 9 farms were in the hands of their families: 114 were fully stocked and efficiently worked, and over 700 acres (out of 2,929 acres) was tilled. In addition the colonists had taken 240 acres of conacre for crops and 388 acres for grazing, at a cost of more than £5,000 a year. No other scheme has been so large, and the more usual form of re-settlement is to take a few families together, or even a single family. The chief remaining problem of the Land Commission lies in the 7,000 holdings, mainly in the western counties, which require rearrangement, enlargement or other improvements before they can be vested in their tenants as potential owners.

The achievement of the past hundred years has been remarkable, as all save a small fraction of the land in the Republic has been transferred from the landlords to purchasers who are becoming owners. Holdings have been improved in countless cases, and a more economic use is made of some of the demesnes that existed a hundred years ago. The unification of all the various agencies concerned in land purchase and improvement into the Land Commission has helped to bring the schemes nearer completion. The welfare of the rural community, especially of the 'congests', is a matter of constant concern, and here at least economic considerations are not always regarded as necessarily the most important. Given years of reasonable prosperity the schemes should be successful, and evictions through failure to pay annuities may become merely of historical interest. But the possibility of ultimate subletting survives and some of the new holdings allocated by the Land Commission may prove too small for economically successful farming. Between 1923 and 1936 the average size of the new holdings was only 25·5 acres, and the charge has been made at times that some of the rural rehabilitation schemes are merely adding to the number of actual or potential 'congests'. One of the greatest contributions to the welfare of the rural community has been the housing programme inevitably associated with the changes land in tenure which, together with the redistribution of land, has changed the landscape of rural Ireland, and especially of the poorer parts, to an extraordinary degree within the past century.

FARM SIZE AND ECONOMY

Table VII illustrates quite clearly one major point – that rural Ireland before the Famine and after were two different lands. In 1841 the small farms predominated everywhere, and for this very reason the few farms of any size, such as the grazing land of Co. Meath, or the few rich holdings of Co. Kildare, were regarded as expressions of greed in the agricultural landscape. The harvests of the richer east were gathered by men from the west though even in the east no county had even one-quarter of its farms in the 30-acre + category. But by 1851, the change was remarkable, for from the Famine the number of small farms diminished further as holdings were united, generally by inheritance.

There are three classes of farm commonly recognized in Ireland, small, medium and large, consisting respectively of those having 1–30, 30–100 and over 100 acres. This classification will seem strange to readers outside

TABLE VII

Farm holdings in Ireland, 1841–1960

Acres	1841	%	1851	%	1881	%	1901	%
1–30	636,997*	93	421,208	74	332,091	69	305,325	66
30–100 } Over 100 }	48,312	7	149,040	26 {	117,275	24	121,515	26
					32,864	7	34,424	8
TOTAL	685,309	100	570,248	100	482,230	100	461,264	100

Acres	N. Ireland, 1960	%	Republic, 1960	%	Total, 1960	%
1–30	41,300	57	144,083	50	185,383	52
30–100	26,490	37	116,265	40	142,755	39
Over 100	4,130	6	29,960	10	34,090	9
TOTAL	71,920	100	290,308	100	362,228	100

* This figure may be too high: it is possible that in the recording Irish acres (= 1·6 statute acres) were used in some cases.

Ireland, accustomed to expect larger units, yet the proportion of small farms remains high though the figures in Table VII overstate the number. In 1951, the number of farmers recorded in the Republic was 235,000 but by 1961 the number had fallen to 210,000. Figures for the Republic show that the number of male workers on the land has fallen from 380,000 in 1961 to 309,900 in 1967: of these 300,200 in 1961 and 254,200 in 1967 were members of farming families and the rest employed labourers. In Northern Ireland there has been a similar decline: from 1961 to 1966 there was a drop from 114,700 to 89,300, of whom 47,700 in 1961 and 38,600 in 1966 were owners of farms. The number of family workers assisting farmers declined from 50,400 to 34,300. The general trend has been for the number of small farms to decline by union into larger holdings: but local figures are available only for 1936. Fig. 32, though it exaggerates the number of small farms, gives a general view of the prevalence of each farm type through the country, and has been so compiled that the uninhabited areas are left blank as far as possible. In the Republic the 1965 statistics of the number of farmers of the small, medium and large units have been plotted and in Northern Ireland the 1964 figures have been used. One difference is that in the Republic the basis is

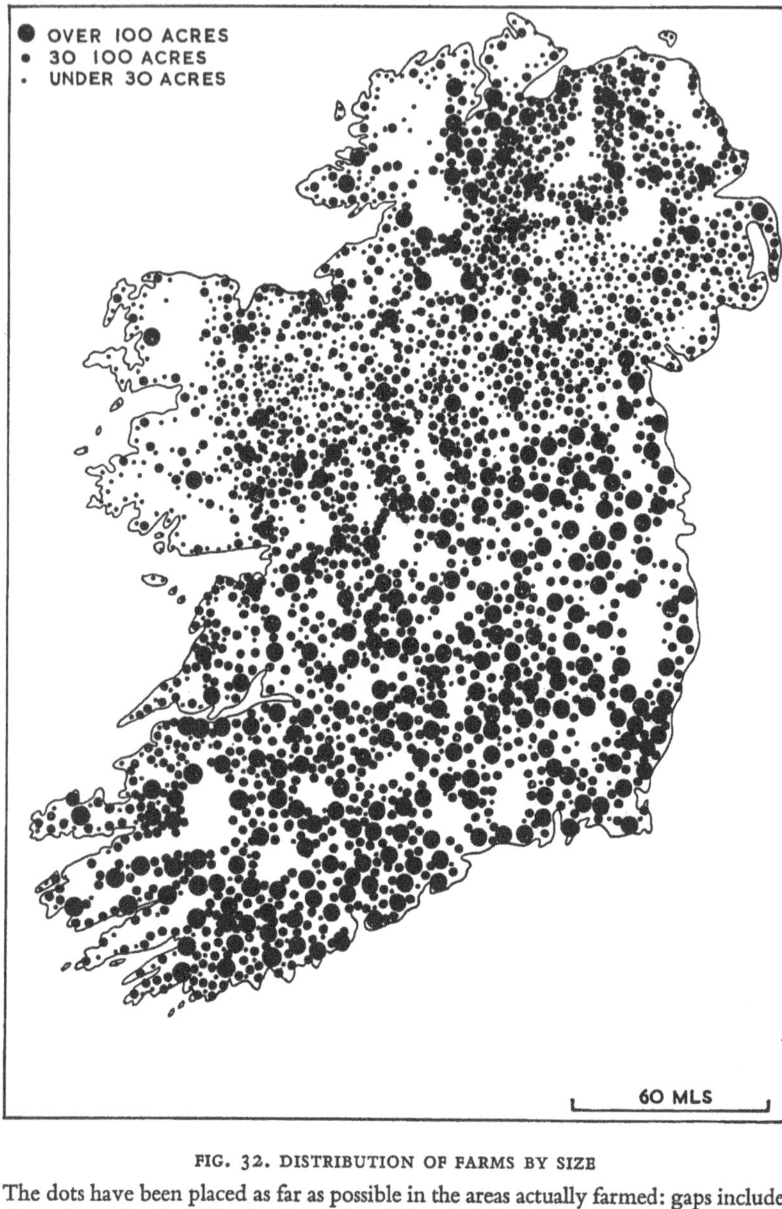

FIG. 32. DISTRIBUTION OF FARMS BY SIZE

The dots have been placed as far as possible in the areas actually farmed: gaps include the major uplands and lowland peat bog. Note that there are statistical discrepancies between the Republic and Northern Ireland. Figures are drawn from the *Ulster Year Book* and the *Statistical Abstract* of the Republic for the nineteen-sixties.

the area farmed by holders, and in Northern Ireland the actual holdings, more numerous than the farmers. But allowing for these inevitable limitations, certain general conditions emerge: first, the prevalence of small farms in the north and west of the country; second, the admixture of farm units of all sizes over the greater part of the Republic south of a line from Galway to Dundalk (Figs. 32 and 33).

The two major areas dominated by small farms, the north and west, differ markedly in their productivity and standard of life. On the one hand, the northern farms are intensively cultivated with as much as half their land in arable crops, such as cereals, grass seeds and potatoes for seed and the market. On the other hand, the farms of the western seaboard and some neighbouring areas such as central Mayo are little more than subsistence holdings on which many workers have to find some additional source of income such as labouring, either locally or elsewhere in Ireland, seasonal migration to various parts of Britain, or remittances from relatives abroad. When the Congested Districts Board began its work and gathered material on the poorest areas of the land, a state of near-subsistence was revealed not only along the Atlantic coasts but also in the almost equally closely settled districts of Co. Mayo, Co. Sligo and Co. Leitrim, all of which still have an unenviable reputation for poverty. Although these are still problem areas, they have been greatly improved by the division into new and larger holdings in some areas, with modern cottages and improved commercial facilities. To avoid, in justice, the word 'subsistence', in Fig. 35 the term 'cottage farming' is suggested for the admittedly poor and somewhat overcrowded districts which correspond to the 'crofting' areas of Scotland.

The prevalence of the small farm in Northern Ireland is partly due to the dominance of domestic industry until the mid-nineteenth century (pp. 228–9) with some home cultivation of flax. Rarely is flax grown now but of 43,000 farming units, 19,000 are so small that they cannot be regarded as modern economic enterprises: even so, about half of them are the sole source of income for their occupiers. The number of small holdings is declining as they are absorbed into neighbouring farms. Some 24,000 farms, mainly of 50–70 acres (excluding rough pasture) provide a living to their occupiers, who in some cases rent extra land on 'conacre'. Letting of land on a permanent basis is not permitted in Ireland, so it is let for a short term, normally eleven months. Many people, including the aged and widows, let their fields in conacre and many farmers are glad to have it to supplement their own holdings. In all, it is estimated that 17% of the farm land in Northern Ireland is let as conacre.

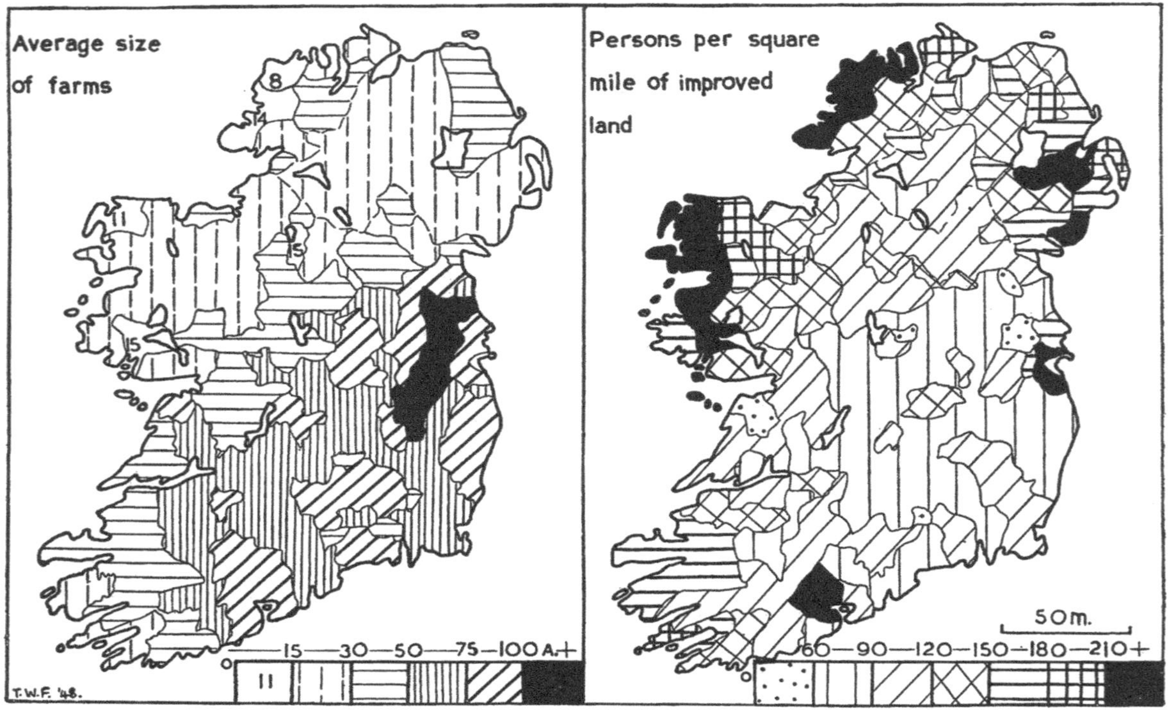

FIG. 33. AVERAGE FARM SIZE: PERSONS PER SQUARE MILE OF IMPROVED LAND, 1936

The general impression is 'the smaller the farm size, the greater the density per square mile'. But there are now signs that on the larger farms the number of workers is slightly increasing and that on the very small farms markedly decreasing: this is a clear inference from columns 7 and 8 of Table VIII. It is also borne out by the stability of population in Co. Meath, contrasted with the marked declines in

Within the Congested Districts, some areas had – even still have – a standard of living humbler than that found anywhere else in the British Isles, except possibly some parts of the Western highlands and islands of Scotland. In 1936 the average farm size was only 8 acres in northwest Donegal and 13 acres in northwest Mayo but since then many farmers have added a second holding to the one they originally had. The few fields are pastures, meadows, or patches of oats and potatoes, and the stock consists of one or more cows and sheep, a few pigs and a surprisingly high number of poultry. All the rough pasture is used as far as possible, even for pasturing cows, and in addition it provides peat for the homestead and heather, bracken and rushes for cattle bedding. The lowland south of the Ox mountains and a belt of country stretching from Donegal bay into Co. Leitrim also have similarly small farms whose holders concentrate mainly on young cattle, poultry, pigs and butter making. The apparent poverty is alleviated dietetically by the heavy consumption of butter and bacon at home, and the prevalence of seasonal migration, to which a certain amount of home industry, chiefly knitting and embroidery, is added in Donegal. But in spite of all the improvements of the past seventy years, which include the enlargement of holdings, the provision of better roads and transport facilities and the increased returns from seasonal labour, the long recognized 'congestion' survives.

Small farms were never as dominant in Kerry as in west Connacht or west Donegal, but yet cottage farming is typical of the entire western seaboard, including Clare. Fishing is only of very local importance, though many farmers on or near the coast fish for their households. In a few places, such as Dingle, Co. Kerry and Valencia island, fishing has at times been prosperous but normally transport costs prove inimical to success. The resources of the shore are generally more important than those of the sea, for the farmers collect quantities of sand and seaweed to manure their lands and carry them to the fields on their backs in basket-work creels or in pannier baskets on a donkey. The amount of labour needed to cultivate even one acre with a spade and a manure supply thus acquired is so great that the decline of cultivation is not surprising. Some of the people living inland have to bring manure so much farther that the struggle is even harder: the decline of the improved area in some western areas since 1871 is as much as one-third, and large areas may legitimately be regarded as submarginal land. The actual declines from 1871 to 1966, in counties along the western seaboard are: Donegal 32%, Sligo 6%, Mayo 18%, Galway 20%, Clare 13% and Kerry 32%. It was around 1871 that most counties recorded their maximum extent of 'improved land', which is officially defined as 'under crops and pasture'. Most of the more favoured counties in the Central Lowland and the south still have as much improved land as in 1871, and it is the less desirable areas that have been abandoned.

Every rural area of Ireland has its small farms, which grade upwards from the holdings of men who are really labourers to those of families who regard themselves as comparatively affluent, even though they cast envious eyes on the larger farms and the demesnes, and hope that their land hunger may eventually be satisfied. Here one may see the gradation into the middle-sized farm group of 30–100 acres which, with an admixture of large farms, characterizes the landscape in most of the Central Lowland south of the Galway bay–Dundalk line and is equally prominent in the mountain-studded lowlands that extend from Wicklow through the south to Limerick. The medium-sized farm is generally worked by the farmer, possibly with some family help though few farmers can afford paid labourers if their holdings are of less than 70 acres. The labour situation has been eased in recent years by the use of tractors, which are loaned out to farmers who do not possess their own. The economy of the medium-sized farm is normally mixed, and the output includes foodstuffs and wheat (except in the west), and feedstuffs such as oats, potatoes, turnips and mangels. Wheat production was greatly stimulated during the 1939–45 war, and the crop is sold to the corn mills. Barley and sugar beet, where grown, find a ready sale with brewers and the four sugar factories located at Carlow, Mallow, Tuam and Thurles. In Northern Ireland, grass seed and seed potatoes are also important: to some extent, these prevail throughout the old Ulster province. The amount of land given to these crops varies from year to year: high during the 1914–18 war, it fell to a low level in the early nineteen-thirties and then rose in response to government propaganda, which became compulsion after 1939, though after the war there was a fall in the acreage cropped. The holder of a medium-sized farm is, in most cases, able to turn his attention to a variety of crops according to circumstance, and is therefore in a central position in Irish farming between the small farm production and that of the large farms, which are in many cases highly specialized, as they prepare cattle for meat markets or provide grazing for stores before trans-shipment. As a group, however, the large farms include a number with a mixed economy, some dairy farms and, especially near the various uplands, some specializing in the rearing and fattening of sheep.

Population distribution in rural Ireland bears a very close relation to the type of farming practised: at one end of the farming scale parts of Meath have only 50 to the square mile, and at the other, parts of Donegal 400 to the square mile. In other words, a square mile of a congested area may have as many as 60–80 farm households, but Co. Meath only 4–6, with a few labourers depending on them. In between lie the middle group, of medium-sized farms, with 8–12 farm households, perhaps labourers on the more prosperous farms, and a population density around 80–120 per square mile (see also pp. 130–4). Every gradation lies between but the correlation between farm size and rural density is generally found in all areas lacking industry or any other cause for the concentration of population. Table VIII shows that in 1961 there were 68 workers per 1,000 acres on very small farms and 14 per 1,000 acres on the largest farms. The table also shows that since 1936 the number of workers per 1,000 acres on farms of all sizes has diminished, but to a greater extent on the smaller than on the larger farms: the decline in the number of farmers from 256,000 to 209,800 in twenty-five years is due to the decrease in the farms under 30 acres from 121,600 to 84,400. In 1936, the labour force on the farms of less than 30 acres numbered 216,200 but by 1961 it had almost halved, with 121,200. Table VIII also shows that the number of farmers having 30 acres or more remained virtually unchanged from 1936 to 1961: the slight increase in the number of farmers having 50 acres or more is due to the union of holdings. Field workers not infrequently discover that a farm of 50–60 acres is made up of what were previously two or three separate farms. Although the number of farmers having 50 acres or more has changed very little, the number of labourers and assisting relatives has declined sharply during this period. In fact the main decline occurred after 1946, as the intensification of farming during the war years and official discouragement of the emigration of farm labourers kept their number, 99,600 in 1946, around the 1936 level. But by 1961 there were only 47,300, half the total fifteen years earlier, and the drop in number of farmers' relatives, from 201,800 in 1946 to 107,800 in 1961, is almost equally striking.

That the process of farm amalgamation and mechanization is still continuing is undoubted. In many parts of the country there are farmers who regard their present holdings as too small to give a reasonable living. It may be possible to supplement their own land by using conacre as even if this hiring of land for eleven months of the year (see p. 177) is comparatively expensive, the use of a few extra acres of grazing or crop land may make it possible to keep enough additional stock to provide an adequate income for the household. An investigation made in June 1965 showed that within the Republic almost 6% of the farmed area was let in conacre (for the position in Northern Ireland, see p. 185). Of this land, 746,000 acres was let for grazing and 84,400 acres for tillage. The area let for tillage was more than 10% of all such land in three counties, Louth, Meath and Donegal, and the area let for grazing was more than 10% of all such land in the three Ulster counties of the Republic, Cavan, Donegal (over 16%) and Monaghan, four of the five Connacht counties (not Galway) and five Leinster counties, Dublin, Longford, Louth (over 15%), Meath and Westmeath.

Conacre land may be available for a variety of reasons. Some farmers, including a number of those in the Central Lowland, may have more land than they can conveniently manage and are therefore willing to let a few acres to others eager to acquire it. In some areas land is let by elderly persons or by those not wishing to farm it themselves. It has long been known that a large number of people in Ireland hold land but were not farmers: a report issued at the 1961 Census showed that they numbered 101,400, of whom 22,100 had less than 1 acre and 46,200 had 1–5 acres of land. Some 14,100 farm labourers had land, with 3,500 in other agricultural occupations (foremen, gardeners, forestry workers and others). But there were also some 13,000 labourers, 6,280 retail traders, 1,400 publicans, 930 teachers and 960 clergy, as well as 3,800 retired farmers, 14,500 other retired men and women, and 14,800 women returned as 'engaged in home duties'. Many of these people will let off land rather than farm it themselves and the concacre system provides a long-established, if not entirely satisfactory, contribution to the problem of supplying people with land.

AGRICULTURAL PRODUCTION

The farmers of Ireland feed themselves and the stock, supply the home market and sell the surplus abroad: estimates for the Republic show that before the 1939–45 war the total agricultural output was absorbed almost equally into these three channels. During the war, however, the export trade fell to one-fifth of the output by value, and the remaining produce was consumed in approximately equal proportions on the farmsteads and in the home market: but afterwards the exports increased and by the nineteen-fifties the produce was divided evenly, as before 1939. No such figures are available for Northern Ireland, though its agricultural exports increased considerably during the war and remained relatively high afterwards. In all parts of Ireland the farm households live reasonably well on a nourishing if not highly varied diet, and this perhaps accounts for

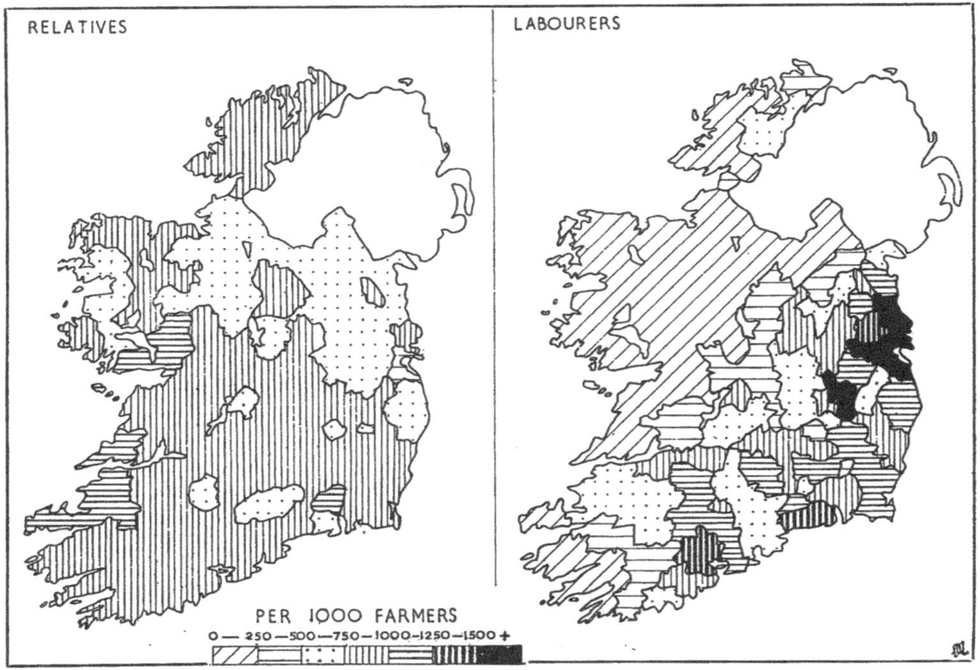

FIG. 34. MANPOWER ON THE LAND, 1946

When labourers were more numerous than now (see text) they were a substantial element of the population in Munster and Leinster but not in the west, except in such areas as the richer farmlands of east Donegal. But at that time assisting relatives were found on farms of all sizes though there were relatively few in much of the Central Lowland and the drumlin-drift belt.

TABLE VIII

Land Distribution and the Agricultural Population in the Irish Republic, 1936 and 1961

Farm size	1–15 acres	15–30	30–50	50–100	100–200	Over 200 acres	Total
Area in acres, 000's	776·1	1,896·7	2,189·6	3,012·9	2,274·5	1,369·5	11,668·8
Percentage of farmed area	7	16	19	26	20	12	100
1936 Farmers	50·3	71·3	55·6	45·3	18·5	5·8	256·0
1961 Farmers	30·4	54·0	53·4	47·2	19·5	5·4	209·8
1936 Relatives	29·7	54·2	49·1	44·6	18·8	5·4	243·2
1961 Relatives	11·2	22·6	27·4	29·4	13·7	3·5	107·8
1936 Labourers	3·6	7·1	13·8	28·6	25·1	21·4	97·2
1961 Labourers	0·9	2·1	5·2	14·1	14·0	11·0	47·3
1936 Total	83·6	132·6	118·5	118·5	62·4	32·6	596·4
1961 Total	52·5	78·7	86·0	90·7	47·1	19·8	334·9
1936 Workers per 1,000 acres	141	82	59	41	28	18	51
1961 Workers per 1,000 acres	68	42	39	30	21	14	29

Figures for farmers, relatives and labourers are given in thousands.

This table shows clearly the decline in the farming population in twenty-five years, especially from the smaller farms. There are some farms for which the area was not declared and whose workers were not included in the above table. The figures given above for labourers do not include those who are unattached to any particular farm, who are not necessarily working on the land for much of the time: the number of such labourers was 127,800 in 1936 and 59,600 in 1961.

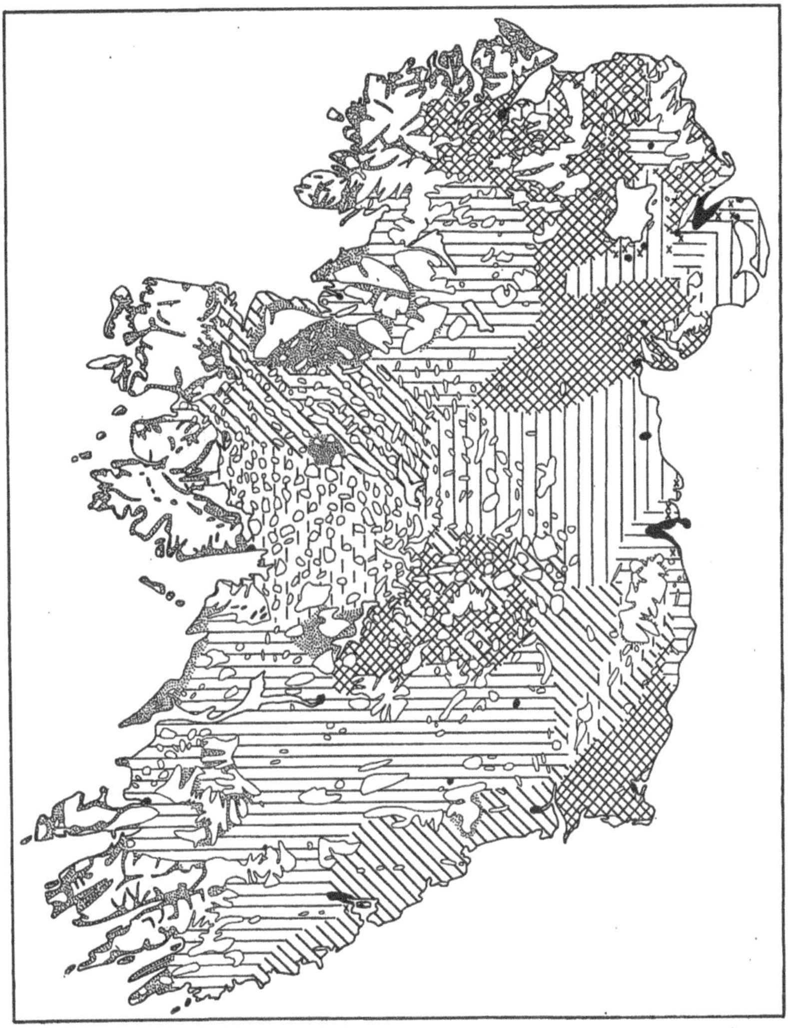

FIG. 35. FARMING TYPES IN IRELAND

- Mixed farming with considerable tillage.
- Mixed farming with stock-raising and feeding for the store cattle trade.
- Grazing and fattening country with some highly specialized large farms.
- Dairying predominant, mainly for creameries, milk supply near towns.
- Sheep raising and fattening with some stock-raising.
- Cottage farming.
- Peat bogs and mountain pastures, mainly uninhabited.
- Market gardening.
- Towns with over 10,000 people.

First published in 'Geogr. J.', vol. 110, 1947, p. 45

the high ranking of the country in international dietetic calculations. The non-agricultural population has increased steadily in Northern Ireland and to a smaller extent in the Republic, and the standard of living has so improved that the consumption in Ireland of meat, butter, bacon, eggs and vegetables has expanded. Yet the prices of vegetables remain high in the Republic and market gardens are surprisingly few in a country well suited to produce bountiful crops. Similarly, tomatoes, apples and soft fruit are grown in very small quantities; only in a few areas around the major cities and in an area to the south of lough Neagh is there any concentration on this type of farming. Individual ventures have proved successful: thus, a jam factory in Dublin, finding it difficult to acquire adequate supplies, established fruit farms near the city, each of which employs several times as many people as the cattle farms that preceded them. In a country adapted to produce excellent fruit and vegetables, the vast majority of the orchards are decayed, vegetable growing is limited to potatoes or cabbage, and soft fruit is scarce. Lack of capital or the diversion of capital from agriculture explains the paucity of these developments, but another factor is transport costs to any export market. The yields per acre have greatly increased since the years around 1850, when the average for Ireland was wheat 13–17 cwt., potatoes 3–5 tons, and turnips 11–14 tons: in 1953–62, the figures for Northern Ireland were wheat 23·4 cwt., potatoes 8·0 tons, and turnips 16·2 tons, and for the Republic (1966) wheat 27·8 cwt., potatoes 9·9 tons and turnips 22·0 tons. The yield of hay remains about 2 tons an acre, now as a century ago.

Cattle production in Ireland shows an economic integration made possible by the strength of the export trade to Britain. Both in Northern Ireland and the Republic the small farms produce more calves than they can bring to maturity, and the surplus is sold to the medium-sized farms, who feed them for a few months and then pass them on to the holders of the larger farms, who prepare them for the meat markets of the Irish towns or for export. Most of the cattle exported spend the last few months of their lives on some of the richer British pastures: Ireland's position in the store and cattle trade is comparable with that of Wales and Scotland, for all supply cattle to replenish dairy herds or to stock fattening pastures. In Northern Ireland the 'producing' and 'feeding' areas have been distinguished in the market reports, which also revealed that most of the cattle born on the smaller farms passed through two or three medium-sized holdings, staying on each for a few months, before reaching the larger farms from which they went to the abattoirs or the ports. The main fattening areas of Northern Ireland are shown on Fig. 35 as the hinterland

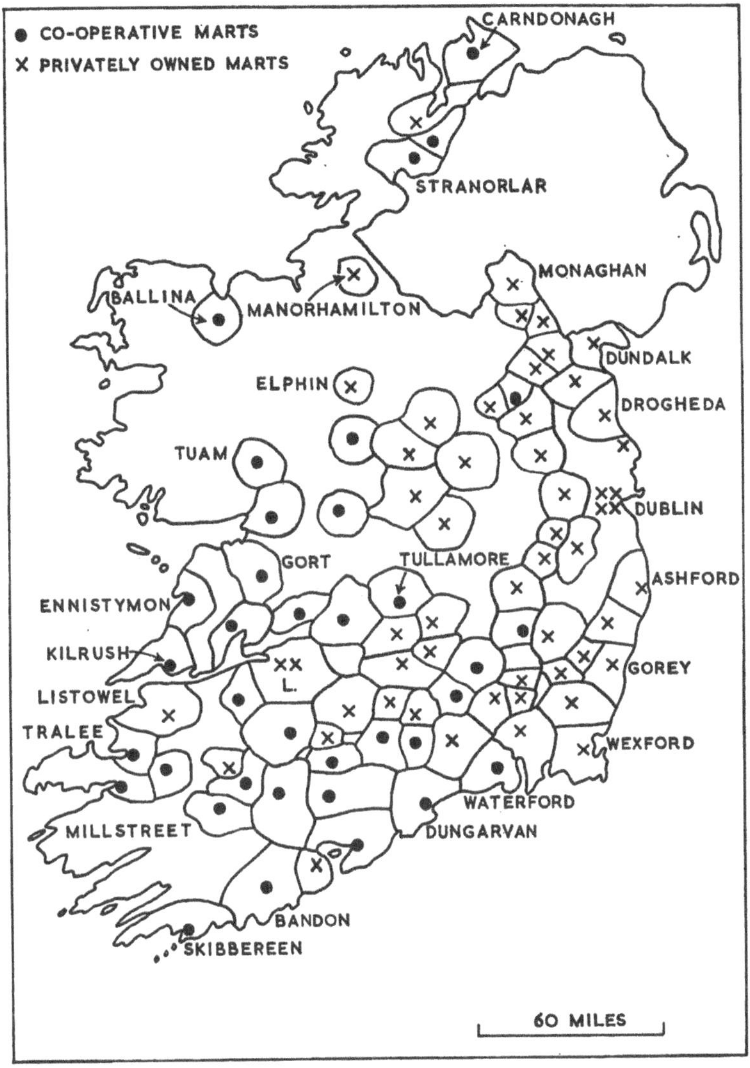

FIG. 36. CATTLE MARTS IN THE IRISH REPUBLIC

This map, redrawn from an original by D. A. Gillmor in *Administration*, vol. 15, 1967, p. 315, shows the co-operative and privately owned marts with the areas from which farmers bring stock for sale. In most of Co. Donegal, a large part of Connacht, and all peninsular Kerry with west Cork except the extreme south, fairs still provide the only outlet for stock sales, except for the comparatively rare cases of direct transactions on the farms. In many towns shown above fairs are still held. Of these the spring and autumn fairs are particularly interesting: at the spring fair in Galway, for example, thousands of pounds change hands and the banks are open from dawn. By the middle of the morning, weary counter clerks have thousands of pounds piled high before them but they are perfectly safe as nobody would interfere with the time-honoured commercial ritual of the fair day. As the day wears on some of the crowd appear to have enjoyed a good deal of liquid refreshment.

of Belfast and the lowlands on the east side of lough Foyle: in recent years, some 100,000 cattle have been imported annually from the Republic for fattening and grazing farms in Northern Ireland. In the Republic the cattle are sent to the large farms of Co. Meath and Co. Kildare and, to a lesser extent, the comparable areas around Cork, Waterford and Limerick. An animal may be sold two or three times in its lifetime at various fairs, in which dealers figure prominently, prices fluctuate sharply and various abuses exist. But from 1940 to 1954, the Northern Ireland Ministry of Agriculture bought all the fat cattle and sheep, though since then official intervention has been limited to the grading and certifying of stock, the supervision of markets and the making of guarantee payments. An increasing number of farmers sell direct to dealers, or at auctions, though fairs survive throughout the country and especially in the west (Fig. 36).

Dairying in Ireland interpenetrates with store cattle raising on many farms, but has become the main specialization of certain areas, notably the Munster lowlands and the lowlands stretching inland from Donegal bay. It is also important around Belfast and Dublin for local milk supplies. Like the store cattle trade, however, dairying is associated with the vast majority of the farms in the country and depends partly on the supply of calves from the small farms. The degree of specialization achieved depends partly on the success of the creameries in the areas they serve. Dairying is closely linked with the export trade in butter, which was already well established in the eighteenth century and was valued at £2,000,000 per annum early in the nineteenth. The original export was of home-churned butter in 'firkins' of seventy pounds weight sent mainly from Cork, but as the firkins were generally filled by several farmers and the quality was unpredictable Dutch and Danish competition was felt very severely in the seventies and eighties. The first creameries were established between 1885 and 1890 in the Golden Vale, and many more were established, both in Munster and Ulster, before the end of the century. At first the flavour of the creamery butter was compared, perhaps maliciously, to 'a raw swede turnip cut in two', but opposition to their development failed. Three types of butter are made in Ireland: farmers' or 'country' butter, factory and creamery; farmers' butter is sold, if at all, locally or to the factories. Factory butter is defined as 'butter which has been blended, reworked, or subjected to any treatment, but which has not as a result ceased to be butter'.

The original inspiration of the creamery movement came from Denmark and is associated with the names of the Rev. Fr Finlay, S.J., and Sir Horace Plunkett, whose work is carried on by the Irish Agricultural

Organization Society. Essentially co-operative in action, the creameries provide a means of processing raw materials, and have therefore carried industry into the countryside. Their strength is greatest in such areas as Limerick, Tipperary, northwest Cork and north Kerry, where almost all the farmers supply them with milk, and their extension eastwards into Kilkenny and Waterford is perhaps a sufficient tribute to the economic advantages they bring to a farming community. In general they have not been conspicuously successful in attracting the allegiance of the small farmer as he has too few cows to guarantee a daily milk supply, though in some areas the travelling creamery, located at some crossroads, has been successful. In Northern Ireland, milk production is now far greater than in 1939, and government purchases before 1955 gave graded prices to encourage a more even production through the year: this has long been advocated by publicists on Irish farming, who wish to encourage winter dairying on Danish lines and hope that milk production will ultimately dominate such areas as the 'tillage districts' of the Midlands (i.e. to the west of Co. Meath, chiefly Co. Leix and Co. Offaly) and also Co. Wexford, which also has a considerable arable acreage.

In spite of all efforts to develop dairying, the store cattle trade remains a main contribution to the export trade of the Republic: in 1966, the stores exported were valued at more than £26,924,000, with fat cattle at £22,912,000. Efforts to develop the tinned and frozen beef trade have had a considerable success, for the exports were worth £21,335,000 in 1966, Meanwhile the butter trade had diminished and then increased to £7,603,000 in 1966: before the war the export value was £1,600,000 but this declined to nothing in the war years and increased only slowly afterwards, to £545,000 in 1955 and £249,000 in 1956. The associated trade in condensed and dried milk, chiefly the latter, increased in value to £3,570,000 by 1966, and the trade in chocolate crumb, made for processing into chocolate from dried milk, cacao beans and sugar, has remained a useful export, valued at £5,658,000 in 1966. All exported foodstuffs of animal origin were worth £110,994,000 in 1966. In Northern Ireland, the number of cattle rose from *c.* 750,000 in 1939 to 1,189,000 in 1966, of which 319,000 were milch cows. From 1940 to 1953, there was a substantial export of liquid milk to Britain, which in some years reached over six million gallons, but when this traffic ceased the factories took the milk for butter, condensed milk, cheese, cream and various other milk preparations. In 1965 the total value of agricultural exports was £78,000,000 which included cattle (almost entirely fat stock for slaughter) at £15,255,000, and sheep or lambs at £782,000: some of the fat cattle are sent to the Irish Republic. Bacon and ham exports were valued at £21,051,000, and other major items included eggs, £12,681,000, milk products £10,079,000 and meat products at £15,891,000. Potato exports, at £2,736,000 included some sent for seed, of which a considerable amount went to Mediterranean countries.

Pig rearing is mainly a small-farm concern in Ireland and is associated with many of the poorest holdings, but also to some extent with the creamery districts. The pigs are fed on skim milk from the creameries or left over from home butter making, cooked potatoes and imported maize meal. As the last was unobtainable in the Republic during the later phases of the 1939–45 war, the number of pigs declined by three-fifths, the bacon factories had no surplus for export and the home market was inadequately supplied. The pre-war bacon trade was, however, expanding both in the Republic and in Northern Ireland, where a successful marketing scheme initiated in 1931 led to an increase in the number of pigs from 220,000 to 627,000 in 1939: imports from the Republic ceased and the export trade with England increased in value to £1,000,000 a year. Feeding difficulties caused a decline to 237,000 in 1944, but the number rose afterwards and by 1966 there were 1,057,000 pigs, more than at any previous time: the pigs are sent to bacon factories, from which 85% of the production goes to Britain, primarily as 'Wiltshire cure' bacon. The fortunes of the pig-rearing industry in the Republic have fluctuated. Up to 1939 the number of pigs was almost a million, but this was halved during the war and recovery was slow though by 1966 the number had increased to 1,013,500. During recent years the export of pig products, chiefly bacon and hams with sausages, has risen from *c.* £4,000,000 in 1955 and 1956 to £11,920,000 in 1966. Irish bacon faces strong and well-organized competition from Denmark. The chief factories are at Waterford, Cork, Limerick and Tralee in the Republic, and at Belfast and Portadown in Northern Ireland.

Poultry rearing both for eggs and table birds has long been regarded as a main support of many farms, especially the smaller ones. In 1939 there were some 20,000,000 poultry in the Republic and 10,000,000 in Northern Ireland: one-third of all the eggs from the Republic and half those from Northern Ireland were exported, and there was also a considerable export of table birds, conspicuously Christmas turkeys. During the 1939–45 war the number of poultry in the Republic declined considerably, with a slow recovery afterwards. In Northern Ireland the story is different: there was a vast increase to 24,000,000 by 1949, but afterwards a decline to approximately half this number. In 1966 there were 10,864,000 poultry in

Northern Ireland, with a considerable export of eggs, valued at £12,681,000 in 1965. During the past few years there has been a development of broiler fowl production, partly for the British market. In the Republic the loss of the British market has been disastrous. By 1966 the exports of eggs had fallen in value to £26,000 and of poultry to £63,000, a small fraction of the figure ten years earlier, and the number of poultry on the farms, having diminished to 12,843,300 by 1961, is still declining. In 1967 there were 10,593,400 poultry.

Sheep are reared in all the uplands of Ireland and are brought to the lowlands for fattening: the fat lambs and sheep are sold in the local meat markets or exported to England, particularly (before the 1939–45 war) to the industrial towns and the holiday resorts during the summer months. In Northern Ireland the number of sheep was increasing rapidly up to 1939, when it approached one million, but during the war and for a few years afterwards, cattle rearing was favoured and the number of sheep declined. In recent years, the number of sheep has risen above the pre-war levels, with 1,054,000 in 1966. In the Republic sheep have fluctuated around three to four millions but increased to 4,662,000 by 1966: the exports included a few fat sheep, many stores and lambs valued at £979,000, mutton and lamb at £4,575,000 and wool at £2,881,000. The main fattening areas of Ireland are the lowlands and valleys of East Ulster, including those around the Mourne mountains and the Antrim plateau, and also the hilly drumlin and drift country extending from Co. Down to Co. Meath. Damp lowlands, such as the Bann and Shannon valleys and many parts of the Central Lowland, have very few sheep, but in East Connacht, where the Carboniferous limestones are thinly covered with drift and even form actual karsts, sheep are reared in considerable numbers: the pastures developed on the thin limestone soils are nutritive, though the frequent outcrops of rock give the country a somewhat desolate appearance. In all the areas where sheep rearing and fattening figure prominently most of the farmers also possess cattle and grow some crops for the support of their household and stock.

As a whole, Irish farming is a mixed economy in which most of the farmers have a small arable acreage but a considerable area of pasture and meadow, and depend chiefly for their living on some form of livestock, with occasional cash crops. The store cattle trade and the creamery dairying have become well-recognized specialization of certain areas, but the core of the agricultural system is the economy of the medium-sized farms, whose holders have a variety of stock and change their production from year to year according to circumstances. With these medium-sized farms must be grouped some of the smaller farms in Northern Ireland on which production has been greatly increased during and since the war years. On the other hand, the 'cottage farms' have so limited a quantity of arable land that they remain dependent on pigs, poultry, calves and sheep: lying outside the creamery area, they cannot profit from daily milk sales. The most specialized of all the Irish farms are the large holdings which, in Co. Meath and Co. Kildare, provide the final grazing and fattening pastures for stock drawn from the medium-sized farms. Hardly less specialized are some of the medium-sized farms, especially in Co. Limerick and Co. Tipperary, where some of the farmers normally have less than one-tenth of their land in crops and rely on the rich pastures and meadows.

The very richness of the Irish pastures, especially on good drift soils with a lime admixture under heavy rainfalls, combined with the possibility of having cattle in the fields for as much as ten months of the year, has militated against the intensification of agriculture on Danish lines, for the farmer has no need to grow large quantities of feedstuffs to support his stock as the grass grows so luxuriantly. In Denmark, three-quarters of the land is tilled: in Northern Ireland one-seventh, and in the Republic less than one-eighth. It is frequently argued that if more crops were grown in Ireland, more stock could be kept and the supply of milk kept more uniform throughout the year. Yet in Northern Ireland the ploughed land declined from 1,117,300 acres in 1861 to 471,000 acres in 1939, and rose to 851,000 in 1943, only to fall again to 324,000 acres by 1966. Wheat was grown on only 3,000 acres in 1966, and flax growing appears to have ceased. In the Republic, the story is broadly similar: the proportion of the improved land ploughed rose from 13% in 1939 to 21% in 1945, but declined to 11% by 1966. Crop production has declined, even for sugar beet and wheat of which the former covered 53,500 acres in 1966 (but 63,900 acres in 1967) and the latter only 131,300 acres in 1966 (but 188,800 in 1967) compared with as much as 300,000 acres in many years since 1945. Both these crops have been strongly favoured by successive governments to obviate the need for imports. But Northern Ireland relies on imports for its grain and sugar, though there is a useful export of potatoes and grass seeds. The essential point remains – Irish farming is primarily grass farming, with stock as the main cash return. In the Republic this is primarily store and fat cattle, with milk products, eggs and bacon as subsidiaries, but in Northern Ireland the farm profits are now acquired largely from milk, eggs, bacon, meat products, potatoes (partly for seed), as well as fat stock: in other words, the grading and processing of agricultural staples has been carried out with at least some measure of success.

CHAPTER 9

Trade and Industry

THE AGRICULTURAL BASIS

The Irish landscape is characteristically agricultural, yet the country has its own well-established manufacturing tradition, dating back far beyond the Industrial Revolution and canalized into the craftsmanship of the workshops in the towns and villages. To this day, many of the factories conceal themselves so successfully that in Dublin and other towns it is not difficult to capture the atmosphere of an eighteenth century when broad streets and stately homes made the foreground of the picture and the workers hid themselves and their activities somewhere in the background. The industrial town or village, utilitarian rather than beautiful, is a phenomenon almost unknown to the majority of Irish people outside the northeast: many a young Dubliner, viewing Manchester or Leeds for the first time, wonders how people could live happily in such conditions, though the visitor from Belfast would find an English industrial city not unlike his home environment.

This contrast between Ireland and Great Britain has led many people to assume that the two kingdoms were designed by nature to be complementary, the one agricultural, the other industrial, but it is now realized that Britain must maintain her agricultural production and that Ireland must develop her industrial life. A hundred years ago, Sir Robert Kane, in his *Industrial Resources of Ireland*, said that the Irish

> dazzled by the wondrous facilities for industrial activity which the structure of the sister kingdom presents, had gradually sunk under the stupefying impression that Ireland was not suited for any manufactures, and the phrase currently ran, 'We are an agricultural people, the English are a manufacturing people: our soil and climate fit us for producing corn and cattle, whilst her mines of coal and iron, the source of her machine power, make her the workshop of the world. The position of the two islands is, therefore, correlative and mutually advantageous: she sends us clothing, and we send her food: from her crowded factories we receive all the products of complex manufacture and in return she takes our corn and cattle, the raw production of our soil.

Only one year after Sir Robert Kane forecast a rapid improvement in the condition of Ireland through the application of science to agriculture with a strong measure of industrialization, the Famine spread through the country and, to quote Kane in 1885, 'made such a profound alteration . . . in the condition of the people that the subject of industry fell out of view'.* But in spite of the emigration of millions, political unrest and years of economic stringency when famine appeared to be imminent, the social and financial desirability of industrialization was constantly urged in Ireland during the latter half of the nineteenth century. Hopeful signs were the developments in the northeast, where flax processing had become a factory industry having an increasing hold on the export market, and shipbuilding based on imported coal and iron ore was acquiring fame: on a smaller scale, the creameries were spreading through the dairy country through the efforts of the Rev. Fr. Finlay, S.J., and Sir Horace Plunkett.

In 1841, 66% of the Irish population lived by agriculture and 24% by trade and manufactures, many of whom were on the land though practising domestic industry in their homes. In 1961, 40% of the insured population was in manufacturing occupations in Northern Ireland, and 17% of the employed population in the Republic, and the number in commerce was 13% in Northern Ireland against 10% in the Republic.† In 1936, the twenty-six county Census showed an increase of 40,000 in the number of industrial workers over 1926, and the rise continued to 1939, when it was partly checked by the difficulty of acquiring raw materials. But there was a further rise after 1945 to a level higher than that in 1939. The re-industrialization of the Republic is discussed at some length in the following pages because it has brought significant changes to the country's life and outlook and has carried manufactures into places where they were unknown before. Northern Ireland's industries are far more deeply rooted: and during the 1939–45 war, as during the 1914–18 war, the linen mills, the shipbuilding yards and heavy engineering works, and the aircraft factories all had their national value: the agricultural production was greatly increased during these years. On the other hand, the northeast is extremely vulnerable to slumps in international trade, and in the early thirties as much as one-quarter of the entire insured population was unemployed: in 1932 and 1933 almost three-fifths of

* The remark of Kane in 1885 was made to the Select Committee on Irish industries; the passage quoted on the relation of Ireland and Britain is from Kane, R., *Industrial Resources of Ireland*, 1844 and 1845, p. 251.

† The figures are not strictly comparable as the insured population in Northern Ireland excludes farmers and family workers on farms.

the shipyard workers were without jobs and only six vessels, of some 20,000 tons, were launched. Resembling Clydeside in its industry, the northeast has needed and still needs new industries, especially as linen manufacture, though important in the export market, is now declining.

Irish industry is rooted in various crafts carried on in the homes of people, and even the great linen trade has its origin in domestic manufacture. There are still some traces of home industry in Co. Donegal, where several small factories provide woollen goods, including knitted garments and tweeds, and the old 'sprigging' (hand embroidery) is continued in workshops making altar cloths and other articles; to some extent also, the factories act as collecting centres for work done in the homes of the people. In many cases the profit of home industry was so meagre that it was, in effect, sweated labour, but the factories are subject to inspection and wage regulation, so the change seems socially healthy. Various small industries have been established in the Congested Districts as part of the effort to provide work, and in recent years the products of the Gaeltacht industries have been given considerable publicity in Dublin. Some of these works are apparently incongruous, such as toy-making in Connemara, but the tweed-making has proved a successful development of a traditional craft and some of the knitwear is attractive in design.

Agriculture is a natural basis for various industries in Ireland, and bacon, milk, cereals, potatoes and cattle may all be regarded as raw materials for factories. But most of the cattle are exported as stores, and the comparison, or more accurately the contrast, between Ireland and Denmark is that Danish live cattle exports to Britain were prohibited in 1892, with the result that the Danes were obliged to concentrate on butter, eggs and bacon, the 'breakfast table staples'. The Irish creamery and factory development has been far less complete than that of Denmark, though it has some elements in common, for many of the bacon factories have been associated with export trade since their foundation early in the nineteenth century. The creameries were the answer to the challenge of the Danish and Dutch competitors in the last quarter of the nineteenth century: previously the Irish butter was sent out in 'firkins' of 70 lb. weight from Cork or 'crocks' and 'lumps' of small content from Northern Ireland. Agriculture is also the basis of the famous distilling and brewing enterprises, which use home-grown as well as imported barley and support a number of malt works. There is every reason to believe that the processing of agricultural raw materials can be extended further, though the Irish producer has to meet the competition of very efficient

British firms and continental and dominion producers. In Ireland, leather-making has developed considerably since the nineteen-thirties and potatoes are used for the manufacture of industrial alcohol.

Supply of the home market has always been an important feature of the industrial life. The modern industrial programme in the Republic was planned primarily for the home market but in recent years there has been an increase in the exports of manufactured goods: in 1962 32% of the exports were of this type but by 1966 40% (not including food products). When Kane wrote his book in 1844, there were nearly 7,000,000 people depending directly on the land, but the majority were backward, illiterate and living at a level hardly raised above subsistence, with a very slight purchasing power. Fifty years later the inspectors working for the Congested Districts Board found that the majority of the people had very small cash incomes and in 1891 as in 1844 the essential need was to increase the buying power of the population, which could be made possible only through opening up the country to commerce by road, canal and railway. While the population has halved in a hundred years, the home demand for manufactures has increased, for once a farm family rise above the mere subsistence level they become potential consumers of goods produced in workshops and factories. Machine-made ropes replace those of straw, iron ploughs those made largely of wood, and the manufactured churn or the wooden beehive supplants the home-produced article: cattle remedies and patent medicines, sheep dips and potato sprays appear in the chemists' shops, and town-manufactured clothes are sold with the resultant decline of home spinning and weaving. There is more sale for groceries and drinks, and even bread is bought by many farmers as their incomes increase. Today there is a considerable sale of goods even among the holders of small farms, and naturally much more among those who have medium-sized and large farms, who buy more groceries and even luxuries: their state of comparative affluence helps to explain the commercial strength of many provincial towns and the prosperity of Irish manufactures such as clothing, footwear and agricultural implements.

THE LOCATION OF INDUSTRY

Minerals, especially coal and iron, were the essential attraction for industrial enterprise in the nineteenth century, but in Ireland the search for deposits, zealously carried out by the Geological Survey since its inception, has not revealed any resources sufficient in themselves for great industrial developments, though some metalliferous minerals such as copper and lead were more extensively worked in the middle of the nineteenth century than now. The lack of minerals is frequently mentioned as one of the causes of the slight industrial developments in Ireland, yet the northeast managed to attract both the coal and the steel necessary for its shipbuilding, and the long-established industries of Dublin also depended largely on imported coal. But the electrification programme of the Republic has been remarkably successful, not least in making possible the building of new factories in remote places. Electricity is now available virtually everywhere in the Republic. The first station, on the Shannon, was built from 1925 to 1929, and hydro-electric plants have been added on the Liffey at Poulaphouca and Leixlip, on the Lee at Inniscarra, on the Erne at Cliff and Cathaleen's Fall, and on the river Clady in Co. Donegal. There are peat-fired stations at Allenwood, Rhode, Portarlington, Ferbane, at Lanesborough and Shannonbridge on the Shannon, and at five places in the west – Gweedore, Co. Donegal; Bellacorick, Co. Mayo; Screeb, Co. Galway; Milltown Malbay, Co. Clare; and Cahersiveen, Co. Kerry. At Arigna local coal is used and there are coal- and oil-fired stations in Dublin and Cork. Normally at least half the power is derived from home resources, of which hydroelectric stations generally provide about 25%. Peat normally supplies 25–30%, native coal some 2% and the balance is acquired from imported fuels at the Dublin and Cork stations. In Northern Ireland the cities of Belfast and Londonderry have their own generating stations, but all the remaining areas are served by the Joint Electricity Authority, which has generating stations at Ballylumford on Island Magee, near Larne, and at Coolkeeragh, near Londonderry. Both these stations are now being extended.

From 1932 the Dublin government tried to encourage industrial development by legislation and loans: this policy, popular with the majority of the population, was inspired by several motives, of which one was to provide work in country towns and so to save them from decay and decline by emigration. Every town wants an industry; many towns have tried to raise capital to invest in a new factory. At first, economic independence of Britain seemed to some possible even though many raw materials must be imported; and during the 1939–45 war it was felt to be a happy circumstance that the country had its own electric supply, sugar-beet works and clothing factories. Northern Ireland's position is different. During the 1939–45 war, the shipyards, flax mills and shirt factories all made contributions to the needs of Britain but the present need is to attract new industries to combat unemployment.

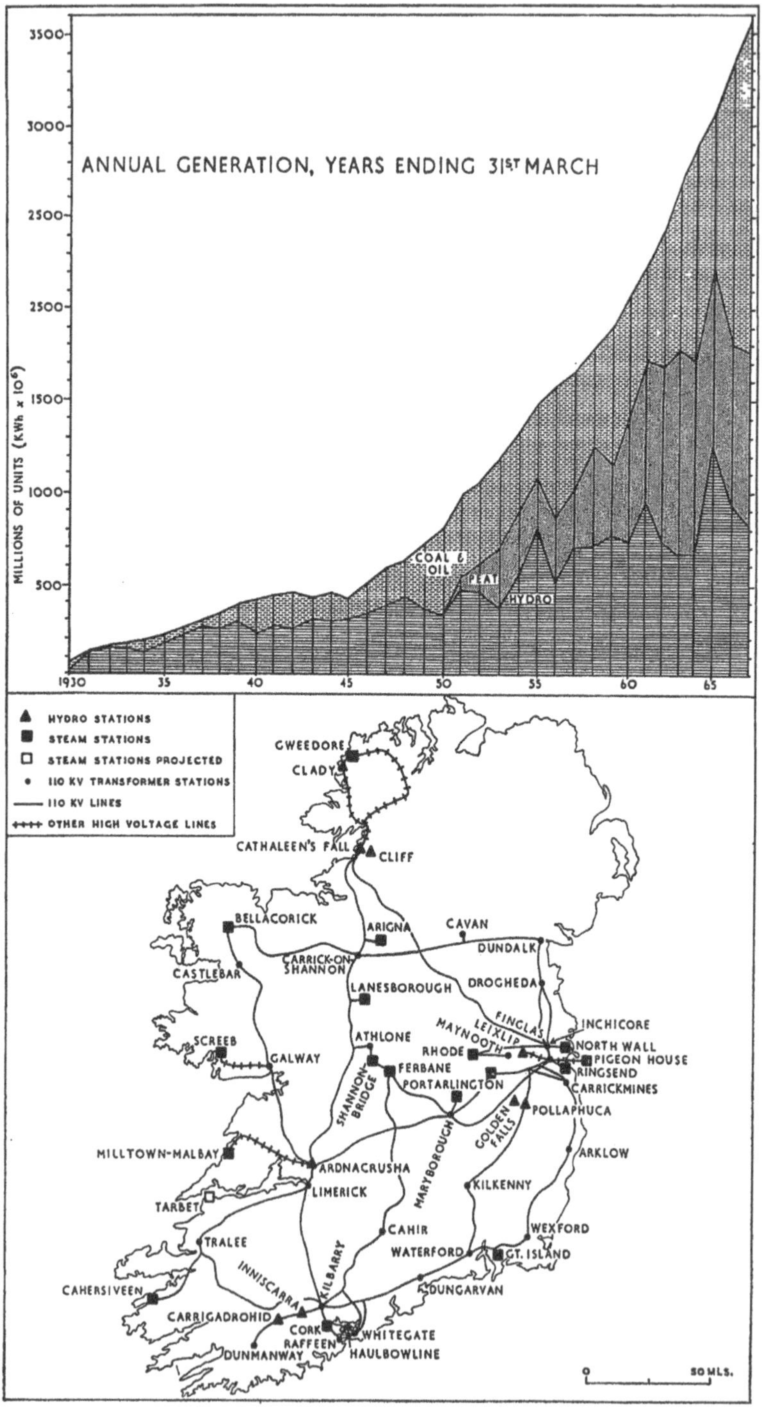

ANNUAL GENERATION, YEARS ENDING 31ST MARCH
MILLIONS OF UNITS (KWh x 10⁶)
3500
3000
2500
2500
1500
1000
500
1930
35
40
45
50
55
60
65
COAL & OIL
PEAT
HYDRO
HYDRO STATIONS
STEAM STATIONS
STEAM STATIONS PROJECTED
110 KV TRANSFORMER STATIONS
110 KV LINES
OTHER HIGH VOLTAGE LINES
GWEEDORE
CLADY
CATHALEEN'S FALL
CLIFF
BELLACORICK
ARIGNA
CAVAN
DUNDALK
CARRICK-ON-SHANNON
CASTLEBAR
LANESBOROUGH
DROGHEDA
FINGLAS
LEIXLIP
MAYNOOTH
INCHICORE
NORTH WALL
PIGEON HOUSE
RINGSEND
SCREEB
GALWAY
ATHLONE
RHODE
FERBANE
PORTARLINGTON
CARRICKMINES
SHANNON-BRIDGE
GOLDEN FALLS
POLLAPHUCA
MILLTOWN-MALBAY
ARDNACRUSHA
MARYBOROUGH
ARKLOW
LIMERICK
KILKENNY
TARBET
CAHIR
WEXFORD
TRALEE
WATERFORD
GT. ISLAND
INNISCARRA
KILBARRY
DUNGARVAN
CAHERSIVEEN
CARRIGADROHID
CORK
WHITEGATE
RAFFEEN
HAULBOWLINE
DUNMANWAY
0
50 MLS.

In the Republic, government planning has achieved a scattering of industry that is even more marked than anything contemplated or forecast in post-war England. Having few strongly industrialized centres apart from the larger towns, but an abundant labour supply almost everywhere, social desirability can weigh heavily in the location of factories producing goods for the home market and heavily protected from foreign competition. If the choice of location had been left entirely to the owners of the new factories, almost all would have gravitated to Dublin, Cork, Limerick, Waterford or a few other major towns such as Drogheda and Dundalk. Instead, many small towns have been revitalized, some conspicuously so: not all the factories have been successful, but a considerable proportion are already well established. Some of the places chosen for new industries, such as Tullamore, a country town in the Central Lowland off the main railway lines, already had successful factories developed by local enterprise. The triumph over obstacles by various able families has been a remarkable feature of Irish industrial life for more than a century, in some cases for at least two hundred years, and the foresight of nineteenth-century business men in Belfast is matched by that of men who made the city and county of Cork famous for its woollen mills, Dublin renowned for its stout, biscuits and poplin, Limerick the centre for bacon curing, Clara the home of jute manufacture, or Tullamore a town supplying whiskey. No single controlling influence has ever operated in Irish industrial life, and therefore the modern planning is perhaps not as incongruous as it might seem, as it is based partly on the desire to give employment that was so manifest in both the works of Kane in 1844 and the Reports of the Congested Districts Board fifty years later.

THE REPUBLIC OF IRELAND

IMPORTS AND EXPORTS

The overseas trade of the Republic has recently shown a marked increase, and the country's policy of industrialization inevitably depends

---

FIG. 37. ELECTRICITY SUPPLY IN THE IRISH REPUBLIC

The graph shows the vast post-war expansion and the use of peat supplies to an increasing extent. As noted on pp. 53–4 the availability of peat and water fluctuates from one year to another. Imported fuels must be used to meet this rising demand for electricity.

NOTE – Cork power station is fired by oil and coal, Arigna by coal. Of the three stations in Dublin, the Pigeon House (only a 'standby') uses coal or oil, North Wall oil and Ringsend coal or oil. The other fuel stations are fired by peat.

largely on the import of raw materials. It is estimated that over three-fifths of the imports are materials for further production, and one-eighth are capital goods ready for use: approximately one-fifth are consumption goods ready for use. To assess the overseas trade by value, the government statisticians have taken 1953 prices to represent parity: in that year the exports were valued at £114,097,000 and the imports at £182,480,000. The adverse trade balance of £68,383,000 was met by invisible exports of various kinds, notably the profits of the tourist trade, remittances from emigrants, pensions from abroad, chiefly to ex-service men, and the interest on stocks held abroad by Irish residents. Especially important in this invisible exports category is the tourist trade, which has developed considerably since the war of 1939–45, and rose steadily in value to a record level of £67,700,000 in 1965. This is a far higher contribution to the national income than the £14,000,000 supposedly sent from emigrants, but this figure may be an underestimate.

From 1925 to 1932, the volume of exports by relative value was similar to that of 1953, but the imports were less, about 80–90% compared with 1953, so the import excess was slighter; nevertheless, money from abroad was then as now indispensable to economic stability. From 1933 to 1939, there was a decline both in exports and imports, to 70–80% of the 1953 volume. This was due in part to the general decline in world trade of the time, and also to the deliberate government policy of trade restriction in the interests of home industries. An idea fashionable in some political circles during the 1930s that a measure of autarchy was desirable even led to the suggestion that some home substitute such as a herb beer should be found for tea, but after the 1939–45 war these views melted away, and the Republic wisely became interested in trading with various international groupings such as the European Free Trade countries, as well as with the dollar countries. There is every wish to deal with a wide range of countries and not with Britain alone, even though it is recognized that Britain must be the main market for exports and the chief supplier of imports. During the war years, the overseas trade sank to very low proportions: compared with 1953, the imports dropped to a minimum of 21% in 1943, and exports to 41% in the same year. Widespread unemployment was alleviated by the heavy outward migration of the time, and the lack of imports gave the country a 'favourable' trade balance in 1941, 1943 and 1944. In fact, economic circumstances were far from favourable.

After the war, the revival of trade began fairly slowly but by 1950 the import surplus began to cause alarm: in that year (1950 prices) the exports were valued at £81,520,000 and the imports at £204,596,000. The wide range of commodities available in the shops was regarded by some as a tourist attraction. Legislation was introduced to restrict imports and the encouragement of home industry continued. Even so, in every year there was the normal feature of an import surplus, ranging from £52,831,000 in 1957 to £80,986,000 in 1961 and £99,700,000 in 1962 (at contemporary prices in each case). Using the 1953 value of money as a standard, the 1966 imports were 76% greater by value and the exports 90% greater than in 1953. At the 1966 prices, the exports were valued at £234,422,000 and the imports at £363,625,000. Annual estimates of the balance of international payments show that the tourist trade is highly significant though the £65,100,000 produced in 1966 was partly offset by £31,600,000 spent abroad by Irish residents. The income from investments abroad was £45,129,000, though £28,814,000 was sent from the Republic to investors overseas. The general rise in overseas trade is rightly regarded as a sign of strength and the increasing export of industrial products is especially welcome.

Exports from the Republic are mainly of agricultural origin, though in recent years there has been an increasing overseas sale of manufactures, notably of clothing and footwear with an export trade valued at £10,423,000 in 1966. But in such vital industries as chemicals and engineering, Ireland's contribution to export trade is very small, though the current governmental policy is to attract foreign firms by credits for buying land and buildings and tax concessions on profits arising from new or increased exports. From 1958 to 1963, 160 new firms were induced to settle in Ireland, mainly from Britain, West Germany and the United States. The aim is to increase industrial productivity by approximately 4% annually, which would give a rise of employment of 7% and substantially reduce, or even perhaps in time virtually eliminate, the present heavy emigration. The main question mark, however, is the degree of success that can be attained in finding export markets.

In 1966, live cattle valued at £54,716,000 formed one-quarter of the export trade: these included stores valued at £26,924,000 and fat beasts, chiefly bulls and bullocks, at £13,912,000. Most of the store cattle, in 1966 numbering 394,000, went to Britain and many of the rest to Northern Ireland. Other stock exported were valued at £9,379,000, 4% of the whole: this included 12,900 horses worth £4,835,000 and sheep or lambs at £979,000. Greyhound rearing still persists, but on a smaller scale than in some previous years: in 1966, 7,230 greyhounds worth £1,100,000 were exported, chiefly to Britain. Foodstuffs of animal origin accounted for almost one-quarter of the exports in 1966 (£56,278,000 out of £235,271,000, excluding re-exports). Significantly, these foodstuffs are now a larger item than the export of live animals even though the overseas sales of eggs and poultry, which at one time accounted for one-third of the foodstuffs group, have virtually ceased. The main increase has been in beef sales, valued in 1966 at £19,705,000, of which £11,433,000 was for fresh or chilled meat and £8,271,000 for frozen meat: most of the former but little of the latter is bought for Great Britain. The tinned beef trade, from which goods at £1,631,000 were exported (mainly to Britain) in 1966, has been far less successful than many of its exponents hoped. There is a considerable export trade of other meats, including mutton and lamb at £4,575,000 and pork at £2,754,000 but the best established meat export is of hams and bacon, at £9,166,000 in 1966. This is clearly competitive with Denmark, and so too is butter, of which the export value in 1966 was £7,603,000. The increase of Irish butter sales in Britain owes much to efficient advertising and packaging. Other dairy products include cream, dried or powdered milk with an export value of £4,501,000 in 1966, and (Cheddar!) cheese with £3,388,000-worth sent overseas in 1966.

Other exports of agricultural origin include various articles of food valued at £19,678,000 (8%), of which the largest single item is chocolate crumb, used in the later manufacture of chocolates, at £5,658,000. There is an export of chocolate at £1,639,000, most of which goes to the British market. Useful recent additions to the exports include dried or dehydrated vegetables, which sold overseas for £1,064,000 in 1966. Of the drink exports, stout accounts for over nine-tenths of the £8,213,000 (4%) received, as the overseas trade in Irish whiskey,* 172,300 proof gallons worth £381,900, is small: in fact the import to Ireland of whisky,* almost entirely Scotch, is valued at £522,900. Another export of some interest is seed potatoes, valued in 1966 at £926,000, almost all of which went to Britain: there is also a small export of ryegrass seed. The export of fish from Ireland at approximately £2,000,000 may seem small. It includes some interesting items, such as £121,000 for eels, virtually all of which are sent to London, and £200,000 for lobsters and crayfish, mainly collected at various places on the Atlantic coast and sent to the French market. The export of salmon has declined in recent years, largely because it is sold increasingly on the home market, particularly during the tourist season.

* Note the different spellings for Irish whiskey and Scotch whisky. It is claimed the soda water was invented by a Dublin firm. Yet it is the custom to drink Irish whiskey with plain water (and generally not much of it).

The hide, skins and leather group of exports, worth £2,372,000 (1%), includes some materials of agricultural origin, notably undressed cattle hides. This group of industries has made a steadily increasing contribution to the export trade, and so too have the clothing and footwear trades, whose exports reached £10,423,000 in 1966. Women's garments and shoes were the main contributors to this total, and the exports are sent mainly to Britain. Equally significant, textile yarns and fabric products to the value of £10,423,000 were exported in 1966, including some carpets of attractive design and colour. The products are of wide range, and many are obviously surpluses from the needs of home production such as, for example, ropes, cables and even floor coverings. Although many manufactured goods appear among the list of Irish exports, in most cases – indeed in almost all – they represent only a small return of money to the country. But, as noted above, every effort is now being made to expand overseas trade, partly by increasing the range of goods manufactured in the country.

THE INDUSTRIAL EXPANSION

The main development of the past forty-odd years in the Republic has been increasing industrialization, seen most fully in the number of people working in factories and workshops. Fig. 38 shows these divided into four main categories listed below: it does not include those in building and

TABLE IX

Main Employment Groups, 1926–61

Year:	1926	1931	1936	1943	1946	1952	1961
Transportable goods	57,768	62,608	97,152	92,982	109,987	139,829	167,279
Building and Service trades	44,847	37,980	56,736	47,812	57,433	81,466	63,283
TOTAL	102,515	110,588	153,888	140,794	167,420	221,295	230,562

Note:

(1) For 1939 complete figures are not available, but the number in the 'transportable goods' class was 104,510.

(2) The occupational groups included under the building and service trades in 1961 were building and construction, 20,193: employees of local authorities and government departments, 23,947: canals, docks and harbours, 729: railways, 2,955: electricity, 7,867: gas works, 2,170: waterworks, 1,427: laundry, cleaning and dyeing industry, 3,987.

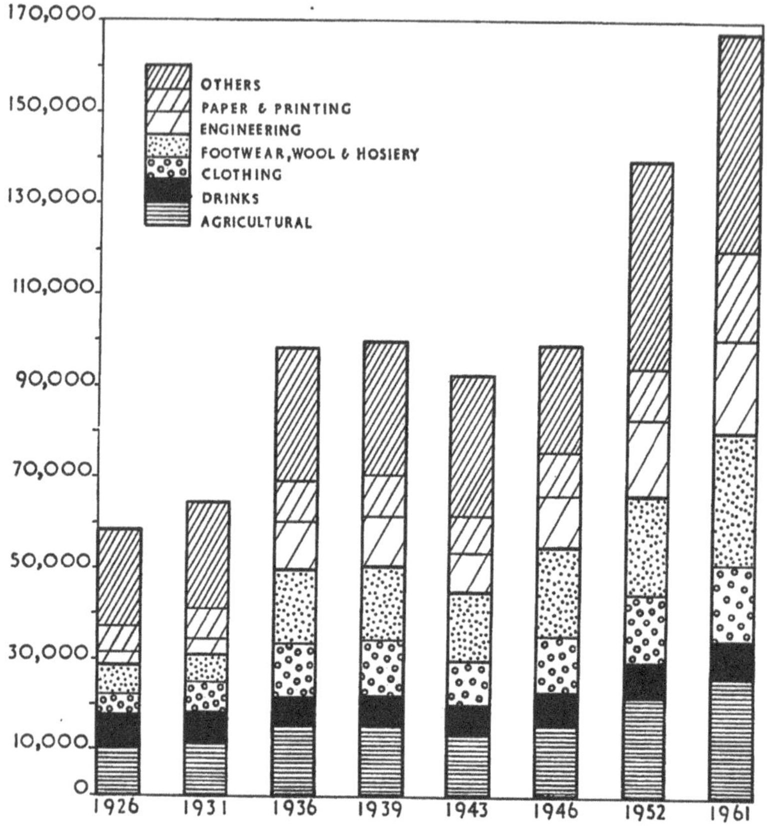

FIG. 38. INDUSTRIAL EMPLOYMENT IN THE REPUBLIC

This diagram, based on data in the *Irish Industrial Yearbook*, Dublin, annual, shows the number of persons working in factories; the industrial groupings shown here are indicated on p. 213. Of the four classes no. 1 is agricultural, no. 2 drinks, no. 3 clothing, footwear, wool hosiery, and no. 4 engineering, paper and printing and various other industries.

service trades. In Table IX, these are added to the number of persons making 'transportable goods', that is the group of workers represented in Fig. 38.

A first sign of the industrial drive was an increase of some 40,000 in the number of factory workers, who numbered 100,000 by 1939. But this advance was impeded by wartime scarcities, best expressed not in the total employment figures used in Table IX, but in the quantitative output. If that for 1936 was taken as 100, the factory production was 64 in 1926, 104 in 1939 and 80 in 1943: the cash value of the output increased steadily during the war, but the output per man declined. Unemployment was less severe than expected, largely due to the abundant opportunities for emigration to Britain. After the war, the number of factory workers rose steadily so that during the earlier 1950s it fluctuated around 140,000, with a welcome rise to 150,000 by 1958 and to 167,000 in 1961. Although this is highly significant, a considerably greater expansion of industrial employment would be necessary to absorb the numbers of people who now became emigrants, even allowing that the general industrial expansion provides employment in clerical and other occupations.

In Fig. 38, employment in the Republic's industries is shown. They fall into four classes:

1. Processed agricultural staples; grain, bacon and other meats, milk, butter, cheese and sugar.
2. Drinks, some of which are made from agricultural produce: distilling, brewing, malting, cider and mineral water manufacture.
3. Clothing, hosiery and footwear industries.
4. Industries mainly for home supply. Among these are mining and extractive activities (including peat collection), engineering, chemicals, paper and printing, wood and furniture, bread and confectionery and miscellaneous industries.

Fig. 38 shows that some industries have shown marked recent growth, notably clothing, engineering with allied trades, paper and printing. Only to a limited extent, however, is this due to changes in the type of export trade, though as shown on p. 211 there has been an increase in the export of clothing. Engineering and vehicle manufacture so far makes only a small contribution to exports, but their increase in employment is indicative of the increasing mechanization of farming and probably of a rise in the standard of living, as well as of the home production of goods formerly imported. The paper and publishing industries have increased markedly, partly as a replacement of imports.

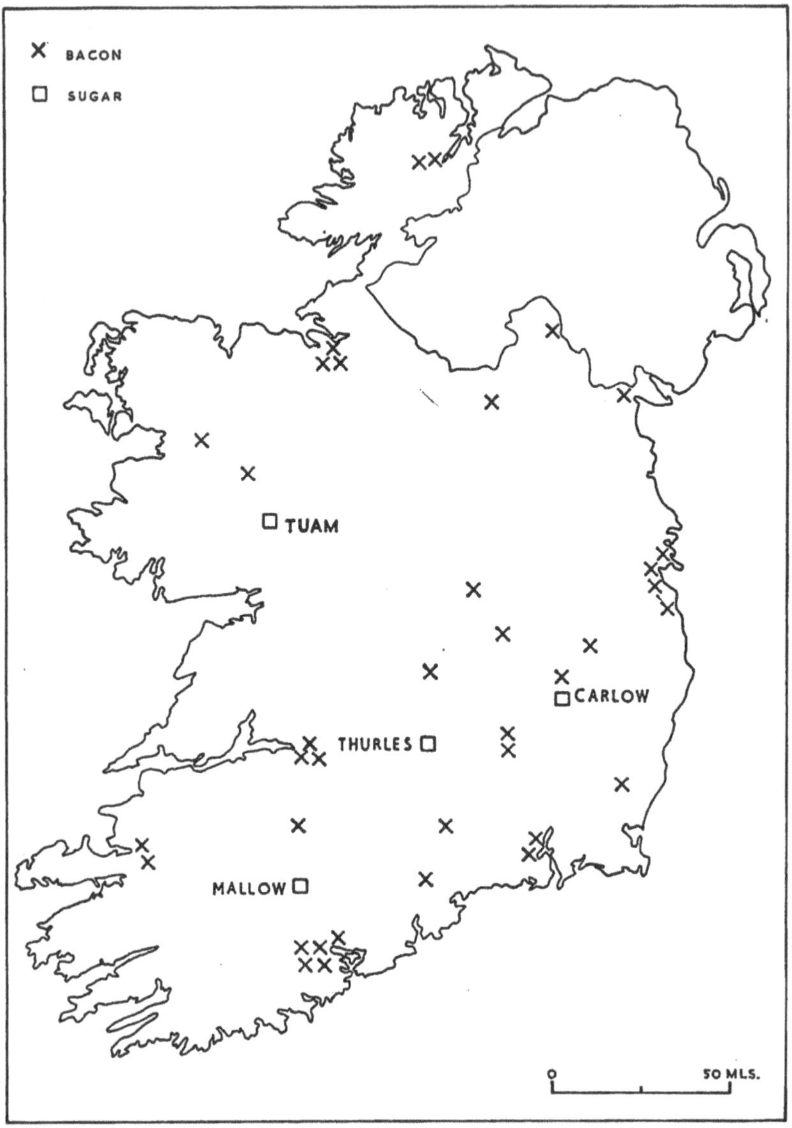

FIG. 39. AGRICULTURAL INDUSTRIES

The four sugar-beet factories are at Carlow, Thurles, Mallow and Tuam, and bacon manufacture is scattered through the country.

1. *Agricultural staples.* Only a small part of the agricultural produce enters into industry, as live cattle remain the chief staple of the country's farms. Factory production is connected mainly with dairying and its natural associate, pig production. There are bacon factories in the four chief towns, Dublin, Cork, Limerick and Waterford, and also in a number of provincial centres. Some of the larger bacon factories date back to the early part of the nineteenth century, and are well-established concerns having various sidelines such as meat and vegetable canning, soup manufacture, or sausage making. Efforts have been made to spread bacon factories throughout the country in order to provide a market for pigs, which are important sources of profit to masses of the smaller farmers. During the war years the scarcity of imported grains, such as maize, so reduced the pig population that the export trade was extinguished and bacon and ham were scarce even in Ireland. In the thirties, however, some £4,000,000 worth of bacon and ham was consumed at home, and the remainder, valued at £1,600,000 to £3,000,000, exported. Gradually since the war this export has revived (p. 198). But the export of bacon and hams is now less significant than that of meat, fresh, chilled or frozen. In 1966, the exports of fresh or chilled beef were valued at nearly £11,500,000 of which, apart from a small quantity to Germany, virtually all went to Northern Ireland or to Great Britain. But of the frozen meat, valued at £8,271,000, nine-tenths went to the United States.

Traditionally Cork was the centre of the dairying industry with the main butter factories of the country, and a general trade in the provisioning of ships crossing the Atlantic. But all the factories have gone now, and the butter is collected in cold stores in Cork and Dublin. The export butter now comes mainly from the creameries, which also provide some cream and dried milk for the export trade: for example in 1966, the cream exported – all to Britain – was valued at £931,000 and the dried or powdered milk at £3,570,000, of which less than half went to Britain: there was in fact a large export of dried milk to Venezuela. Cheese making is a sideline at various creameries and at the large farm produce factory in Mitchelstown, Co. Cork: recently there has been a marked increase in the export of cheese (p. 210).

Sugar-beet growing has been fostered to supply the works at Carlow, Mallow, Thurles and Tuam. This crop was only introduced in 1925, and it was usual during the war years to be thankful for its existence. Since the war a limited amount of sugar has been imported, but there has been a useful development of the jam, confectionery and chocolate industries, chiefly in Dublin and Cork, mainly for the home market. The most spectacular recent development has been the export of chocolate crumb, valued at £5,658,000 in 1966, which uses milk, sugar and cocoa. In the sugar, chocolate and sugar confectionery industries over 7,000 people were employed in 1961.

Fig. 38 shows that the number of workers in the agricultural industries increased considerably from 1926 to 1961. Of the various groups those in creameries and butter factories have changed least, from 3,350 to 4,600, the highest ever recorded: grain milling has increased from 3,050 to 5,350, due largely to the availability of foreign grain for both human and animal food. The main increase has been in the sugar industries, from 2,200 in 1926 to over 7,000 now, due in part to the chocolate crumb industry which also uses large quantities of milk. The bacon factories have about double the employees they had in the 1930s, and there are now some 2,500 workers in meat factories. Equally striking is the growth of the canning industry, with over 3,000 workers in 1961, and deep-frozen vegetables now have a profitable home and export sale. Small as many of the numbers may seem, they indicate an expanding use of agricultural produce for industrial processing.

2. *Drinks.* In the seventeenth century, Dublin is said to have had 1,200 ale-houses and 91 breweries, many of them in housewives' kitchens: now Dublin has one brewery, Guinness, but its proprietors say that it is the largest in Europe, exports more beer than any other brewery in the world, and operates three ships of its own to carry its beer (mainly Guinness stout) to Britain, for shipment to 126 more distant markets. The term 'beer' includes stout and porter, which became the popular drink in Ireland from the middle of the nineteenth century. The Guinness firm have an adequate tenure of their 65 acres, or at least of the part of it which was leased to Arthur Guinness in 1759 for 9,000 years. Other breweries in Ireland include Cherry's in Waterford, which provides ales, and the lager brewery of Dundalk, both of which are owned by Guinness. A brewery at Dundalk is owned by Ind Coope of Burton. Smethwick's in Kilkenny dates back to 1710 and has always relied largely on local barley: another old firm, at Rathdowney, provides ales widely sold in the Irish midlands, but is now allied to Cherry's of Waterford, and therefore to Guinness. In Cork there are two well-known and long-established breweries, Ladyswell ('Murphy's Stout Gives Strength') and Beamish and Crawford, recently taken over by the Canadian Breweries company. Both these Cork breweries have some 'tied houses' in the district, and Guinness has recently entered the retail trade, but on a limited scale. The Republic now

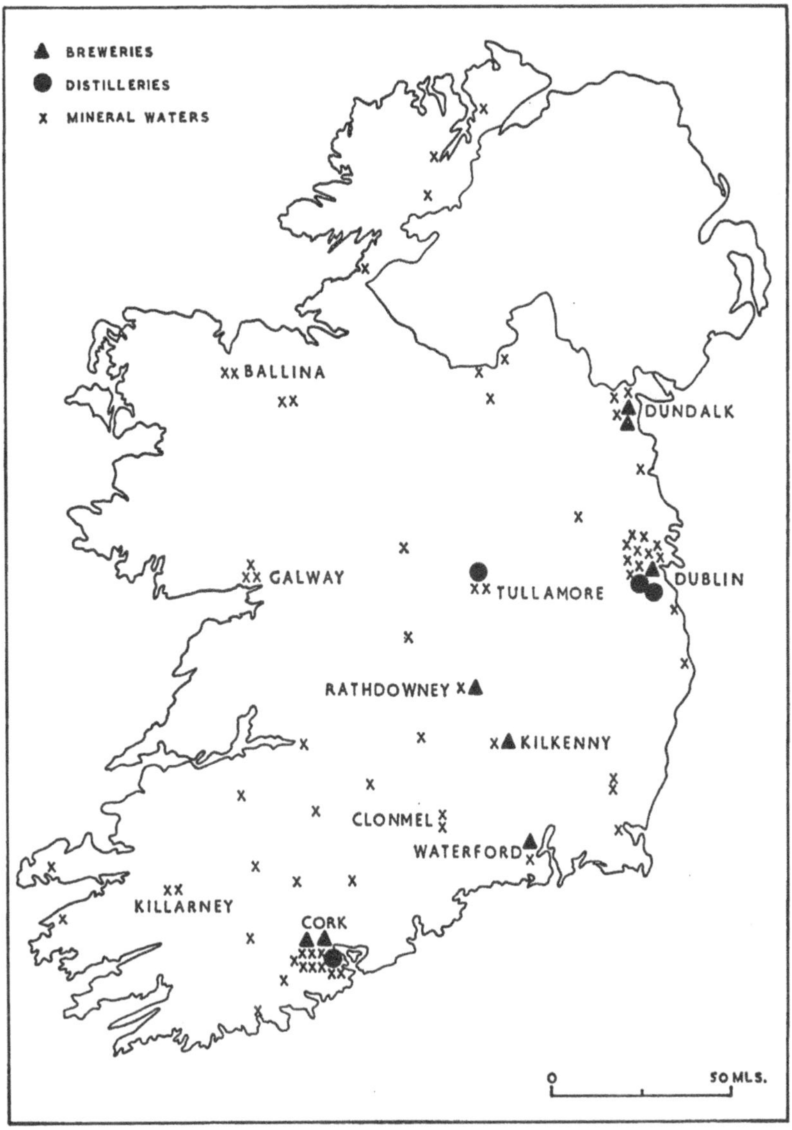

FIG. 40. MANUFACTURE OF DRINKS

Dublin, Cork, Dundalk and Tullamore are the main centres: mineral waters are manufactured in various towns.

has only four distilleries, two in Dublin, one in Cork and one in Tullamore. Those in Dublin date from 1780 and 1791, but one of them has recently spread its activities to distilling gin and 'Saratov vodka'. The large Cork distillery originated in its present form through the union of several smaller concerns in the mid-nineteenth century: it too provides gin and 'Nordhov vodka'. The distillery at Tullamore makes various whiskeys and a liqueur called 'Irish mist', based on old Irish whiskey and honey. In all, the number of people directly engaged in the drink industry is small – in 1961, 665 in distilling, 803 in malting (some of them only seasonally) and 4,730 in brewing. Yet the following quotation from a pamphlet issued by the distilleries shows that the whole story is more complicated: incidentally, practically all this quotation could apply to brewing, except that hops are imported, largely from Britain.

> The industry indirectly gives employment in every corner of the country. It embraces farmers, maltsters, corn factors, coopers, bottlers, case-makers, skilled wood-workers, carriers, wholesalers, and retailers. All the barley and other grains used are the finest grown on Irish farms. All packaging materials are also Irish. In fact, the only non-native part of a bottle of Irish Whiskey is the cork!

Lastly, there are approximately eighty firms through the country providing soft drinks of which one, at Clonmel, also manufactures cider, primarily from local apples.

3. *Leather and Clothing*. This group of industries had only 10,000 employees in 1926, but 45,500 in 1961. The leather industry showed an advance from 1,000 to 8,250 workers, hosiery from 900 to nearly 6,000, fabrics from 1,600 to nearly 8,000, woollen and worsted spinning and weaving from 2,400 to 6,800, and clothing from 4,300 to nearly 17,000. Of these industries woollens and tweeds had the background of domestic spinning and weaving in the west of Ireland, of which there is still an interesting survival in Co. Donegal: in addition some mills in the Cork area had preserved a tradition of supplying fine cloths for suitings. By contrast, the leather industry had by the 1920s become virtually extinct, largely through the failure to meet British competition.

In 1923 only one large tannery survived in Ireland, at Limerick, though there were a few small concerns providing harness leather: there were very few footwear factories, apart from some small ones in Dublin, with some long-established firms run by families in Cork and Killarney. Several more have been established, some of them with British affiliations: at Dundalk, for example, there are large firms dating from 1928 and 1932,

FIG. 41. COTTON AND WOOLLEN INDUSTRIES

Dublin, Cork and Co. Donegal are prominent, but numerous other towns and villages have woollen industries. Tullamore (see p. 277) has a large yarn factory.

with 2,000 employees between them and another smaller factory opened in 1952. Drogheda also has two major footwear factories. In Munster, Cork's main footwear factories date from 1880 and 1925, and others date from 1926 in Tralee and 1933 in Limerick. Clonmel and Kilkenny also have footwear factories, and there are others in Ballinasloe, Co. Galway, Birr and Edenderry in Co. Offaly and Bailieborough in Co. Cavan. In several instances these firms have developed with the aid of local capital, and have acquired premises cheaply, even including the old workhouse at Birr. The first post-1922 growth of the footwear industry preceded the foundation, or in some cases the re-establishment, of tanneries. Many are placed in rural areas, close to rivers for a water supply: steadily the varied types of leather required for footwear have been produced as new tanneries have been added. Those established in the pre-war period included establishments at Portlaw, Co. Waterford and New Ross in 1935, at Dungarvan and Gorey in 1936, at Carrick-on-Suir in 1938, with Portarlington in 1940. All these were producing sole and upper leathers, but after the war the range of materials was widened to include, for example, chamois leather and shoe linings at a Dublin tannery opened in 1948, and a variety of leathers including calf skins, suede and glove skins at Kenmare, 1953. Other post-war tanneries include one in an old malting mill at Milford, Co. Carlow, 1948, Cork 1949, with Tralee, a branch of the Limerick tannery in 1952 and Ballybay, Co. Monaghan, 1952. At Ballybay local people raised a large sum of money to open the tannery, largely because there was no other industrial employment available.

Textile industries have an honourable tradition in Ireland in woollen goods, but the linen industry was virtually extinct by the 1920s: unfortunately there is no prospect of any expansion in linen weaving in the Republic for, as noted on p. 477, it is one of the quality industries destined to decline. But cotton spinning and weaving has been successfully established since the 1930s. Historically, the woollen trade was developed especially in Co. Cork, with mills at Blarney 1824, Rosscarbery 1847, Araglin 1853 and Douglas 1889. All these occupied premises vacated by other industries, including flax mills at Rosscarbery and Douglas, a flour mill at Araglin and a linen mill at Blarney. In Kilkenny, also, a woollen mill of 1905 used a former linen mill. One enterprise of special interest is the Foxford mills, Co. Mayo, opened in 1892 by nuns anxious to alleviate the poverty of the district and, with the help of efficient management, rapidly made successful: in this case an old corn mill was acquired and the old mill wheel used to drive machinery.

By 1922 there was a sound basis for an expanding textile industry, for

FIG. 42. LEATHER INDUSTRIES

In these industries there has been a marked advance during the past twenty years not only in the major towns such as Dublin, Cork, Drogheda and Dundalk, but also in many smaller towns and villages.

apart from the firms mentioned above others were well established and successful, for example at Ardfinnan, Co. Tipperary, Athlone, Co. Westmeath and Convoy, Co. Donegal. The factories of Co. Donegal developed partly from the home weaving tradition: in 1950 a yarn factory also making tweeds was established at Kilcar, in southwest Donegal, and in 1952 a hand weaving tweed factory was built at Downings in north Donegal. Much has been done in Donegal to preserve this craft: several factories have been successful for many years in the northwest (p. 452). In the city of Galway, there is a survival of hand weaving, partly for local tourist sales, and the mill at Avoca, Co. Wicklow, has some fame for its interesting designs. But the main growth since 1922 has been in worsted spinning and weaving, with the Tullamore factory of 1938 in the old prison as the largest spinning plant, employing 600 people (1965): weaving plants exist at Maryborough 1937, Cork (two, 1938 and 1946), with firms having both weaving and spinning in Midleton 1946, Wexford 1950 and in the Dublin area. At Cootehill, Co. Cavan, the old workhouse has been used since 1949 for a shoddy mill.

The great advance of the hosiery industry, with knitwear, was in part due to its marked success in the Cork area, where two British firms combined to set up a large factory in 1928, and later established a woolcombing plant in Midleton, and two mills at Youghal, one in 1946 for synthetic fibres and another in 1952 for fine cotton yarns. There are also several firms in the Dublin area, with others in various provincial towns, including Dundalk, Kilkenny, Wexford and Galway. This industry has had some success in Co. Donegal, notably at Buncrana and Letterkenny.

Cotton spinning and weaving has developed under tariff protection during the past thirty years. The cotton thread factories established in 1934 in Dublin and Westport, Co. Mayo, were regarded as revolutionary growths at the time: a cotton braid and tape mill at Ennis, Co. Clare, of 1935, had a location almost as remote as that of Westport. The site was chosen in response to the governmental policy of industrial dispersion, the availability of labour, and local encouragement. In 1935 a cotton weaving mill was set up in Athlone, to which spinning was added in 1946: in all this mill now has 800 workers. A large weaving mill was added at Slane, Co. Meath, in 1936, with weaving mills at Naas in 1938 and 1952: this town was chosen as there was little employment locally. In 1950 cotton spinning began at Ardee, in association with the older surviving cotton and linen mill at Dundalk, and another spinning mill was opened at Loughrea in 1956, with a weaving mill at Portumna in 1954 and a textile printing works at Galway in 1957. Other weaving plants were established in the Dublin area, including Balbriggan, and in 1950 at Longford, a town which also has a ribbon factory.

The large increase in the number of workers in clothing factories is mainly indicative of an increasing home demand, satisfied by home production. The imports of clothing in 1965 were valued at approximately £3,260,000, but the exports at £7,863,000, among which shirts and women's coats were prominent items. Naturally, there are still imports of materials and of luxury articles already made up, but the expansion of the clothing industry has been planned to meet the needs of all types of trade. The industry is prominent in Dublin and Limerick, with hosiery manufacture in Cork and district: shirt making is chiefly a Donegal industry.

4. *Industries for Home Supply*. A wide variety of industries intended for the home market employed 87,000 people in 1961 compared with 30,400 in 1926. These fall into six main groups: (*a*) extractive, chiefly peat cutting and quarrying for road materials and cement; (*b*) engineering, including vehicles; (*c*) chemicals, notably fertilizers; (*d*) paper and printing; (*e*) wood and furniture; (*f*) bread, confectionery and tobacco. Of the six groups, the last named has shown little change in employment since 1926, when there were 8,100 compared with 9,450 in 1961. The wood and furniture trades show a comparatively small increase, from 4,400 in 1926 to 7,000 in 1961. Far more remarkable is the rise in the engineering and vehicle trades from 3,600 in 1926 to 26,400 in 1961, and even the growth of paper and printing from 5,700 to 14,700 in the same period. The small chemical industry increased from 1,400 to 5,600. In chemicals, fertilizers, machinery and electrical goods, even vehicles or parts of cars and many other manufactured commodities, semi-manufactured goods or raw materials, Ireland depends largely on imports.

Mining hardly exists in Ireland, except for workings in the Castlecomer field and at Arigna, in the latter case for power production. But several thousand men are employed in digging peat for the power stations and for domestic consumption. Bricks are made from Triassic clays near Kingscourt, and from glacial and other clays in the Dublin area. There are earthenware potteries at Arklow and at Carrigaline, Co. Cork: the most notable artistic industry is the Waterford glass manufacture, recently revived after an interval of several decades.

Engineering industries have expanded considerably during the past thirty years, and a recent exotic addition is the crane factory, of German origin, near Killarney. Contrary to many statements, it is not a blot on the landscape as the factory is of pleasing design: most of the parts are imported and the finished products are exported. Only since 1938 has steel been made in the Republic, on Haulbowline island in Cork harbour, near Cobh: this island has deep-water berths available at all states of the tide. Cork has had the Ford motor works since 1919, the vast Dunlop rubber factory, and several other engineering works. Many small engineering factories are scattered through the country, and old enterprises include the making of agricultural machinery at Wexford. Of the scattered industries, examples are the making of razor blades at Carlow and Enniscorthy, of aluminium utensils at Nenagh, of cutlery at Newbridge, electric bulbs at Bray and electric cables at New Ross. The industrial dispersion may create difficulties: the Ford plant in Cork uses heavy springs from Wexford, glass from Templemore, upholstery and plastic leather from Trim, Co. Meath, Westport thread, sealing solutions from Belturbet, Co. Cavan, batteries, bulbs, sparking plugs, door panels and seat springs from Dublin. But the paints, enamels, tyres and tubes are available in Cork.

Chemical industries are relatively minor, with fertilizer works at Dublin, Drogheda, Wicklow and Cork. The paper and printing industry includes some 5,000 people making paper and paper products and nearly 10,000 working in printing, mainly in producing newspapers, journals and other ephemeral literature, but there are no large-scale publishing houses. An increasing array of 'miscellaneous' industries employs some 14,000 workers, more than twice the number in 1936: many of these are in small consumer goods factories.

NORTHERN IRELAND

In 1966, Northern Ireland had 511,000 insured employees, of whom 192,590 (38%) were in manufacturing industries and 247,170 (48%) in service trades and professions. Some 51,090 were in building and other constructional occupations, and only some 4,030 (less than 1%) in mining and quarrying. These figures include the number unemployed in the summer of 1966, in all 6% of the whole, but though they include 13,600 labourers in agriculture, forestry and fishing, they do not include the farmers and family workers. In all there were 89,300 agricultural workers in 1966, of whom 38,600 were farm owners, 16,400 wives of owners working on the land, and 20,700 other family workers. In 1961 there were 114,700 people on the land, including 47,700 farmers and 50,400 family workers: the number engaged in agriculture has shown a continuous decline since 1952, when the special census recorded almost

145,000 of whom 57,000 were owners, 66,000 family workers, 13,000 full-time and 9,000 part-time workers (p. 183). The fall in the number of workers in agriculture is characteristic of the whole of Ireland, but as an economic unit Northern Ireland differs from the Republic in its degree of dependence on industry, and especially on three industries which together employed 126,130 people in 1966. These are engineering, with 43,960 workers of whom 13,880 were in shipbuilding and marine engineering; textiles with 55,170 workers; and clothing 27,000, including 9,000 workers in the shirt-making industry, of whom approximately half are in Londonderry. In appearance, however, Northern Ireland is primarily an agricultural area in which the Industrial Revolution has overlaid a veneer of manufacturing and town life, and the produce of the farms was the basis of the flax industry, organized first in the homes of the people, and later in the factories of the towns. Belfast, like Manchester, began as a market town; it gradually became the economic hub of a large part of Ulster, having in time no effective rival as port, factory town, commercial centre or, in short, regional capital.

An analysis for 1960 of all manufacturing firms having twenty-five or more employees shows that of the 204,500 workers in their factories, 56% were in Belfast with the new industrial estate at Castlereagh, another 12·5% in Co. Antrim, and 8·5% in Co. Down. The city of Derry had nearly 10,000 industrial workers, almost 5% of the total for Northern Ireland. An earlier analysis by K. S. Isles and N. Cuthert, using the figures for all the insured persons, that is for the whole of Northern Ireland, 470,000 in 1955 (511,000 in 1966, see above), gave comparable results: they showed that three-fifths of all the insured workers were in Belfast itself, and seven-tenths in an area covering the city, the Lagan valley and the upper Bann, including Portadown and Banbridge. Indeed, three-quarters of the workers were within a radius of approximately twenty-five miles, and four-fifths within a radius of some thirty miles of Belfast, which would include Portadown, Newcastle and Ballymena. Obviously the main industrial concentration is in Belfast and its hinterland. Belfast also handles the great bulk of the overseas trade, with Larne as a second port of consequence: taken by value the share of the 1960 import trade was Belfast 69%, Larne 17%, Derry 4%, and of the export trade Belfast 60%, Larne 31%, Derry 4%.

The figures of exports and imports, from 1924 to 1965, include a certain quantity of goods for re-exports (partly to the Republic), and except in 1944, 1945 and 1946, show a slight adverse balance of trade, made up by the 'invisible exports' of the tourist trade, remittances from those overseas, dividends on investments in Britain and abroad, and pensions paid from Britain. In 1965 the imports were valued at £512,968,000 and the exports at £458,852,000, three-quarters of which was for manufactured goods valued at £354,000,000, including textile yarns and fabrics valued at £108,268,000. The remaining exports were food and drink, at £75,592,000, including meat and meat preparations at £38,250,000 and dairy produce with eggs and honey at £23,276,000. The export of live animals, valued at £18,652,000, has been declining for several years.

Imports in 1965 valued at £512,968,000 consisted mainly of manufactured goods at £322,803,000 with another £62,615,000 worth of basic materials for manufactures. Textile fibres and waste imports were valued at £15,218,000, fertilizers and other minerals at £3,575,000. Among the manufactured goods, major items were chemicals at £19,943,000, paper and paper products at £15,879,000, textiles at £50,163,000, metals, machinery and electrical goods at £122,316,000, vehicles including aircraft and boats at £56,624,000, with clothing and footwear at £16,620,000 (offset by an export of £33,608,000). In the food, drink and tobacco group, valued at £117,317,000, the major items were cereals at £29,585,000, feeding stuffs at £12,860,000, and tobacco at £26,484,000. There was also a small import of live animals, at £10,234,000, partly from the Republic. A study of this catalogue shows first, that the total value of Northern Ireland's export trade is far greater than that of the Republic (p. 208). Secondly, in the economic structure of the province, the profit from the linen factories and the shipyards is of very great importance – unfortunately the export value of the shipbuilding trade is not given as the yards are controlled by one firm. Thirdly, the imports in the food, drink and tobacco group are greater than the exports. Fourth, Northern Ireland has no pretensions to autarchy, and imports almost all its supply of wheat and even, in recent years, all its flax. The re-industrialization programme is similar to that of the Development areas of Great Britain and bears no resemblance to the politically inspired programme of the Republic.

The main concern of the post-war years in Northern Ireland has been the high unemployment rate, and for some years there has been vigorous propaganda to attract new industries, especially those producing consumer goods. New factories established or planned since the war produce nylon stockings, artificial silk, hosiery, drawing instruments, wireless and gramophone apparatus, carpets, toys, prams, optical goods, furs, vacuum cleaners, electric cells, jewellery, office tabulating machines, rock-drilling bits and turbine equipment. Such a list might well come from a trading estate in Great Britain. There have been various Acts to foster industrial development from 1932 onwards, when Northern Ireland was suffering heavily through the Depression. 60,000 new jobs are in enterprises begun or expanded since 1945 and it is expected that these concerns will eventually provide another 6,000 jobs. There are two main aims – first the industrial diversification of Northern Ireland, and second the provision of employment for male workers. Of the 60,000 new workers, over half were males: many British trading estates have factories staffed largely by women in areas of heavy industry, but in Northern Ireland female employment is relatively high.

Although many people think first of Northern Ireland in terms of its great industries, the largest of these, textiles, drew its initial strength from work on the farms. Flax is no longer a main crop in Ireland – indeed it was grown only on 9,250 acres in 1955, far less than in the years to 1940, so over nine-tenths of the raw material for linen production had to be imported. By 1958, only 1,000 acres were sown, and none at all has been grown recently. That it could be grown widely is shown by the 1944 figure of 125,000 acres, when supplies from Belgium and Holland were not available. Even in the eighteen-forties there were complaints that too little flax was grown in Ireland. Of other agricultural produce, the Babington report of 1946 commented that 'if methods of processing agricultural produce can be developed successfully, a whole range of employment hitherto unknown will be opened up'. Significantly they add that 'of no country is it truer to say that agriculture constitutes the basis of economic life'. Some success has already marked such industries, as a famous milk firm has established itself and the meat trade has developed considerably, along with canning of fruit and vegetables. In 1966, nearly 31,000 people were employed in the food, drink and tobacco industries and it is estimated that some 20,000 people are employed in work ancillary to agriculture, including the manufacture of fertilizers, and feedingstuffs, fatstock marketing, meat processing, egg packing and canning.

As long ago as the seventeen-seventies, Arthur Young looked forward hopefully to the days of linen mills in towns, but it was the cotton industry that first used steam power when in 1789, only a year after the first steam engine reached Lancashire, one was installed at a mill in Lisburn as no water was available. The imports of cotton wool and cotton yarn into Ireland showed a steady increase from 1783 onwards, and by 1811 there were, in the neighbourhood of Belfast, fifteen cotton mills driven by steam, six by horses and hand, and twelve by water. Apart from the workers in the mills, there were numerous outworkers who were either employees of the factories or of agents from Scottish firms who took advantage of the lower wage rates paid in Ireland. Not a few of these home workers were linen-weavers who changed over to the easier task of cotton-weaving; and some changed from one to the other as the demand for each warranted. In the early part of the nineteenth century, cotton-weaving spread through a large part of northeast Ireland, but the slump after the Napoleonic wars led to its restricted localization in Belfast and neighbouring centres such as Bangor and Donaghadee, and others in the Lagan valley, such as Lisburn and Hillsborough, though the final decay came only in the eighteen-sixties.

The linen industry of the northeast was largely an eighteenth-century growth, though its foundations were laid by Scottish pioneers who settled during the late sixteenth and the seventeenth century, and by various Huguenot refugees who came during the seventeenth century. Ireland already had a reputation for fine spinning by the eighteenth century and the farmers of Ulster, more secure on their holdings than those in the other provinces, launched out into the domestic production of linen. This linen, once spun and woven, was sold to bleachers, at markets in the streets of various towns, or, in the more important Ulster towns, in the linen halls that survive to this day. The Ulster Custom, by which a tenant had security on his holding, is often regarded as the main reason for the success of the linen industry in the northeast, but was only one of several contributory causes which included the presence of a naturally industrious people, partly Scots in origin, the steady increase in the number of towns and market villages founded by landlords eager to encourage industry, and finally the provision of effective marketing and transport facilities by canal, road and later railway, with suitable ports such as Newry, Belfast, Coleraine and Londonderry.

Domestic in origin, linen weaving remained a home craft in some cases throughout the nineteenth century and even into the twentieth century, but domestic spinning did not survive long after 1825, when James Kay of Preston discovered that a thorough soaking in cold water made the fibres sufficiently slippery to be drawn by machinery into a really fine yarn. Already in 1820, the famous York Street mill in Belfast was founded, first for cotton and then for linen. The English triumphed in cotton, but the Irish in linen, fortunately for Ulster, as it provided a sure source of employment for which the people, having so long a tradition of flax working, were eminently suited: the 'middlemen' bought yarn from the spinners in bulk and supplied it to the weaver, then collected the webs, which were sold in the linen markets. As the production increased, handloom weaving declined though the final decay was a gradual process lasting from *c.* 1825 to 1865.

Outside the northeast, flax-spinning in Ireland gradually declined, as it never reached the factory stage except on a small scale at Dublin, Drogheda and Dundalk. Everything combined to favour Ulster, for a regular cargo service from Belfast to Liverpool was established in 1824, and within a few years there was a substantial import of coal from Ayrshire and Cumberland for the new steam-driven mills as well as for domestic consumption. Belfast harbour had been steadily improved from the eighteenth century, but the first major deep-water channel was completed in 1839 and the first iron ship was built in 1844. Economic growth from this time gave a demand for male labour in shipbuilding and female in the mills. Belfast, with its satellite towns, was able to attract labour in abundance from the surrounding countryside, as the marked drift to the towns or overseas had begun in earnest about the middle of the nineteenth century. There was a great development of banking, especially between 1825 and 1865, and many of the bleachers, who had risen to power as buyers of brown linen, became powerful manufacturers and exporters, steadily accumulating more capital, which has often been difficult to acquire for Irish industrial developments. The earlier advance of cotton to the factory stage left, in its decline, a useful reserve of capital, buildings, and industrial tradition. Although many towns in Northern Ireland have a sad legacy of mid-nineteenth century artisan housing, the rise of industry in the northeast caused a population growth in Belfast and district which was unique in Ireland after the Famine.

In 1960 the number employed in linen manufacture was 42,000, but the industry is now greatly diversified by the use of other materials. In 1912, the output of linen and union cloth was 226 million square yards, but in 1961 it was less than 60 million, though the inclusion of cotton with cotton–rayon mixtures (27 million) and rayon (35 million) with other materials (10 million) brought the yardage to 132 million. Northern Ireland has plants built by five of the major fibre-producing firms and other textile manufactures include both tufted and woven carpets, warp knitting and a variety of hosiery products. There are also factories for twine, nets and ropes, and the famous ropeworks in Belfast is said to be the largest in the world. Clothing includes the shirt, collar and pyjama industry, for which Northern Ireland gives one-fifth of the United Kingdom output.

Shipbuilding is now entirely in the hands of the famous Harland and

Wolff firm, dating from 1863, whose yards were built on reclaimed land in the Lagan estuary. Before the 1939–45 war, shipbuilding and engineering normally employed some 23,000 people, but there were times of heavy depression, notably in 1931 and 1932 when almost three-fifths of the workers were unemployed. In 1937, aircraft construction began in Northern Ireland. During the 1939–45 war, as during the 1914–18 war, shipbuilding yards were vigorously employed, but there have been periods of anxiety since, though large liners are still made in the Belfast yards, which provide a fine spectacle for those sailing up the lough on passenger boats from British ports.

Northern Ireland has both industrial strength and economic problems. The unemployment is greater than that in Great Britain as a whole, or in any of the Ministry of Labour's separate administrative regions: since the 1939–45 war, as for several years before it, the government has attempted to attract a wider range of industries, to create new factories through the six counties, and to strengthen existing industries. Assistance of some kind has already been given to two hundred firms, and sixty more have submitted modernization schemes. In fact, Northern Ireland has become industrialized through the enterprise of its citizens: as its major trades cannot be expected to expand to any marked extent, the hope for the future lies in the attraction of new light industries similar to those found on the trading estates of Great Britain: a further hope is that more of the agricultural output will be processed at home rather than sent out as raw materials. And there are signs of at least partial success in both these aspirations.

CHAPTER 10

Communications

The basis of the Irish communication system is the road, and the modern highways are fed by a network of lanes which reach back to remote farmsteads. The main roads exist for the traffic between towns and the minor roads for the day-to-day movement of the farming population, who travel by farm-cart or donkey trap to the shops, fairs and churches. The dispersion of settlement, the small size of many holdings, and the dependence on peat supplies by large numbers of the agricultural population all combine to multiply the number of tracks through the countryside, but the footpath through fields is less frequent in Ireland than in England, though the walker can generally go straight through the fields except at times when precious crops have to be guarded against cattle, and the hedges are reinforced with gorse, hawthorn and similar deterrents. Ireland's road pattern is the accretion of many generations, with a veneer of modernity in its trunk roads, which are of excellent quality in most counties but not all: except in a few areas, mainly around Dublin, and in the most densely peopled parts of Northern Ireland, they carry very little traffic.

River navigations and canals were constructed from the eighteenth century and have not entirely outlived their usefulness though many waterways have been abandoned recently and the fate of others is in the balance. Railways spread through the country from the eighteen-thirties and were regarded as a fine contribution to the country's economic troubles though not as rivals to the canals. Since the 1939–45 war the railways have lost so much money that they have been integrated into government-supported combines, Coras Iompair Eireann, which controls most of the road, railway and canal services in the Republic, and the Ulster Transport Authority, which has similar functions in Northern Ireland. The air routes reaching Ireland centre on three main aerodromes – Aldergrove (Northern Ireland), Collinstown, near Dublin (an airport of fine design) and Rineanna, on the Shannon. There are regular services to London and major British cities including Glasgow, Liverpool, Manchester and Birmingham, and some continental services run direct from Ireland or with a stop at London or Manchester. In summer services are augmented and air travel has made a useful contribution to the tourist trade. Rineanna, on the Fergus estuary close to the Shannon, has trans-Atlantic services. There is also an airport at Cork.

THE IRISH ROADS

Once off the main roads of Ireland it is not difficult to see the relics of a more leisured age than the mid-twentieth century, as most of the traffic is by farm-cart and bicycle. In 1776 Arthur Young commented favourably on most of the secondary roads, though not the main turnpikes:

> For a country so very far behind us as Ireland, to have got suddenly so much the start of us in the article of roads, is a spectacle that cannot fail to strike the English visitor exceedingly. It is a common complaint that the tolls of the turnpikes are so many jobs, and the roads left in a state that disgraces the kingdom.

The turnpikes were not numerous, though they included such highways as that between Dublin and Belfast, along which various old tollgates may still be seen. In Young's day the turnpikes were worn down by coaches with teams of several horses, and Young advanced the argument that one-horse carts were more suited to the country's roads than the large coaches.

In 1760 all compulsory road work was removed by law, and the building of 'presentment' roads was made possible: any person wishing to make such a road presented maps, plans and estimates to the magistrates, who passed them to the Grand Jury, and the road was then built by the petitioner, who was reimbursed out of the County Cess. The roads so built had to be at least 21 ft. wide, of which at least 14 ft. was to be stone and gravel on an earthen foundation. Naturally it was mainly the gentry who planned such roads, sometimes to provide work for their poor relations or to make a profit for themselves. Arthur Young thought well of the presentment system, and notes that it was an effective means of producing roads, however great the possibilities of graft. The roads generally spread like rays from the demesnes, but yet

> in a short time, those rays, pointing from so many centres, met and then the communications were complete . . . . I found it perfectly practicable to travel upon wheels by a map, I will go here; I will go there; I could trace a route as wild as fancy could dictate, and everywhere I found beautiful roads without break or hindrance. . . . What

a figure would a person make in England who should attempt to move in that manner, where the roads . . . are almost in as bad a state of repair as in the time of Philip and Mary.

Not all the evidence is quite so favourable, but at least the presentment roads made a substantial contribution to the communications of the country. In time the Board of Works took over their powers and built such famous highways as the Antrim coast road in the middle of the nineteenth century, which was seen by W. M. Thackeray: mention must also be made of the employment of many men on roads in various relief schemes, especially in the Famine years of the eighteen-forties.

The crux of the road problem in Ireland has always been to provide access to remotely placed farmsteads and to peat bogs. Most of the older roads go over a hill rather than round it, and any smoothly graded trunk road in, for example, the drumlin country, is a recent addition. Many roads follow the outer walls of a demesne, so that travellers are taken along two sides of a triangle. Some roads were built for strategic reasons, notably the famous Military Road in the Wicklow mountains, constructed after the Rebellion of 1798, which runs through the heart of the mountains between a series of derelict barracks: having an altitude of more than 1,200 ft. for fifteen miles, it is the longest stretch of high-level road in Ireland and, recently re-surfaced, provides a fine tourist route. Several factors have contributed to the growth of the present road pattern but the country lane remains the dominant element. In Northern Ireland, for example, there are 13,915 miles of public road, of which 385 miles are trunk roads (including motorways), 972 miles are class 1 roads connecting major towns or having a large volume of tourist traffic, 1,735 miles are class 2 roads connecting less important towns with some holiday traffic, and 2,864 miles are class 3 roads acting as feeders to other roads. In addition there are 7,959 miles of unclassified roads. The motorway from Belfast to Londonderry through Omagh will be completed to Dungannon (38 miles) by the end of 1968: in addition sections of the motorway to Coleraine and Londonderry are finished or under construction.

Long-distance buses now supplement railway travel in Ireland, but the network of services is far less complete than in Britain and few Irish towns have a network of bus services fanning out along the country roads. The nearest approach to this is seen in the northeast: over most of the country the farmers depend on cars, carts and bicycles, though lorry services penetrate much deeper into the countryside than passenger transport. In the nineteenth century there were stagecoaches on the major trunk roads, which, then as now, were focused on Dublin so that Ireland's communications resembled 'a wheel with a hub and spokes but no rim'. Some of the traffic was carried by water, but this was slow, both for passengers and goods. Clonmel, described by Wakefield in 1812 as having 'more internal commerce than any other town in Ireland', was linked to Waterford by long river-boats which carried flour, corn, pigs and firkins of butter to be exported, and brought back coal, sugar and tea. These boats took a day and a night to reach Waterford. It was at Clonmel that Charles Bianconi, an Italian print-seller, began to operate his 'cars' known as the 'bians'. In 1815 he used the jaunting-car, with one horse, to link Clonmel with Cahir, eight miles away, and within ten years he was operating services on more than 1,000 miles of road, at first in the lowlands of Munster, but in time throughout the land. By 1835 the 'bians' ran on more than 2,000 miles of road, and by 1845 on more than 3,000: meanwhile the seating capacity of the cars increased from six to ten, and later to twenty, as the original two-wheeled vehicle was transformed into the 'long car' with four wheels. The passengers sat back-to-back facing the hedges, unprotected from the weather or from the beggars roaming the countryside, but contemporary accounts suggest that the journeys were enjoyable, at least to those possessing a sense of humour. When Bianconi began his services in 1815, artillery horses left over from the Napoleonic wars were available, but later two horses were needed to do the same amount of work. Throughout the period of the 'bians' smaller cars were kept for less important routes. There were other coach proprietors also: by the eighteen-thirties Purcell's services ran from Dublin to 'major' provincial towns, including Belfast, Londonderry, Dungannon, Cork, Clonmel, Kilkenny, Waterford and Wexford: this firm also ran services from Waterford to Cork and Limerick. Gosson's coaches were also based on Dublin, and travelled to a number of towns in the Central Lowland with terminal points at Cavan, Drogheda, Kells, Navan, Killashandra, Granard and Enniskillen.

In the middle of the nineteenth century, Bianconi's cars extended to Sligo, Clifden, Tralee, Wexford and Longford (see Fig. 43), filled gaps in the stagecoach routes, and carried mails beyond the places reached by the Post Office. But the railway era was beginning, though with several false starts, and Bianconi recognized that 'it was as foolish to try to prevent the establishment of railways as to try to stem the Liffey'. He therefore bought railway shares and became a Director of the then Waterford and Limerick railway, and also adapted his 'bian' services to feed the railways so that as late as 1865 he still had cars running on some

2,500 miles of road, over three-quarters of their extent in 1845. The 'bians' survived longest in the extreme west of Ulster and in Connacht, where they remained until the dawn of the petrol age at the end of the century. In all, the work of Bianconi was of inestimable benefit to the country, for his cars helped to remove the isolation of many rural communities. Railways in Ireland are no longer profitable, and a less expensive form of transport, such as the 'bians' a century ago and the bus or lorry now, is needed to fill up the gaps in the transport system.

Many interesting forms of transport are still found in Ireland, but it is impossible to study each in detail here. At the poorest end of the social scale people carry seaweed, sand or peat on their backs in a basket-work 'creel' not unlike that used by the Swiss peasantry on their journeys up and down the mountains. A similar contrivance is the pannier baskets fitted on either side of a donkey, who can negotiate poor tracks and is entirely composed on the seashore. Both the creels and the panniers are common in the extreme west of Ireland, especially among the poor communities who depend for their manure supplies largely on materials collected from the seashore. The slide-car, which survives in the remoter parts of Ireland, has no wheels but two shafts connected by cross-bars on which an open box made of wood or basket-work is supported, dragged along by a horse or donkey: it is so ruinous on the roads that it has been banned by Act of Parliament, though it is of particular value on the bogs and heathlands. The farm-carts are of various sizes and colours, according to local custom, but a general feature is detachable sides, which are made of slats not unlike a gate. These can be added when peat, calves, pigs or sheep are carried, and removed at other times. When the railways began to carry peat during the war of 1939–45, a wagon with side-pieces like those of the farm-carts was designed for the purpose. Some of the farm-carts have the surprising addition of pneumatic tyres taken from a worn-out car. Traps are also a frequent sight on the Irish roads, especially on Sunday mornings; then, too, several jaunting-cars will be seen, though they are not as frequent in the cities as visitors suppose, as they appear with the tourists and disappear as abruptly. Motor-cars are becoming more numerous, but many of the poorer farmers now have no means of travel except bicycles as horses need not be kept for ploughing if tractors can be hired.

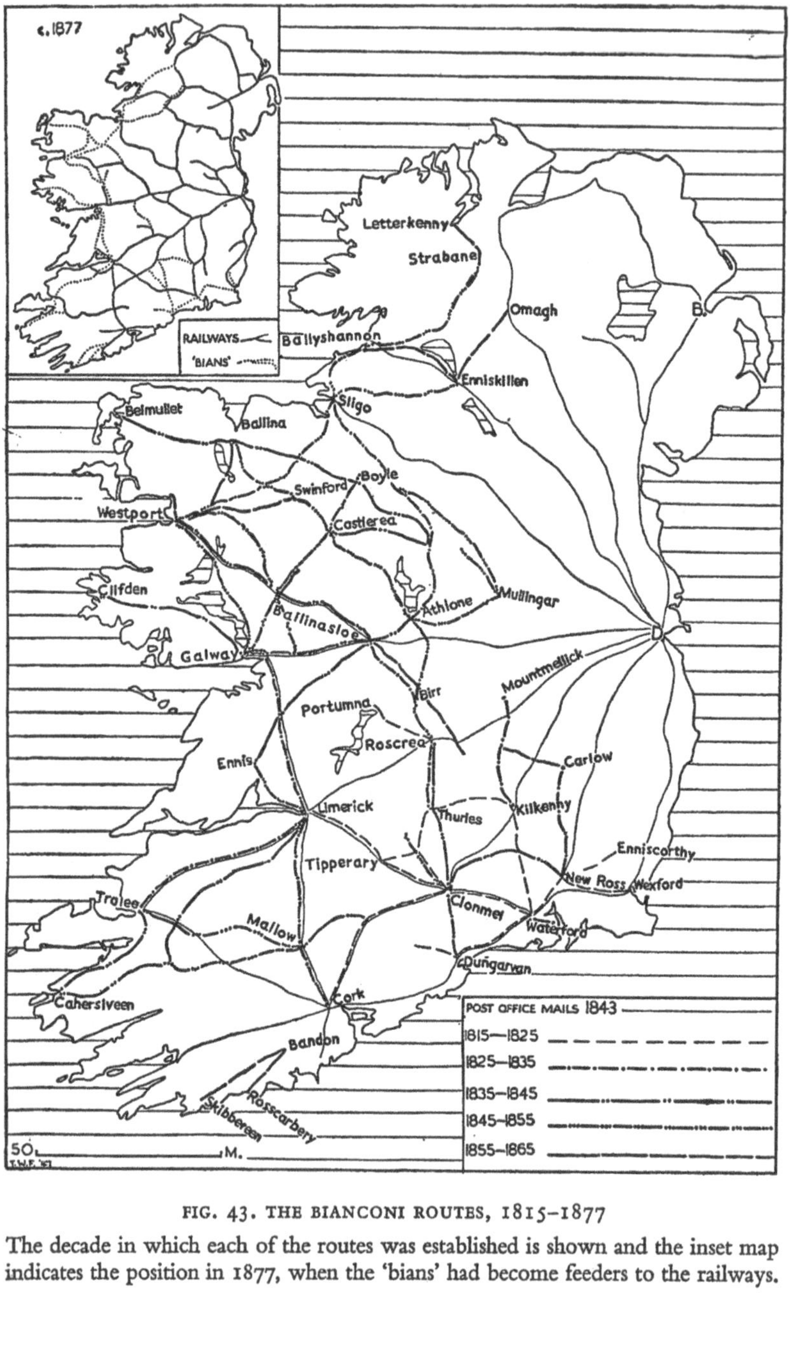

FIG. 43. THE BIANCONI ROUTES, 1815–1877

The decade in which each of the routes was established is shown and the inset map indicates the position in 1877, when the 'bians' had become feeders to the railways.

CANALS

Canals in Ireland seemed to offer economic salvation in the middle of the eighteenth century, as it was hoped that the barges would bring in the raw materials needed for industry and commerce and take away manufactured goods to Dublin and the export markets, all at very low cost. The country's Central Lowland, together with its ramifying lowland corridors between the hills, offered gently graded courses for the new waterways, and in addition many of the major rivers, such as the Shannon, the Boyne, the Suir, the Nore, the Barrow, the Bann and the Foyle, had long navigable reaches which could be incorporated in a system of water transport. The optimistic views expressed around 1750 were hardly borne out by experience for the canals took much longer to build, proved more costly, and attracted less traffic than was anticipated. The Grand Canal from Dublin to the Shannon, with a branch to the Barrow, was begun in 1756, reached Athy on the Barrow in 1791, and the Shannon in 1804. Until recently, this canal still had a considerable barge traffic. Not so the Royal Canal, which runs from the north side of Dublin to Mullingar and Termonbarry and Longford: built between 1789 and 1817, this canal was later acquired by the Midland and Great Western Railway Company, whose line runs beside it to Mullingar. Both these canals had passenger services from 1780 to 1852, and 'express' boats, drawn by teams of galloping horses, reached a speed of 10 m.p.h. Along the route a number of handsome hotels were built, for example at Portobello Bridge (Dublin), Sallins, Tullamore and elsewhere, and these, with the dignified warehouses beside the various quays, still show that the prosperous days of water transport coincided with an age of good building. The Grand Canal was obviously far more likely to succeed than the Royal, as it connects Dublin with both the Shannon and the Barrow and was, in the pre-railway age, an important link between various provincial towns and ports including Waterford and Limerick; but the Royal Canal runs through thinly populated country and touches few villages and towns, except Mullingar and Longford.

Waterways in Northern Ireland include the Bann and Lagan navigations, the Strabane canal of five miles from the Foyle to the town, a ship canal from Newry to Carlingford lough, the Coalisland canal, and a link from lough Neagh to the Erne. The Newry canal is of special interest and of early date, as it was completed above the town in 1741 and below it in 1761, and the town has some substantial mansions and warehouses dating from the eighteenth century, when the town was an important commercial centre: by 1834 there were exports to Liverpool and Glasgow of linen cloth, butter, wheat, barley, oats, flour, oatmeal, flax, eggs, cattle, pigs and horses, and imports of tea, sugar, iron, salt and general merchandise. All the waterways were constructed before the railway era, but as long ago as 1837 the cost of transport on the Lagan canal, due to high tolls and its winding course, was reported to be greater than that of road transport. Only the lower Bann navigation, from Coleraine to lough Neagh, is open – and traffic is declining. The Strabane canal is disused; the Newry Inland canal was closed in 1949 and 1956; in 1954 the Coalisland canal, the Upper Bann and part of the Lagan were closed and in 1958 the rest of the Lagan from Stranmillis to Lisburn was closed.

Various other small canals were made between the middle of the eighteenth century and the Famine period of a century later, when the railways were hailed as the source of economic and social salvation in Ireland. Galway was linked to lough Corrib by a channel three-quarters of a mile long, and a second 'canal' connected lough Corrib to lough Mask, but only in theory, as it was constructed through porous limestone and holds no water at all. Various other schemes have been put forward at different times for Irish rivers, though few have been effectively canalized and in general the financial story is a sorry one. The Royal Canal was inherited by Coras Iompair Eireann along with the railway beside it. The Grand Canal, also owned by C.I.E., was profitable up to 1949: since 1954 there has been no barge traffic at all on the river Barrow, south of Athy and from December 1959 C.I.E. ceased to operate barge services on the main canal. On the Shannon, barge traffic ceased in 1960 but since that time the use of the river, and also of the Grand Canal, for pleasure craft has greatly increased. C.I.E. also runs services by passenger launches as a tourist attraction, and the Tourist Board is spending £140,000 on the improvement of waterways.

RAILWAYS

In Ireland, as in England, the forerunners of the railways were in the timber trackway laid down to transport coal and other minerals from the mines and quarries to the roads and ships: these were found at collieries in Armagh and Tyrone, and at Dun Laoghaire, where the granite blocks used to construct the piers were brought from Dalkey hill along a route still known as the 'metals'. Dun Laoghaire, then Kingstown, became the packet-station used by ships from Holyhead in 1826, and was connected to Dublin by rail in 1834. This, the first railway in Ireland, was rightly hailed as an epoch-making addition to the neighbourhood, though some of the inhabitants regarded it as an expression of the increasing vulgarity of the times: not so the *Dublin Penny Journal*,* which welcomed it in the following words:

> Hurried by the invisible but stupendous agency of steam, the astonished traveller will now glide, like Asmodeus, over the summits of the houses and streets of a great city – presently be transported through green fields and tufts of trees – then skim across the surface of the sea, and taking shelter under the cliffs, coast along the marine villas and through rocky excavations, until he finds himself in the centre of a vast port which unites in pleasing confusion the bustle of a commercial town with the amusements of a fashionable watering-place.

Meanwhile, Belfast commercial interests had sensed the possibilities inherent in railways and in 1839 the Ulster Railway Company opened the line from Belfast to Lisburn, but not without opposition to the running of Sunday trains in which, it was said, 'souls are sent to the devil at the rate of sixpence apiece' (first class, one shilling), and 'every shout of the railway steam whistle is answered by a shout in hell'. Capital for extensions appeared reluctantly and the railway promotors were obliged to seek government aid, with which the line was extended to Portadown by 1842. The line from Dublin northwards was begun in 1840 and reached Drogheda by 1844, having skirted Dublin bay on an embankment and crossed wide estuaries at Malahide and Donabate: not until 1852 was there a complete railway link between Dublin and Belfast, and even then passengers had to walk through the town of Drogheda as the viaduct bridge, almost 100 ft. above the river level, was not built till 1855.

Two major controversies had marked the earlier years of the railways, the course of the Belfast–Dublin line and the gauge to be used. Brunel favoured the 7 ft. gauge used by the Great Western Railway in England, and Stephenson the 4 ft. 8½ in. standard which eventually triumphed on the British Railways, but Ireland adopted its own gauge of 5 ft. 3 in. in 1846. The Dublin–Belfast coastal route was fixed by a parliamentary commission in 1836, but there were strong advocates of an 'inland' route through Navan and Armagh. In the discussions it was said that the population of the seaboard was 'almost industrial in its density' by some and 'merely that of a few villages and fishermen's huts' by others. Of the inland line, some said that there were 'a large number of communities,

* Quoted in Joyce, W. St J., *The Neighbourhood of Dublin*, Dublin, 1912, 1939, p. 45.

small individually, but in total substantial', while others said that the route must fail as there was 'a view of beggary on every side', with 'thatched cabins and nothing but misery'. The inland route was never completed.

The year 1852 saw the completion of the Belfast–Dublin line, except at Drogheda, and the lines from Dublin to Cork and Galway were already completed. The Midland and Great Western Railway had bought the Royal Canal, and constructed their line beside it as far as Mullingar, where the line leaves the canal on its journey to Athlone, an easy crossing-place of the Shannon with 'dry ground' on both sides. Having crossed the Shannon at Athlone and the Suck at Ballinasloe, the line runs across undulating country to Galway. The line to Cork is one of the most attractive in the country, as it passes through the lowland corridors between various Armorican ranges and finally crosses a ridge between Mallow and Cork through a narrow pass at 460 ft. O.D.: the gradual rise from Dublin to the 200 ft. level so widely developed in the Central Lowland emphasizes the prevalence of this peneplane level. Two other lines were also completed by 1852, those from Limerick to Clonmel, and from Kildare to Kilkenny: Clonmel was a town of some note in the estimation of a century ago and Kilkenny also figured as one of the major provincial towns.

Railway building continued actively in the years following 1852, and by 1863 many of the larger country towns were connected with the trunk railways to Dublin and Belfast. Waterford was linked with Limerick through the Suir valley, and also with Kilkenny; the important branch line from Mallow to Killarney and Tralee was also built, partly in the expectation of a considerable tourist traffic to Co. Kerry. The Galway line was joined by two important affluents, one to Castlebar, Co. Mayo, and the other to Sligo from Mullingar. At Inny Junction, a few miles north of Mullingar, one of the most desolate railway stations in the world, a line diverged to Cavan and Clones. A line from Athlone to Maryborough passes through Clara and Tullamore. In Co. Meath the Drogheda to Oldcastle line was completed in 1863. It took from 1845 to 1859 to build the line from Dundalk to Enniskillen, but only from 1847 to 1854 to reach Enniskillen from Londonderry. The two main lines from Londonderry to Belfast were completed by 1863; the first to Portadown, Omagh and the Foyle valley, and the second through Antrim and Ballymena to Coleraine and around the edge of the basalt scarps to the Foyle. Each line serves a number of important provincial towns, and this is the only example of competing routes in Ireland.

After 1863, the main development on the railways was the extension to a few additional towns, such as Bundoran, Ballina and Westport on the Atlantic seaboard, the provision of a few more branch lines, and, in effect, the completion of the railway network already foreshadowed by 1863. Not all the branch lines were successful and two survived only for ten years, the line from Portumna to Birr, 1868–78, and that from Armagh to Castleblayney, 1914–24. The railways were comparatively prosperous immediately before the 1914–18 war, and the system reached its zenith by 1920, since when a variety of circumstances have led to its decline. The political disturbances of 1921 and 1922 interfered with the running of the trains, and in places tracks were cut, but a more permanent menace to the railways came with the motor-buses and lorries of the middle twenties, though the railways were authorized to run bus services in the Republic in 1927 and in Northern Ireland in 1935. During the war of 1939–45 the railways of Ireland prospered, even though passenger services on the Great Southern Railway were cut to one daily and even, on some lines, to four, three or even only two trains a week. But road transport was so curtailed that people were obliged to travel by train, and there was heavy military and industrial traffic in Northern Ireland, with peat transport in the Republic. After the war, however, conditions became financially so serious for the railways that government help was sought to keep the railways running.

In 1944, the Dublin government set up a new company, Coras Iompair Eireann, to acquire the Great Southern Railway and the Dublin United Transport Company: in 1950 the Grand Canal Company was added. Later, in 1948, the Ulster Transport Authority was constituted to operate the railways from York Street station, Belfast (which had belonged to the London, Midland and Scottish Railway under the Northern Counties Committee), the Belfast and County Down Railway, and the services of the Northern Ireland Road Transport Board. It was originally intended that the Belfast transport services and the part of the Great Northern Railway within the six counties should be included also, but Belfast retained its own bus services. The Great Northern Railway endeavoured to remain independent but impending bankruptcy compelled its joint nationalization by both governments in 1953. At the end of the war, this railway had 332 miles of its total length of 562 miles in Northern Ireland, and the lines crossed the Border at fifteen separate points, with seven customs stations. Finally its services were taken over by Coras Iompair Eireann and the Ulster Transport Authority in 1958, so that virtually all the rail and road with most of the lorry services are now in the hands o these bodies. As in other countries many railway lines have been closed and bus services substituted.

LIGHT RAILWAYS

Light railways appeared to offer great economic advantages to Ireland in the latter years of the nineteenth century, and at the maximum development there were eighteen such lines extending over 562 miles: the last three to survive were the West Clare, the Cavan-Leitrim railway and part of the County Donegal Joint Committee line. Mixed motives of desirability and economic advantages actuated the promoters of light railways: it was declared that such lines would remove the isolation of many communities in the west, and the government were prepared to guarantee as much as half the cost, and finally all of it, in the various Light Railways Acts passed between 1860 and 1896. The gauge used was 3 ft. and the rolling stock was in many cases of strange design, though adequate for its purpose. On some western railways, occasional gales were a severe menace. In Donegal on January 30, 1925, a train was blown off the Owenacurra viaduct and most of the passengers were either killed or injured. Subsequently, an anemometer was installed at Dunfanaghy Road station and train services were suspended when high gales were recorded. Similar precautions were taken on the West Clare railway, which has an anemometer at Quilty (see also pp. 50–1).

Seven of the light railways were in Northern Ireland, which had a heavier network of railways than the Republic. Two of these, from Ballymena to Larne and from Ballymena to Retreat, were originally built in the eighteen-seventies to carry iron ore from mines in Co. Antrim, and the Ballymena–Larne line was also used by passengers from the Stranraer boat. The Retreat line never had much traffic, and was closed in 1930, and the Larne line was closed for passengers in 1942. The Ballymoney–Ballycastle line was closed in August 1950, and in the previous September services closed on the Giants' Causeway tramway: hardly a railway in the technical sense, this line was opened in 1883 and was the first electric tramway in the world. It ran from Portrush through Bushmills to – disappointingly – a point one mile from the Causeway, and its power was acquired from a waterfall at Bushmills, supplemented in summer by steam power. Another short length of tramway, built in 1885 and operated until 1948, connected the town of Newry with Bessbrook, when the mills employed 3,000 people: the power was drawn from a fall 29½ ft. high on the Camlough river.

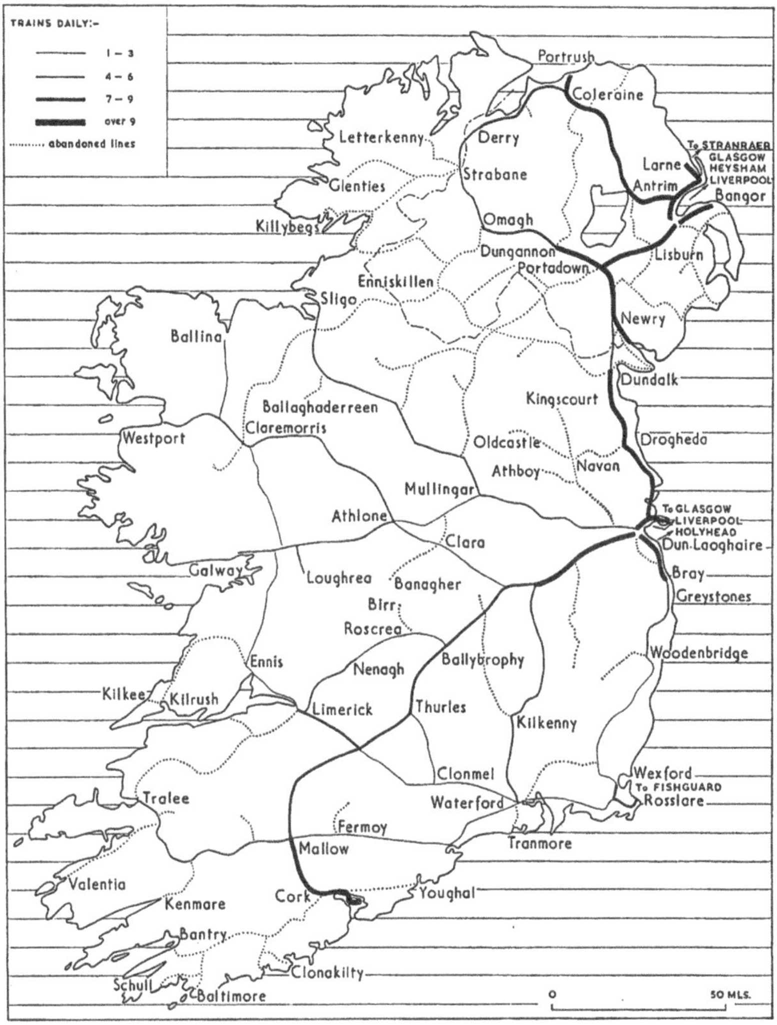

FIG. 44. RAILWAY SERVICES, 1963

This map's main interest is that it shows the last phase of reasonably adequate train services. Since 1963 passenger services have ceased on the line to Newry and on the Portadown–Omagh–Derry line; and also from Waterford to Mallow and from Thurles to Clonmel. A few lines are still open for goods traffic, for example, Limerick to Tralee, Claremorris to Sligo, Drogheda to Navan and Kingscourt, Waterford to New Ross. There are also a few passenger trains from Cork to Youghal.

The Clogher Valley railway, opened in 1887 and closed in 1941, ran through agricultural country with small market centres such as Aughnacloy and Fivemiletown, each of which has a population of less than a thousand, in Co. Tyrone and Co. Fermanagh. It was connected at either end with the Portadown–Enniskillen railway and followed the Belfast–Enniskillen road for most of its course, though in places it ran through the adjoining fields. Worked by steam locomotives, it resembled many continental tramways: its progress through Fivemiletown was celebrated in the verses sung at fairs and markets, of which one may be quoted:

This new tramway three times a day she runs both up and down
The most mistakes I'm sure they make is at Fivemiletown.
She clears the streets of all she meets, it is the whole vexation,
Going up and down, through the town, to and from the station.

By 1930 the financial position of the railway had become precarious and the deficits were increasing annually owing it was said to the unfair competition of the Northern Ireland Transport Board, which ran three buses daily from Belfast to Enniskillen and also conveyed livestock in lorries. About 1935, however, the cattle and pig trade on the railway increased considerably around Augher, which is a short distance from the border, here lying on the Slieve Beagh, a desolate line of uplands, but more efficient patrolling of the frontier led to a decrease in the number of illegally imported animals. The line was closed in 1941.

The light railways of Donegal and the Foyle valley date back to 1863 and were gradually extended so that by 1905 they provided at least a skeleton network for transport in Co. Donegal. There were two separate railways, the Londonderry and Lough Swilly in the north, and the County Donegal Joint Committee in the south. The L.L.S.R.'s first line ran from Londonderry to Farland Point, where ferry boats crossed lough Swilly, in 1863, and reached the town of Buncrana, also on lough Swilly, in the following year. In 1883 the line to Letterkenny was opened; in 1901 the extension to Carndonagh, a small market centre in the north of the Inishowen peninsula, was added, and in 1903, that from Letterkenny to Burtonport. In all, the L.L.S.R. had 100 miles of track and was regarded as an asset to the countryside, but the financial returns were inadequate to maintain the line on a commercial basis, and gradually the company developed bus services of considerable efficiency. The line from Buncrana to Carndonagh was closed in 1935 and passenger services on the Buncrana to Londonderry line ceased at the same time though freight trains ran until 1941. In this year, owing to the shortage of fuel for buses, the train service was restored to Buncrana, and continued until 1953. The line from Londonderry to Burtonport was closed in 1940, but the last railway of this company, from Londonderry to Letterkenny, was closed only in 1953.

In South Donegal the light railways survived into the 1950s. The lines of the County Donegal Railways Joint Committee were owned partly by the British Transport Commission and partly by the successors of the Great Northern Railway – that is partly by the Minister for Industry and Commerce in Dublin and partly by the Ulster Transport Authority! The C.D.R.J.S. began its lines at Strabane, and reached Stranorlar in 1883, Donegal in 1889, Killybegs in 1892 and Glenties in 1894. Later extensions included a line to Londonderry in 1900 parallel with the main line from Dublin but on the other side of the river, the Donegal–Ballyshannon branch in 1905 and the Letterkenny line in 1909. The railway was run with considerable enterprise and – what is by no means universal – courtesy and consideration for the passengers. Railcars were run by one man who issued the tickets, dealt with parcels and stopped for passengers to alight or enter virtually anywhere, though in 1944 this privilege had become so thoroughly used that some limitation of stops had to be imposed. The railway bought redundant stock from extinct lines, such as the Clogher Valley, the short Castlederg–Victoria Bridge line (in the Foyle valley, closed in 1933), and the Blessington–Dublin railway, which existed from 1882 to 1932. It owned motor vehicles, run with difficulty on poor roads, but the increasing costs of rail services and the road improvements combined to make the railway working unremunerative, so the lines were gradually closed. The line to Glenties was closed for passengers and goods in 1947, though occasional trains for peat and livestock ran for a few years afterwards and the final closure was dated March 1952. The Strabane–Londonderry section was closed at the end of 1954, and all the remaining sections at the end of 1959.

The Cavan and Leitrim railway served the upper part of the Erne and Shannon basins, one of the less-visited parts of Ireland, and was used chiefly for the transport of coal from the Arigna mines near lough Allen. In 1887 the line from Belturbet to Dromod was opened and in the following year the section from Ballinamore to Arigna was completed, but the line never carried more than three trains a day, and after 1921 the traffic declined sharply. But the Arigna extension mineral line, added in 1920, saved the railway from extinction and even in the 1950s as much as 1,000 tons of coal a week were carried, most of it used at the Arigna power station though the line was closed in 1959. As one of the last survivors of the light railways, this line had rolling stock from several extinct lines.

The West Clare railway has almost all the attributes of an Irish light railway: it is celebrated by a song 'Are ye right there, Michael?' describing the leisurely progress and frequent crises of the trains; it has an anemometer, a speed limit of 25 m.p.h., was never remunerative at any time, and lived in perpetual danger of extinction. From 1952, however, diesel railcars were used, at first for passenger and later for goods services. The passenger trains in each direction ran more speedily than ever before but the experiment was short-lived, for all services ceased in 1961. Perhaps this railway has gathered more lore than any other. In its early days, animals strayed on the line and were occasionally run down by a train, so their infuriated owners retaliated by placing stones on the rails or even hurling them through the carriage windows. The first section of the line, from Ennis to Milltown Malbay, was opened in 1887 and the remainder, to the towns of Kilrush and Kilkee, was completed by 1892. Until 1914, a steamer ran from Limerick to Kilrush, but this service was not resumed after the 1914–18 war.

Two light railways in Co. Kerry are of special interest.

(*a*) The Listowel–Ballybunion 'monorail' line was opened in 1888 and ran on a single rail, 3 ft. 6 in. off the ground, on either side of which a carriage was suspended. There was also a double engine to draw the train, the wheels ran on the central rail, the noise was deafening, and as the weight on either side had to be equalized the actual arrangement of the passengers and goods was complicated. The hopes of transforming Ballybunion from a small seaside resort into a busy port never materialized, and the traffic receipts in winter never paid the expenses, so the line was closed in 1924.

(*b*) The Tralee–Dingle railway covered the thirty-two miles between the two towns and had a branch line to Castlegregory, a village on the north side of the Dingle peninsula. The main difficulty was the steep gradients, in one case 1 in 35–40–31 on the climb through the Anascaul pass, and as a result of accidents the speed on this section of the line was cut to a minimum. The railway was opened in 1891 and closed for passenger traffic in 1939, though one freight train ran daily between Dingle and Tralee until 1953.

Cork city had its light railways, built to attract the outer suburban and tourist traffic but now dismantled. The earlier in date was the Cork–

Passage railway, of which the first section was built in 1850: this line was worked in connexion with steam ferry services on Cork harbour. In 1902–04 the line was extended to Crosshaven, a small holiday resort near the outer entrance to Cork harbour, but the traffic declined even before the 1914–18 war and, after a number of vicissitudes, the whole system was closed in 1932. The Cork–Muskerry railway, on the west side of the city, was constructed between 1887 and 1893, and penetrated the pleasant hilly country north of the Lee valley which, it was hoped, would become residential in character, and it was also expected that there would be tourist traffic to Blarney and other villages. The city did not grow as fast as the builders of the railway hoped and so on some parts of the line, such as that to Donaghmore, there was never much traffic; it was closed in 1934. Mention must also be made of the Schull–Skibbereen railway, in the extreme southwest of Co. Cork, which was opened in 1886, but never profitable. In its heyday it carried two 'mixed' (i.e. for both passengers and goods) trains a day, with an extra service on market days: the line was closed in 1947. Though not a light railway, the Galway–Clifden line was constructed under the Acts named earlier, and was operated from 1895 to 1935.

The short life of most of the light railways is only too clearly apparent: like the canals, their day appears to be finished, though in one case, south Donegal, enterprise and adaptation to changing circumstances at least prolonged the period of usefulness of a light railway. To some extent the story of light railways is one of broken hopes, in which schemes such as the Shannon navigation by steamers, or possible new ports, figure prominently: hardly less significant is the failure to begin many industries that, it was said, would arise once a railway was built. The position of these light railways was financially precarious even before the buses took the roads, though they were an essential public service. Once road transport could replace them, their work was done. Even so, their extinction has caused regret and even disturbance, not least in north Donegal in 1940, where people attempted to prevent the removal of the rails for scrap. The closing of branch lines on the broad-gauge system generally raised a comparable storm.

CURRENT TRENDS IN IRISH TRANSPORT

The derelict railway track is a familiar feature of the Irish rural scene and in effect only the main trunk lines now survive. In the extreme west, railways reach the coast only at Tralee, Galway, Westport, Ballina and

Sligo for the whole of Co. Donegal, a large area of west Mayo and west Galway. Co. Clare west of Ennis, the peninsulas of Kerry and West Cork and all the coast to the southwest of Cork city now have no railways. Some lines have been closed for passenger traffic and left open for freight a few years longer, but under such conditions maintenance of the track proves too costly for the line to survive. The 1957 Transport Commission in the Republic produced a report that seemed as drastic as the Beeching recommendations in Great Britain: there would be lines radiating from Dublin to Wexford and Rosslare, to Waterford, Cork and Cobh, Tralee (through Killarney), crossed by a line from Rosslare harbour to Limerick. To the west, there would be lines to Galway, Westport, Ballina and Sligo, and the number of stations would be thirty-three, excluding those on suburban lines in Dublin and Cork. These stations would be approximately twenty-five miles apart: between Dublin and Cork, for example, there would be stations at Kildare, Maryborough, Thurles, Limerick Junction (where the Rosslare–Limerick line crosses the main line) and Mallow. The line from Limerick to Sligo, it was suggested, might survive: in fact there is still one train a day from Limerick to Ennis, Tuam and Claremorris but the line from Claremorris to Sligo is no longer used for passenger services. In short, the 1957 report suggested the closing of half the mileage of the railways and of two-thirds of the stations. In fact many lines have been closed though some small rural stations still survive.

Much of the old Great Northern railway is now extinct. In 1957 all the lines from Dundalk to Enniskillen, with the connexion from Enniskillen to Omagh, were closed with the line from Portadown to Armagh and Tynan. From Enniskillen the independent Sligo, Leitrim and Northern Counties Railways was closed in the same year: the result has been to leave a large area of Ireland far from a railway as the distance between the lines from Dublin to Sligo and from Dublin to Dundalk and Portadown is forty to fifty miles. Some interesting minor lines have been closed, notably the light railways (*v.s.*) now all extinct. The Dundalk, Newry and Greenore railway, which ran on each side of the Carlingford peninsula, maintained a somewhat Edwardian air through its long period of decline. The section from Dundalk to Greenore was completed by 1873, and the Newry section by 1876. From 1874 to 1926 passenger and cargo services were run to Holyhead, and a large hotel at Greenore ministered to travellers. Cargo services were continued to the end of 1951 but when they were abandoned the fate of the railway seemed certain, and it was closed in 1957. Within Northern Ireland, the only railways still carrying passengers are the main line from Belfast to Dublin, the lines from Belfast to Bangor, Larne and Londonderry, with the branch from Coleraine to Portrush, which has services in summer only. The line to Bangor is the last remnant of a once-extensive system serving Co. Down, from which most services were withdrawn in 1950 though trains ran to Newcastle until 1957. The lines from Goraghwood to Newry and Warrenpoint, and from Portadown to Londonderry, were closed in 1965.

The two national combines have done much to integrate transport services, and on some routes an inclusive fare covers both rail and bus travel. Some notable attempts have been made to provide quick rail services, such as those from Dublin to Cork in three hours and the 'enterprise' expresses from Belfast to Dublin in slightly over two hours. Although the loss of a railway link is regretted everywhere, and fought with vigour in many places, the bus and lorry services have been co-ordinated with the rail services, and in many places the buses run from the railway stations and use their facilities. But one main cause of the decline in public transport has been the increase in the ownership of private cars, which in Northern Ireland (1966) totalled 234,000, one to every six persons, and in the Republic over 281,000, one to every ten people. These proportions are lower than those in many other European countries, but the ownership of cars is increasing rapidly and the roads have been greatly improved during the post-1945 period. Except in Dublin and Belfast, virtually all the roads seem empty to British visitors and with some justice tourist agencies can claim that in Ireland motoring is still a pleasure.

From this chapter, readers may discover that Irish transport is hardly a favourable investment, for so much of the story is of broken hopes and great but unrewarded expectations. The railways have been largely extinguished by the bus, the private car, the lorry and the van, but a still more fundamental point has been the inadequacy of traffic on all except a few lines such as those from Dublin to Belfast, Belfast to Londonderry, Dublin to Cork. But the rapid services from Dublin to Cork and Belfast have proved competitive with other forms of travel – even, in the case of Belfast, with air travel. As Cork now has an airport, it may be possible to travel there from Dublin more rapidly than by train, allowing for the not inconsiderable time consumed in reaching the city centres to and from the airport. On the tourist side, both C.I.E. and U.T.A. have provided excellent day and longer tours, mainly by bus, though from Dublin there are train day trips to Killarney and Connemara which are a source of great satisfaction to American and other visitors. Recently the Shannon has been re-opened as a tourist attraction with steamers able to take large numbers on excursions: some private companies are providing launches for hire on the Shannon and also on the Grand Canal, due to the increasing popularity of cruising holidays. Outside the two combines, the Londonderry and Lough Swilly Railway, and the County Donegal Joint Committee still provide bus and lorry services for Co. Donegal. It is perhaps strange that the former still keeps the word 'railway' in its title for it is now many years since its last line, from Londonderry to Letterkenny, was closed. But at Letterkenny the buses pull up in the town square, and then at the old station on its edge, where the buildings are used as offices and stores and the scene is a mixture of decaying railway solidity and modern motorized efficiency.

A survey made in 1962 by D. J. Reynolds shows that in the Republic the road system is less used by vehicles than that of any other comparable country, that is any having a good road system, in Europe, even though the number of vehicles registered doubled from 1951 to 1961. Not unexpectedly, vehicles per head of population, including tractors, motorcycles, commercial conveyances, as well as private cars and taxis, are almost twice as numerous in the richer counties of the east and south compared with those in the west. The use of private cars, taxis and hired cars is increasing rapidly, but there is only a slight increase in the use of buses, even though many bus services have replaced railways. On the railways fewer passengers are making longer journeys than formerly, largely due to the closing of stations and branch lines and the development of express services. But in 1960 it was estimated that only 7% of the state's inland passenger transport, and only 20% of the freight, went by rail.

PART II

Regional Geography

CHAPTER II

The Regional Divisions of Ireland

Ireland is itself a natural entity, differing in its social geography from even its nearest neighbours, in spite of its long-continued contacts with them through invasions, trade and population movements. The clearest claim to uniqueness that Ireland possesses is its population history for in a century of vast world population increase, the number of its people at home has fallen almost to half. At present the changes are slight in total numbers, but emigration continues steadily and the heaviest outward movement is from the rural areas of the Republic, but all through Ireland the farm population declines. As the countrysides lose their people, more land is available for those who remain, and the farmed landscape is being altered by the union of holdings, a process which is likely to continue for a long time. In the chapters on regional geography an attempt is made to assess the regional divisions as they now appear with the main stress on the conditions of today, though with some veneration for a past which lives on into the present.

There are wide contrasts of scene, and of types of life in Ireland: quite simply, the journey from east to west is a study in increasing poverty of resources, for example from the rich cattle-pasturing lands of Meath to the poorer and smaller farms of the country across the Shannon to the western seaboard of Connacht, where the dwarf holdings still present an economic and social problem by no means solved yet. Even in a single county such as Donegal there is a visible contrast from east to west, with the *soignée* farms of the Lagan at one end and the few-acre holdings of the almost-barren Rosses at the other. Few if any geographers would maintain that the life of a community can be totally explained by the area occupied, though it is clear that life in the Rosses must differ from life in a fertile and well-farmed lowland somewhere in the east of Ireland. There are differences of soil, of climate and weather, of accessibility and transport costs, to name only a few. Much has been done to raise the standard of living in the poorer districts of Ireland, with some success, but all the social and economic enterprise conceivable could not remove a contrast that is inherent and inevitable. Even so, may one repeat emphatically that the changes have been vast since the Famine and that poor as many parts of

Ireland may now seem, the changes have been good. As a preparation for the regional treatment, a summary is given of the main features that emerged in Part I, with a brief indication of the quality of each region discussed in Part II.

THE BASIS OF REGIONALIZATION

Structurally Ireland is the western continuation of Britain and possesses clearly marked Caledonian and Armorican fold mountains, each with their distinctive trends. Caledonian mountains and hills, aligned from northeast to southwest, are a striking feature of Donegal and neighbouring parts of Tyrone, of West Connacht, of the 'Newry axis' country and of the Leinster chain. The Armorican east–west lines appear in the long ranges of Munster, which reach the coast of Kerry in mountainous peninsulas separated by rias, the drowned ends of deep valleys. Central Ireland, between the Caledonian north, the Armorican south and the Caledonian southeast, has a number of ranges consisting of Devonian and Carboniferous beds folded over a Caledonian floor. Armorican in structure, they show the Caledonian trend; many of them are insignificant as features of the landscape and fail to break the generally undulating lowland surface, but others, such as the Slieve Bloom, stand up as ranges comparable with those of Munster.

More than half Ireland is covered with Carboniferous beds, varied in composition but including large areas of limestones. The older Carboniferous beds generally form lowlands, particularly the great Central Lowland but also the many valleys and corridors that provide easy passageways between the Armorican ranges. Some of the Carboniferous uplands, which may in general be termed residual plateaus, bear some resemblance to the English Pennines. Formed of a wide variety of rocks, they rise in some cases to 2,000 ft., and include most of the country's meagre coal resources. Conspicuous among them are the high plateau-like hills of Sligo and Leitrim, the lower, desolate uplands that extend northward from Killarney to the Shannon estuary and on into west Clare, and the Castlecomer coalfield. But over most of Ireland the rockfloor of lower Carboniferous strata, chiefly limestone, though folded into numerous anticlines and synclines, exists as a smoothly planed surface beneath a variety of drifts and glacifluvial deposits.

These three morphological elements, the Caledonian and the Armorican mountains, the Carboniferous lowlands and residual plateaus, have been etched out by long ages of denudation. They present a fundamental basis of regionalization as they are definite features of the landscape. In addition, ice has been a moulding influence: in the west, a wide belt extending north and south from Galway bay is practically driftless and possesses limestone surfaces of a karstic character, but in the east the drifts reach a thickness of as much as 200 ft. Swarms of drumlins are developed across the whole lowland from Dundalk in the east to Clew bay in the west, and also in many parts of the northeast. Again, in the heart of the Central Lowland there is a magnificent series of ridges which are generally termed 'eskers' but may include a considerable number of moraines. Finally the endless variety in form and content of the ground moraine has produced innumerable local contrasts in soil and drainage. In most of the lowland corridors of the south, the drift is a thin rind through which the underlying limestones appear here and there; in the Central Lowland well-tilled patches of ground moraine alternate with flat expanses of bogs through which eskers stride. While therefore the major structural regions stand out with some emphasis, the local diversities are so considerable that they form one basis for a more detailed local regionalization.

Climatically, Ireland has the epitome of what is commonly called the West Maritime type. Sufficiently far from continental Europe to escape many of the cold spells that mark most winters in Great Britain, Ireland also fails to experience many of the heat waves that extend to the English Lowland in summer. The country's weather is dominated by depressions, which bring liberal supplies of rain throughout the year with a winter maximum that is universal but more marked in the west than in the east. Relief rain adds to the total precipitation in the mountains, where some falls of 100″ per annum are recorded, with higher falls – most probably – only awaiting measurement. There is a gradual climatic transition from the mean winter temperatures of 44° in the west to 41° in the centre and east; in July the averages vary from 61° in the south to 57° in the north. The rainfall on the lowlands is 50″–60″ in the western peninsulas, but less than 30″ in a lowland belt between Dublin and Dundalk; while the rainfall decreases gradually from west to east, the number of raindays everywhere is high. The contrasts in natural vegetation and agriculture are considerable: in the southwest an evergreen flora, both native and introduced, is luxuriantly developed especially in the valleys that are sheltered from the powerful winds which, along some sections of the coast, prevent trees from growing at all. The east of Ireland has splendid stretches of deciduous woodland, seen especially in some of the demesnes around the Leinster chain and in Meath. The main crops of Ireland, oats and potatoes, are grown throughout the country; while the east is more suited than the west to arable cultivation, the over-riding influence of the cattle trade is such that Ireland lacks anything comparable with the arable farming belt of the east of England. Climatic factors affect the agricultural life closely but somewhat indirectly; they are only one criterion among many in any delimitation of agricultural regions.

Non-agricultural land in Ireland consists largely of woodland and peat bogs. In spite of the efforts of the Forestry Commissions, Ireland is still the least wooded country in Europe: only in a few areas, chiefly mountainous, are 'natural' or planted woods an important feature of the landscape. The bogs, however, still survive in spite of centuries of active cutting for fuel and are in fact of such extent that until recently one might reasonably have regarded them as a permanent feature of the native scene. But during and since the 1939–45 war there has been so vigorous a use of peat both for domestic fuel and to supply power stations that hopes turn once more to the land 'reclaimed' when bogs of several square miles have been cleared away. In parts of the west, little will be found except bare rock, but in some of the lowland bogs there may be possibilities of agricultural settlement, given control of drainage; but this is almost entirely of the future rather than the present. Little impression has so far been made on the vast 'blanket bogs' in the western peninsulas. In the centre of Ireland the 'raised bogs' are seen in hollows separated from one another by 'dry ground', which frequently consist of drift-covered Carboniferous limestone. In the east and south lowland bogs are few and small. But both the blanket bogs of the west and the raised bogs of the centre are sufficiently extensive to be a vital feature in the building of a regional scheme.

The land use pattern in Ireland has been worked out by a population who have always lived primarily by farming: even now the only large urban concentrations are those around Dublin and Belfast. Negative areas of settlement such as peat bogs and mountains are used as rough pastures or for turf supplies over the centre and west of Ireland: in the east and south there are continuously occupied lowlands in many areas, without even belts of woodland between the fields. The limit of agricultural occupation has not been constant: it has been pushed up the mountain sides, then allowed to retreat once again, and recession is typical of the present age. There is no single statistical limit of cultivation: south of Dublin cultivation climbs above 1,000 ft. on fertile drifts, but in Connacht most hills have never been settled at any time. Nor is the present limit a straight line, but rather a series of farm boundaries oscillating around a level that is, in effect, the limit of human patience. But on the lowlands land is added to the farms rather than subtracted from them, as some of the soils revealed by bog clearance will support crops of potatoes or oats. In addition, some of the demesne lands have been divided between new holders who are settled in houses built under a succession of government schemes. Land use is one key to a regional scheme, though under Irish conditions farm size is of particular significance.

Heavy decreases in population have made possible many adjustments that seemed improbable a century ago. The agricultural Ireland of 1845 was an overcrowded country, and almost as soon as the heavy Famine emigration ceased, the farms were enlarged by unions. This process has continued ever since and the farms are now cared for by their owners and a much-diminished force of labourers and family helpers. Over much of the country the holdings are, by English standards, small, and they are still, in the main, larger on the better than on the poorer lands. This factor enters into the regional scheme; there is an obvious difference between the farmer of 100–200 acres in Meath and the holder of 5–10 acres in Donegal. The former is a producer of cattle for slaughter or a grazier of cattle for export, but the latter is little removed above the subsistence level, though he may sell a few calves, some butter, eggs or poultry. The two extremes still remain, with endless economic shades in between them: and in Irish farming no uniformity exists. Local specialization rises out of a mixed farming economy, with some arable cropping, but a prime dependence on pastures and meadows. The land division into farms of different sizes is, therefore, rather more important in a regional classification than the actual land use which, in the Republic, is largely a revelation of the local form of a cattle trade that is essentially a unity over the whole country. Northern Ireland's recent agricultural history has given its farming some significant differences from that of the Republic; for this and other reasons it has proved useful to treat the two political units separately.

The present population distribution in Ireland is most remarkable for its wide disparities between one area and another. Leaving aside the few patches where town populations are overflowing into the surrounding countryside, the density of population per square mile bears a direct relation to the average size of the farm unit: some areas in Meath have a mere 50 per square mile as compared with 100 or more in areas of smaller farms. If the densities are assessed on a basis of improved land only, some western districts have as many as 400 people to the square mile. Nothing could be more fallacious than to regard population density in rural Ireland as an indication of the relative wealth of the soil in each area; rather is it an expression of the particular type of farming evolved. The rural density of population is a valid criterion in the regional division of the country; great as the adjustments of the past fifty years have been, one may still recognize congestion, or something not very far removed from it, in parts of the west. This book contains, in Chapters 12–22, dot maps of population drawn from data in the 1961 Census. These maps replace those in the two earlier editions (1950 and 1960) based on the 1936 Census for the Republic and that of 1937 for Northern Ireland. The marked decline of the rural population from 1936 to 1961 is best seen by comparing Figs. 21 and 22, pp. 131 and 133.

In the actual distribution of the rural population many interesting contrasts emerge. The western counties have a great concentration near the coasts, the lowlands west of the Shannon a patchwork of cultivated areas and bogs, the fertile farmed lowlands of east and central Cork an even spread of population that is rare. Everywhere, from north to south, from east to west, dispersion prevails. Another feature of some importance is the relative strength of town and rural populations throughout the country: some of the poorer regions have little more than one-tenth of their numbers in nucleated settlements, but the richer regions have larger market towns, of which some have industries. Historically, Ulster's linen industry developed from the flax growing, processing and spinning on individual farms, but for over eighty years this industry has been concentrated in factories. There are a few survivals of home industry in Donegal, but as a general rule there is a primary dependence on agriculture everywhere. There is little fishing for sale, though in the most densely settled areas most farmers have some resource apart from their land, such as local labouring, emigrants' remittances, seasonal migration or even doles. Without such extras, the population would decline still more. Although the density of population is indicative of regional differences, some account has to be given of these qualifying factors.

Administrative divisions in Ireland, civil and ecclesiastical, provide at least some evidence of natural barriers within the country, though their evolution has been conditioned by the relative strength of various native and immigrant rulers at different periods. The modern counties developed gradually through the Anglo-Norman penetration, which reached its maximum strength in the fourteenth century, and the Tudor and Stuart periods, when large sections of the country were granted to English and Scottish settlers, the 'planters' of historic and sometimes hated memory. The Tudors extended the English Pale to the Shannon and settled the counties now called Offaly and Leix but formerly named King's and Queen's after Philip and Mary. Scots settled in Armagh and Down, and, under James I, Derry became Londonderry through its plantation by the Livery Companies of London. Cromwell finally systematized the county organization, which had grown slowly through five centuries of conquest, and established the landlords for the next two hundred years. Geographically, the most interesting features of the counties as a whole are first, their relation to some of the major topographical lines in Ireland, and second, their grouping into four 'provinces', which are still recognized in Ireland to this day.

The Shannon, like the Thames, is a county boundary for almost the whole of its course; unlike the Thames it is still a barrier marked by bogs and water meadows (callows) with infrequent bridges. The Bann, with its wide floodplain, is also a county boundary except near its mouth, and the Lagan still divides Down and Antrim, even in the centre of Belfast. In the south the Barrow and the Suir are boundaries for several miles above the tidal reaches, but the Nore valley between them is the heart of Co. Kilkenny. One could multiply such illustrations, but these will at least serve to show the varied nature of the counties as they exist today. The grouping of counties into provinces implies a broad recognition of fundamental regional differences in the country as a whole. Ulster covers the north, Leinster the east and southeast, Munster the southwest and Connacht part of the west. The Ulster province is essentially the four river basins of the Foyle, Bann, Erne and Lagan, together with the Tirconnail country in Donegal; Connacht is the country of mediocre resources west of the Shannon; Munster the uplands and lowlands that stretch from Waterford to Kerry and Clare; and Leinster the country around the great mountain chain that extends in unbroken Caledonian line from Dublin southwestwards. Meath, a fifth but now vanished province, is the Mercia of Ireland, a kingdom that spread through the east part of the Central Lowland almost to the Shannon, but was ultimately truncated and absorbed into Leinster.

Even more venerable in origin, the dioceses of Ireland are exactly equal to the counties in number but sharply different from them in shape. Several of the older dioceses, still surviving in the Church of Ireland, are linked with remote places of monastic sanctity, though all the major Norse settlements, Dublin, Waterford, Cork and Limerick, became cathedral cities. The dioceses were delimited at the Synod of Kells, 1152, and in spite of numerous changes since that date their boundaries still show significant geographical influences. The Shannon is a diocesan divide except at Killaloe; the Meath diocese reveals in its history the same lack of natural frontiers that has been noted in the history of the Meath kingdom, and is still much larger than the county of the same name; the Ferns diocese corresponds almost exactly to the county of Wexford, the fertile lowland of the southeast between the Leinster chain and the sea. In 1152 the country was also divided into four ecclesiastical provinces, ruled from Armagh, Dublin, Cashel and Tuam; these are comparable but not identical with the political provinces of Ulster, Leinster, Munster and Connacht.

While both the counties and the dioceses, combined respectively into civil and archiepiscopal provinces, have originated to some extent from the natural regional framework of the country, their full interpretation must take account of the life of the centuries in which they were formed. They are more closely linked with the regional geography of, say, the fourteenth century than with that of the twentieth, even though they still remain part of the fabric of Irish life. While the four provinces still have their distinctive characteristics, the one administrative boundary now of consequence is that between the six Ulster counties and the twenty-six of the Republic. The Border, a series of county boundaries running mainly through cultivated lowlands, is an economic divide between the areas focused on Belfast and Dublin respectively; in some ways the duality of economic organization has been accentuated, but if Britain enters the Common Market the Republic is likely to do so and to make economic adjustments. While the Border is in no sense a morphological divide over most of its course, the regional geographer cannot neglect this separation within Ireland as he makes his division of the country into a number of units.

Ideally, natural regions are clearly marked entities, each possessing distinctive qualities that mark them off from their neighbours. These qualities are derived from the morphology, climate, past and present space relations, natural vegetation, land use, industry and population distribution. Uplands and lowlands, cultivated farmlands and bogs, form contrasting environments in every part of Ireland, often within a few square miles; even the Central Lowland is saved from monotony by the presence of low hills, many of them miniature Armorican ranges, which break the generally even line. But this is only a recognition of an inevitable fact – that many areas are far from homogeneous and exhibit contrasts within quite short distances. On the other hand, one can travel for over sixty miles through the farmed coastal fringe of Connemara and see little change in the immediate scene, or for thirty miles through bogs interrupted only by a few patches of farmland, on the road to Belmullet. Small-scale differences are frequently much more obvious than the larger ones, yet there is, for example, a marked general contrast between the east, with its cattle-grazing farms and its industrial towns, and the west, still partly congested and revealing a small farm economy little removed from the subsistence level. Again there are marked differences between those areas which were effectively planted by immigrant settlers and those only partially occupied or ignored. And much of the history is expressed in the division of the land among its holders.

Two factors stand out as of supreme importance: first, the physical features; and second, the type of farming evolved through the centuries. The first is in the widest sense the natural physical environment, including climate, soils and natural vegetation, and the second the use made of that environment by the people who have lived in it. But at this moment in time, after more than a century of heavy emigration and land agitation, never was so much land farmed by so few.

THE ELEVEN REGIONS

All the eleven regions show some distinctive traits, yet Ireland does not lend itself readily to such division. Why then try? Simply, because there are differences between one area and another, even if a striking landscape contrast may be rare. Even the observant traveller may find that there has been a change of scene, but so gradually that he could hardly say where. Nevertheless, the contrast between lowland and upland Kerry, or between the Antrim uplands and the Bann valley, stand out clearly; and if one travels through the town of Galway, the difference between the area to the east and that to the west of the town is sharp. In the following chapters, the boundaries of the regions and subregions will at least help the reader to locate himself with the aid of Folder II at the end of the book. The six regions (Chapters 12–17) cover most of Leinster and all Munster, the next two (Chapters 18–19) two contrasting but related sections of Connacht, and the last three the old Ulster province and its borderlands (Chapters 20–22).

1–6. LEINSTER AND MUNSTER

1. *Dublin and the East Central Lowland.* In the east there are large farms specializing in cattle grazing to the virtual exclusion of arable cropping, though there is dairying near Dublin, and a limited development of market gardening. Farther west, cultivated land and bogs are mingled in the landscape, the farms are smaller and have more tillage than in the east, but cattle, received from districts west of the Shannon, are fed for a year

or more and then passed on to their final destination in Meath. The rural density of population is remarkably low in Meath and higher towards the Shannon; there are several prosperous market towns, some of which have industries of considerable significance. Dublin, the capital of a thousand years, is considered in this chapter.

2. *The Leinster Chain.* This Caledonian range is surrounded by drift-covered lowlands, divided mainly into medium-sized and large farms concentrating on cattle production, with sheep rearing in the uplands.

3. *The Suir, Nore and Barrow Basins.* These basins, mainly lowlands, are diversified by Armorican ranges and Carboniferous plateaus. They have mixed farming and medium-sized holdings with varied types of production, but an increasing emphasis on dairying to the west. Waterford is a port, industrial centre, and to some extent a minor regional capital.

4. *The Cork Uplands and Lowlands.* Composed of parallel east–west ranges and valleys of Armorican structure, this region has medium-sized farms specializing in dairying: Cork, its regional capital, is a port with a variety of industries.

5. *Kerry and West Cork.* The Armorican ranges terminate in peninsulas, and there are Carboniferous plateaus in the north. The cultivable area is limited by blanket bogs on the lowland, and the farms, though generally small, are larger than those of West Connacht or Donegal, and the region is therefore less congested. Young cattle, butter, eggs and poultry are the main products, though creameries have spread into parts of the region from the great Munster dairying belt to the east.

6. *Limerick and Clare.* Carboniferous limestone lowlands, partly bare rock in Clare but drift-covered in Limerick, provide varied soils used mainly for pasture: the Carboniferous beds form desolate expanses around 1,000 ft. in the west, and Armorican ranges of equally dreary aspect around lough Derg in the east. The creameries of Limerick use almost all the available milk supply, but in Clare the farmers' main products are butter, poultry, eggs and young cattle. Limerick, a bridgehead town and regional capital, is the one important industrial centre.

7–8 CONNACHT

7. *East Connacht.* This lowland to the west of the Shannon has small farms, very few towns, hardly any industries, but so high a density of population in 1891 that considerable areas ranked as congested. The main development since then has been heavy emigration, the enlargement of farms and an increasing specialization in sheep farming in the bare limestone areas, with cattle rearing elsewhere.

8. *West Connacht.* Though primarily a lowland, West Connacht has some magnificent lines of hills composed of quartzites or of Ordovician and Silurian beds, strikingly exposed in bluffs around Killary harbour. The lowlands are covered with blanket bog except where the inhabitants have made farms, mostly of a few acres only, along a coastal fringe one to three miles deep. There is very little settlement away from the coast with its many islands and peninsulas.

9–11. ULSTER AND ITS BORDERLANDS

The chapter on Northern Ireland has been placed last not for any motives of disrespect, but to keep together the treatment of the Republic. A separate chapter is given to Donegal, and another to the borderland country, part of the great drumlin-drift belt, which lies both in the old Ulster province, and also in neighbouring areas of Leinster and Connacht.

9. *The Drumlin-drift Belt.* Drift, generally of considerable depth and frequently moulded into drumlins, covers most of the lowlands north of a sinuous line drawn from Dundalk to Clew bay. Small farms are characteristic of this belt (and of its natural continuation across the border in Northern Ireland), and it may be conveniently divided into two main sections.

9*a*. *Leitrim and Sligo.* Plateaus of Carboniferous strata and the Ox mountain ridge of Caledonian structure break up the lowlands, which have cultivated land on drumlins and ground moraine, but also considerable stretches of peat bogs: large parts of the lowland have poor soil and were regarded as congested in 1891. The loss of population has been remarkably high.

9*b*. *The Eastern Hills and Lowlands.* Hilly country of Ordovician and Silurian strata prevails in this eastern region, though lowland corridors with a Carboniferous rockfloor lead through the Erne valley and through central Monaghan. The farms are generally small, but the soils are more fertile than farther west: home flax weaving was widespread in the pre-factory stage of the linen industry. The rural population has declined very heavily, but the market towns have remained stable: the largest of them, Dundalk, has a number of important factories, some of which have been established quite recently.

10. *Donegal.* The west of Donegal is an environment comparable to that of West Connacht, with intense occupation of restricted areas but a more active development of home industries and a greater emphasis on seasonal migration. In the east the farms are medium-sized and given

mainly to cattle production with home growing of feedstuffs, most of which are milled locally. There are no large towns.

11. *Northern Ireland.* In the first edition of this book, this chapter was divided into two, using the Bann as a divide, and to some extent this fusion into a single chapter is a concession to political geography. More and more the six counties are welded into one economic entity even if political life remains unchangingly tense in its division of sympathies. From Belfast, the economic and political capital, an easy way opens through the Lagan valley to the Bann, and thence to the Erne and the Foyle, though the Border runs through these valleys, a series of county boundaries elevated in status to a frontier. There are many interesting contrasts of physical features and economic life in Northern Ireland, though most of its industrial activity is in the east, with Londonderry, half circled by the Border, as a plantation outpost city with modern manufactures.

CHAPTER 12

Dublin and the East Central Lowland

DUBLIN

The city of Dublin is the natural approach to the Central Lowland, of which the east section is considered in this chapter. The built-up area of the city and its environs covers some forty square miles, but any road or rail journey out of the city shows a rapid transition to the essentially agricultural landscape that stretches away to the Shannon and beyond it to the Atlantic coast. Inspection of any road or railway map of Ireland will show the concentration on Dublin of routes from all parts of the country, and throughout its history it has been a bridgehead town looking both landwards and seawards. In this it is comparable to London, but the Romans gave London to England and the Norse people gave Dublin to Ireland. Its name, Dubh-linn, 'the black pool', recalls a Norse harbour, but the postmark and Irish name, Baile atha cliath, 'the town at the ford bridged by hurdles', probably refers to an early river crossing near the present centre of the city. The two main factors in the site are expressed in these names, for it was at once bridgehead and port, and remains so to this day. From the beginnings, which probably antedate the Norse arrival, it was one of the few sheltered havens on the east coast of Ireland, and the first evidences of settlement suggest occupation of two patches of dry ground, of which the more important was the spur on which the castle was built near the main ford of the river: the other was the low gravel ridge at Ringsend.

Close to the river, the first walled city was established by the Norse people or 'Ostmen', but houses spread beyond the walls and an early suburb on the north side was known as 'Ostmanstown', a name which still survives as Oxmantown. In 1170 the Ostmen succumbed to Strongbow, and in the thirteenth century the castle was built by the Normans: it remained a centre and symbol of English power until the twentieth century. The thirteenth century saw the addition of St Patrick's Cathedral outside the walls, only a quarter of a mile from Christ Church: both are Church of Ireland today, but Christ Church is the cathedral of the dioceses of Dublin and Glendalough, while St Patrick's is a national

FIG. 45. THE SITE OF DUBLIN

Most of the dock and industrial land has been reclaimed since the eighteenth century: elsewhere limestone boulder clay occurs except in the valleys. The oldest part of Dublin lies around the Castle and Christchurch, and the city is now centred in streets planned in the eighteenth century, near Trinity College and the Customs House.

Based on the Ordnance Survey, by permission of the Minister for Finance of Ireland. Drawn by J. P. Haughton and reproduced from 'Irish Geogr., vol. 1, p. 54

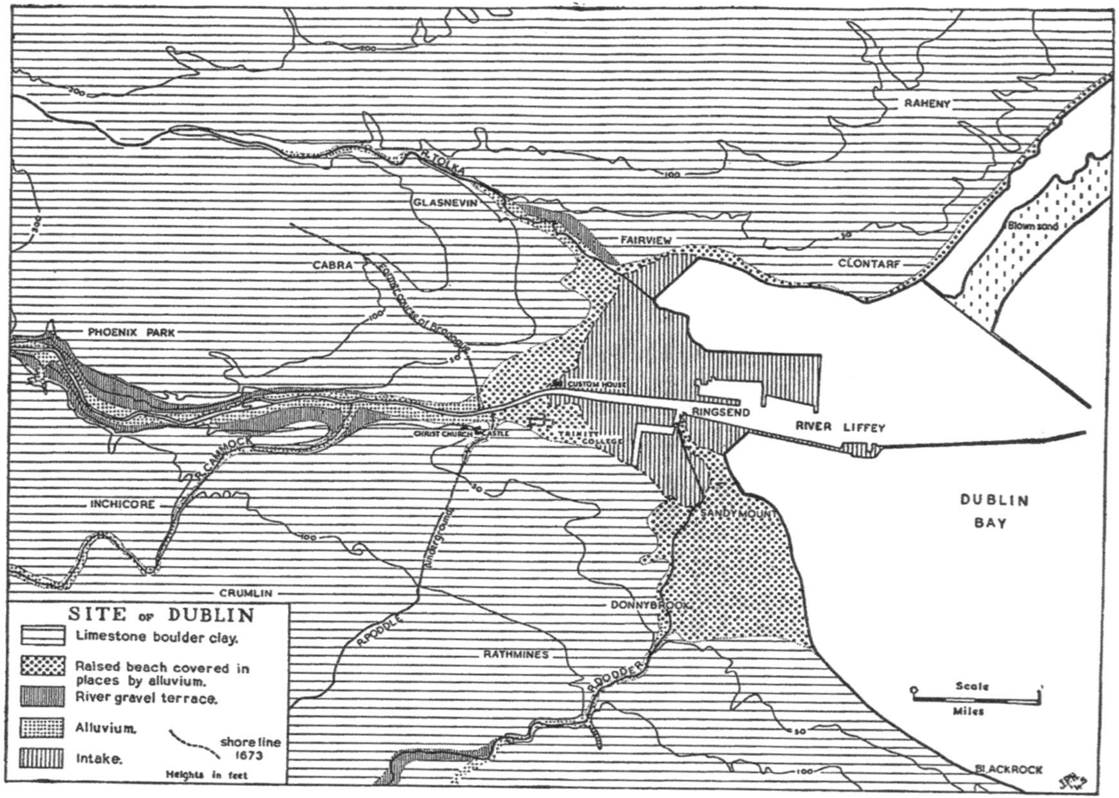

cathedral. Medieval Dublin was a walled city surrounded by the 'Liberties' outside the walls with a northern suburb in Ostmanstown, and the bay was much larger than now. Trinity College, at its foundation in 1591, was given land 'juxta Dublin' on the site of the demolished monastery of All Hallows, but it is now in the very centre of the city with shopping, commercial and industrial quarters close to its walls. In the college, as in the main part of the city, Georgian architecture figures prominently. This phase of Dublin's history is strongly impressed on the city partly because many new roads, some as much as forty yards broad, were planned by the Wide Streets Commissioners, and numerous celebrated architects of the day were engaged to design homes for the wealthy families associated with the then 'second city of the British Empire'.

Dublin, like other cities, has incorporated various outlying villages in its growth, such as Sandymount, Donnybrook, Rathmines, Crumlin and Inchicore on the south side of the river, or Glasnevin, Clontarf and Raheny on the north side. Its expansion was mainly around the bay and towards the mountains, and there is now a continuous line of settlement from the city to Howth and Dalkey, which guard the entrances to the bay, though Dun Laoghaire is still a separate borough extending from Blackrock to Killiney. The large increase of population in Dublin and district since 1922 has its expression on the ground in many square miles of new suburbs, both on the outskirts and in the areas left between various old villages which had been suburbanized. Many of the older suburbs have a spaciousness that impresses visitors; the reasons for this include the survival into the nineteenth century of the Georgian tradition of making wide roads and building long terraces of houses with large gardens in the rear and sometimes also in front, and also of building large houses surrounded by a few acres of grounds within two or three miles of the city centre. Many of these are now surrounded by smaller houses, but a considerable number are still occupied either by families or by various institutions such as monasteries and convents. The most famous of all the open spaces in Dublin is the Phoenix Park, an expanse of almost three square miles to the west of the city which is used by thousands for recreation and has within its bounds the former Viceregal Lodge (now the home of the President of the Republic), the Papal Nunciature, the American legation, the Ordnance Survey department, a military barracks and the zoological gardens.

The 'functional areas' of Dublin have been distinguished by J. P. Haughton whose map is published as Fig. 46. Many areas close to the centre of the city are still strongly residential, though Dublin shares with other large towns the phenomenon of a declining population in central

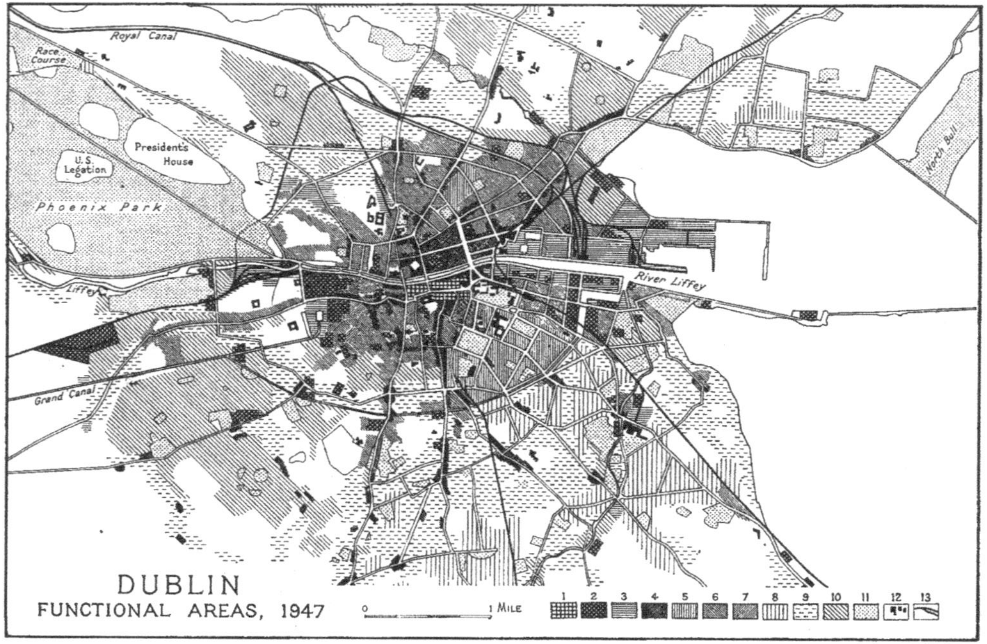

FIG. 46. DUBLIN, LAND USE IN 1947

This map, by J. P. Haughton, was based on street-by-street observations. There is a well-defined commercial core close to the shopping district but the industries are scattered. Both the decayed and the prosperous city residential quarters exist and the suburban spread has been least marked on the west due to the fortunate barrier of the Phoenix Park.

Key: 1, commercial core; 2, factories; 3, transport and warehouses; 4, shopping centres; 5, substantial city residences; 6, old tenements and slum areas; 7, new tenements; 8, expensive suburban houses; 9, medium-sized suburban houses; 10, cottages and corporation housing estates; 11, recreation, including parks and playing fields; 12, public buildings; 13, railways.

areas. The commercial core of the city, the headquarters of banks, insurance companies and the like, lies between Trinity College and the Castle: there are also many offices in other streets close to, and interpenetrating with, the major shopping quarters of the city. In this central area too there are hotels, theatres, cinemas, restaurants and churches, and as a whole it is busy but attractive in general appearance: through the heart of it runs the Liffey, and beside it is the splendid small park of St Stephen's Green. Two main stations on the margins of this central belt – Pearse and Connolly, are connected by an ugly but much-used bridge across the river. Heuston, the terminus for many trains to the south and southwest, is inconveniently located near the Phoenix Park.

Factories are spread through the city and only near the docks is there any marked factory-warehouse belt; even here the mixture with housing will be noted as a feature in Fig. 46. Guinness, the vast brewery, is to the west of the commercial centre: its barges on the river go to the firm's wharf on the quays for export. Most of the goods produced in factories are intended for the home market, as in various other capital cities, and a trend of recent years has been to establish factories outside the city altogether, in such places as Finglas to the north. Maritime trade has been the basic feature of Dublin's commercial life for a thousand years, and the reclamation of large areas of land along the estuary began in the eighteenth century, when the river was confined between the two parallel lines of quays which are one of the main features of the city to this day.

In 1961, 31% of the employed population of Dublin were engaged in industry, compared with 17% in administration and professions (including clerical staff) and 22% in commerce and transport. For the borough of Dun Laoghaire, largely suburban in character, the figures were industry 27%, professional 24%, commerce and transport 26%. The large number in commerce, administration and transport gives a clear indication of the metropolitan status of Dublin. Its dominance as a shopping centre was shown by the 1960 Census of Retail Trade, which showed that the total for the whole country was £212,631,000, of which £78,075,000 was spent in Dublin and its suburbs, or in Dun Laoghaire and suburbs, 37% of the total. These figures have confirmed the general impression that Dublin is a shopping centre of immense strength and attraction. The city has many contrasts, the stately squares and streets of eighteenth-century houses, modern suburbs of middle-class houses that could exist anywhere, and slums close to the finest streets, including tenements in decayed Georgian houses on the north side of the river. To many it seems that in Dublin one half of the world does not know how the other half lives, for as recently as 1946, 27% of the city's population lived at a density of more than two persons per room. The equivalent figure in 1936 was 35%, but in 1961 was down to 14%. The city has considerable areas of new tenements: some of the old Georgian streets have been reconditioned and their population substantially reduced, and the city's housing estates have considerable populations – for example, in 1966, Cabra 34,000, Kimmage 22,200, and Crumlin 33,700. These have declined slightly in population from 1956, but in Ballyfermot the population increased from 13,200 in 1951 to 31,900 in 1961 and 33,900 in 1966. In short, the housing areas built twenty or more years ago are now virtually stationary in population, and the areas at present expanding include Finglas, with 31,800 people in 1961 and 43,300 in 1966 and Coolock, with 8,100 and 15,600. At present the Dublin area is in a phase of active expansion (see pp. 125, 137).

THE EAST CENTRAL LOWLAND

The scenery of the Central Lowland is varied; north of Dublin there is the drift-covered limestone country of Meath, the richest pasture land in Ireland, but a westward journey of only twenty miles reaches the first of the many bogs that divide the farmlands into cultivated islands of firm ground, and some fifty miles from Dublin the traveller reaches the Shannon tributaries, which flow through bogs and widen into shallow lakes. The Shannon forms a suitable western limit for this region as it has remained a barrier rather than an artery of communication throughout most of its history; and the fringe of the great drumlin belt is a convenient line of demarcation on the north. On the south the boundary is drawn through the lowlands which encircle the Slieve Bloom, the Castlecomer plateau and the Leinster chain. So defined, the region covers some 4,700 square miles, and is in some ways as amorphous a unit as those parts of the English Lowland through which the kingdom of Mercia expanded and contracted at different times. In this region the Kingdom of Meath rose and fell: here also was the English Pale, perpetually menaced by the hillmen from the Leinster chain; here, too, were the successive invasions and plantations of which the county names, King's and Queen's, now Offaly and Leix, remained to tell the tale, just as the surviving Meath and Westmeath speak of the ancient kingdoms. Today this region is merely one part of Dublin's hinterland. In it the landscape contrasts, though considerable, emerge so gradually that a traveller across it may observe with surprise that the fertile farmlands of Co. Dublin grade into bog-strewn country around Kildare: it is in fact a country of transitions, but on its margins the Leinster upland, the Castlecomer plateau and the Slieve Bloom possess a clarity of line that rarely occurs in the lowland itself. Yet there are contrasts of great importance between the continuously farmed lands of Meath and the bog-divided agricultural landscape near the Shannon, though the transition from one type of area to another is gradual. On the whole the lowland is not thickly populated, though it has a number of small towns, several of which have recently acquired some new industry, though their main functions are to provide market facilities for a rural community of relatively high purchasing power.

LANDFORMS AND LAND USE

This lowland has been called a limestone peneplane from its widely developed Carboniferous limestones. But it does not appear to be a single surface of erosion, for though the level of 200 ft. is widely represented there is a gradual rise to 400 ft. And the rockfloor is of complex structure, with many anticlines and synclines, apparently moulded to a peneplane surface by long-continued denudation. Commonly regarded as 'monotonous', this lowland is broken by small isolated fold ranges of Old Red sandstone with Silurian and Ordovician beds, all of which form hills rising to 600 ft. or more; one of these folds appears first in the Hill of Allen range and then remains hidden beneath the Carboniferous cover until it is exposed at Portrane, north Co. Dublin, and in Lambay island. Even more significant is the emergence of Silurian and Ordovician strata along a line from Balbriggan to Longford: here shales, slates and sandstones form hilly country rising to 600–900 ft. near Oldcastle and to 912 ft. in Corn Hill, north of Longford. The most important topographical feature, of close relevance to agricultural study, is the widespread glacial and glacifluvial deposits: few areas of solid rock are exposed.

In this lowland the influence of the Quaternary glaciation and glacifluvial action has been felt in three main ways. First, the relief has been altered and generally smoothed out by the filling of valleys and the covering of lowlands with sand, gravel and drift. The pre-glacial limestone peneplane has an abrupt edge in the neighbourhood of Dublin, but this has been obliterated by the various glacial and glacifluvial deposits. In places the drift cover is thick, with a record of 234 ft. in the centre of the Curragh, where it exists as rolling ridges of morainic material. A second vital effect of the Ice Age was the entire reorganization of the drainage: meltwater channels in the lowland are naturally less striking than those on the flanks of the Leinster chain (see pp. 36–7), but they are immediately obvious to the field worker in such areas as Co. Meath south of the Boyne: one inevitably sees the present drainage as the humbler descendant of a much more powerful meltwater system. Thirdly, the decay of the ice-sheet provided opportunities for the development of large lakes and fens, many of which have become bogs in the present landscape. These vary in size from a few acres to as much as twenty square miles, but an unobservant traveller could cross the lowland from Dublin to the Shannon and remain almost unaware of their existence, partly because the roads keep whenever possible to the drifts and eskers that form the dry ground (see pp. 283–4).

Two main types of river form are seen in the lowland. The middle and lower courses of the Boyne and Liffey and the Barrow below Athy are incised streams cut in drift or, more rarely, solid rock, but the upper courses of the Boyne and Barrow lie in bogs, where the drainage is so badly organized that few clear water-partings exist. One classic case is the Barrow and middle Liffey watershed which runs through the Bog of Tankardsgarden: this bog, resting on 30–40 ft. of drift, and therefore masking the pre-glacial landforms effectively, drains to both the Barrow and the Liffey: if the Liffey waters rose 15 ft. they would be carried to the Barrow. The greatest of all the Irish rivers, the Shannon, has the least attractive course, for it passes through extensive bogs and widens out into lakes that are generally ascribed to the solution of Carboniferous limestone: what part, if any, was played by meltwaters in their formation remains unknown, but the lakes still remain one of the most significant features not only of the Shannon but also of its tributaries, the Inny and the Brosna.

The remarkable series of eskers in the lowland is comparable with that in Finland or North America. The three main types recognized in Ireland include ridges, isolated mounds or hummocks, and broad flat-topped masses. The most characteristic development is the long ridge, persisting as a natural passage-way for as much as twenty miles across the lowland. Two such ridges are shown in Fig. 48: it will be seen that both provide routes across the bogs. Especially prominent towards the Shannon, and visible in some cases for many miles, they are locally obvious farther east. Rarely of much agricultural use, many of the eskers are covered with heathland or trees; some are used for road metal such as the few close to Dublin It is unlikely that all the ridges are eskers: they are formed generally of stratified sand and gravel though boulders, blocks of limestone and strikingly laminated clays and silts have also been found. In form they are varied: some of the ridges consist of a single embankment 30–100 ft. high, and some of several parallel ridges, which may be connected by transverse ridges and vary in width from a narrow causeway to a ridge hundreds of feet wide. Both the ridges and the mounds, hummocks and masses of similar materials show very little weathering, possibly because their sides generally slope at 25°–30°, which is near the angle of repose for such materials. Some of these forms are perhaps more truly kames with kettle-holes, and many have cores or nuclei of bedrock.

Little trace of the natural vegetation remains in the east part of the Central Lowland: even the bogs have been grazed for centuries and their margins cleared for fuel. The deeper layers provide the best 'turf', and the present active exploitation is bringing their ultimate exhaustion within measurable time, though the supply still seems almost unlimited. The cleared sections generally reveal a clayey soil that will grow potatoes successfully, and has, in fact, been specially recommended for the growth of seed potatoes. Some bogs that have ceased to grow are already colonized by sub-spontaneous pines and birches, and a few bogs are under the care of the Forestry Commission and planted with conifers. Natural fens in the Central Lowland are found only in small areas around the bogs and on the periodically flooded land around the rivers: these 'callows' are particularly well developed around the Shannon and the upper Barrow. The major part of the lowland consists of farmland with few patches of woodland except in demesnes, which are most numerous towards the east coast, especially in Co. Meath. Some of the demesnes possess exceptionally fine woods with magnificent beeches, but several have already been taken over by the Land Commission, and divided into small farms, some of them for Irish-speaking migrants. Other demesnes have been taken over for plantation by the Forestry Commission and some large forests now exist in the Slieve Bloom; but there are no state forests in Co. Meath. In all, woods cover hardly 1% of the whole area, and the landscape in many parts of the lowland seems almost treeless, though not everywhere, for some roads have lines of trees with occasional small groves and shelter-belts around the farms. In spite of the region's present paucity of woodland due to its long-continued use for farming, it is not difficult to imagine it as one large forest broken only by bogs and grass-covered eskers some two thousand years ago.

Farmland under crops and pasture covers 95% of Co. Meath and 80–90% in most other districts of the Central Lowland east of the Shannon: only in the heart of the bog country around Edenderry and towards the Shannon is the proportion of improved land substantially less, and even there it generally covers at least two-thirds of any area of, say 50–100 square miles. The trend of the past hundred years has been to add to the improved land by drainage and bog clearing; very little land is removed from agricultural use by afforestation or by building development, except in the immediate vicinity of Dublin. Thus Co. Longford had 75% of its entire area under crops and pasture in 1851 and 78% in 1961; Co. Westmeath had 81% in 1851, 82% in 1891 and 85% in 1961. Lowland areas are reclaimed or maintained as farmland, but upland areas are more readily withdrawn from agricultural use, either by the desertion of farms or by the sale of land to the Forestry Commission.

The eastern part of this lowland region is climatically the most suitable area in Ireland for arable cultivation, and it is commonly assumed that such counties as Meath were at one (generally unspecified) time devoted mainly to arable farming. This was not the case, at any rate during the past hundred years, for in 1851 Meath had a mere 31% of its farmlands under crops; at the same date Westmeath had 26%, Offaly 26%, Dublin 33% and Longford 37%. It has been noted that the new holdings of the immigrants had a relatively high percentage of tilled land, yet this was hardly one-fifth of the whole. Pastoral farming was prominent in this lowland long before the Famine, though on the smaller holdings much of the land was used for crops. The general organization of agriculture in this region lends itself to concise summary. Meath is the heart of the cattle-fattening country in Ireland, and in some parts of the county as many as two-thirds of all the cattle are more than two years old. The fattening belt extends into Westmeath and southwards through Co. Dublin into Co. Kildare, but in these areas the specialization on fattening is not so fully developed. The immediate hinterland of Dublin has a dairy-farming belt similar to that around most large cities in the British Isles, but it extends only for a short distance beyond the outer suburbs, and merges into the fattening belt, where dairy cattle generally number less than 20% of the herds, and in some cases less than 10%. All these areas, with their obvious specialization, have less than one-tenth of their land under arable crops; the cattle live on pastures and hay, supplemented mainly by imported feedstuffs in normal times. Outside the fattening and dairying areas the farming assumes the mixed character which is more typical of Irish agricultural conditions; dairy cattle, store cattle, fat cattle, all figure in the economic organization in varying degree on different farms, and arable crops cover as much as 20% in some districts, especially in Co. Leix, in south Kildare and in Carlow, though pastures cover most of the farmland and hay is an important crop everywhere.

Changes in cropping during the past hundred years include a marked decline in the acreage of wheat, oats and roots with a temporary rise in cereal cultivation during the 1914–18 war and an even greater rise during the 1939–45 war. But during the inter-wars, and in recent years, wheat was grown on less than 1% of the farmed area and the acreage of oats fell to one-third or less of its 1851 level: meadows and pastures have taken the place of arable fields. Sheep rearing is subsidiary in this region, but the well-drained soils of the hilly country in Meath and Westmeath are good sheep pastures, and there is a certain amount of fattening on the drier ground for the Dublin and overseas fat sheep markets. Poultry are kept everywhere, but are of most commercial significance on the smaller farms. As yet, the farm specializing exclusively in poultry is rare, even in the immediate neighbourhood of Dublin. Pig production is generally allied to dairy farming, and is more usual outside the cattle fattening belt in the area of mixed farming.

This eastern part of the Central Lowland has a large number of excellent farms, including some famous studs. Around Dublin the holdings average 80–100 acres and there are a number of large and highly specialized dairy farms intermingled with much smaller concerns, but the Meath–Kildare fattening belt has a mean farm size of over 100 acres, which represents a considerable number of large holdings interspersed among smaller ones. Large holdings specializing in cattle grazing and fattening support a relatively small farm population, and houses appear so rarely in the landscape that people journeying rapidly by road and rail have been known to ask who worked the apparently deserted but productive land. Such areas are essential to the cattle trade in its present development, for in them the cattle received from the extensive hinterland are prepared for slaughter. The fattening belt passes into Westmeath where the medium-sized farm, with an average of 70–80 acres, is more usual. Towards the Shannon, the farm units become smaller on the whole, with an average of 40–80 acres. There is a gradual transition to the more fragmented agricultural landscape characteristic of the lands west of the Shannon and the drumlin belt to the north, but to the south the medium-sized farm remains the chief type.

INDUSTRIES

Although Dublin has a large proportion of the newer industries, many of the country towns of this lowland possess several attractions for the proprietor of a factory: land is plentiful and cheap, labour generally abundant (especially in days of bicycles and bus services), certain raw materials are available locally, and there are good rail and other transport facilities to and from the port of Dublin. It is therefore not remarkable that the towns of this region should have developed some industries before the minor industrial revolution of the post-1932 era began. The measure of industrialization achieved in the towns is varied. Some, like the towns of Co. Longford, are market centres with a few factories while others, such as Clara and Balbriggan, are dominated by a single industry. Several have a number of small industries with one large factory, generally of recent establishment. In Drogheda, the largest town (17,900), factories line the banks of the Boyne and use the port: they include a large brewery, shoe factories, a chemical manure works, various mills producing foodstuffs and feedstuffs, a cement plant and, most interesting of all, linen mills, one of which dates back to the beginning of the nineteenth century. Balbriggan has woollen mills employing several hundred workers, and may perhaps be regarded as an industrial satellite of Dublin. The towns of Co. Meath have a number of small factories which include those for clothing, farm tools, carpets and furniture at Navan, boots at Kells, and cotton goods at Slane. Obviously one could search in vain for any geographical influence in such a factory location, especially when carbon papers, typewriter ribbons and similar articles are made in Trim, an ancient market town of 1,900 people. In Co. Westmeath post-1932 factories provide cotton goods and a variety of woollen materials at Athlone. Until recently there was a pot-still distillery at Kilbeggan, a village of some 700 people, where whiskey was made by a process used since 1757.

The market towns to the west and southwest of Dublin have a wide variety of industries, many of which have been conspicuously successful. In many cases, inquiry reveals the enterprise of a few families. Maryborough has a large flour mill with branches at Sallins, Co. Kildare, Portarlington, and St Mullins on the river Barrow. Again, several hundred people are employed by two firms in Tullamore in a kind of composite industry based on the needs of licensed grocers' shops: the drinks supplied range from whiskey through wines and beer to mineral waters and tea; seed potatoes are sold locally and exported: and finally, each firm controls a chain of grocers' shops. Tullamore also has a distillery whose products are widely advertised, and an associated company manufactures a liqueur that rejoices in the name of 'Irish mist'. But a creamery here was unsuccessful and turned to bacon products. Such industries have grown to success from the ordinary market-town trade of fairly prosperous countryside. They are of indigenous and not civil service origin, but they do not complete the present industrial scene for there are several new concerns each employing at least a hundred workers. Newbridge has rope and cutlery works; Naas, cotton mills; Kildare, a wallpaper factory; Athy, an ironfounding and general engineering plant; Carlow, beet sugar, shoe, engineering and mineral water factories; Birr, footwear; Maryborough, woollen mills; Portarlington, sports goods, leather and wooden articles of endless variety. The majority of these factories, but by no means all, are post-1932, but in some cases an old industry has been expanded in scope; for example, the new wood-working industry at Portarlington is based on a century-old sports factory and the furniture industries of Edenderry and Leixlip are based on long-established saw-milling.

Two towns, Tullamore (6,900) and Clara (2,400) in Co. Offaly, set in the heart of rural Ireland on branch railway lines, show a remarkable industrial development that calls for special mention. In neither case does the actual population of the town reflect its commercial importance, for labour is drawn from the adjacent rural areas. Tullamore has the chain of food and drink-producing enterprises mentioned above, but in addition there is a Co-operative Society employing over 100 people mainly in bacon curing and the butter, egg and poultry trades: the old prison grounds gave a site for a woollen and worsted factory with some 600 employees. Altogether, Tullamore provides industrial employment for at least 1,500 people, who are drawn from a radius of approximately ten miles. Clara has a jute factory, with as many as 600 workers at times, and the firm responsible for this extraordinary development has branches at Waterford, Cork, Limerick, Dublin and Donegal with a subsidiary company at Slane, Co. Meath, for the spinning and weaving of cotton and rayon, employing 250 people.

This review of some of the industrial developments in one part of the Central Lowland shows the unexpected and haphazard nature of its economic life. Market towns differ so strangely in their material prosperity that no general ruling is possible for some are decaying and others prosperous, yet most live primarily on their trade with the factories as a supplementary source of employment. The old-established industries are deeply entrenched in the economic life of the country, but the new ones have had chequered fortunes in some cases, especially during the war years when imported raw materials were scarce, though their sales in the home market are guaranteed so long as tariffs remain. As a result, knives come from Newbridge rather than from Sheffield, cotton textiles from Athlone rather than from Lancashire, and sugar from Irish-grown beet refined at Carlow and not from imports. Industrial decentralization has been made possible by the use of electric motive power in the factories, and by the conscious planning since 1932: without such planning many of the new industries, if they ever arose at all, would almost certainly have gravitated to Dublin and its immediate surroundings.

POPULATION

Excluding the Dublin conurbation, the population of this region was 516,000 in 1891, 430,000 in 1951 and 420,000 in 1966: the towns had 98,700 people in 1966, about 5 per cent more than at the 1961 Census, though in 1966, another 11,000 people were in the suburbs of a few towns. Since 1936 the rural population has declined slowly, except in Co. Longford, where it has decreased from 37,800 to 29,000. The density of population is high, *c.* 150–250 per square mile, throughout Co. Dublin, and in the extreme northeast of Co. Kildare around Celbridge and Leixlip: such densities are not typical of essentially rural areas in the east of Ireland and indicate the presence of urban works in some strength. Many of the rural areas not dominated by Dublin show a remarkably low population density: in Co. Meath the average density, towns included, is 72 per square mile. In general the areas dominated by large cattle-grazing farms have low densities, but to the west, where the farms are generally smaller and mixed farming prevails, the density increases to 80–120 per square mile of cultivated land. The gaps in the settlement pattern (Fig. 47), apart from the Slieve Bloom and a few other uplands, are mainly the uninhabited bogs which extend in a discontinuous series from the neighbourhood of Edenderry to the Shannon. The settlement pattern is noticeably thin outside the towns in Meath and Kildare but much denser in the occupied areas elsewhere, and relatively heavy in Co. Longford, where the average size of the farms is less than 40 acres – i.e. one-third to one-quarter that of the Meath–Kildare fattening belt.

Throughout the lowland, which includes some of the most productive agricultural land of Ireland, there has been a slow but continuous fall in the numbers on the land. In Co. Meath, however, this has been less marked than in Co. Longford, which had 8,610 people working on the land in 1951, but 6,700 in 1961, almost 23% less. In Co. Meath during the same ten years there was a decline from 15,260 to 13,300, a loss of 13%. In Longford the number of family workers decreased by 16% from 7,040 to 5,900, but in Meath the same group declined by only 7·5% from 8,870 to 8,200: among both the permanent and temporary labourers there was a loss of almost one-fifth in Co. Meath, which had 4,430 permanent and 1,970 temporary farm workers in 1951, but 3,500 and 1,600 respectively in 1961. In Longford the labourers diminished from nearly 1,600 to 800, almost by half.

Most of the towns in this lowland are within six to fifteen miles of another town, and the market villages and periodic rural fairs fill gaps in the distribution pattern. Apart from Drogheda (17,900), only six other towns had more than 5,000 people in 1966 – including the suburbs; of these Athlone (11,000) and Carlow (9,800) are major route centres serving large areas and possessing industries; Mullingar (7,900) is an important route and commercial centre and Tullamore (6,900) has seen the interesting industrial growth mentioned on p. 277. Maryborough (5,900) is at once a market, industrial and administrative centre; Navan (5,900) is the major town of Co. Meath. Most of the remaining small towns have declined within the past seventy years but some are floating on a rising tide of prosperity, due to modern industrialization.

SUBREGIONS

This region has many contrasts, some essentially local and others covering large areas: the former include the inter-mingling of small bogs and carefully farmed lowland, the latter the essential differences between the lowlands of Meath and those nearer the Shannon. A division into subregions may be sought in the nature of the present agricultural landscape, which is the result of choice and tradition among the inhabitants: such a soil and such a climate as that of Meath would not preclude and might even favour arable farming, with enclaves of market gardening and fruit growing. Again, the naturally rich pastures are equally suitable for dairy cattle as for fat cattle, and yet the latter reign supreme, except in the immediate vicinity of Dublin. Beyond the Meath–Kildare grazing belt, mixed farming is more usual, but the store trade is still dominant over dairying. In so far as there is any evidence of 'geographical control', one may note that large cattle-fattening farms could develop successfully only in an area unbroken by bogs such as Meath, while smaller mixed cattle-breeding farms are more suited to a bog-interrupted landscape such as the lowland near the Shannon. Different as the farming types may be, they depend on one another, for the breeding and feeding areas supply the fattening areas, which in turn rely on the demands of Dublin and the British market.

1. THE MEATH–KILDARE CATTLE-FATTENING BELT

Economically this belt corresponds fairly closely with those areas having an average farm size of more than 100 acres: physically, it is undulating

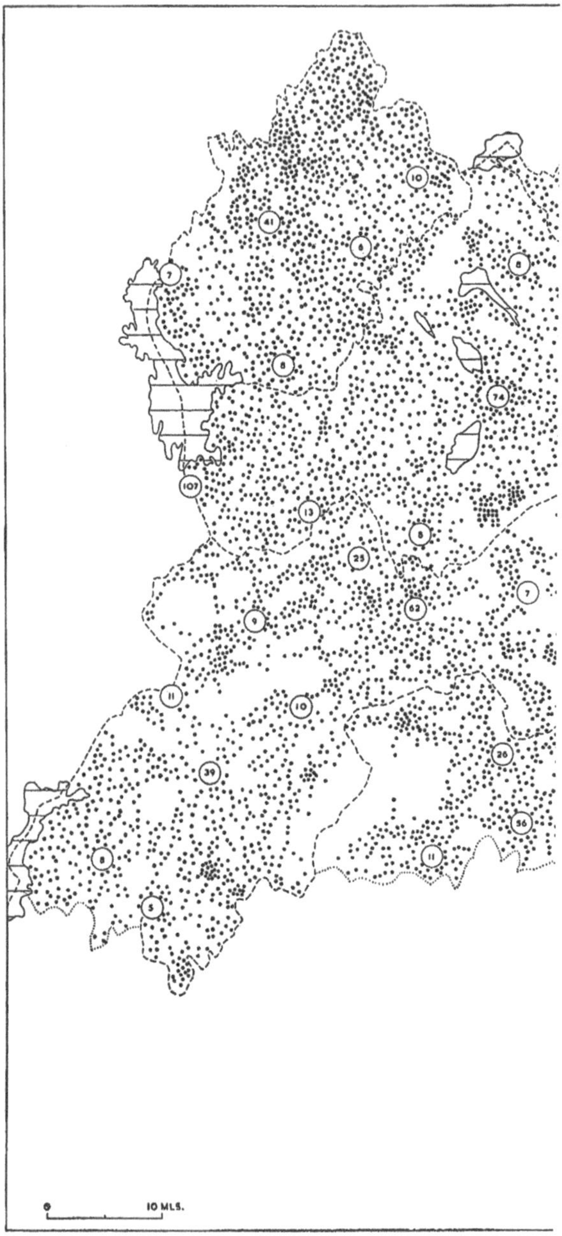

FIG. 47. POPULATION DISTRIBU

One

A great variety of settlement types will be noted. The inf
Meath the sparsity of occupation is greater than anywhe
farmland gives an intermittent pattern typical of many p

CENTRAL LOWLAND

rsons

ıblin extends for some miles around the city, but in Co. land. Farther west, the interpenetration of bogs with entral Lowland.

country rising up from *c.* 200 ft. to 400 ft., with some isolated hills over 500 ft., including the historic hill of Tara (512 ft.). Covered largely with glacial drift, it has soils varying from heavy clays through marls to the light sands of the Curragh – which from its military use invites comparison with Aldershot. Near Dublin there is some fruit growing near Rush and some interesting bulb farms and market gardens, but these are not considerable and the dairying belt is restricted though there have been signs of expansion during the last twenty years. The natural wealth of the cattle-farming belt is undoubted and it remains famous for its fine grass pastures.

2. THE MIDDLE BARROW LOWLAND

Lying between the Castlecomer plateau and the Leinster chain, this lowland is drained by the middle sections of the Barrow and by the Slaney in sharply incised courses. Drift-covered Carboniferous strata, Ordovician slates, and granites form country around 200–400 ft., varied by hills of upper Carboniferous limestones near Stradbally, and of granites near the Barrow south of Carlow. The drift, which occurs almost everywhere, contains lime and varies in depth from a thin rind to as much as 100 ft. Eskers are few and bogs occur only on the surprisingly level country to the west of the Blackstairs mountains. Agriculturally the Barrow lowland lacks the specialization of the Meath–Kildare cattle belt, and shows variety rather than uniformity. About half the farms are of medium size but both large farms, especially those of 100–200 acres, and small holdings of less than 30 acres are also represented. Generally one-fifth of the farmland is given to arable crops, chiefly cereals and roots, with sugar-beet growing for the Carlow factory. Cattle are reared both for dairying and stores, and there is neither the fattening emphasis of Kildare nor the creamery dominance of Kilkenny. Until recently, there was considerable barge traffic along the river but lorry services have replaced the barges though the railway to Kilkenny and Waterford still exists. Carlow is the largest of the market towns, which occur at an average distance of ten miles one from another, and the general density of the rural population is around 80 to the square mile, or 100 per square mile of improved land, a figure typical of many mixed farming areas in the south and east of Ireland.

3. THE BOGLANDS EAST OF THE SHANNON

Some 1,000 square miles east of the Shannon consists of carefully farmed drift-covered lowland interpenetrated by extensive bogs. This belt of country, which extends to within twenty miles of Dublin, is drained by the headwaters of the Barrow, the Boyne and two Shannon tributaries, the Brosna and the Clodiagh, though there are many areas of indeterminate drainage. Skirted rather than crossed by main roads and railways, this area is long-settled and of considerable agricultural wealth rather than a desolate wilderness: few parts of it covering even 100 square miles have as much as half the surface in bog. Most of the area is 200–300 ft. high, rising gently to 400 ft., but varied by higher ground such as the Allen range near Kildare with four hills around 700 ft. high, and the igneous Croghaun, 769 ft. in the middle of vast bogs west of Edenderry. In general the landscape is a drift-covered lowland crossed by a wonderful series of eskers, whose human importance lies in their value as routeways, especially in those areas where the boglands are most widely developed. The Shannon crossings at Athlone, Shannonbridge and Banagher are all associated with eskers, and the ancient cathedral centre of Clonmacnoise, with its seven churches, stands on an esker overlooking the extensive bogs around the Shannon. Clonmacnoise, however, though so ancient and reached by a 'pilgrims' road' along the esker has no bridge across the Shannon, for there is no firm ground on the west bank. Fifteen miles separates the bridge of Athlone from Shannonbridge, and the river flows for another eight miles to reach the crossing at Banagher. It is not therefore remarkable that the Shannon should be a county boundary for almost the whole of its course. There is sufficient 'dry ground' in the Central Lowland for the roads to avoid peat but the canal and the railways cross through the heart of the bogs.

The distribution of settlement (Fig. 47) shows very clearly alternation of farmlands and boglands. In the extreme east of Co. Offaly, Edenderry has only 56% of its area in crops and pasture; farther west, Birr I rural district has 64%. These figures contrast sharply with the 94% of districts in Co. Meath, yet farming is the main resource, and as much as one-fifth of the settled area is given to arable cropping in some parts. The cash resources are varied and mixed farming is general: cattle are reared for milk production, stores, and, in some cases, fattening, but very few sheep are kept. The small type of holding becomes increasingly prevalent towards the Shannon, and the average farm size decreases steadily from 100 acres in the east, around Edenderry, to 45 acres near the Shannon. Several of the market towns in this area are excellent trading centres with some useful industries. Athlone is the largest town on the Shannon above Limerick and the meeting place of numerous roads: the smaller towns include Tullamore, Maryborough (5,900), Birr (3,900), Edenderry (2,900), Mountmellick (2,600) and Portarlington (2,800).

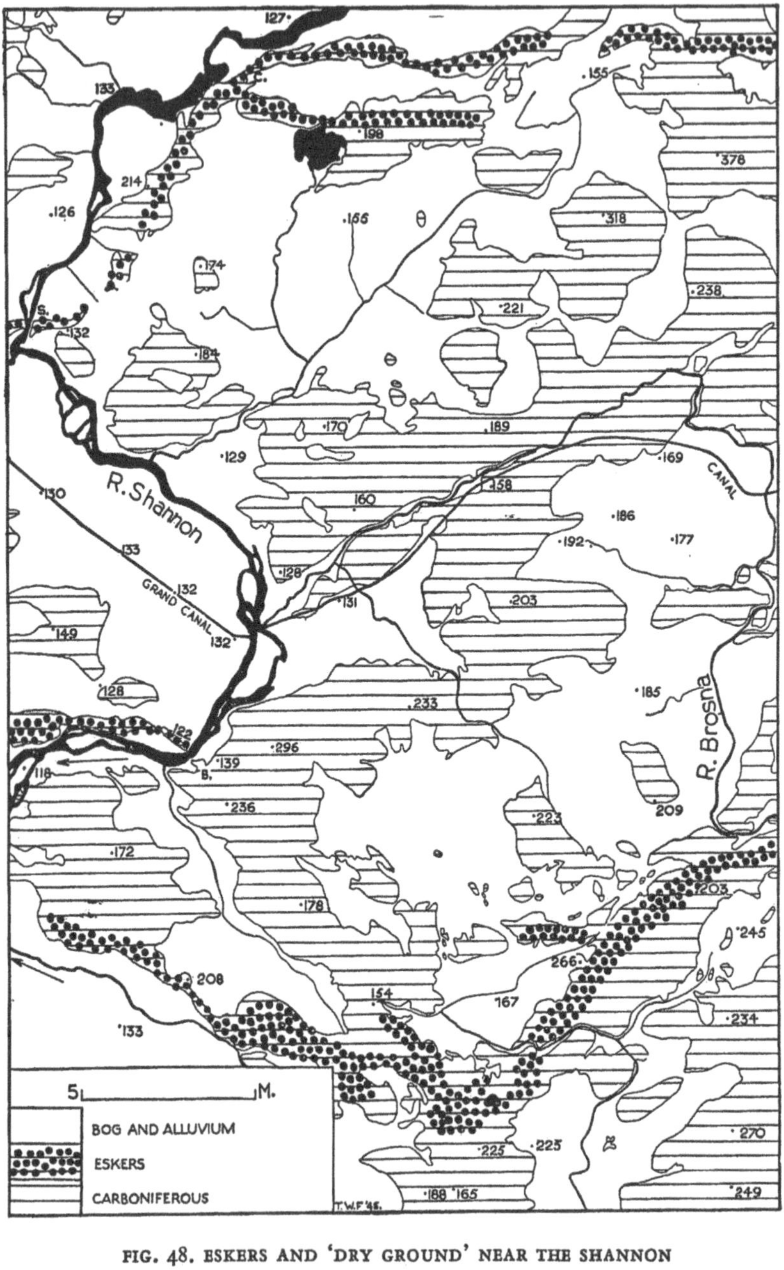

FIG. 48. ESKERS AND 'DRY GROUND' NEAR THE SHANNON

The bog and alluvium is not improved land but the remainder is settled: eskers wind across the lowland. The mixture of the bog and dry ground is typical of the country near the Shannon. C. (near top) denotes Clonmacnoise; S. Shannonbridge; B. Banagher.

4. THE LOWLAND OF WESTMEATH AND LONGFORD

This subregion is the agricultural annexe of the Meath cattle-fattening belt and in some ways similar to it, though its farms are smaller, its pastures more broken up by bogs and its general topography more varied. Much of it consists of an undulating Carboniferous-floored and drift-covered surface around 200–400 ft., but the structure is complicated by numerous folds and the lower Carboniferous limestone rises into hilly country more than 600 ft. high to the west of Mullingar. Other hills 600–900 ft. high are formed of unremoved middle and upper Carboniferous beds or are anticlinal ridges of pre-Carboniferous strata, such as the isolated hill at Ardagh, 659 ft. high, and the Slieve na Calliagh near Oldcastle, 904 ft. These hills diversify a landscape that becomes dreary and monotonous in such areas as the flat-floored and bog-covered Inny valley, in which the river falls only 9 ft. in a stretch of twenty-four miles. Towards the Shannon bogs become more frequent, but even so they rarely cover as much as one-quarter of any area of as much as 100 square miles.

The settlement pattern of this subregion is less broken than that of the area to the south and less continuous than that of Co. Meath to the east. The density of population in the rural areas is 40–90 per square mile, and 50–100 if uninhabited areas are excluded. The relatively low densities may be correlated with an emphasis on cattle rearing, both for stores and fattening, though towards the Shannon mixed farming becomes more general, and in the north of Longford dairying becomes steadily more important with the transition to the drumlin-drift country.

Everywhere, however, the farming is pastoral in emphasis and only one-tenth of the farmland is given to arable crops at the expense of pastures and meadows. Young cattle move into this sub-region from areas west of the Shannon, and nearly half the herds of Westmeath are beasts more than two years old destined for final fattening in Meath. The farms are smaller than those of the more specialized areas in Meath; the average of 80 acres in Westmeath falls to 35 acres in Co. Longford, where most of the farms are of less than 30 acres. The towns consist of market centres such as Mullingar, Longford (4,100) and Granard (1,040): Mullingar has a variety of small industries, including knitwear, and Longford now has cotton and rayon weaving, with a considerable trade in meat partly for the export market. The towns are widely separated one from another, and the trade of the countryside is drawn into various smaller centres that are mere villages. As yet, this subregion shows little traces of that intermingling of agriculture with industry which may be seen in some parts of the Central Lowland east of the Shannon.

CHAPTER 13

The Leinster Chain

From the southern suburbs of Dublin there is a fine view of the northern edge of the Leinster chain, which runs in a southwesterly direction for over sixty miles from Dublin to the neighbourhood of New Ross, and is an anticlinal fold having a granite core, with varied Cambrian and Ordovician rocks penetrated by igneous intrusions on its margins. Best developed as a mountain range in the north, it narrows farther south into the sharp ridge of the Blackstairs mountains and finally appears only in isolated hills of which the chief is Mount Brandon, a few miles north of New Ross. The coastal lowlands are nowhere more than five miles wide between Dublin bay and Wicklow head, south of which they widen into the fertile lowland of Wexford, sometimes called the 'garden of Ireland'. As a barrier the Leinster chain is important only to the north of the Slaney, which rises on the west side of the mountains and then cuts through them in a well-defined gap near Newtownbarry. The dominance of the Caledonian trend is the salient topographical feature of southeast Ireland: it is seen in the lines of hills and valleys and even in the Wexford lowland, and is followed in the pattern of roads and railways, settlements and farms.

Much of the topographical detail is derived from the moulding influence of icesheets and mountain glaciations, with important glacifluvial phases. The soils are varied, but among them are many that are well drained and of considerable fertility, developed from the morainic accumulations of hummocky drift and the lake-clays or sands. The southeast is mainly an agricultural area, economically well placed with relation to Dublin's market and port facilities: there is sheep farming on the mountains and cattle rearing for stores, meat and dairying on the lowlands. The land use is varied, but approximately seven-eighths of the farmland is in grass near Dublin, though arable crops are more prominent in Wexford. There are several market towns such as Wicklow, Gorey, Enniscorthy, New Ross, Arklow and Wexford, and a number of market villages which help to canalize the economic life of the region. In the north the outer suburban fringe of Dublin stretches along the coast to include Bray and Greystones.

LANDFORMS AND LAND USE

The great Caledonian anticline is a granite bathylith (or laccolith) flanked by lower Palaeozoic sediments with associated igneous rocks. The major topographical features in the core of the mountains are rounded granite domes with wide open valleys modified in places by glaciers to U-shaped trenches. The mica-schists form sharper hills along the margins of the upland, but the Ordovician slates and grits have weathered into plateau-like expanses with a few quartzite knobs, notably Bray head, the Little and the Great Sugar Loaf. The Wicklow mountains are nowhere crossed at an altitude of less than 1,000 ft. and the Military Road through the mountains, built after the 1798 insurrection in an effort to control the inhabitants, runs at a height of more than 1,000 ft. for over twenty miles: it has recently been opened up as a tourist route and forms a fine mountain drive. Remnants of former surfaces of erosion are an impressive, if as yet incompletely interpreted, feature of the Wicklow mountains. Two of the most striking are the Vartry peneplane on the east and the Blessington basin on the west, both of which slope from *c.* 1,000 to 500 ft. The former is underlain by Ordovician and Cambrian grits and slates, the latter by Ordovician beds and granites. The recognition of the various surfaces is made difficult by the presence of glacial and glacifluvial deposits that vary from a thin rind to a covering as much as 200 ft. deep, but there are indications of surfaces around 1,300–1,400 ft. and perhaps also at 2,000 ft. in the mountains; the subdued and generally rounded form of the granite hills themselves suggests yet another surface around 2,500 to 3,000 ft. At altitudes below 1,000 ft., there appear to be erosion surfaces at about 800 ft., 600 ft., 400 ft. and 200 ft.

South of Lugnaquilla the Leinster chain ceases to be a significant barrier, but the granite core of the anticlinal ridge appears in several isolated hills between Lugnaquilla and the Slaney and appears also in the Blackstairs mountains between the Slaney and the Barrow. On the east side of the range Ordovician slates and grits, which have a continuous extension from Killiney bay to Waterford harbour, are pierced by two lines of igneous intrusions of varied character and hardness which form two lines of hills. The first of these extends from a point southwest of Wicklow town through Croghaun Kinsella (nearly 2,000 ft.) into the northern part of Co. Wexford, the second extends from Arklow head southwest for eighty miles across Wexford and Waterford to the coasts near Dungarvan (p. 308). Where these intrusions form hills, they break the monotony of the lowland and imprint upon it the Caledonian trend in unmistakable emphasis.

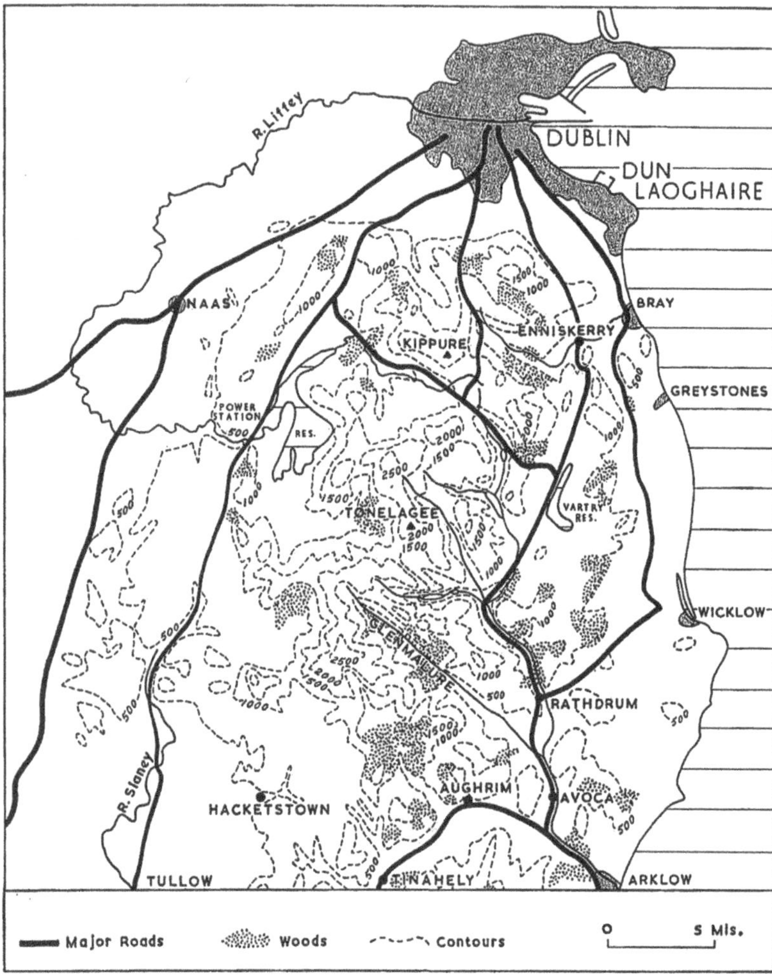

FIG. 49. THE WICKLOW MOUNTAINS

This is the northern part of the Leinster chain, though Three Rock and other hills are commonly called the Dublin mountains. The limit of cultivation and settlement ascends in places to 1,000 ft. and woods are prominent in the valleys.

Based on the Ordnance Survey by permission of the Minister for Finance of Ireland.

Two general icesheets from the north have left limestone-bearing drift, of which the older is generally a tough clay with a varied stone content, including rocks of Scottish origin, though limestone pebbles are the most frequent of all. The newer drift penetrated some of the Wicklow valleys to an altitude of at least 1,800 ft.: its morainic deposits are also a very clear feature of many parts of the coastal lowland, especially to the south of Cahore point, where there is a jumble of hills and ridges, with many kettleholes, rising to a height of 350 ft. and extending to a point only three miles north of Wexford harbour. The deposits of these icesheets have proved vital influences in the land use of the region. In some valleys, such as those in the northern part of the Leinster chain (locally known as the 'Dublin mountains'), the past or present limit of improved land corresponds almost exactly with the upper limit of the drift. In general, the moraines themselves and the flanking outwash deposits provide varied conditions of soil and drainage, but there are some particularly fertile soils on natural loams. In the mountains there are some fine glacial features, including the remarkable moraines left by two glaciations on the Vartry peneplane, and some U-shaped valleys on the east side of the mountain crests, notably in the four valleys converging at Laragh and also in Glenmalure. Striking morainic accumulations occur in many of the valleys, especially around Annamoe at the southern end of the Vartry peneplane, around Rathdrum, and also in Glendalough and in Glenmalure. Some of these valleys are spanned by terminal moraines and have almost undisturbed lateral moraines covered in places with indigenous oak-birch woodland or with new forestry plantations. There are also some corries with block moraines; two very fine corries at the head of the Glencree valley are divided one from another by an arête-like wall, and occupied by lakes (upper and lower lough Bray) dammed by block moraines.

The present rivers are moulding their courses in valleys that are mainly plastered with drifts and lacustrine deposits and have reached solid rock only in a few places. As a whole, they are in a youthful stage of development, contrasting strongly with the mature outlines of the granite mountains. In many of the valleys erosion is particularly active and almost every flood brings some readjustment both on the valley floors and on the hillsides. At present the peat is accumulating, if at all, at so slow a rate that a remarkable natural wastage is in progress and all the rivers are coloured brown. When the granite is reached, it is turned into the gravelly deposit so frequently seen in the channels intersecting the upland peat bogs. Further, the clearing of many upland peat bogs for wartime fuel needs will accelerate erosive processes by the removal of a temporarily protective cover; on the other hand, the afforestation of many Wicklow hillsides will help to preserve them against erosion though there are many very steep and unstable slopes above the forest level on which soil creep is very active. The Wicklow valleys have varied exposures to insolation and other elements of climate. The southward-facing slopes have generally less steep sides than those facing north and there is a widespread tendency to asymmetrical forms. The settlements are mainly on the south-facing slopes, as would be expected, but it is not clear why the drift deposits should be thicker on the same side.

Approximately one-half of Wicklow, but little more than one-tenth of Wexford, is rough pasture or non-agricultural land. There is a wide stretch of open country in the Wicklow mountains, where the 1,000 ft. contour encloses an area of 205 square miles, and similar but much smaller moorlands exist on the Blackstairs mountains and on the many uplands that flank the Wicklows or break up the Wexford lowlands. The sand dunes and shingle ridges along the coast at intervals from Greystones to Bannow bay (Co. Wexford) are also negative areas for agricultural settlement: some of these dunes enclose freshwater or salt marshes, such as the Murrough to the north of Wicklow town. The main agricultural resource lies in the soils derived from the glacial and glacifluvial materials on the lowland, and to some extent also in the uplands.

The frontier between the farmlands and the moorlands has oscillated during the past hundred years through an intermediate or hill-pasture belt, mainly between 600 and 1,000 ft., which has gorse and bracken plentifully spread through it. The low-growing *Ulex gallii*, dominant on the higher ground, is moulded into rounded bosses separated by sheep walks; the tall *Ulex europaeus* is found mainly on the lower ground. Comparison with conditions in 1904 shows that several farms have been abandoned and the hill-pasture belt therefore extended. The belt is, in fact, a frontier of agricultural settlement in which the enemies, gorse and bracken, are perpetually on the offensive and already possess a 'bridgehead' on the margins of fields now ranking as improved land. Broken stone walls are widespread features of this marginal belt, but its decline in status is not necessarily recent, for many upland farms were abandoned from 1870 onwards. Nor is the belt useless, for the present farmers graze their sheep wherever they can. The field worker is constantly shown fields at an altitude of as much as 1,000 ft. which 'grew wheat' a century ago, and while the past productivity is perhaps exaggerated, the frontier of settlement has been pushed back, though one family will maintain a high-lying farm and another will not: in many cases, a remote upland farm is abandoned on the death of the holder, especially if all the sons and daughters have emigrated and the family is, to quote an old woman living in Glencree, 'worn out'.

Although the hill-pasture belt has been closely studied only at the northern end of the Leinster chain, it occurs at similar altitudes in other Irish uplands. The tide of decay has been arrested in some places by the planting of woods, some of which reach an upper limit of 1,000–1,500 ft., which must correspond fairly closely to the climatic limit of trees in this area. Already the afforestation has altered the appearance of several valleys, generally by the addition of contrasting colours, especially in the larch belts, and many of the former farmsteads have now been entirely obliterated by the new plantations. As yet, the afforestation has not submerged all traces of the sub-spontaneous or natural woodlands, chiefly of oak and birch, sometimes with holly, which clothe many of the steeper slopes in the mountain valleys. The treeless hillsides in many of the valleys are also of interest, particularly in April and May, when they have colourful sheets of bluebell, violet, germander speedwell, lesser celandine and primrose on areas covered later in the season by bracken. Such vegetation probably represents the lower layers of the former forest community, now dominant. Bracken is not only able to hold its own against gorse, but even to push back its rival, especially on east-facing slopes sheltered from the westerly winds: in some hollows and glens the east-facing slope is covered with bracken and the west-facing with heather. Heather moors are widely developed throughout the Leinster chain, where the drainage is good, and in many places extend continuously from the hill-pasture belt to the tops of the mountains; the heather is as much as 2 ft. high at the lower levels, but dwarfed at higher levels. Generally the *Scirpus* or *Eriophorum* (cotton-grass) associations are developed on the higher and more level areas, with some *Rhacomitrium* moss that is intermingled with *Sphagnum* and associated with dwarf *Calluna*, *Scirpus*, *Erica tetralix* and *Juncus squarrosa*. The three associations marked by *Scirpus*, *Eriophorum* and *Rhacomitrium* are almost certainly forming bogs at the present time, but the accelerated cutting of many *Scirpus* and cotton-grass areas will probably result in an extension of the heather moors.

Agricultural life in southeast Ireland is closely influenced by the proximity of the Dublin market. The proportion of dairy cattle in the herds increases from one-quarter in Wexford to one-third in Wicklow; most of the lowland farmers sell the milk to wholesale dealers, who collect it by lorry from platforms outside the gates of the farm or by direct rail services. In Co. Wexford a few co-operative creameries have been established, but direct sale to dealers is more usual. Cattle are reared for the local meat trade or for export through the ports of Dublin, Waterford and Wexford. Some of the salt marshes are used as pastures for cattle, sheep and horses: one part of the marsh between Wicklow and Greystones is

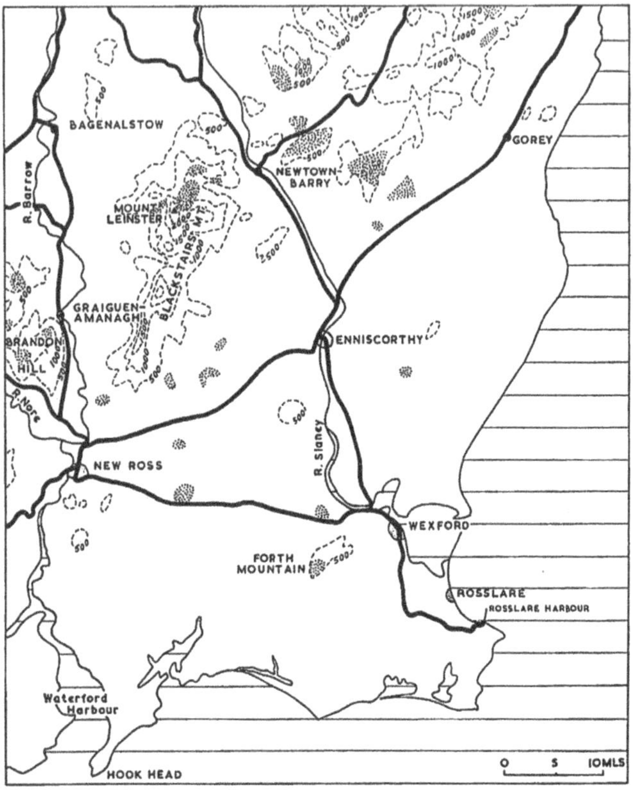

FIG. 50. THE SOUTH PART OF THE LEINSTER CHAIN

The Slaney river makes the only lowland route through the Leinster Chain.

Based on the Ordnance Survey by permission of the Minister for Finance of Ireland

traditionally supposed to cure sick horses. The sheep are of the Wicklow mountain breed, a cross between the native strains and the Cheviot breed: in winter they are brought down to the lowlands in very large numbers, and they are reared both for meat, which is sold at local fairs or to dealers, and wool, which is sold co-operatively through the Wicklow Mountain Sheep Breeders' Society. Pigs and poultry are reared on most farms, but are relatively of greater importance on the smaller farms. In recent years some poultry farms have been established, but this is rather the effort of a few enterprising people than any widespread development.

Mixed farming with a pastoral emphasis is the most typical form, and approximately two-thirds of the improved land is in grass. The arable cultivation of the region is generally planned to meet the needs of livestock and the farm household, but Co. Wexford still has more of its land in crops than any other part of the Republic, except for the completely different country in the west of Donegal; before the post-1932 propaganda drive, it had more wheat cultivation than any other Irish county except Cork, with substantial acreages of oats and barley. Sugar-beet has been grown here in steadily increasing quantities and Co. Wexford normally supplies about one-tenth of the national output. Half of the farms of south-east Ireland are of 30–100 acres, and the holding of more than 50 acres is widely represented in most lowland landscapes. One-fifth of all the farms are even larger, though only 5% are of more than 200 acres. The standard of farming varies widely: it would be possible to tour a dozen farms of 100 acres, and find magnificent standards of work with splendid farm buildings, adequate machinery and a well-kept house. It would be possible to tour another dozen farms and find a different picture. There are also numerous small farms of 30 acres or less, interspersed among the larger farms of the lowland: in some of the upland valleys, even within a few miles of Dublin, holdings not unlike those of the west of Ireland may be seen. The houses range from the thatched and whitewashed two-roomed cottage to substantial dwellings of granite, some of which date back to the eighteenth century, of austere but attractive architectural features.

INDUSTRIES

The few small industries that exist in the region include several enterprises supplying agricultural needs or making articles from agricultural produce, together with a few new concerns located here for no particular geographical reason. The latter group includes a pottery at Arklow, a razor-blade factory at Enniscorthy, a works making insulated wires and cables at New Ross and enterprises at Bray making inks and polishes and also the entire home supply of electric light bulbs. The industries allied to agriculture are of much greater interest and are generally of older establishment. They include fertilizer factories at Wicklow and Kilcoole, maize, wheat and oat mills (some of which are quite small enterprises) in various towns and villages, bacon factories at Enniscorthy and Hacketstown, tanning in Gorey and malting at Wexford and New Ross. Successful small enterprises include hand-weaving mills, which produce tweeds of attractive colour and design, at Avoca in the Vale of Avoca, and a woollen mill near New Ross making blankets, tweeds, flannels and rugs. Wexford has iron foundries which provide a large variety of goods including agricultural machinery and employ several hundred people. Creameries have been successfully established in Co. Wexford, though those in the north of the county have to meet the competition of the prices offered for milk by the Dublin wholesale merchants. There is no widespread industrialization: the towns exist primarily as market centres, but have also attracted a few concerns such as the pottery at Arklow and retain some older enterprises such as the Wexford iron manufactures, which supply most of the agricultural machinery produced in the Republic.

POPULATION

The population of this area was 182,000 in 1891, *c.* 156,000 in 1936, 159,000 in 1951 but 151,000 in 1966, 105, 87 and 85 to the square mile respectively. The rural population has declined by one-third from 149,000 to 98,000, but the town population has increased from 33,000 to 50,000: there are also some 19,500 people in small towns and villages counted as rural population. The outer suburban limit of Dublin is reached in Delgany and Greystones, which in 1966 had 3,600 people: Delgany is an old village, but Greystones grew from a tiny fishing hamlet into a residential centre with 500 people in 1918 and several times as many now. Bray, five miles nearer Dublin, a market, light industrial and residential town, increased steadily in population from 7,000 in 1891 to 13,400 in 1966. The smaller places include some built by the gentry, such as Enniskerry and Blessington, which have a dignified appearance, and others not so attractive. Many new cottages have been built in the past twenty years, some of them close to existing villages. The pattern of rural settlement is remarkably even over most of the region and gaps are either moorlands, peat bogs or sand dunes and shingle ridges along the coast. The general density of population, 50–70 per square mile of improved land, is typical of areas with relatively large farming units. Some areas have densities of less than 50 to the square mile (see Fig. 22, p. 133) with large farms. So far the use of the coast for holidays has brought little permanent settlement.

SUBREGIONS

The main characteristic of the southeast is the intermingling of upland and lowland, with uplands more prominent in the north and lowlands in the south. As a whole, the Leinster chain and its bordering lowlands is relatively easily divided into smaller units, and many of the subregions are fairly clearly delimited by physical features and possess definite characteristics in their life, so that they may be regarded as small *pays*.

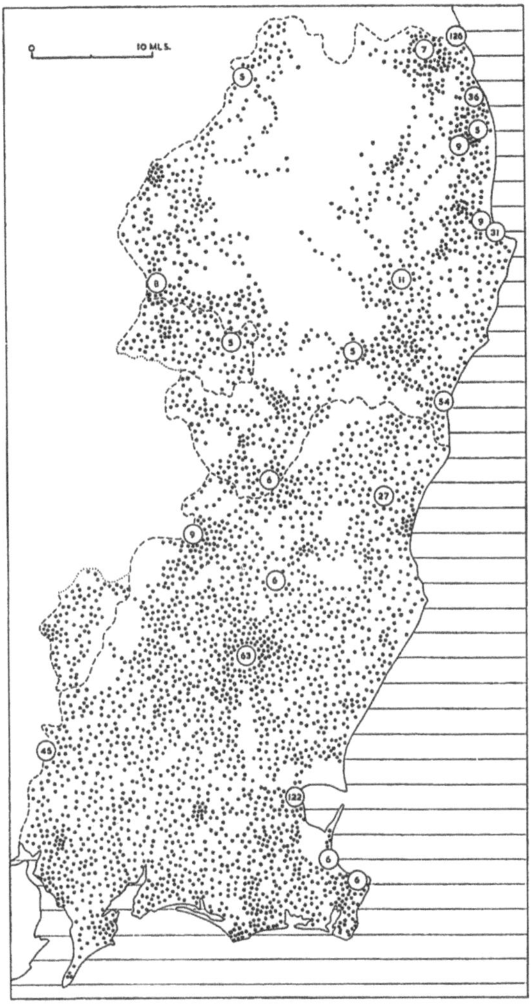

FIG. 51. DISTRIBUTION OF POPULATION, LEINSTER CHAIN

One dot to 25 persons

South of Greystones, the population density is not heavy; it is typical of many areas dominated by medium-sized farms.

1. THE CENTRAL MOUNTAIN BELT

The central part of the Wicklow mountains, about ten miles wide and twenty-five miles long, consists of rounded granite uplands with summits of 1,500–2,000 ft. in the north and more than 2,000 ft. in the south, finally reaching 3,039 ft. in the schist-capped Lugnaquilla. The junction of the granite with the mica-schist is seen on the east as a line of sharp-featured hills from Knockree to Maulin and Djouce, and thence as a hilly ridge intersected by deep gorge-like valleys to lough Dan and Glenmalure. On the west side the mica-schist line is less clearly defined, though it forms several hills and ridges along the upland edge. The granite upland is mainly bog-covered moorland intersected on the east by valleys that are normally wide and open on the higher ground bu t narrow and steep-sided where they were occupied by glaciers. On the west side the valleys are usually mature in outline and less sharply glaciated than those of the east. The rivers on the east are downcutting rapidly, and in places dramatically, and have waterfalls or torrent courses where the lower youthful reaches are cutting back into the mature upland surface around and above them: the rivers on the west side flow mainly through lowlands covered with drift and glacifluvial materials.

Land available for cultivation and settlement is restricted to valley floors and a few of the hillsides; improved land covers less than 10% of the whole area. The valley floors have small floodplains of alluvial material, much of it re-sorted drift, which forms good cattle pastures or arable fields for oats and potatoes. In addition, a few of the moraines provide good and well-drained soils, but the many gaps in the settlement pattern include marshy areas, some with willow scrub, or moraines and outwash deposits of coarse sand and gravel left in forest, some of which is oak-birch woodlands, with holly in places. The hillsides are equally varied in their present land use, for sheep farms on well-drained drift deposits alternate with land never cleared for agriculture and covered in part with some relics of the original woodland, seen most clearly in Glendalough. But the large plantations, of which some cover at least two square miles, are adding a new element in the landscape, and thereby improving or desecrating the scene according to the taste of the observer. The mountains have obvious tourist attractions but the one place visited by large numbers is Glendalough, visited for its Seven Churches, the Round Tower and the fine U-shaped valley with its two lakes separated by the delta of a stream flowing from a hanging valley (Fig. 52).

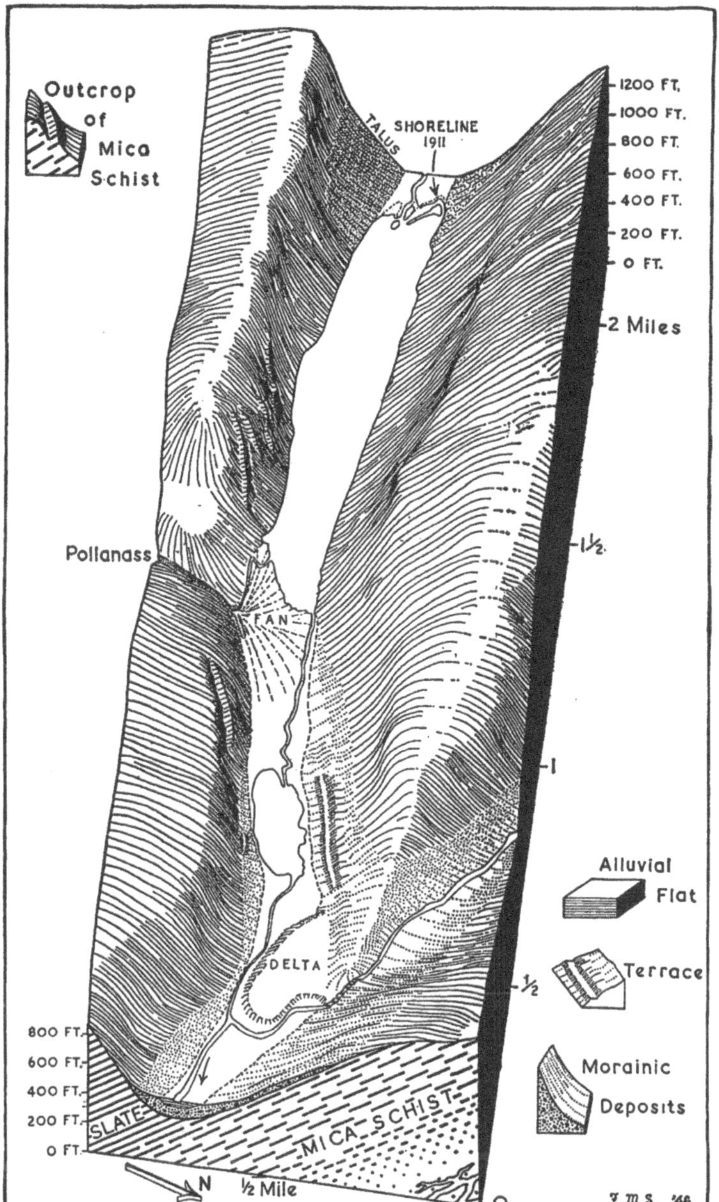

FIG. 52. THE GLENDALOUGH VALLEY

This non-perspective diagram shows the deeply glaciated Glendalough valley; morainic deposits cover part of the hillsides and the valley floor and the height of a former lake level is shown by the delta and terrace.

Constructed from field observations and from the Ordnance Survey, 6″ to 1 mile, sheet 23, Co. Wicklow, Dublin, 1911, by permission of the Minister for Finance of Ireland.

Drawn by F. M. Synge and published in 'Irish Geography', vol. 1, p. 60

2. THE BLESSINGTON LAKE-BASIN

The name of this subregion combines recognition of both the glacifluvial lakes that filled it to a height of 900 ft. at its greatest extent and the modern reservoir constructed with a water level around 615 ft. to augment the water supplies of a rapidly growing Dublin, and to provide electric power (Fig. 53). The new reservoir was constructed in 1938–40 and submerged for ever some remarkable morphological features in an area of several square miles, where the Liffey and its tributary, the King's river, were cutting a series of remarkable meanders in the old lake bed and exposing varved deposits; the soils of the basin varied from humified peat to fen-covered clays and well-drained gravels. The new lake is scenically attractive, as it reaches up into various valleys in long fingers of water separated by farmed ridges. The whole lake basin was worked out in preglacial times through the capture of the 'Liffey' (1 and 2 in Fig. 53) which found a course westward through the Slievethoul ridge. This outlet was closed by glacial deposits and the new Poulaphouca gorge made: the tributaries began to dissect the basin, a process still proceeding outside the area occupied by the new lake.

The farms around the lake include several of considerable size, some of more than 100 acres, but there are also a number of smaller holdings, especially at the upper levels: in places settlement reaches 1,000 ft. The quarries of Ballynockan have supplied large quantities of granite for buildings in Dublin. Blessington, a village of late seventeenth-century foundation, with a granite church built in 1683, now has some 550 inhabitants and is a small market centre. The road from Blessington to Dublin had a steam tramway from 1882 to 1932, but almost the only relic of its track is a series of memorials to those killed along the route An extension of this tramway ran from Blessington to the Poulaphouca gorge from 1890 to 1927.

3. THE VARTRY PENEPLANE

Though possessing a strong resemblance to the Blessington area, the Vartry peneplane has an entirely different surface cover which consists mainly of piedmont moraines from ice tongues that pushed out of the Annamoe river valley and spread almost to the eastern border of the area. These moraines cover the older drift, which may now be seen as a very much weathered deposit; its thin leached soils support only bogs and heather moors. Most of the fields are on the moraines or on the floodplain of the river Vartry. The whole area, which now appears as a peneplane sloping from 1,000 ft. to 600 ft., would remain mature in outline even if

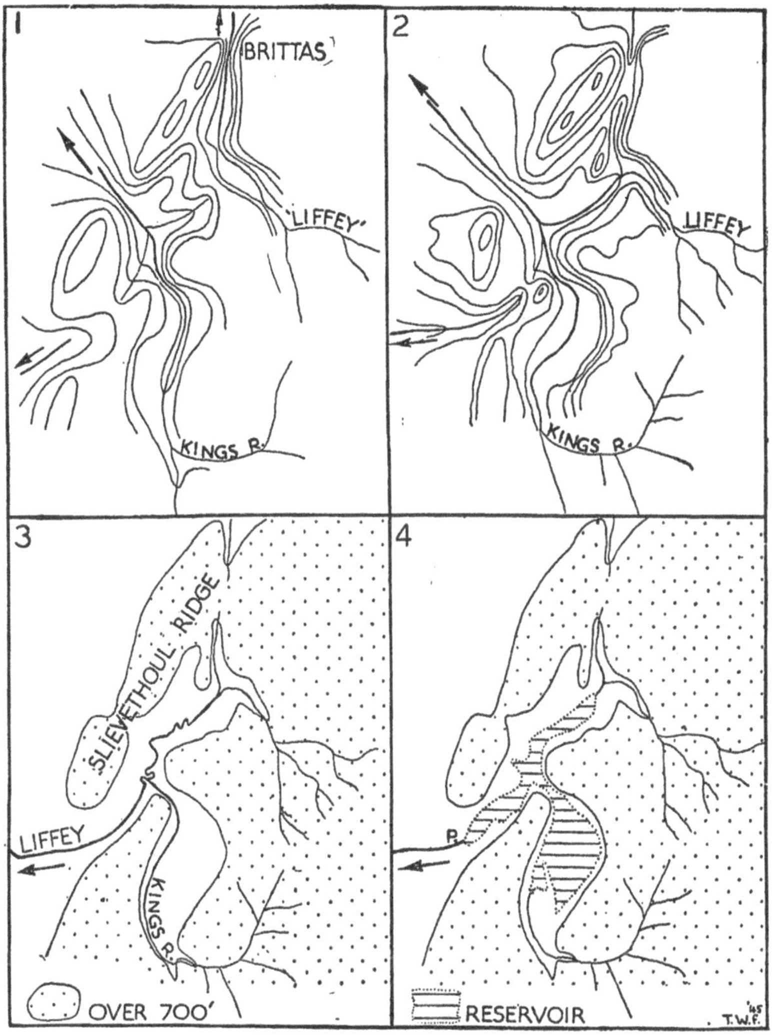

FIG. 53. THE UPPER LIFFEY VALLEY

The four stages are discussed in the text, p. 298.

the shallow drift were removed. The Vartry river, now cutting vigorously into the surface deposits, finally escapes to the coastal lowland through the Devil's Glen, a deep gash used by glacifluvial waters. Some four miles wide and ten miles long, this peneplane is clearly bounded by the quartzite Great Sugar Loaf, 1,654 ft. on the north, by knob-like hills of quartzite reaching 1,000–1,200 ft. on the east, and by the hillfoot of Djouce and other mica-schist hills on the west.

The general upper limit of cultivated land is 700 ft., with local extensions to 1,000 ft. through a well-defined hill-pasture zone dominated by gorse and bracken. Cotton-grass moors are developed with some interesting gradations into heather and bilberry moors as natural drainage improves, but trees are restricted mainly to windbreaks around some of the farms. In the northern and higher areas around Calary there are farms of 50 to 70 acres, many of which are settled by people of English descent settled here by their landlords, but farther south, in the Roundwood area, the farms are much smaller and the holders Irish. The farming is mixed, with as much as one-quarter of the land given to oats and potatoes, and sheep rearing is predominant, though there is also cattle rearing, some horse breeding, and poultry and pig raising. The cattle and sheep are mainly stores sent to the lowlands and the area produces butter rather than fresh milk. An interesting special development is seed-potato production both for the local trade and for export. Thinly peopled, with less than 30 to the square mile, this area has a small market village in Roundwood (220), and is served by a bus from Dublin to Glendalough.

4. THE BRAY–WICKLOW LOWLAND

This lowland lies between three prominent quartzite hills, the Great Sugar Loaf (1,654 ft.), the Little Sugar Loaf (1,120 ft.) and Bray head

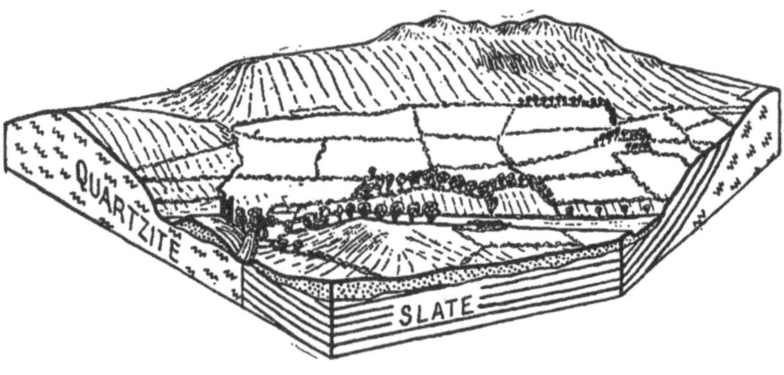

FIG. 54. THE LITTLE SUGAR LOAF

Quartzite stands out in Bray head, the Little and the Great Sugar Loaf above the surrounding country, here of Cambrian slates: most of the low-lying area is covered with drift.

Drawn by F. M. Synge

(795 ft.) on the north, and Wicklow head on the south; it extends on the west to the edge of the marked break of slope formed by the edge of the Vartry peneplane. It is one to five miles broad and fifteen miles long, and is an important agricultural area just penetrated on the north by a residential development that shows the extreme limit of the Dublin suburban spread. The town of Bray, on the north side of the head, belongs essentially to the Dublin complex and has become its Southend and Blackpool since the railway was built in 1851. Previously there was 'no esplanade . . . nothing but the heaped sand, shingle and low sand dunes, along which a rude raised pathway led to the Head, interrupted by numerous hollows and irregularities caused by the digging and carting away of the sand by farmers and others in the neighbourhood'.*

Greystones stands on a small rocky promontory made up of Cambrian slates and grits with some quartzite, all under sharp attack in winter storms along the coast. A holiday and residential centre of infinite respectability, it is in part a model suburban development of the pre-1914 era, with tree-lined roads, substantial houses, spacious gardens and a pleasant lack of uniformity. Fig. 51 shows that the density of population in the coastal lowland decreases sharply to the south of Greystones.

Coastal erosion of some severity has occurred between Bray head and Greystones, and the railway has twice been re-laid farther inland. Between 1905 and 1928 the loss of land was estimated at 4 ft. a year and storms have removed houses from the North Beach at Greystones, and almost demolished the harbour. To the south of Greystones a ridge of shingle and sand forms the coast for more than ten miles and carries the railway to a point one mile from Wicklow town. Wave attack has been arrested by planting marram grass and laying down enormous concrete blocks, each marked with the date. Immediately behind this ridge there is a belt of salt marshes with lagoons, notably in the Murrough area immediately to the north of Wicklow town, where the Vartry river finally reaches the sea. The coastal marsh belt is everywhere less than a mile wide, and the greater part of the low land consists of limestone drift of hummocky appearance rising to about 400 ft. at the hillfoot and penetrated here and there by small quartzite hills and isolated bosses. The evidence of glacifluvial drainage is seen in a remarkable series of channels, most of which run parallel to the hillfoot, and also in the deep cuts of the upland edge: these include the Glen of the Downs, which is 400 ft. deep, about 450 yards wide, and one and a quarter miles long, and the Devil's Glen. Some of the channels on the lowland are followed by roads, but though their mark on the

* Joyce, W. St J., *The Neighbourhood of Dublin*, 1912, 1939, p. 83.

landscape is indelible, they are almost ignored by the present drainage, which consists of short and inconspicuous streams running from the edge of the Vartry peneplane to the sea.

Agriculturally the lowland is divided into large farms averaging 100 acres but varying in size from 40 to 300 acres. About one-fifth of each farm is sown to crops, especially oats, potatoes, roots and greens intended for animal feedstuffs, though small quantities of wheat and barley are grown and there is sometimes a surplus of oats sold for milling. The farmers breed store cattle and sheep, and receive young stock from such neighbouring areas as the Vartry peneplane. The milk trade with Dublin is merely one branch of a characteristically mixed farming. The well-wooded appearance of the landscape is due mainly to trees in hedgerows and windbreaks, though there are also plantations laid down by the Forestry Commission, particularly on hillslopes, for example at Delgany and Rathnew. There are now no native woods in the area, but the oak-hazel-holly woods clothing slopes in the Devil's Glen and the Glen of the Downs are apparently subspontaneous. The density of population is 50–80 per square mile, and the area has a number of villages which are partly residential and partly market centres, with labourers' cottages recently added. Wicklow (3,300) is a market town, with a port much used in the mid-nineteenth century for the shipment of copper and lead mined at Avoca.

5. THE GLENEALY–SHILLELAGH CORRIDOR

This subregion, eight miles wide and over twenty miles long, consists of a corridor of Caledonian trend limited on the west by the Wicklow mountains and on the east by a line of hills, chiefly made up of igneous intrusions, of which Croghaun Kinsella, 1,993 ft., is the most significant. The Caledonian line is also emphasized by hills 500–900 ft. high within the corridor itself, but its main feature is a large expanse of land around 400 ft. cut by a number of river valleys. Those in the northern part are the middle reaches of the rivers draining the highest parts of the Wicklow mountains, here flowing in youthful valleys some 200 ft. deep; to the south there is a long wide subsequent valley drained by the Derry river to the Slaney, and by the Derry water to the Aughrim river. The soils of the area derived mainly from drifts or form some of the striking moraines deposited by ice-tongues from the uplands. There are also some alluvial fans at the junctions of the tributary streams, and only in the Derry river and Derry water valleys are there marshes of wide extent. The farms average 70–80 acres and the farmed land generally extends to a height of about 700 ft. with some extensions to 1,000 ft. through the transitional hill-pasture belt.

Some two-thirds of all the improved land is in pasture and nearly one-fifth in arable cultivation with oats normally predominant, followed by turnips and potatoes. The farmers rear store cattle and sheep rather than dairy cattle. There are no towns in this area, but the market villages include Rathdrum (1,000) and the smaller Aughrim, Tinahely and Shillelagh.

Mining. Auriferous gravels exist in several valleys between Croghaun Kinsella and Shillelagh, including the Gold Mines valley which joins the Vale of Avoca at Woodenbridge. No efforts at exploitation have ever proved a commercial success, though the deposits were worked by the government from 1796 to 1803, and at various times people have found and sold pieces ranging from minute grains to a nugget of 22 oz. discovered in 1795. The mother lode, however, has never been found, though the association of the mineral with tinstone and wolfram may indicate that it may exist at the margin of the Leinster granite. The Avoca mines lie in an extensive mineral belt stretching along both sides of the Avoca valley. The ore, a complex mixture of sulphides of copper, lead, iron and zinc with a small content of precious metals, was worked at the end of the eighteenth century for copper and during the middle part of the nineteenth century for sulphur, but the competition of purer continental pyrites ores made the enterprise uneconomic, though there was a revival of working during the 1914–18 war, and since 1939.

6. THE UPPER SLANEY VALLEY

The Slaney is the only river which cuts through the Leinster chain; from its main source in the North Prison, a corrie on the west side of Lugnaquilla, it descends in a steep torrent course to the Glen of Imail, a basin cut in granite at 500–700 ft. and surrounded by mica-schist and granite hills rising to more than 2,000 ft. Having left the glen, the river assumes a north–south direction with several sweeping bends across the granite lowlands, and swings east near Tullow. The Slaney gap has the Blackstairs mountains to the south and various isolated hills to the north; in it the river is entrenched some 50–100 ft. below the 200 ft. level; and the latter level is widespread in the gap. Though a connecting link between the lowlands of Carlow and Wexford, this gap has never had a railway and has no towns but merely villages at the river crossings, of which Newtownbarry (900) is the largest. Entirely agricultural in its life, this area consists mainly of medium-sized farms averaging 70–80 acres, with a few larger holdings and some small ones, especially near the limit of cultivation, which ranges from 600 to 1,000 ft. Store cattle and sheep are more important than dairy cattle, and two-thirds of the improved land is in pasture, with the remainder divided almost equally between meadowland and arable crops, especially oats. The density of population is around 70 per square mile, and nearly 80 per square mile of improved land. Most of the settlement is dispersed, though there are a number of small clusters of farms, especially at the higher levels occupied. Nearly one-fifth of the population live in market villages such as Baltinglass (800) and Rathvilly (300) on the Slaney and similar centres scattered through the lowlands at intervals of a few miles. The only town is Tullow (1,800), on the Slaney.

7. THE BLACKSTAIRS MOUNTAINS

This striking range divides the lower part of the Barrow valley from the Slaney lowlands in Co. Wexford and extends for nearly twenty miles in a south-southwest direction. It has a maximum width of six miles, but is generally a ridge only one to three miles wide, though it maintains a summit level of at least 1,500 ft. and reaches 2,610 ft. in Mount Leinster, and 2,409 ft. in Blackstairs mountain. Between these two summits the deep Scullоge gap carries the one route through the range at slightly over 600 ft. The range consists mainly of steep, heather-covered slopes rising continuously from the lowland level of *c.* 400 ft., and some of its valleys show traces of a mountain glaciation, probably to be correlated with the last of the three in the Wicklow mountains, but much less forceful in its action on the landscape. Comparatively little land is farmed, though there is a thin ribbon of settlement through the Sculloge gap, and there are scattered farmsteads in the valleys and on the hillsides to a height of 800–1,000 ft.

8. THE WICKLOW–WEXFORD LOWLANDS

Wicklow head stands at the northern end of an extensive lowland that is one of the most climatically favoured and fertile parts of Ireland. Developing first as a series of small river basins between low hills of igneous rock around Wicklow, the lowland widens near Arklow and stretches for over twenty miles between the Blackstairs mountains and the sea in Co. Wexford. Everywhere the lowland is saved from monotony by scattered hills of igneous rock, aligned parallel to the Blackstairs mountains. The coast mirrors the Caledonian trend between Wicklow and Wexford, but is varied in features: between Wicklow and Arklow it consists of several bays separated by rocky headlands with one splendid belt of sand dunes, three miles long and backed by marshes, at Brittas bay. South of Arklow there are wide sweeping bays with rocky headlands, some sand dunes and some boulder clay cliffs, of which the finest are those just north of Wexford, where the moraine of the last general glaciation truncates the coast in a jumbled mass of kames and kettleholes. Between Wexford harbour and Waterford harbour there are signs of east–west Armorican features: Carboniferous beds have been laid down in a belt from Ballyteige bay to Wexford harbour. The south coast of Wexford has an interesting series of bays that are submerged parts of short river valleys and are now almost, or in one case (Lady Island lake), completely closed by bay bars.

Almost the whole of these lowlands consists of farmlands, and the only gaps in the remarkably even spread of settlement are a few heather-covered hills, rare patches of lowland bog and some sand dune belts along the coast. The soils are varied types of drift, generally reasonably well drained, and the farms average about 70 acres, with more than three workers per farm. Only 13% of the farms are of less than 15 acres, and only 4% of more than 200 acres. Mixed farming is general, with rather more arable cultivation than is usual in Ireland. Even so, less than one-quarter was cropped before the 1939–45 war, though there was a rise to more than 30% during the war years, mainly through a considerable expansion of the wheat acreage; over two-thirds of the cropped land is used for cereals, with barley and oats rather than wheat as the main crops. Turnips are more generally grown than mangels or potatoes. The chief reliance is on store cattle, exported mainly through Dublin though Waterford is also used: dairying has been encouraged by the creameries, and cheese manufacture has been developed in Wexford. Here also malting is of considerable significance.

Co. Wexford has a general population density of 90 per square mile, with 70 to 90 on improved land: the main market towns are Arklow (6,100), Gorey (2,900), Enniscorthy (6,300), New Ross (4,600) and Wexford (12,700). As a group of market towns they are spread at intervals of ten to eighteen miles, and the essential market functions of the countryside are met to a certain extent by the various villages that have failed to grow into towns. The fact that two-fifths of the population live in towns and villages is itself remarkable: it means that there is more purchasing power in this part of Ireland than in many others, and that commercial life is more thoroughly developed than elsewhere. This lowland may be regarded as a farming area where a mixed economy prevails with both arable crops and stock, having rural creameries and in the towns mills and processing factories for agricultural staples.

CHAPTER 14

The Suir, Nore and Barrow Basins

These rivers are known in Ireland as the 'three sisters', for they eventually join and reach the sea in the long drowned estuary known as Waterford harbour, after flowing between ranges of Caledonian and Armorican structure. This landscape of hills and lowlands belongs essentially to the south of Ireland: most of the lowlands are floored by Carboniferous rocks covered with drift that is thin or even non-existent in places and the soils, generally fertile and well drained, are industriously farmed, mainly in medium-sized holdings. Both store and dairy cattle are found on most farms and creameries have spread through the region from their original home in Tipperary and Limerick to Waterford and Kilkenny. In the west towards Limerick dairying predominates; in the east towards Wexford arable cropping is more significant, and as a whole the agriculture is prosperous and the rural scene made attractive by the care lavished on the land. The towns include such historic centres as Waterford, Kilkenny and Clonmel, of which the first-named is a natural regional capital combining the functions of a river port, a sea port, a major market centre and an industrial town. There are also many small trading centres, a few of which have a variety of industries.

LANDFORMS AND LAND USE

The meeting of Caledonian and Armorican structural elements is one of the main features of this region (Fig. 55). To the east the Leinster chain, making its last appearance in Mount Brandon, is aligned from the northeast to southwest; to the west, the long east–west parallel ranges and valleys extend for over 100 miles into Kerry and the extreme west of Cork. The Armorican ranges within this region are the Comeraghs with the Monavullaghs (in effect one range), the Knockmealdowns, the Slievenaman and the Galtees, all of which run east–west. Towards the Central Lowland, however, there is a change, for the ranges, though still Armorican in structure, assume a trend from southwest to northeast, probably over a hidden Caledonian floor: this trend appears in the Slieve Ardagh ridge and the Castlecomer plateau, both of which are composed of

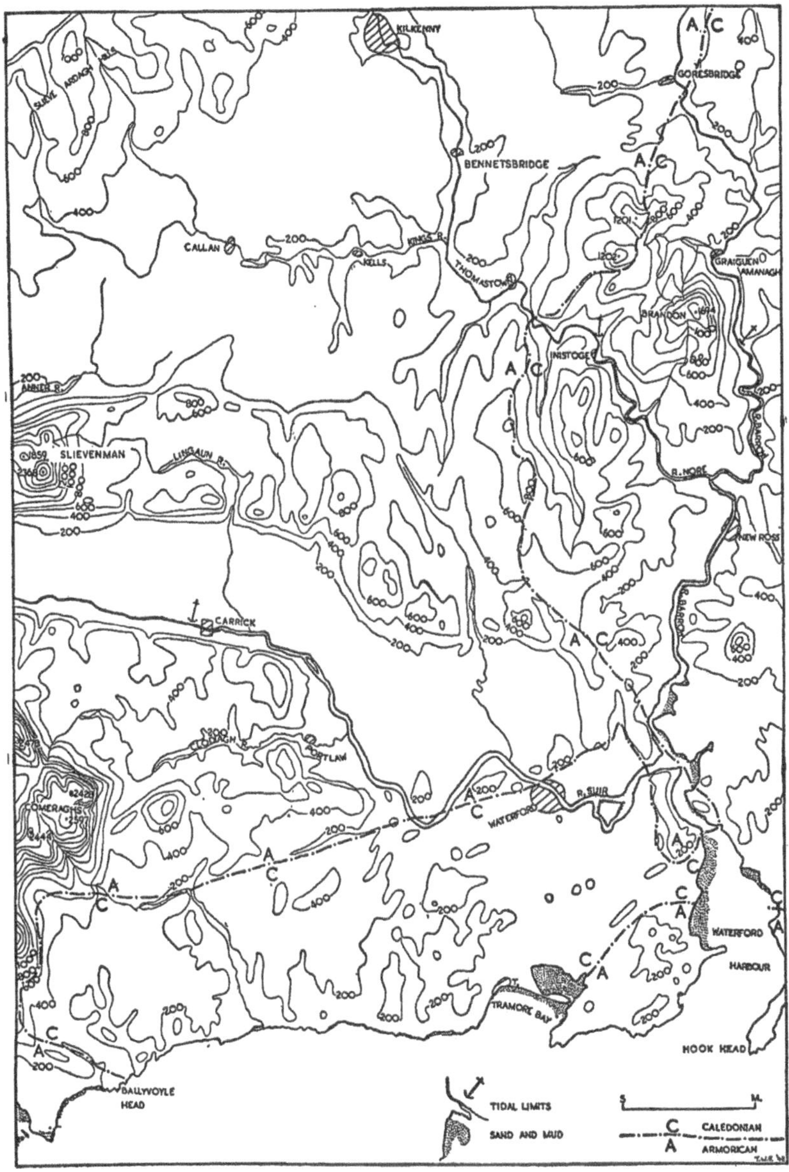

FIG. 55. CALEDONIAN AND ARMORICAN AREAS IN THE LOWER SUIR, NORE AND BARROW BASINS

The northeast trend of the Leinster chain and the east–west trend of the Slievenaman range meet in this area: Slieve Ardagh is of Armorican structure but Caledonian trend. This map also shows the cliffed coast of south Waterford, the lowlands between the various ranges, and the long estuarine reaches of the three rivers.

Carboniferous rocks, and in the long line of hills that extend from the Slievefelim, near Limerick, to the Slieve Bloom, and are made up of Devonian and earlier rocks. Where the Old Red sandstones survive there are hills reaching altitudes of between 2,000 and 3,000 ft., or crag-like eminences of lesser height but hardly less striking form: where the Devonian strata have been removed, the older Silurian and Ordovician strata have been eroded into depressions, many of which are ringed round by hills of Old Red sandstones. Between these hills lie lowlands, generally several miles wide and linked one to another by easy passage-ways, so that the whole landscape has an alternation of uplands and lowlands typical of the southern sector of Ireland.

Although it is in the hills that the Caledonian and Armorican trends may be most easily seen by the informed traveller, the lowlands also show the imprint of each great structural phase. In south Waterford, and also around the lower Barrow and the lower Nore, the lowlands are floored by Ordovician, Silurian and Devonian strata with varied igneous intrusions, and belong to Caledonian Ireland: elsewhere the lowlands are floored mainly by Carboniferous rocks and belong to Armorican Ireland. But the most remarkable feature is that the 200 ft. peneplane truncates all these varied rocks, so diverse in hardness and resistance to erosion. Nowhere is this better seen than in south Waterford, where the 200 ft. level extends for several miles as an even-surfaced lowland, and the rivers flow in wide, open and marshy valleys with every appearance of senility until they plunge through gorges near the cliffed coast. Viewed from Waterford harbour, this coast is as remarkable an expression of the 200 ft. level as its analogue in South Wales, for there is a line of cliffs extending for some twenty-five miles from Creedan head to Ballyvoyle head, near Dungarvan, with only one major bay, at Tramore. In this peneplane the rivers are incised, in many places with magnificent entrenched meanders and long gorge-like reaches. Above this level there are signs of levelling around 400 ft., especially in the lowlands some distance from the coast, for example around Thurles and Tipperary. Again, there are traces of a surface at about 800 ft. around various uplands.

Glaciation has left its clearest traces within the region in the great corries of the Comeraghs and Monavullaghs, the Galtees and the Knockmealdowns, all of which are the product of mountain glaciers that never reached the lowlands. In the southern lowlands the earliest, or Eastern General glaciation, left deposits of 'limestone gravel', a strongly weathered drift, but the later, or Midland General glaciation, extended only to a line running from the north side of the Galtees through Cahir, along the north side of the Slieve Ardagh ridge, then into the Nore valley near Thomastown and the Barrow, south of Borris (Fig. 4). North of this line, limestone drifts occur generally as hummocky ground moraine interspersed with bogs where the drainage was interrupted and lakes or fens formed: associated with these features is the presence of a large number of bogs, some very small and others covering several square miles, which have developed on a fen base between the ground moraine. The major feature of the detailed regional geography of such areas is the alternation of these two elements, bogs and ground moraine, as the farmlands are found on the gently rolling country and the bogs occupy the hollows, their brown and purple colouring contrasting pleasantly with the green of the pastures and meadows or the changing colours of the crops through summer months.

Agriculture in this region shows careful farming and long occupation, and the proportion of improved land is remarkably high, in Co. Kilkenny 87%, and in Co. Tipperary 80%. Most of the non-agricultural land is in the mountains, notably the Comeraghs, where there are very obvious signs of withdrawal of land from cultivation. A small part of the non-agricultural land is forested, but the area so far developed by the Forestry Commission covers little more than 1% of the region. Woodland is found mainly in demesnes and on some of the steeper hillsides, though there are also some interesting patches of scrub, with oak as a main component, in the gorges of the three main rivers. With the demesnes and hillside woodlands as a basis, the Forestry Commission has established some large plantations, notably on the northern slopes of the Comeraghs, around Clonmel, on the north side of Slievenaman, and on the Old Red sandstone ridges south of Carrick-on-Suir and Portlaw. The slopes of the Galtees and of Slievenamuck, on either side of the Glen of Aherlow, have also been planted, and there are several new plantations on the north side of Mount Brandon. Two demesnes in the Nore valley, near Thomastown and Inistioge, have been afforested, but as yet few plantations are found in the lowlands.

Most of the farms are in the medium-sized range of 30–100 acres, with a large number over 50 acres: 17% are of more than 100 acres, and approximately one-quarter of less than 30 acres. Farming is generally mixed: during and since the war, the cropped acreage increased and in some years Co. Kilkenny has had one-tenth of the improved land under wheat. But stock farming is the main type, both for meat and milk. The creameries have spread eastwards from Co. Tipperary into Waterford and Kilkenny: they are supplied by 88% of the cows in Co. Tipperary, 85% of those in Co.

Kilkenny and 82% in Co. Waterford. Valuable as the milk cheque is to many farmers, less than one-third of the cattle are milch cows, and the main cash resource of most farmers is store cattle. The arable cultivation is mainly intended for feedstuffs, and the proportion of land sown to crops is higher in the east of the region than in the west: normally oats, barley, potatoes and turnips are the main crops, with a smaller area under mangels and sugar-beet. The amount of land in pasture varies from two-thirds in Co. Kilkenny to as much as three-quarters in some parts of Co. Tipperary and Co. Waterford. In fact, it is particularly high in the Suir valley and throughout Co. Waterford, where cattle are reared for the export trade and for the canning industry. Pigs are also exported or sold to the bacon factories. So much of the local produce goes to creameries or factories that the region has an agricultural economy in which production, processing and sales are reasonably well organized, and the farmers with their typically varied production, are able to alter the relative emphasis on particular staples according to the trend of prices.

INDUSTRIES

Agriculture is the main resource of this region and employs half the working population. And the number of people who depend indirectly on the land is large, for the towns and villages are well provided with shops and furnish various services that give employment and indicate a not inconsiderable purchasing power. The agriculture is itself industrialized to some extent, as there are about fifty creameries scattered through the region, some of which have branch separating stations. Creameries are found in almost all the towns, in most of the villages and also on the roadsides away from any definite centre of population. Most of the creameries are quite small, with less than twenty employees, though a few have added shops and sawmills; one such, in the farmed country between the Comeraghs and the Knockmealdowns, is in effect a rural department store and supplies groceries, drapery, footwear, feedstuffs, machinery, hardware, even coal. Milling is based on numerous small concerns serving a limited area, to which farmers brought their wheat and oats for grinding: many old mills are seen beside rivers. Various brands of porridge oats, wheat flour and feedstuffs, including maize meal, are produced at the larger mills, notably those at Waterford, Carrick-on-Suir and Cahir. Bacon factories exist at Kilkenny, Waterford, Roscrea, Clonmel and Callan: there are factories for canned meat in Roscrea, Clonmel and Waterford, and other industries of directly agricultural origin include wool preparation and fellmongery at Clonmel, a condensed milk factory at Tipperary, with tanneries at Carrick-on-Suir and Portlaw. One industry giving direct service to agriculture is the works at Roscrea providing biological, medicinal and veterinary products. There is a chocolate crumb factory at Carrick-on-Suir.

Industrial enterprise in this region has spread far beyond the processing of the output from the farms. Clothing industries have achieved some success, and include woollen mills in Kilkenny, Cahir and Ardfinnan, a village in the centre of the Suir lowland. Waterford has, in addition to its flour mills, tinned meat and bacon factories, boot and shoe works, iron foundries, a works making sacks and bags, printing works, a brewery and a nursery providing seeds, bulbs and trees. But of all the city's industries, the most remarkable is the glassworks, which in 1947 revived a craft industry famous in the eighteenth and early nineteenth centuries. Clonmel has a large number of small industries, including a cider works and firms making motor-car bodies, perambulators, mineral waters and footwear. The Clonmel cider works is mainly supplied with apples from orchards on farms south and east of Cashel, especially between Clonmel and Clogheen: some are grown in the north of Co. Tipperary, but everywhere the tradition of home cider making has ceased. Linoleum is made in Tipperary, and safety glass at Templemore. Kilkenny has a large brewery and furniture works as well as its clothing and footwear industries. This dispersion of industry through a large number of towns is by no means new, though there are several factories that date back only to the post-1932 period. It is largely a question of individual enterprise, often of particular families: for example, a group of farmers established the canned meat industry in Waterford and a family established a chain of mills in various towns. Some towns have been left behind in the struggle, while others have prospered.

POPULATION

This region had 315,000 people, 125 per square mile, in 1891, 254,000 in 1936, 243,500 in 1951 and 243,000, 96 to the square mile, in 1966. The population in the larger towns has increased from 70,000 to 81,000 with 32,000 in smaller towns and villages; excluding these the rural density is now 56 to the square mile. The rural population has diminished but this region has a large number of small clusters of houses, of which the humblest are little more than a few farmhouses lining a road with possibly a church and a few shops. The dot map of population distribution shows a generally

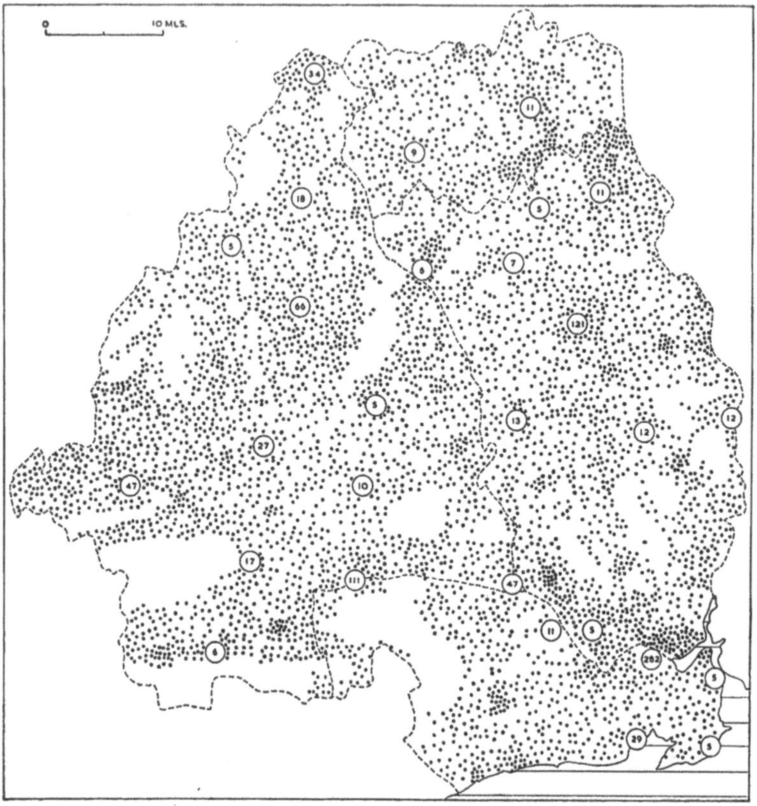

FIG. 56. POPULATION DISTRIBUTION, SUIR–NORE–BARROW REGION
One dot to 25 persons

The settlement pattern is remarkably even: in the south the main gaps are such uplands as the Comeraghs, Slievenaman, the Knockmealdowns and Galtees, but to the north lowland bogs interpenetrate with farmlands.

even pattern if the various mountain masses such as the Galtees, the Knockmealdowns, the Comeraghs, and Slievenaman are excluded. The only lowland gaps of consequence are found in the scattered bogs towards the north of the region. In a few places, notably east of Thurles, it is possible to distinguish ribbons of settlement that stretch along roads in some bogland areas, either because the road follows a patch of better-drained ground moraine or because small clearings have been made (Fig. 56).

The towns fall roughly into four groups (Fig. 57). Waterford is the main port, market and industrial centre, and in effect a regional capital: its population was 28,900 in 1956 and 29,800 in 1966. Next come Kilkenny (12,000) and Clonmel (11,500), two of the larger inland towns of Ireland, both of which are industrialized to some extent and serve large areas as markets. A third group of towns is represented by Tipperary (4,500), Carrick-on-Suir (4,900) and Thurles (6,900), which are typical of many moderate-sized Irish market centres that exist primarily on the trade of the countryside and generally lack any considerable industry. Thurles alone of these three towns has grown within the past sixty years, or more accurately the last thirty, partly through the establishment of a sugar-beet factory. The fourth urban group consists of the small Irish towns that are very closely allied to the rural areas and seem in many cases to be little more than overgrown villages; Roscrea (3,500) and Cashel (2,700) are the largest of these but the group also includes Templemore (2,000), Cahir (1,700), Callan (1,300) and Fethard (1,000). Except at fairs, most of these places are quiet if typical country towns. Fig. 57 shows that the towns are generally eight to sixteen miles apart, but that in some areas – for example between Kilkenny and Roscrea – the commercial life is gathered entirely in villages of a few hundred people.

SUBREGIONS

Some parts of this region could be subdivided into a very large number of small units, while other parts are more or less homogeneous over considerable areas. On the one hand, the complex arrangement of hills and valleys in the Comeraghs, the Slievenaman ridge and the country on both sides of the Nore between Thomastown and New Ross gives a landscape of contrasts in which almost every bend of the road reveals something different. On the other hand, the long traverses across the lowlands that extend from Bagenalstown to Clonmel or from Clonmel to Tipperary and Roscrea are monotonous in comparison, and the chief scenic variety in them is provided by the surrounding hills. If the whole region were submerged to a height of 500 ft. above sea level, the lowlands would appear as a sea surrounding a number of hilly islands, which would fall into three main groups: first, the Knockmealdowns, Comeraghs, Slievenaman and Mount Brandon; second, the Galtees, Slieve Ardagh and Castlecomer plateau; and third, the Silvermines and neighbouring hills including the Devil's Bit and the Slieve Bloom. The lowlands are richer in geographical interest than in scenic attraction, for they show a puzzling interplay of surfaces around 200 ft. and 400 ft., a river system of complex

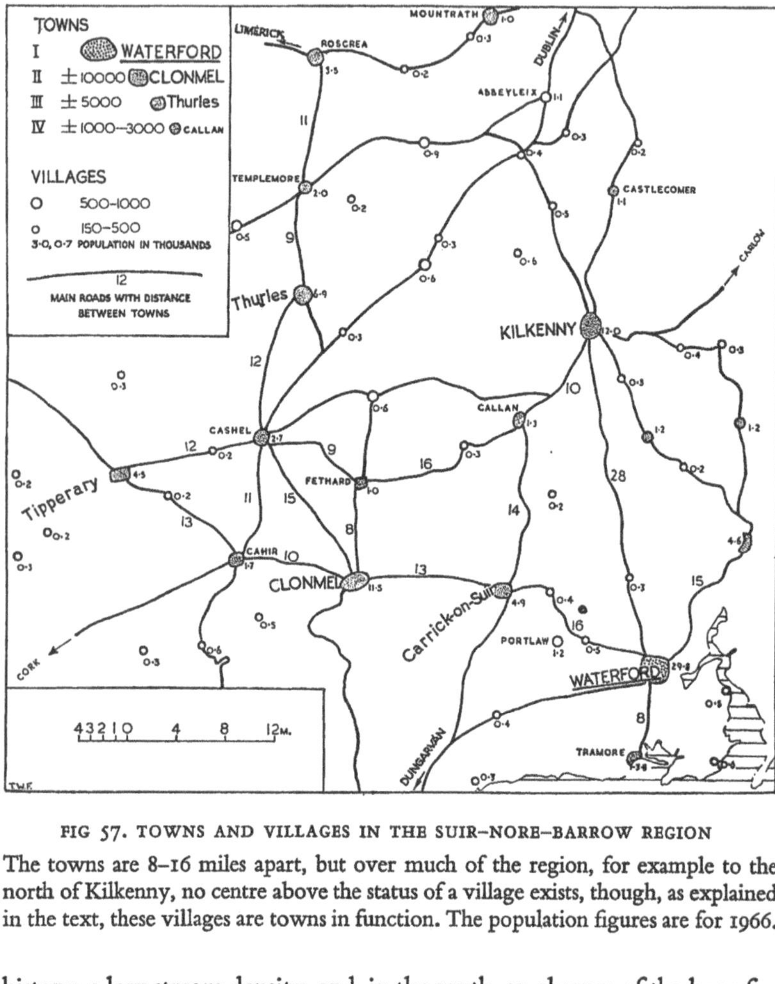

FIG 57. TOWNS AND VILLAGES IN THE SUIR–NORE–BARROW REGION

The towns are 8–16 miles apart, but over much of the region, for example to the north of Kilkenny, no centre above the status of a village exists, though, as explained in the text, these villages are towns in function. The population figures are for 1966.

history, a low stream density, and, in the south, an absence of the bogs for which so many Irish lowlands are justly famed. Almost continuously farmed and in the main fertile and prosperous, these lowlands have many signs of long settlement in ancient castles, churches and abbeys. The soil is the chief resource, and the agricultural life shows both variety and resilience to changing economic circumstances; the facilities for processing the farmers' staples are a particularly useful feature of the industrial life.

Eleven subregions are distinguished here. Four of the regions are wholly lowlands, but none are wholly uplands, for all the mountain groups are penetrated by significant lowlands and valleys. The first three regions (A1–A3) belong to Caledonian Ireland, the remaining eight (B4–B11) to Armorican Ireland.

A1. THE WATERFORD LOWLANDS

The lowlands around the city of Waterford show a remarkable development of the 200 ft. surface, with hills rising to 400 ft., truncating various Ordovician rocks with igneous intrusions. This level is particularly clearly seen along the coast, where the cliffs are broken only by the bay at Tramore, a town on a hillside rising to the 200 ft. level. The Suir valley is outlined by slopes rising, with varying steepness, to the coastal peneplane, and the river has a magnificent series of meanders with small floodplains. Agriculture is the main resource of the subregion, which is divided into medium-sized and large farms, many of which profit from the proximity of Waterford with its canning and bacon factories, store cattle exports, and local meat trade: in addition a number of farms supply milk to creameries. There is a dual emphasis on store cattle and dairying, though fat cattle normally account for about one-fifth of the exports. There is very little rough pasture, though marshes have developed in some stream valleys on the 200 ft. surface, and there are a few small bogs. The rural population has declined by more than one-third during the last sixty years.

Waterford city developed around the creek provided by the St John's river, and is the first crossing place of the river from the sea: it possesses a mile of quays on the south side of the river, and can be entered by vessels drawing 20 ft. of water. Most of the shipping is coasting steamers, whose main export cargo is cattle: for some years passenger steamers from Fishguard came to Waterford, but they now use Cork and Rosslare harbours. The suburb on the north side, Ferrybank, has the main Rosslare–Cork railway on a narrow strip of river-bank, backed immediately by a steeply sloping rise, but Waterford city rises more gently from its quays to reach the 200 ft. level. The St John's river, which traverses the alluvial flat on which part of Waterford is built, flows through a marshy valley. The most widely known of all the city's industries is the revived Waterford glass manufacture, but there are also flour mills, bacon factories, a tinned meat factory, iron foundries, printing, a works making sacks and bags; near the town is a nursery specializing in bulbs, trees and shrubs. The population of Waterford, which has remained stable for many years, was 29,800 in 1966, a small increase since 1961. Tramore (3,300) is a small market centre, holiday and residential town, whose fortunes are closely tied to those of Waterford.

A2. THE LOWER NORE AND THE LOWER BARROW BASINS

The Nore and the Barrow unite two miles above New Ross and flow southwards as the Barrow to the point where they are joined by the Suir, six miles east of Waterford. They cut through the 200 ft. surface in valleys with incised meanders which are generally at least 100 ft. lower: only in a few places, such as the area north of New Ross where the two rivers join, is there a floodplain of considerable extent. New Ross climbs from the river by steep streets and staircase-like alleyways to the 200 ft. level though to the west there is a gradual rise into rolling hills about 400 ft. high. Similarly to the north of the Barrow the main road from New Ross to Graiguenamanagh rises steeply to 400 ft., only to descend again on to a very clearly defined patch of 200 ft. surface between the Barrow gorge and Mount Brandon, which rises from it without any higher break of slope and provides a scenic focus for the whole district. But on the west side of Mount Brandon, the 800 ft. ridges are very clearly defined, and possess a north–south line that is accentuated by the valleys between them (Figs. 58, 59).

The variety of aspect in this subregion is considerable, and two-thirds of the land is farmed: the remaining third is rough pasture of four main types – first, the heather-covered Mount Brandon; second, the gorse- and heather-covered upper parts of the 800 ft. ridges, partly decayed pastures; third, a few flat and ill-drained parts of the valleys in the ridge country, and fourth, the intractable flooded areas around the Nore and the Barrow. The limit of cultivation is 600–800 ft., and the farming is mixed, with 10–15% of the land under arable crops, and the remainder in meadow and pasture. There are a large number of very small farms at the higher levels, but as a whole the medium-sized and large farms are dominant. Creameries are now well established in this area, and the market centres are New Ross (4,600) and Thomastown (1,200), a typical small country town found round a mill and a river. Inistioge (220) is a small demesne village around an ancient abbey at the first crossing place of the Nore above its confluence with the Barrow: the large demesne has fallen into decay and is now taken over by the Forestry Commission.

A3. THE GRAIGUENAMANAGH BASIN

This small subregion of some fifteen square miles (Fig. 59) is worthy of special consideration as an exceptionally clearly defined basin, which is very carefully cultivated and possesses a market town in Graiguenamanagh (1,200), placed where the Barrow gorge opens out slightly as several streams join the main river after flowing through trenched valleys divided

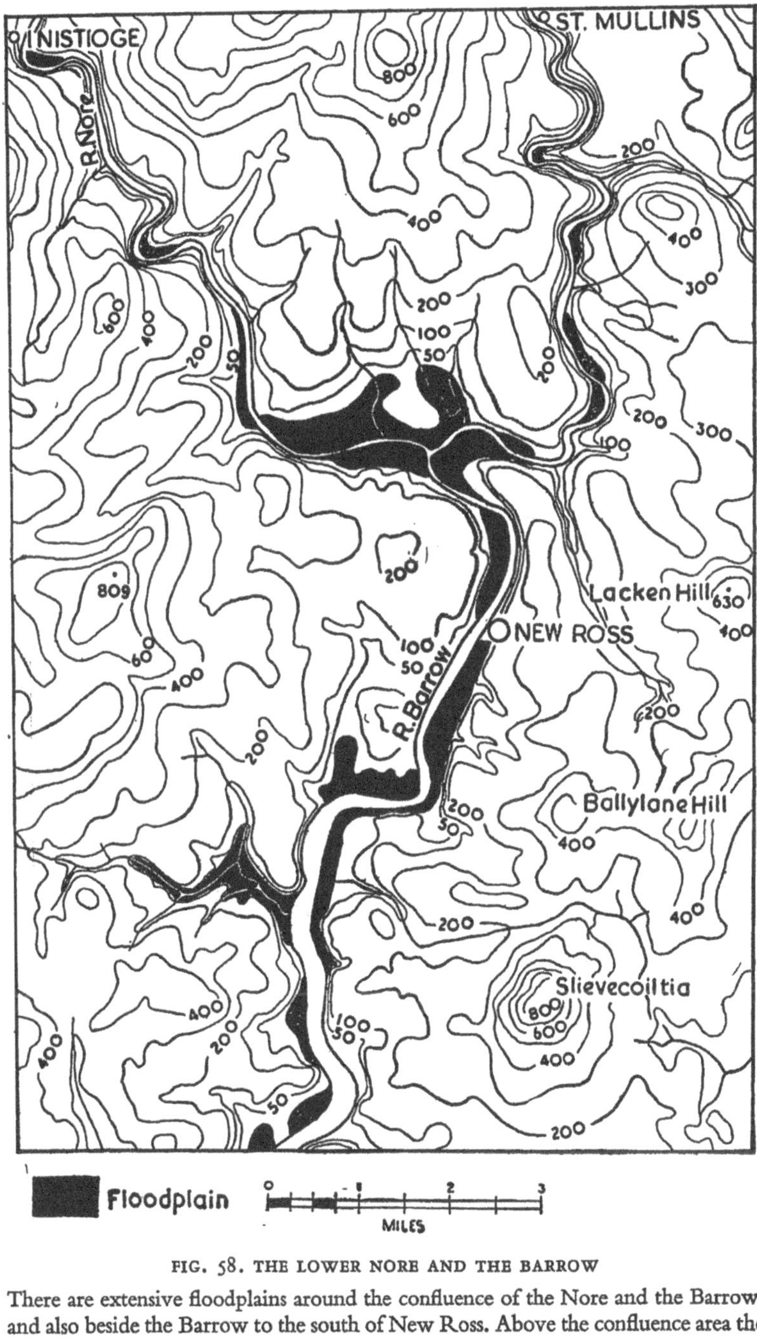

FIG. 58. THE LOWER NORE AND THE BARROW

There are extensive floodplains around the confluence of the Nore and the Barrow, and also beside the Barrow to the south of New Ross. Above the confluence area the floodplains, though smaller, are developed at various points: the river has some gorge-like reaches and incised meanders. The 200 ft. peneplane is widely developed, and a second surface around 400 ft. may exist: the 800 ft. level is also clearly seen in this area.

Map drawn by W. E. Nevill

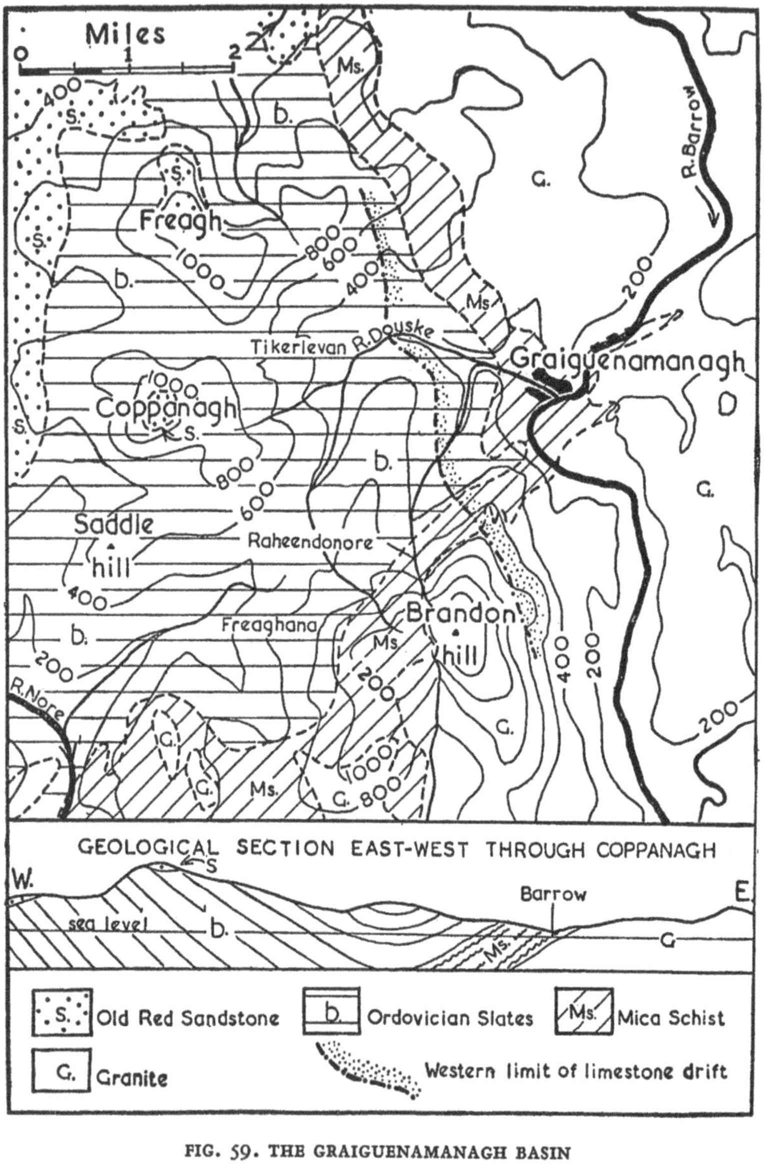

FIG. 59. THE GRAIGUENAMANAGH BASIN

The Leinster granites and Ordovician slates are separated by a band of mica-schists: over the slates lie caps of Old Red sandstone. The 200 ft. peneplane is extensively developed here and the Barrow is incised, with some striking entrenched meanders. The limestone drift, mapped by A. Farrington and J. P. Haughton in 1945, was previously supposed to have ceased some miles farther north.

Drawn by J. P. Haughton and reproduced from 'Irish Geography', vol. 1, No. 3, 1946, p. 62 from one another by long sloping interfluves. Mount Brandon defines this subregion on the south, and two flat-topped hills, Freagh and Coppanagh, both 1,200 ft. on the west and north: these hills are capped with Old Red sandstone and probably mark the beginning of Armorican structures in this part of Ireland. But the basin is by no means isolated, as apart from the river crossing at Graiguenamanagh into Co. Carlow, there are passes between the various hills and the 200 ft. peneplane is continued northwards on the west side of the Barrow gorge. The limestone drift of the second general glaciation penetrated the lower part of the Graiguenamanagh basin, but otherwise the soils are chiefly of local origin and derived from the granites, mica-schists, red sandstones and Ordovician slates of which the area is composed.

Well-drained fields occur on the long sloping interfluves of the basin, and have pastures, meadows, and arable crops, including wheat, barley, oats, potatoes and roots. The most remarkable feature is the aggressive attitude of the farmers towards the encroachment of rough pastures: a survey made in 1945 shows that many fields had been wrested from gorse and bracken, since the 6″ to one mile map was published in 1903, and that very few fields had degenerated to rough pasture during the same forty years. Individual stories of the clearing of particular fields showed how hard the struggle for reclamation may be and the enemies, gorse and bracken, are perpetually on the watch, for they line the hedgerows waiting to advance into the fields. In this area the general limit of cultivation lies around 800 ft., though it is lower on the steep slopes of Mount Brandon and higher on the gentle slopes of Coppanagh and Freagh, with some fields reaching 1,100 ft. The farming is characteristically mixed, with sheep, store and dairy cattle as the main livestock. There is a separating station at Graiguenamanagh, and churns of milk are brought in by the farmers or collected by lorry from platforms beside the farm gates. Graiguenamanagh is an ancient town living mainly by retail trade and monthly fairs for the neighbourhood: its narrow street is lined with extremely well-stocked shops. Houses and gardens have submerged the old abbey, whose cloisters and outbuildings are hidden away in gardens, though part of the old abbey church has been used by the Catholics since 1813. The town has a woollen mill and a sawmill.

B4. THE COMERAGHS AND THE KNOCKMEALDOWNS

The long line of hills that includes the Knockmealdowns and the Comeraghs forms a rampart on the south side of the Suir valley for more than thirty miles between Mitchelstown and Carrick. Seven of the peaks

FIG. 60. LAND USE IN THE GRAIGUENAMANAGH BASIN

1, Arable land; 2, Meadow; 3, Permanent pasture; 4, Rough pasture; 5, Coniferous and deciduous woods; 6, New plantations. Contours at intervals of 150 ft.

The basin is cultivated to a level varying from 450 ft. to 850 ft.: the forests on Mount Brandon reach 1,200 ft.

Drawn by F. M. Synge and originally published in 'Irish Geography', vol. 1, No. 3, 1946, p. 65

in the Knockmealdowns are of more than 2,000 ft., one of them 2,609 ft., and the range has an abrupt slope to the valley of the Tar, a stream tributary to the Suir: a famous tourist road from Caher to Lismore climbs the slopes through a series of hairpin bends and achieves its summit of

1,114 ft. at a point called the Gap. The Comeraghs and Monavullaghs are really a single mountain range extending from Clonmel to the lowlands around Dungarvan, and now appearing as a north–south line, though belonging structurally to the east–west folds of southern Ireland. Like the Knockmealdowns, they have several peaks above 2,000 ft., and reach 2,597 ft. at one point, but they are much more sharply glaciated, and their stream-cut valleys, which must have been impressive before the mountain glaciation, are hollowed out into splendid cirques, most of which enclose small lakes overhung by craggy precipices. Coomshingaun, one of the finest corries in the British Isles, is walled in on every side except the east by almost perpendicular cliffs that rise at one point to 1,288 ft. above the surface of the lake, which is dammed by a block moraine. A few of the other corries are almost equally impressive, and the range shows a clearness of glacial sculpture that is rivalled in few parts of Ireland except perhaps in the Dingle peninsula, in the Reeks of Kerry, and in some of the Leinster chain valleys. East of the Comeraghs there are hills 500–1,200 ft. high, formed of Old Red sandstone around a lowland, or in appearance an upland basin 200–500 ft., floored by an inlier of Silurian strata. Here the anticline has been denuded by the removal of the Devonian beds and the softer Silurians have been worn down into a basin consisting of comparatively fertile land drained by the Clodiagh river to Portlaw.

In this subregion the landscape is completely dominated by the Knockmealdowns and the Comeraghs, both of which are mainly covered with a heathery vegetation. The lower hills of Devonian strata beside the Suir valley have some old woodlands, and are now being intensively forested. The basin formed of Silurian beds is farmed and settlement penetrates the Comeraghs in various remote valleys on the western side to as much as 1,000 ft. in places, but on the steep eastern side cultivation generally ceases at the hillfoot. There is also continuous agricultural settlement in the threshold between the Comeraghs and the Knockmealdowns, but in this range cultivation ceases at the hillfoot almost everywhere. The farming is mixed with sheep, dairy and store cattle, and normally about three-quarters of the land is in pasture, with the remainder divided fairly evenly between meadow and arable crops.

B5. THE SUIR LOWLAND

This lowland lies between the Comeraghs on the south and the Slievenaman range on the north, and is mainly between 100 ft. and 200 ft. high. It is formed of the lower Carboniferous limestone covered, generally thinly, by limestone drift of the earlier glaciation, and is a continuously farmed area laid out in medium-sized or large farming units, with mixed farming, including dairying, as the usual type. As a whole, it is fairly typical of the Carboniferous limestone lowlands that lie between Armorican ridges in the south of Ireland. North of Clonmel, there is a rise of land to 300 ft. and even 400 ft. in places, with a few patches of karstic limestone scenery. The Suir, which flows in a well-defined valley at the southern fringe of this lowland, possesses a floodplain which is narrow at Clonmel, but widens out towards Carrick and becomes a marshy flat more than a mile wide at Portlaw. The river is tidal to Carrick and willow growing for baskets and hampers is a traditional industry, for which 200 acres of ground is used. Ridges 3½ ft. to 4 ft. wide are made in summer, and planted with cuttings from November to March: the rods are cut annually and stored in low damp pits under refuse. They are generally peeled and dried in the sun to bring out a white colour but some are boiled to give a reddish-brown shade. The willow beds are partly covered at low tides, and completely submerged at high tides and during periods of heavy rain. All the main north–south roads descend into Clonmel or Carrick, the two market towns situated at the principal crossing-places of the river. Both these towns are market centres for considerable areas, including those around the Comeraghs; but Clonmel is the larger, more prosperous and more industrialized. Chocolate crumb is now manufactured at Carrick.

B6. THE SLIEVENAMAN RIDGE

This east–west ridge extends for some twenty miles on the north side of the Suir lowland and is structurally comparable with the Comeraghs and their environs described as region B4. Old Red sandstone forms Slievenaman, a peak 2,368 ft. high, at the western end of the ridge, and two lines of craggy hills 700–1,000 ft. high on the north and south of an upland basin, 400–600 ft. high, worked out in Silurian slates. The slates in this basin, which is drained by the Lingaun river, were formerly quarried for roofing materials, but only the scars of quarrying and a vestigial industry remain today. East of the Lingaun basin the Devonian sandstone forms a series of north–south ridges, all around 800–1,000 ft. These ridges are continued to the Nore between Thomastown and New Ross, but the Old Red sandstone fades out along a line running due south from Thomastown, which must be regarded as the present eastward limit of Armorican structures (Fig. 55). The Lingaun basin is farmed over its whole extent, and cultivation also extends on to the north–south ridges to the east, interspersed with rough pasture lands. But Slievenaman is a definite gap in the settlement pattern, though it is farmed on its gentler eastern slopes to 1,250 ft. in places and has a few farms on its steeper western slopes. The farming of the area is mixed, and there are several co-operative creameries in this subregion and immediately outside it. The Forestry Commission has planted large areas on the slopes of Slievenaman, and there are also woodlands on the hills surrounding the basin.

B7. THE SUIR–NORE LOWLAND

This lowland extends for more than fifty miles in a northeasterly direction from the foot of the Knockmealdown mountains to the Barrow valley at Bagenalstown. It is defined, not too clearly, by discontinuous lines of hills on either side, and is in fact a great natural route leading from the Upper Barrow to the Mitchelstown gap and the upper Blackwater, with easy connexions to all the lowlands in the south of Ireland along the valleys of the Suir, Nore and Barrow. But it has not become a major artery of trade and commerce, for roads and railways cross it rather than follow its central axis. It would appear to offer the easiest route from Dublin to Cork, but both the road and railway traverse the parallel lowland on the northwestern side of the Slieve Ardagh ridge. The two chief towns associated with this lowland, Clonmel and Kilkenny, are on its margins, and as a whole it preserves an almost exclusively agricultural character with a quality of remoteness that is, in all the circumstances, surprising.

Carboniferous strata underlie this lowland, and the striking mountains that rise abruptly from it in the south, Slievenaman, the Comeraghs, Knockmealdowns and Galtees, are formed of Old Red sandstone and the underlying Silurian strata. To the north the Slieve Ardagh ridge develops much less conspicuously in a series of low hills, some little more than 400 ft. high, around Cashel and Fethard. These hills are formed of the younger members of the Carboniferous system, while the older members, such as the lower limestones and the lower limestone shales, form the prevailing level around 200 ft., rising in places to 300 ft. and falling, especially where shales are developed, to 150 ft. or less. At a few points in the lowland there are small eminences due to the preservation of younger strata, but generally the surface is undulating rather than flat. Drift deposits, usually ascribed to the older of the two general glaciations, cover most of the surface, but many of the soils are extremely thin, and probably derived directly from the underlying limestones; in some cases, notably to the south of Kilkenny, rock outcrops are not infrequent, and the stream density is remarkably low over the entire lowland.

This lowland is continuously farmed but relatively thinly populated: it now has 50–60 persons per square mile and the population has declined by more than one-third in the past seventy years. Very little land is unimproved or has reverted to rough pasture, though there are marshy pastures around the stream courses on the 200 ft. surface where the rivers are not incised. The landscape is dominated by medium-sized farms and the average farm size varies from 55 acres in the upper Suir lowland to 80–90 farther east. On the average about one-sixth of all the farms are of more than 100 acres. The farming is mixed and permanent pastures cover two-thirds of the farmland in the east, and as much as three-quarters in south Tipperary, with the remainder divided between meadows and arable crops. Co-operative creameries have now been established throughout this area and wooden platforms for milk churns are seen at most farm gates: while the immediate market for milk provided by the creamery is of particular importance, store cattle raising is also general. As already noted, the main towns serving this subregion lie outside it in such centres as Clonmel, Cahir, Cashel and Kilkenny. Two very small towns, Fethard (1,000) and Callan (1,300), are typical of the Irish towns that, though long established, have not risen in status beyond minor market centres. Fethard still preserves its ancient walls, and was given a charter by Edward III, and Callan is almost equally venerable. But although Fethard, at least, has a stately eighteenth-century main street, both towns now lie off the major trunk routes and are commercially subsidiary to Clonmel and Kilkenny respectively.

Kilkenny. The historic city of Kilkenny stands on either side of the Nore, but, unlike Clonmel, it turns its back on the river. Kilkenny, the see of the Ossory dioceses, with two cathedrals, one of them dating from the thirteenth century, and a castle that is said to be on the site where the kings of Ossory lived, has been a significant centre for more than a thousand years. It is placed on the Nore between the Slieve Ardagh ridge and the Castlecomer basin. Today the city has a dominant commercial position over an area covering the whole of Co. Kilkenny, and there is no town of even 5,000 inhabitants within a radius of twenty miles. Its importance was recognized by the railway builders in the nineteenth century, and it was linked to Dublin through Kildare by 1852. The lines to Waterford and Maryborough were built during the next thirty years. However, in spite of Kilkenny's continued prosperity as a market centre and its maintenance of a number of industries through all crises, the town declined steadily from 1841, when its population was 19,100, to 10,300 in 1946, though in 1961 it had 12,100, and 12,000 in 1966. Its main industries are brewing, furniture, woollen textiles and footwear. The fame of Kilkenny marble is considerable: it is a greyish crystalline limestone which becomes deep blue and black on polishing. Much of it was taken from a quarry near the town to which everyone had access, and the stone has been used for building, and even for paving streets.

B8. THE SLIEVE ARDAGH

This ridge appears as a number of hills around Cashel and Fethard, and extends in a northeasterly direction for more than thirty miles to the Nore valley. In trend, as in the strike of its rocks, it is Caledonian, but in structure it is Armorican, for it consists entirely of Carboniferous strata synclinally folded. There is a definite scarp on the northwest, but a gentler fusion of upland and lowland in the southeast. The ridge rises to 1,100 ft., is pierced by numerous gaps, including that followed by the railway from Thurles to Clonmel, and is not a barrier to agricultural settlement, as most of the hills are farmed to their summits, and the density of population varies from 50 to 80 per square mile, with farming similar to that of the lowlands to the south. The coalfield has been exploited at various times for anthracite. There are eight or nine known seams in synclinal basins, but the upper and more accessible ones have been exhausted though many scars of former workings are to be seen.

B9. THE CASTLECOMER BASIN

The Castlecomer plateau, or basin, which covers an area of more than 200 square miles between the Nore and Barrow valleys, is one of the most clearly marked *pays* in Ireland. It is surrounded but not crossed by important trunk routes, followed by roads and railways, and it is seen from almost any of the surrounding areas as a plateau ringed round by an even ridge-like skyline about 900–1,100 ft. high, with well-marked scarps. A synclinal basin, with sandstones (flagstones) and shales on its margins and Coal Measures in the centre, it is drained by one main river, the Dinin, which joins the Nore a few miles above Kilkenny. This valley gives easy access to the basin and was followed by the railway line built during the 1914–18 war from Kilkenny to Castlecomer, but elsewhere roads climb the scarps with steep gradients from the surrounding lowlands. The Castlecomer basin with the Slieve Ardagh area discussed above, forms what is rather grandiloquently called the Leinster coalfield. It is the most significant Irish coalfield and provides anthracite: in recent years there have been about 400 workers. The coal has been mined for two or three hundred years, at first along the outcrops of the synclinally disposed seams, later by driving in adits from the edges of the basin, and more recently by working inclined adits inside the basin, which have reached the water level so that pumping is necessary. There are many evidences of the three phases of mining in the basin. During the war of 1939–45, the peacetime output, which rarely reached 100,000 tons per annum, was considerably increased, but home supplies of coal cannot meet the Irish demand.

Agriculture employs many more people than mining in this plateau-basin, and in Castlecomer rural district, the miners number less than one-fifth that of the workers on the land. There are mining settlements, notably Castlecomer (1,140), and over 30% of the population lives in villages, a somewhat unusual feature that is obviously related to the mining, which is in many cases combined with the tilling of an acre or two of land. There are few gaps in the farming settlement, and 93% of the Castlecomer area is improved land, in spite of the altitude of as much as 1,100 ft. and the fact that the sandstones and shales weather into acid soils supporting rushy pastures. Mixed farming with a decided emphasis on cattle production for stores and dairying is general, with co-operative creameries. Some two-thirds of the improved land is in pasture, one-tenth under crops and the rest in meadow, and about half the farms are medium-sized, with an average of 60 acres. The density of population per square mile was 127 in 1891, and 76 in 1966, a decrease of two-fifths, and the basin is essentially an agricultural area with a development of coal mining that has left its inevitable imprint on the landscape and is likely to provide employment for at least some time to come.

B10. THE TIPPERARY–THURLES LOWLAND

This lowland lies between the Slieve Ardagh on the southeast and the hills that extend from the Roscrea gap to Tipperary town on the north-west. The lowland is generally about fifteen miles wide and merges at its southern end into the Golden Vale: through the lowland runs the main Dublin–Cork railway line like a central artery, with branches to Roscrea, Clonmel and Cashel. The lowland is entirely floored by Carboniferous strata, covered almost everywhere by drift, and crossed by morainic belts that mark stages in the recession of the ice from its greatest forward spread in the second general (Midland) glaciation. Bogs are increasingly prevalent away from the outer morainic margins: around Cahir and Tipperary there are no bogs, between Cashel and Thurles they appear as a minor element in the landscape, but east of Templemore they cover many square miles of land drained by the upper Suir and the upper Nore. A 'level' of 400 ft. is widespread in this lowland, but there is an extremely gentle slope southwards towards the 200 ft. level recognizable in the Suir lowland around Cahir and Clonmel: in addition various hills, which may be monadnocks, rise to 600 ft. or even in one case 775 ft. at Killough hill, between Cashel and Thurles. These hills are not numerous, but in an undulating and in places almost flat landscape they are conspicuous: some appear to be due to differential erosion in locally more resistant strata, as in the case of a line of hills almost 600 ft. high formed of crystalline-dolomite to the south of Rathdowney.

Agriculture is the main resource of this lowland and the medium-sized farm is the most characteristic unit, with an average farm size of 60 acres, though there are more small farms than large farms. Permanent pastures cover at least two-thirds of the land and most of the remainder is meadow. Fodder crops, such as potatoes, mangels and oats, are grown, with some wheat and barley, but the building of a large sugar-beet factory in Thurles has stimulated the growing of this testing crop. In 1933, a low-water mark of arable farming in Ireland, as little as 5% was under crops around Tipperary. Cattle production is still the main resource, and the areas around Tipperary and Thurles are usually called 'dairy farming' belts. Even so, three-fifths to three-quarters of the cattle are stores and the farm economy remains mixed rather than specialized. Pig and poultry production is widespread in association with milk, and there are numerous creameries. The towns are market centres with small industries processing the agricultural output: Tipperary and Thurles have creameries and Roscrea has bacon and meat factories. Thurles (6,900) has grown considerably in recent years, but Cashel (2,700) has remained stationary. All these, with Templemore (2,000) and Cahir (1,700), and a number of small centres that can only be called villages, are indispensable markets of the lowland, having in all approximately one-third of the population.

BII. THE GLEN OF AHERLOW

This glen, an offshoot of the Suir lowland, lies between the Galtees and the Ballyhoura hills on the south and the Slievenamuck range on the north. The Galtees are a typical Armorican range rising to several points above 2,500 ft., culminating in Galtymore, 3,018 ft. high. These mountains have some fine corries with lakes and block moraines: having persisted for some fifteen miles they break down into an upland basin worked out in Silurian slates and surrounded by Old Red sandstone ridges – a structural feature similar to that in the Comeraghs and the Slievenaman range. On the other side of the valley the Slievenamuck range is formed of Old Red sandstone and rises to 1,216 ft.: its northern edge is a fault scarp which forms a very clear limit to the lowlands around Tipperary. The Glen is 200–300 ft. high, from one to two miles wide and some ten miles long: it stands out as a continuously farmed area, with the main emphasis on dairying for the co-operative creameries. The hills on either side are heavily wooded, partly with old woods and partly with new plantations laid out by the Forestry Commission.

CHAPTER 15

The Valleys and Uplands of Cork

The valleys and uplands of Cork are the heart of the old Munster province, and have a southern softness of climate and verdancy of growth that adds to the attraction of the scenery, which is generally pleasant if rarely striking. The countryside is mainly well farmed, and the holdings are larger than those of Kerry: the great success of the creameries since the eighteen-eighties has meant that dairying still predominates with the result that the city of Cork is, now as for centuries past, the centre of the butter trade. A cathedral city, with industries and wharves beside a tidal river, Cork is the natural heart of this region and the one major city of the south-west. The agricultural landscape is dominated by medium-sized farms averaging 75–85 acres around Cork and Mallow, but only 40–50 acres farther west. As a result of the farmers' main reliance on a return from cattle in some form, the actual land use is pasture first, meadow next and arable crops last and least. Even the arable crops are intended chiefly as cattle feedstuffs or as foodstuffs for home use. Industry, more deeply rooted in this region than in many other parts of Ireland, is based mainly on the processing of agricultural raw materials. Cork city, however, has recently acquired some heavy industry, but the economic pattern is in general agricultural, and the towns are primarily market centres with a stationary or diminishing population.

LANDFORMS AND LAND USE

Nowhere else in the British Isles is the east–west trend of Armorican structural elements as well developed as in this region, which consists of a series of anticlinal ridges, of varying altitude but including some peaks over 2,000 ft., and synclinal valleys drained by streams flowing east, some of which have a right-angled bend before they reach the sea. The country south of a line from Killarney to Dungarvan has the east–west Devonian ridges very perfectly developed: north of this line, the Armorican ranges rise out of a lowland composed mainly of Carboniferous limestone. This limestone disappears near Carrigaline, and is replaced by Carboniferous

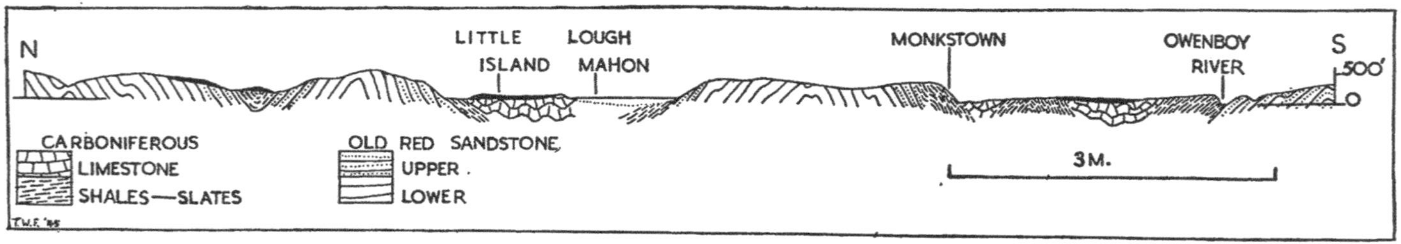

FIG. 61. SECTION DRAWN APPROXIMATELY NORTH–SOUTH THROUGH LOUGH MAHON

The anticlines and synclines have a sharp development, and the dips are at high angles. The vertical scale is three times the horizontal, but the even nature of the upper surfaces appears clearly.

Copied from the Geological Survey, 1" to 1 mile Cork District, drift series, Southampton, 1905

slates and Coomhola grits: this change is of great topographical significance, as in south Cork both the Old Red sandstones and the Carboniferous strata, especially the grits, form hilly country, though the hills are nearly as much as 600 ft. high and are cut by an intricate system of valleys (Fig. 62). Farther north the correspondence of river valleys with synclines and ridges with anticlines is much clearer; these are:

1. *The Cloyne Syncline*, which extends from Ballycotton bay to Cork harbour, where the Carboniferous limestones appear in two well-known islands, Spike and Haulbowline.

2. *The Knockadoon–Great Island Anticline*, a ridge with summits of 200–400 ft. in the east, and 600–800 ft. in the west.

3. *The Cork Syncline*, a lowland stretching from Youghal bay to lough Mahon, through Cork city and west into the Lee valley.

4. *The Boggeragh–Nagles Anticline*, in its full extent reaching from the borders of Kerry to the Drum hills, south of Dungarvan. The general level of the sharply dissected ridge is 600–800 ft., but the Nagles rise to 1,400 ft. and the Boggeraghs to 2,100 ft. In addition there are small synclinal depressions in this area, notably those around Riverstown, Blarney and Dripsey, north of Cork: in the east, the river Bride flows through a synclinal valley to join the Blackwater. This, the greatest of the Cork rivers, has a long west–east course from near Millstreet to the right-angled bend at Cappoquin, where it begins a gorge-like course to Youghal.

Two levels are widely seen in these ridges; the higher is sometimes called the *South Ireland Peneplane*, around 600–800 ft. It is well developed in the long ridge lying to the south of the Blackwater valley from the estuary to the Boggeragh mountains, which has swelling crests around 600–800 ft. It is also seen in the Drum hills and on the south side of the Knockmealdowns, which appear to rise above it like a hilly island in a sea. Various ridges and hills south of the river Lee are also around the 600 ft. level. The lower peneplane, called Coastal or – better – Lowland, is at 200–400 ft. Widely represented in many parts of Ireland, it is seen with wonderful clarity on the coast here, as everywhere from Waterford harbour to the extreme southwest of Co. Cork. The drift cover is generally thin or even non-existent in some places, such as the areas around the Boggeraghs, Nagles and Knockmealdowns. In the main the drifts are sandy or gravelly, and thick only in a few river valleys, though some of the river gravels and alluvial fans such as those in the Lee valley and near Midleton give particularly fertile soils. In places where the drift is very thin or lacking there are sinkholes and caves, as around Midleton and Mitchelstown, and on most of the limestone lowlands the stream

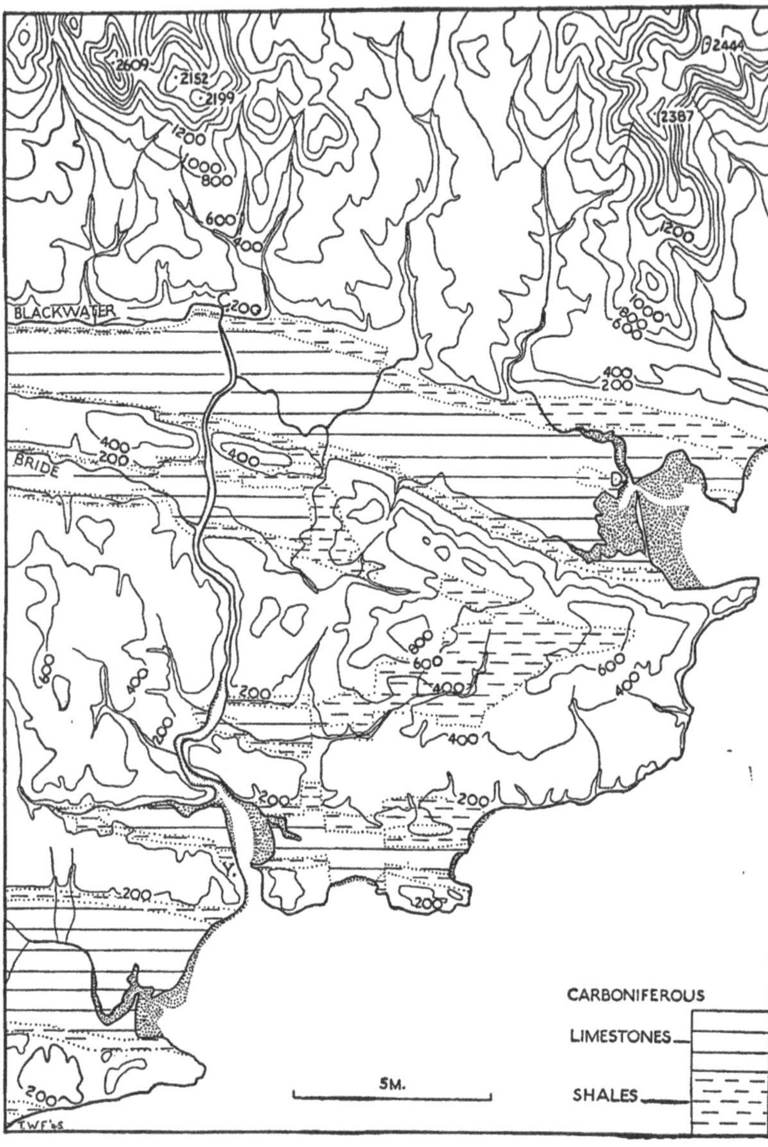

FIG. 62. THE LOWER BLACKWATER VALLEY

C, Cappoquin; Y, Youghal; D, Dungarvan

The bend at Cappoquin inspired the researches of J. B. Jukes: this map shows the exposures of Carboniferous limestones and shales in the lowlands: the remaining area is made up of Old Red sandstones, which form the Knockmealdowns and the Comeragh–Monavullagh range, and also extensive areas around 600–800 ft. and 200–400 ft.

density is low, so that the main rivers must be fed from subterranean sources.

The climatic circumstances of winter mildness, heavy rainfalls and infrequent frosts favour the growth of the various exotics that are also associated with Kerry. The richness of gardens, the density of woodlands, and the abundant wild flowers or garden escapes on waste ground are notable features. Peat bogs are developed only to a slight extent on the lowlands, though they normally cover the higher parts of the uplands. It seems fairly clear that blanket bog does not accumulate on lowlands as easily as in Kerry, Mayo or Donegal, probably because this region has a lower rainfall and better natural drainage. The valleys west of Mallow show a transition to the Kerry landscape as the pastures become more and more sedge-filled, and patches of bog appear in a steadily increasing degree. Here the proportion of non-improved land is nearly half the total, as against less than one-fifth in east Cork.

Demesne woodlands are an attractive scenic feature of this region, notably in the Blackwater valley. The Forestry Commission has acquired over twenty areas for afforestation, covering some 140 square miles, of which over nine-tenths can be planted though most of it was previously scrub and heathland. Most of the new forests cover several square miles and the largest, near Tallow in the Bride valley, covers nearly 12 square miles, most of which is already planted. There are also large forests on the Ballyhoura hills, at Kilworth to the north of Fermoy and near Dungarvan. The forests are located not only on the mountain slopes but also on some of the former rough ground on the windswept ridges so characteristic of the southern part of the region.

Agricultural transitions and contrasts are numerous, and this region has a complex mingling of small, medium-sized and large farms, with the middle group most significant everywhere. The rich heart of the region is the country to the north of Cork city, an area well served by the excellent market facilities of its regional capital. The valuation of land averages £6 per head in the east, and half that amount in the west: typically enough more people are supported on the poorer land which, equally typically, is laid out in smaller farms. Creamery developments have done much to integrate the economic life of the region. In north Cork the proportion of improved land sown to crops has been as low as one-tenth in some years, and all the rest has been used for hay and pasture: but in the south of Cork and the west of Waterford, where cattle production for stores and meat is more significant, the proportion of arable land is generally at least 20–25%. In some form mixed farming is characteristic everywhere, with stock production as the main aim: towards the south dry cattle for the store trade and, less frequently, local meat markets are more important than farther north, where the creameries have brought even the small farms within the commercial network of the dairy industry, which also includes the butter collecting depots of Cork city and the cheese manufacture of Mitchelstown.

More than seven-tenths of this region is improved land and the limit of cultivation is relatively high: the gaps in the settlement pattern are mainly in the higher upland areas. Although a certain number of farms have been withdrawn from cultivation, this region does not show the wholesale advance of rough pasture that may be seen in some Irish uplands. The proportion of improved land in corn crops fell from 19% in 1851 to 9% before the war of 1939–45, doubled during the war years to 18% and declined afterwards to 14%. Since 1851 the area used for hay has steadily increased, and it now covers 16% of the improved land; the most striking development in Co. Cork has been a vast increase in the numbers of cattle to 581,300 in 1961 and 727,900 in 1967. Pig rearing remains profitable, but with fluctuations in numbers (295,000 in 1961 and 207,000 in 1967). Poultry increased from *c.* 600,000 in 1851 to *c.* 2,800,000 before and after the 1939–45 war, but have declined since then to 1,054,000 in 1967. Fewer sheep are kept now than during the latter half of the nineteenth century (233,700 in 1967): the main emphasis is on dairying and the associated store cattle trade. The stresses of the war years demonstrated that farmers were able to increase arable cultivation when necessary, and the main hopes for future agricultural prosperity lie in the extension of creamery production, with more homegrown foodstuffs, and the grading up of pastures that are inherently rich over the greater part of the region. The wealth of farming in this southern area is apparent in its attractive rural landscape.

INDUSTRIES

Co. Cork and Co. Waterford have a balance of agricultural and industrial life that is reflected in the high proportion of people in towns, and in the considerable numbers employed in commerce and distribution. Only 43% of the employed population in Co. Cork and 48% in Co. Waterford is directly engaged in agriculture, as even in the rural areas many people provide goods and services or work in various small industries.* The industrial life is closely interwoven with the agricultural economy. Cork

* 1961 figures, excluding the cities of Cork and Waterford.

city, still the third town of Ireland in size, gathers up the trade of the countryside, exports its products, and provides it with goods and services. If, for example, an optician in a Kerry town wants some new lenses he will turn to some supplier in Cork: similarly its shops and warehouses provide goods for hotels, local traders and others who might look to Dublin to supply their needs. Printing, furniture making, engineering, and even such industries as the manufacture of brushes, paints and carpets are located in Cork. In many ways Cork is overshadowed by Dublin, with its greater port, better communications, vast home market and traditional economic strength but it has been developing economically in recent years and at present is reasonably prosperous. The population within the city, redefined in 1965, was 122,100 in 1966 (112,300 in 1961).

The creameries of Co. Cork carry industrialization into the remotest countrysides: most of them are small concerns employing less than a dozen persons, but a few have already become large enterprises employing as many as 100 workers. One of these at Mitchelstown produces various light cheeses, but while a number of creameries produce cheeses this side of dairying remains of minor significance. Some of the creameries also deal in eggs, pork and such feedstuffs as maize meal. Bacon production is controlled by large factories in Cork, and there is also a small factory in Cappoquin. Cork has now ceased to be a centre of butter manufacture though its merchants control butter exporting, and there is a margarine factory. Milling is a widespread industry in Co. Cork, with a few large concerns in Cork city and many small ones in the various market towns and villages. These mills provide feedstuffs, especially maize meal, and grind wheat grown locally or imported.

Hosiery manufacture is a long-standing development in this region, with large factories at Cork and smaller ones at Bandon and Kanturk, which is one of the few towns in the Republic almost dominated by industry. The trade in woollens and tweeds is also very firmly established, with others in Blarney (over 500 workers), Midleton, Fermoy and Dripsey, the last named a village with a mill employing 200 workers and producing excellent tweeds and woollens, especially blankets. In Cork itself ready-made clothing and weatherproofs are made to supply the large wholesale and retail trade. A fertilizer firm employing over 200 people is also located here. Boots and shoes are also made in Cork and there is a large tannery, established in 1936, at Dungarvan. Almost all these industries, it will be noted, are of agricultural origin, even though they depend to a considerable extent on imported raw materials. Two other special trades are distilling, by a firm in Cork with a branch in Midleton, and brewing in Cork city. Stout and porter are produced, and consumed in various tied houses: this is the one part of Ireland in which the 'tied house' system predominates.

The industries lacking any perceptible agricultural basis are located almost entirely in the city of Cork. They cover a wide range of articles and depend mainly on imported raw materials. Engineering trades include ironwork, electrical equipment and allied heavy steel and motor manufactures. The building trade is supplied by lime and timber merchants, retail concerns by several firms making confectionery and soft drinks, and there are also stationery and printing works. Cobh (Queenstown) lost a great deal of trade through the withdrawal of the British naval dockyard but the Republic's first heavy steel mill was established on Haulbowline island in Cork harbour. Ford cars are manufactured and assembled in Cork and the Dunlop factory produces tyres and rubber articles in the city. These non-agricultural industries have spread outside the Cork focus only to a very limited extent, though mention should be made of an ironwork and general engineering factory at Macroom; and a pottery at Carrigaline with over 200 workers producing earthenware of attractive design and colour.

POPULATION

This region had a population of 409,000 in 1891, 340,000 in 1936, 322,500 in 1961 and 330,500 in 1966. The decline of 20% in population 1891 to 1951 covers a loss of over 40% in the rural areas and a slight increase in the towns as a group, though in fact only Cork gained to any significant extent as many of the smaller towns were stationary or declining in population. All the towns together had one-third of the inhabitants in 1891 but 205,000, about three-fifths of the total population, by 1966. Of the 205,000, 28,400 were in very small towns and villages. Industrial developments in rural areas give a socially healthy balance between agricultural occupations, industry and commerce.

The relatively even pattern of settlement is largely an indication of the size of farms, and outside the closely settled area around Cork harbour where farmlands and outer suburban developments interpenetrate, the density per square mile is generally 50–100 per square mile in the east, about 100 per square mile in the centre, and rather more in the west, especially the northwest. In the actual mapping (Fig. 63) it was found that farms are generally built on roads, and some of the areas apparently unoccupied are in fact covered with fields. Apart from a few hill masses, such as the

FIG. 63. POPULATION DISTRIBUTION, CORK UPLANDS AND VALLEYS

One dot to 25 persons

Cork and its satellite area appears very clearly: elsewhere the gaps are mainly uplands and the settlement is slightly more intensive in the west than in the east. The coasts were formerly very densely populated, but it is not so now.

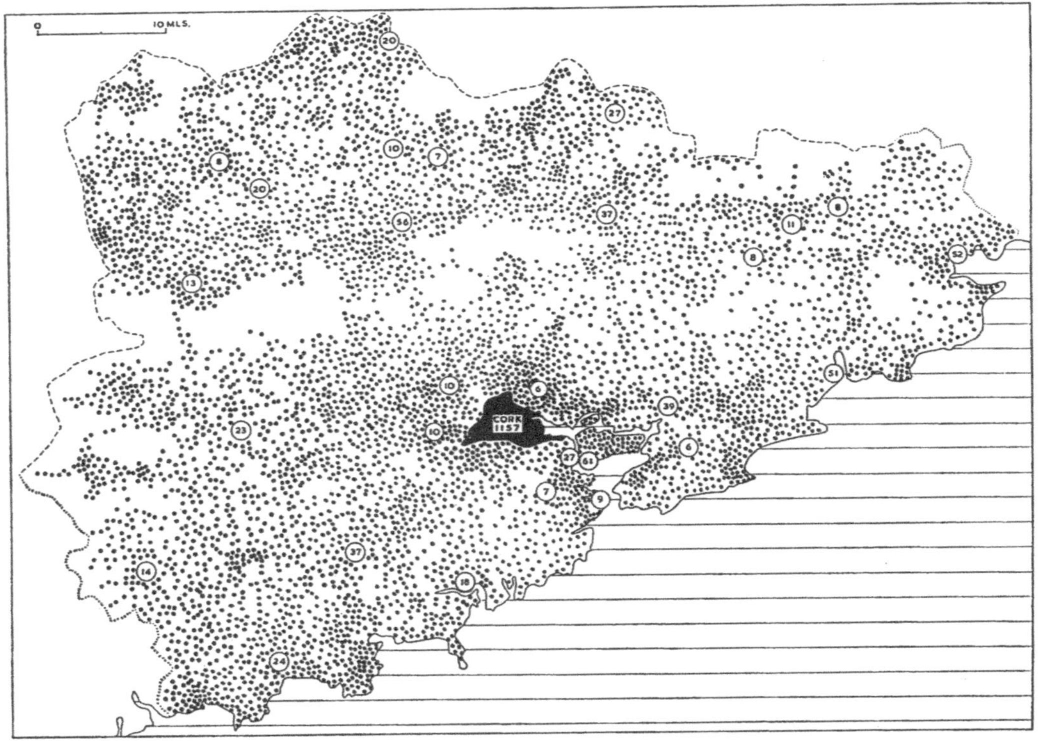

Nagles, Boggeraghs, Ballyhouras and the Knockmealdowns, gaps in the agricultural occupation are infrequent. Only in the extreme west on the border with the Kerry–West Cork region is there any marked population divide, and even there the farmlands of the upper Blackwater valley are linked with those east of the town of Killarney. Ribbons of agricultural settlement penetrate the various uplands, generally along hillside roads, and there is no uniform limit of occupation; the frontier of agricultural settlement lies between 600 ft. and 1,000 ft., and the extent of withdrawal from the maximum occupation depends largely on personal factors. Population decline in the various rural districts between 1891 and 1966 was heavy everywhere, except Cork where the figures are weighted by the outward spread of the city. Of fifteen other rural districts the lowest decreases were in Mitchelstown (−30%) and Bandon (−33%); seven rural districts had decreases of 30–40% and eight −40% or over, including Youghal, Co. Waterford, with −54%. Rural areas along the seaboard show particularly heavy declines of population, which may be ascribed partly to the decay of fishing, a feature common to the whole of south and southwest Ireland (see also pp. 357–8). The main loss, however, is due to an emigration movement from the farms which, though considerable, would have been even more marked but for the successful developments of the creameries.

For many years after the 1922 Treaty the towns of Co. Cork lost population for various reasons: in Cobh the departure of the British forces caused a decline from 8,200 to three-quarters of this figure. In Fermoy, a garrison town, there was a decrease from 6,900 in 1911 to little more than half this figure, with 3,700 in 1966. Comparable decreases are noted in the fishing places, such as Kinsale with 4,000 people in 1911, 2,400 in 1936 and 1,750 in 1966: here a small and specialized tourist industry, with the somewhat exotic sport of shark fishing, has been developed. Several of the other towns have profited from a limited amount of industrialization, and have assumed a more prosperous appearance than formerly. But many of the towns change very little in population from one census to another and none show marked growth.

SUBREGIONS

The structural unity of this region as a whole is one of its most striking features; as a population unit it has few blank patches between its settled lowlands, and there are few lowland bogs to separate one area of fertile soils from another. But it possesses an abundance of extremely small regional entities, such as valleys cut in Carboniferous limestone between the well-defined peneplaned ridges of Old Red sandstone, or again, hidden estuaries with towns and villages around harbours whose significance belongs to the past rather than to the present. Only an infinitely detailed scheme of division could do justice to so varied a landscape, but the subregions used here are at least a convenient descriptive framework and will be found to possess a certain individuality.

1. THE CORK NUCLEUS (Figs. 64, 65)

This includes the city of Cork with the area dominated by it around Cork harbour and the lower parts of the Bride and Lee valleys. The approach to the port from the sea is a mile-wide gap through a ridge of Devonian sandstone; and the great harbour with its extensive anchorage and deep shipping channel is the submerged part of a synclinal valley. The channel cuts through the next sandstone ridge on the west side of Great Island, and enters lough Mahon, a wide sheet of water in the synclinal trough which runs from Cork to Midleton and Youghal. Cork city is placed at the head of navigation on the Lee, and its quays and wharves, which have water up to 30 ft. deep, still receive most of the shipping, though trans-Atlantic liners use Cobh as an outport. Few cities have a more attractive site, for there is great charm in its long winding sea channels with a river divided into two arms around the very centre of the city, and wooded suburbs on higher ground to the north and south. Cobh, in spite of its loss of population, remains fairly prosperous, partly through its use as an American passenger port, though it has also a dockyard, a small tourist trade, employment in the steel mill on Haulbowline and residential attractions for people working in Cork. In Passage West (2,700), on the approach to Cork opposite Great Island, the waterfront and docks decayed, many fine old houses and shops fell into ruins, and some terraced houses were turned into tenements; lately this small town has been much improved. An important development on Cork harbour is an oil refinery at Whitegate, with a jetty, opened in 1959.

Apparently of monastic origin, Cork attracted Norse raiders in the ninth century, but the Danes probably began the settlement on the marshes in the Lee, where the walled town which grew up later became an Anglo-Norman stronghold. It has been suggested that Cork grew around two centres, one on the marshes, and the other on the higher ground to the south of the river. The former is still part of the city centre, but the walls, partly destroyed by cannon in 1690, when the city was still faithful to James II, gradually decayed and were never rebuilt. The process of

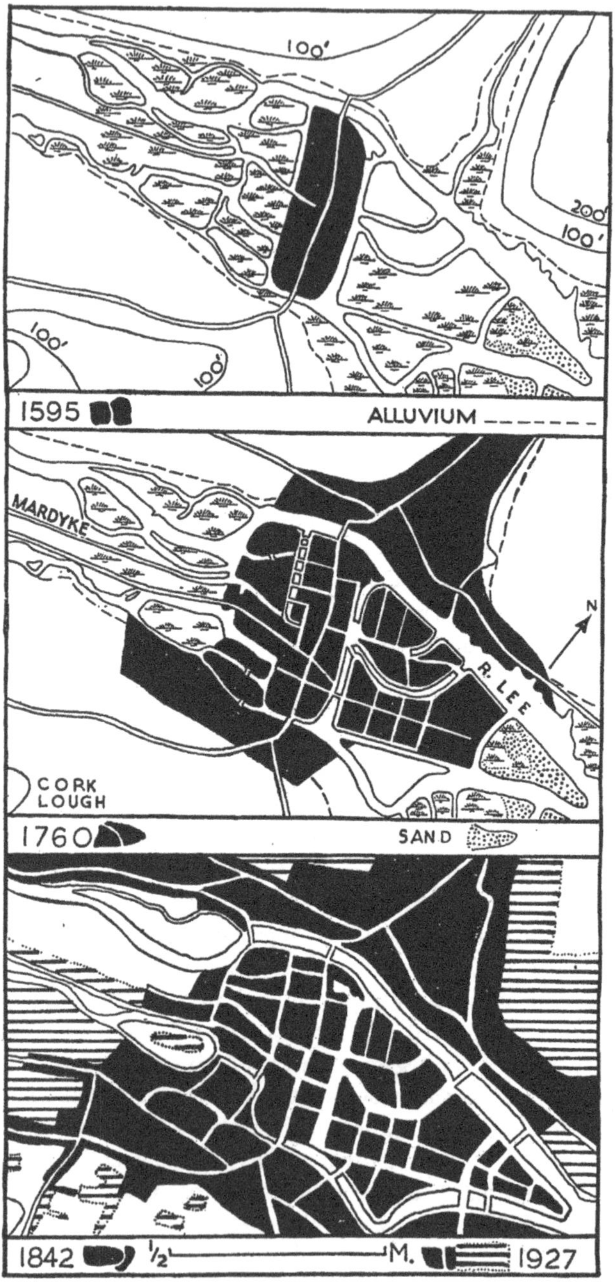

FIG. 64. THE CITY OF CORK

The marshes in the Lee valley were gradually united, and the channels between them were covered in to form the main streets of the city.

Based on maps in the Geological Memoir 'The country around Cork and Cork harbour', Dublin, 1905, and the 6" to 1 mile maps, by permission of the Minister for Finance of Ireland reclamation had begun long before this date, and the channel which ran through the city from east to west had been covered in by the beginning of the seventeenth century. There was already some settlement on the firmer ground to the north and south, but its exact limits are uncertain. By the middle of the eighteenth century reclamation of the marshes had brought the centre of the city to a state at which it bore some resemblance to Venice or Amsterdam of the present day. The old main street through the original walled city survived, as indeed it does still, but buildings covered most of the marshes, and a large number of bridges crossed the channels between them. The Mardyke, a promenade that runs through the west of the city, was laid out in 1720. Island Cork was complete in form by the end of the eighteenth century; its channels were covered and spacious streets, notably Grand Parade, Patrick Street and South Mall, were already laid out. The final developments included the construction of numerous quays, altogether four miles long, of which some are used for shipping.

Cork does not possess the impressive Georgian architecture of Dublin, Waterford and other Irish towns, as most of its eighteenth-century houses are of red brick imported as ballast on ships that maintained the then profitable export trade with the Netherlands. Some of the later buildings have a pleasing mixture of red sandstone and white limestone, but the latter finally prevailed, and Welsh slates gradually became universal. One of the city's major problems is the large number of very small houses that has fallen into decay, but its main streets and quays are dignified and attractive, and the Mardyke on the west runs beside a large public park. Cork is a regional capital serving a primarily rural community; port, industrial centre, cathedral and university town, it is a minor metropolis, but one facing the increasing industrial power of Dublin, with its superior port facilities, its greater natural nodality and its proximity to the British export market. But it remains a strong town: at the 1960 Census, the retail trade of the city and its suburbs was valued at £11,441,000, more than half the total for the county of £22,413,000. In 1964 the port of Cork received 1,128 ships, tonnage 880,177, about one-tenth of the Republic's total. Oil tankers go direct to Whitegate.

2. SOUTH CORK

South Cork is dominated by ridges formed of Old Red sandstone and various Carboniferous slates and grits: the clearest morphological features are the South Ireland peneplane, here at 600–750 ft., and the 200–400 ft. surface, which is best seen in the many cliffed headlands along the coast.

The river system is extremely complicated, but there is an interplay of west–east and north–south stretches seen, for example, in the Bandon valley; the only Carboniferous limestone valley is that drained by the Owenboy river to Cork harbour through Carrigaline (Fig. 65). The rivers, incised into the surfaces, reach the sea through the long winding estuaries that interrupt the cliffed coast, but few ships now reach the little ports, as fishing has declined almost to extinction. South Cork now lives mainly by farming the valleys and uplands, and medium-sized farms are the most usual type. The farming is mixed, with as much as one-fifth of the land in crops, especially oats and potatoes, over most of the area, and some farmers grow sugar-beet for the factory at Mallow. Cattle are reared for the store trade, but also for dairying, including creamery supplies. Sheep are reared, but they are generally of subsidiary importance to cattle, pigs and poultry.

The rural density of population is now 50–100 per square mile, but in 1891 it was 100–150 per square mile. The small market towns, such as Kinsale, Clonakilty, Bandon and Dunmanway, have lost population in harmony with the rural areas for various reasons which include the decline in fishing, now the occupation of only a few men compared with many hundreds before 1914. Kinsale, now having some tourist trade, has declined in population from 4,600 in 1891 to 1,750 in 1966. It has been splendidly described by Lennox Robinson in a book that recalls the social atmosphere of the pre-1914 years, when the Protestant community of south Cork was much stronger and the British garrison still of considerable social significance. In addition, fishing smacks came from far and near to the little harbours along the coast.* Kinsale itself, with its narrow winding streets, had 'gentlemen's houses side by side with shops or small houses, and a street which seemed almost a slum would suddenly display a large house, and behind that house would stretch a very large garden crammed with fruit trees'. Now, however,

> the Kinsale people are fewer and the ancient town is so depopulated, so stagnant as to be almost dead. The mackerel and herring in their queer inconstant way suddenly deserted the Kinsale waters. No Manxmen and Cornishmen, no men from Scotland come any more, and in place of the four or five hundred fishing smacks which once used to moor on a Saturday night round the harbour, so close together that you could walk the whole way stepping from deck to deck, there is not now a single craft bigger than a driever to be seen. The battalion of infantry

* Robinson, L., *Three Homes*, Dublin, 1938, pp. 65 and 134–5.

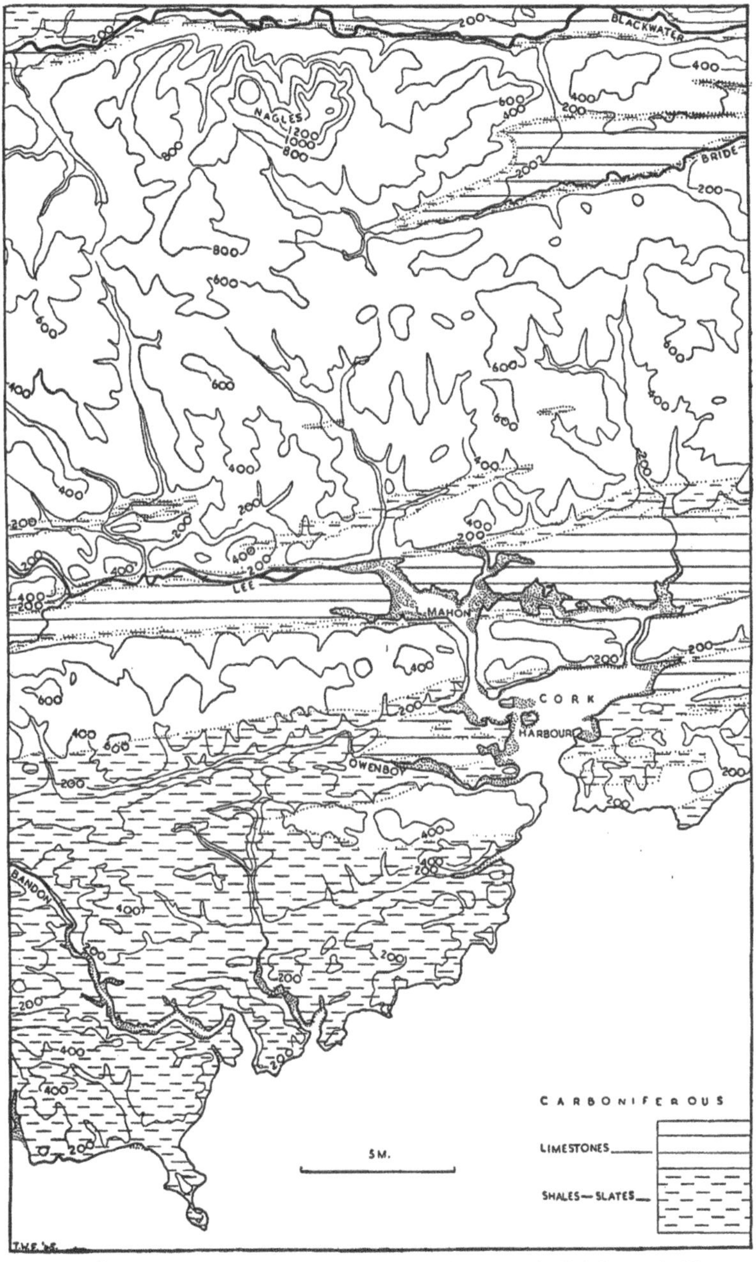

FIG. 65. COUNTRY FROM CORK HARBOUR TO THE BLACKWATER VALLEY

Anticlinal ridges and synclinal valleys appear very clearly: Old Red sandstones form the greater part of the area north of the Lee, though Carboniferous beds occur in the synclinal valleys, such as the Blackwater, the Bride and the minor depressions around Blarney and Riverstown. In the extreme south Carboniferous limestones merge into slates and shales, some of which show a marked resistance to erosion.

was withdrawn after the Anglo-Irish treaty, and in a spasm of upside-down national enthusiasm the empty barracks was burned down, and its empty skeleton broods gloomily over the town. The coastguards went too, and the coastguard station was burned like the barracks . . . the social life sickened and the fine old parish church of St Multose emptied. . . .

3. THE LEE AND BRIDE VALLEYS

The Lee and Bride basins are bounded by ridges 600–800 ft. high and dominated by a girdle of hills that includes the Boggeraghs on the north and the Derrynasaggart and Shehy mountains on the west. On the south, however, there is no clear limit to this subregion. The Lee flows through a deep valley in the Devonian sandstone ridge from its source to a point seven miles west of Cork, where it joins the Bride, which has traversed a wide open synclinal valley floored with Carboniferous limestone. East of Macroom there is almost continuous farmland, but farther west the proportion of rough pasture steadily increases and gaps in the settlement pattern become more frequent even on the lowlands, for bogs develop more readily as the rainfall increases steadily to the west. In effect, this is an indication of transition from the fertile and verdant lands of Cork to the saturated lands of the west. The farms over the whole of the subregion average 50 acres, and the density of population per square mile of improved land was 81 in 1966, compared with 97 in 1936 and 160 in 1891. These figures are for the Macroom rural district which, though not exactly coterminous with the subregion, represents its main features fairly accurately. The largest market centre is Macroom with 2,300 people in 1966 compared with 3,500 in 1891, and over four-fifths of the population lives in dispersed habitations.

4. THE BOGGERAGHS, NAGLES AND MALLOW GAP

The Devonian sandstone ridge between the Blackwater valley and the Cork syncline is mainly 600–800 ft. high, but mountain ranges rise above this level to 1,400 ft. in the Nagles and to 2,100 ft. in the Boggeraghs. These ranges are sufficiently extensive to make gaps in the settlement pattern of Co. Cork, but the gaps are insignificant in comparison with those made by the mountains of Co. Kerry farther west. There is continuous agricultural settlement between the two ranges, in a lowland some eight miles wide, and the limit of cultivation varies from 600 to 1,000 ft. The farming is mixed, with specialized milk production for creameries and for the Cork market. The main railway line from Dublin to Cork

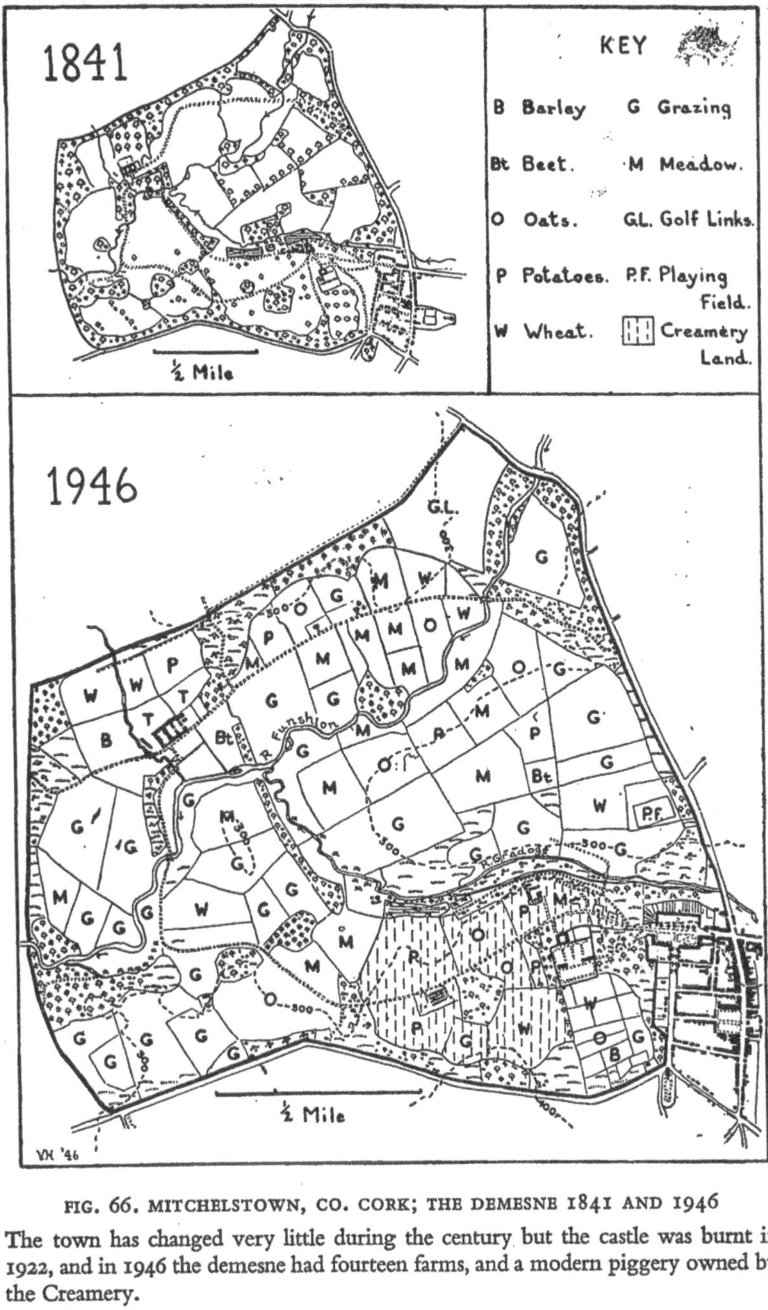

FIG. 66. MITCHELSTOWN, CO. CORK; THE DEMESNE 1841 AND 1946

The town has changed very little during the century but the castle was burnt in 1922, and in 1946 the demesne had fourteen farms, and a modern piggery owned by the Creamery.

The upper map is taken from the 1841 Survey: the lower is based on the map issued in 1937, by permission of the Minister for Finance of Ireland. Drawn by D. Vida Henning, and first published in 'Irish Geography', vol. 1, No. 4, 1947, p. 98 crosses the Blackwater at Mallow by a long viaduct, climbs 300 ft. in seven miles, and then runs to Cork along a track hardly less steep.

5. THE NORTH CORK LOWLANDS

These lowlands are defined by the edge of the Old Red sandstone ridge on the south, by the Carboniferous uplands that extend into north Kerry and Limerick on the northwest, and by the Ballyhouras and Kilworth mountains on the east. Covering an area of approximately 800 square miles, the lowlands are typical of many hill-girdled areas that lie between the lower Shannon, the Barrow and the Blackwater. Wide gaps continue the settled agricultural lowlands into Co. Limerick through Charleville, into the middle Suir valley through Mitchelstown and into the lower part of the Blackwater valley through Fermoy. At Mallow, the natural centre of this basin, the main Dublin–Cork railway meets the line to Killarney and Tralee.

Four-fifths of this lowland is in farms, which are spread fairly evenly across the peneplane surfaces of which it is formed. In the east, around Mallow, Mitchelstown and Charleville, the 200–400 ft. surface (or surfaces) is developed and to the west, around Millstreet and Kanturk, the 600–800 ft. surface appears. The area is drained by various tributaries of the Blackwater, but the stream density is conspicuously low over much of the Carboniferous limestone area, and the drift cover is generally thin or even absent. Around Mitchelstown there are a number of underground caves. In the inter-war period the proportion of arable land was in some years as little as 10%, with as much as 70% of the improved land in pasture and the remainder in meadow. The reliance on dairy cattle is most marked in the west around Millstreet and Kanturk, but it is also considerable around Mitchelstown, Charleville and Mallow. Everywhere the creameries are a dominant influence; at Mitchelstown the creamery is in effect a large factory specializing in butter and cheese, with an egg-collecting station and a pig farm: in all some 200 people are employed. The establishment of one of the four sugar-beet factories at Mallow has stimulated beet growing, but many farmers, here as elsewhere, regard it as a crop demanding far too much attention to be really profitable. The farms are generally medium-sized, of 30–100 acres, with a number of holdings between 15 and 30 acres and some large farms. The population has declined almost by one-third since 1891, and is now 60–100 per square mile of improved land. Mallow (5,500) is the largest of the market towns; Millstreet (1,300), Kanturk (1,900), Charleville (2,100) and Mitchelstown (2,600) are minor centres. Most of the towns and villages, which together

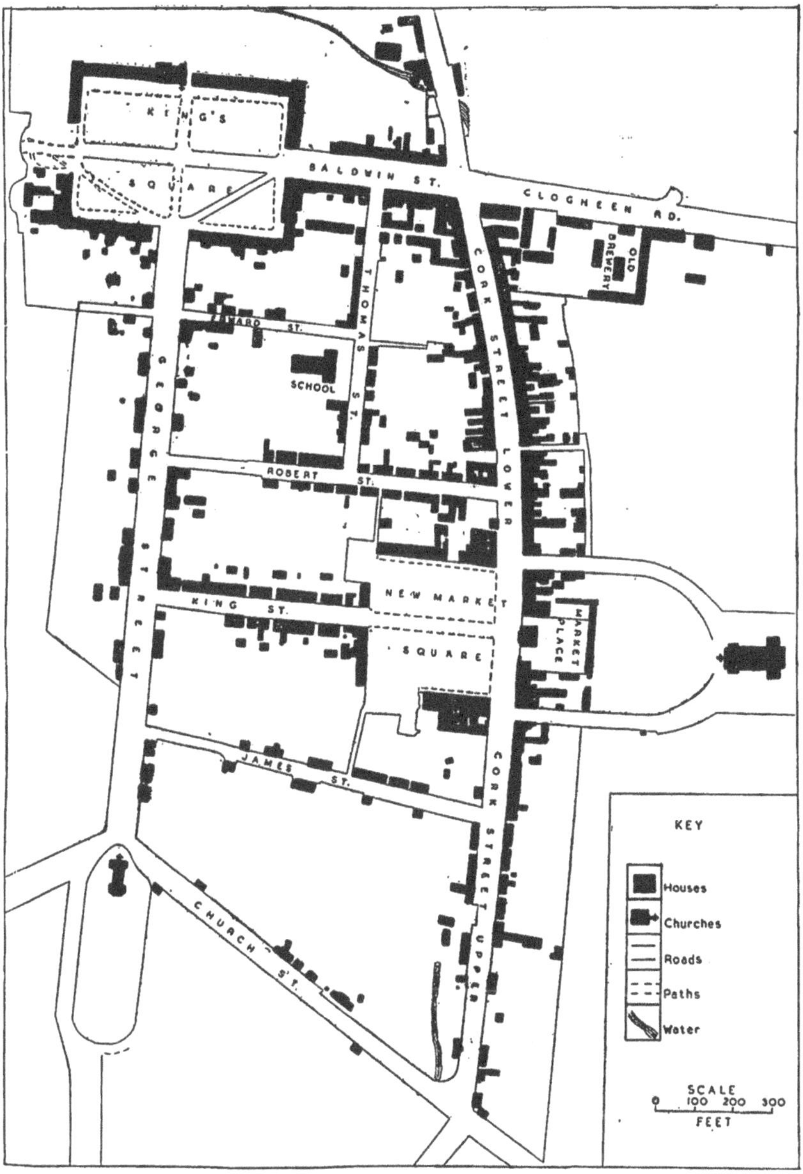

FIG. 67. MITCHELSTOWN IN 1841

This is an excellent example of a planned town, built by the Kingston family: there are two squares, two churches, and an interesting mingling of straight and curved roads. All the houses are on the street and the town is little changed today.

Reproduced from 'Pre-Famine Ireland', Manchester 1957, p. 222: drawn by D. V. Henning and first published in 'Irish Geography', vol. 2, no. 3, 1951, p. 207 include about one-fifth of the population, are merely agricultural centres with a few shops, banks and market facilities, though the creameries have already carried a small measure of industrialization into this dominantly agricultural area.

6. THE MIDDLE BLACKWATER AND BRIDE VALLEYS

The North Cork lowland gradually narrows into the Blackwater valley around Fermoy, and the river flows for nearly ten miles between the Knockmealdowns and the ridge that separates it from the Bride valley. The Knockmealdowns here consist mainly of the South Ireland peneplane surface, *c.* 600–800 ft., deeply incised by valleys now grading their courses to the Blackwater and crowned by lines of hills, nearly 1,200 ft. high south of the Araglin valley and over 2,000 ft. high to the north, where the highest summit is 2,609 ft. The ridge between the Blackwater valley and the Bride has several hills 500–600 ft. high and is broken by a gap south of Fermoy, now a marshy hollow drained by the Shanowennadrimina stream, but probably a relic of the consequent drainage pattern. The Bride valley is a clearly defined synclinal trough with patches of the South Ireland peneplane on either side.

More than two-thirds of this area is improved land: farmlands spread with little interruption from the valleys to the peneplaned surfaces above them, and reach an altitude of as much as 1,000 ft. There are, however, extensive woodlands on the steeper slopes of the Knockmealdowns, on the ridge between the Blackwater and the Bride, and in numerous demesnes along the Blackwater valley. The farms average 65–80 acres, and while medium-sized farms are the predominant type, there is an appreciable number of large farms. Thus in Fermoy rural district nearly 30% and in Lismore rural district approximately 20% of the farms are of more than 100 acres. As the amount of employment given on farms is comparatively slight the density of population per square mile of improved land is relatively low, and averages 50–65; close inspection of Fig. 63 will reveal the sparsity of the population distribution. There has been a decline of more than one-third since 1891 and Fermoy has lost heavily through a variety of causes (p. 338). However, the area is one of considerable agricultural significance, with pasture covering seven-tenths of all the farmlands, and the remainder divided almost equally between crops and meadows. The store cattle trade is generally more important than dairying and creameries are less numerous than farther west.

7. THE DUNGARVAN LOWLAND

The lowland between Cappoquin and Dungarvan is as clearly defined a part of the synclinal Carboniferous limestone trough as the Blackwater valley; on the south, the Drum hills carry the Old Red sandstone ridge to its sharply cliffed termination in Helvick head, and on the north there is a sharp break of slope at the edge of the 600–800 ft. peneplane. This erosion surface forms a saddle cut by several rivers, crossed by various roads and partly occupied by farms between the Knockmealdowns and the Monavullagh mountains. The main morphological interest lies in the continuation of the tectonic line that extends from Dingle bay to its extremity in Helvick head, and in the failure of the Blackwater to continue its subsequent course across this lowland, which is drained by sluggish streams incapable of effecting river capture. Otherwise it is merely an agricultural area of mixed farming generally similar to that described above with one main market and small industrial centre in Dungarvan (5,400) placed beside a sand-encumbered estuary that is the drowned end of the lowland.

8. THE DRUM HILLS–ARDMORE AREA

To the east of the lower Blackwater an intricate landscape of heathery ridges and cultivated lowlands is developed on Devonian sandstones and varied Carboniferous strata. The northernmost ridge runs almost unbroken from Helvick head to the Blackwater and includes a number of summits from 500 ft. to nearly 900 ft.; this and the other ridges, one of which has a summit of 991 ft., probably belong to the South Ireland peneplane. But the 200–400 ft. level of the Coastal (Lowland) peneplane is also well represented in various ridges above the depression worked out in the less resistant Carboniferous beds. The coast from Helvick head to Ardmore bay is a cliffed ledge similar to that farther east in Co. Waterford (pp. 315, 331), on which the streams flow in wide open valleys before descending in gorges to the sea. The southern coast between Ardmore and Youghal has two clearly defined ridges of Devonian sandstone just over 200 ft. high, and separated by Whiting bay, cut in Carboniferous limestone.

The density of population is 50–60 per square mile of improved land, and the decline in numbers has been more than 40% during the past seventy years, due in part to the lessened employment in fishing. Ardmore and Ring are primarily tourist villages: one of the attractions for visitors is the presence of an Irish-speaking community cut off from the main Gaeltacht (pp. 165–70). Mixed farming is dominant, with both stock rearing and dairying for creameries. Threequarters of the farmland is in pasture, some 15% is cropped and the rest is meadow: the medium-sized farms are the most frequent, though nearly one-third are of less than 30 acres.

9. SOUTHEAST CORK

Three peneplaned Devonian ridges and two low-lying troughs filled with Carboniferous limestone form southeast Cork. From south to north there are the cliffed coastal ridge at 200–300 ft., the Cloyne depression, the Knockadoon ridge over 300 ft. high, the Midleton depression and finally a wide ridge that rises to 800 ft. The remarkable east–west folding is well seen on the coast, where the Devonian sandstones stand out as headlands and in cliffs above the sandy bays of the Carboniferous lowlands. Farms cover seven-tenths of the total area, mostly medium-sized, and the density of rural population is approximately 50 to the square mile. The store cattle trade is more important than dairying and approximately one-quarter of the land is normally given to crops, chiefly oats and barley with potatoes and feedstuffs. The rural areas have declined by some two-fifths since 1891. The towns, which have almost half the population, include Youghal (5,200) and Midleton (4,200), the main market centres; in addition Youghal is a seaside resort with some sea and river salmon fishing. At Youghal there are also various industries, including a small stained glass works dating from 1837, a cabinetmaker's works established in 1870, with larger modern factories for cottons and carpets, both employing over 300 people. Cloyne (600) has an ancient cathedral, associated with the philosopher, George Berkeley, its Bishop from 1732 to 1753.

CHAPTER 16

Kerry and West Cork

This region, the extreme southwest of Ireland, is deeply penetrated by the sea in a series of rias with numerous islands: it is at once maritime, peninsular and mountainous. Maritime in its climate of mild moist days with a small annual range of temperature, it has a prodigality of vegetative growth reminiscent of Cornwall or Brittany; peninsular in its penetration of the Atlantic, Kerry is the most westerly part of Europe; mountainous everywhere, it has some of the finest ranges in Ireland through the combined effects of long-continued denudation and glacial sharpening. Some places in southwest Ireland have scenery of international fame, such as the lakes of Killarney, a complicated series of waters in a wide wooded valley surrounded by mountains, or the harbour of Glengarriff, set with many small islands in an amphitheatre of heathery hills having subtropical vegetation on their lower slopes. The charm of the scenery lies largely in the many local contrasts that this region provides by the intermingling of sea and land, hills and valleys, woodland and moorland, farmland and bogs.

Climatically the resemblances between southwest Ireland, Cornwall and Brittany are striking, but there are also similarities with other parts of the western fringe of Ireland. These are partly climatic, partly vegetational but not structural, for the southwest belongs to Armorican Europe, and Donegal and West Connacht to Caledonian Europe. There is an overriding human unity along the western seaboard, most clearly revealed when the investigations of the Congested Districts Board showed that the poverty of Donegal and West Connacht was matched by that of Kerry and the west of Cork, even though the fragmentation of the land into tiny farms was less marked. Southwest Ireland remains to this day an area partly dependent on remittances from emigrant relatives and it is still unable to absorb its increase of population within the region. Some developments that seemed likely in the days of the Congested Districts Board have not materialized, for fishing has decreased, many of the rural industries have declined and specialized vegetable growing is little developed. But the tourist industry is expanding and has contributed to a rise in the standard of living for at least some of the population. There have been agricultural improvements, and the housing is very obviously better than at any previous period.

LANDFORMS AND LAND USE

The main structural lines are clearly seen in the long anticlinal Armorican ranges, primarily composed of Devonian sandstones, separated by synclinal valleys filled with Carboniferous strata. Much of the topographical detail is due to the glaciation: indeed, the whole of this region is a museum of glacial forms. An icesheet stationed to the west of Kenmare overrode the Iveragh peninsula to Dingle bay, leaving some fine moraines (Fig. 70), and moved south across the Caha peninsula to the area around Skibbereen. Evidences of this icesheet are seen in the large areas of bare rock to the south of Killarney on the Kenmare road. But the Reeks, though showing a jagged form on the skyline, were apparently a sharp-peaked range before the Ice Age, though their forms have been sharpened by ice action. The mountains of the Caha and Iveragh peninsulas had unglaciated summits rising as nunataks above the valley and piedmont glaciers. But no two valleys are alike, though almost all in the mountains show the very clear imprint of glaciation. Corrie basins occur singly or in a series; hillsides and valley heads have been made precipitous by glacial erosion and have moraine-dammed lakes; and moraines spread themselves in a bewildering series both in the valleys and on the surrounding lowlands. Although the local detail is highly varied, the major features are clear, even dramatic, for marine submergence has etched out six separate peninsulas with drowned valleys or rias between them; these are, from north to south:

1. *The Kerry Head Peninsula*, which has rugged hills rising to 700 ft.; it is merely a miniature of the Dingle peninsula some ten miles farther south.

2. *The Dingle Peninsula*, which rises sharply from the lowlands around Tralee and runs for nearly forty miles westward to the Blasket islands, which have precipitous cliffs over 900 ft. high. The greater part of the peninsula consists of mountains deeply seared by valleys, and the major lowlands are in the extreme west.

3. *The Iveragh Peninsula*, the greatest of them all, thirty miles long and fifteen miles wide, which has a long line of mountains in the south with twenty summits of more than 2,000 ft., and the Reeks in the north, a sharply glaciated range culminating in Carrauntoohill, 3,414 ft. In the west of the peninsula there are a number of isolated hills overlooking wide and generally bog-covered lowlands: hills dominate the landscape to

the very shores of the Atlantic and perish only in magnificent cliffs several hundred feet high and as challenging as those of the Blasket islands at the western end of the Dingle peninsula.

4. *The Caha Peninsula*, between the Kenmare 'river' and Bantry bay, is of similar length to the Dingle and Iveragh peninsulas, but little more than ten miles broad at its base. It is sharply indented by the sea towards the west, where there are a number of natural harbours and islands such as Bear island, which shelters the magnificent anchorage of Berehaven. The peninsula has a striking mountain range with many summits over 2,000 ft., and the tourist road from Kenmare to Glengarriff crosses the range by climbing with difficulty to over 1,000 ft. through a series of tunnels. Only in the extreme west is there easy passage from north to south, and even there the lowlands are very restricted: the greater part of the coast is sharply cliffed.

5 and 6. To the south of Bantry bay there are two peninsulas which terminate in *Sheep's head* and *Mizen head* respectively. These peninsulas have hills lower than those farther north, with summits mainly around 1,000 ft., but the Armorican line is clearly seen. Lowlands are very restricted in the Sheep's head peninsula, which is ridge-like in form, but the Mizen head peninsula consists of hills and valleys intermingled one with another.

The landforms of north and central Kerry are more subdued, and include the largest continuous area of lowland in the southwest: this is developed on Carboniferous limestones and extends from a point east of Killarney to the inner shores of Dingle bay, around the Dingle peninsula to Tralee and northwards to the Shannon through Listowel. It is covered almost everywhere by glacial drift and has extensive areas of reedy pastures. Sharply delimited by the hills of the Dingle peninsula and the hill edge from Killarney to Glenbeigh, it is also defined by the Carboniferous uplands that cover a large part of east Kerry and extend into Cork and Limerick. These uplands have flat rounded outlines with wide valleys, and are deeply covered by peat resting on drift.

Kerry and West Cork have the oceanic type of west maritime climate developed in its fullest form, and while the climate and weather is in some ways similar to that of the west of Connacht and Donegal, its closest parallels are to be found in southwest Wales, the extreme southwest of Cornwall and the Atlantic coast of Brittany. The winter mean temperatures of 44° F. are similar to those of Falmouth and Penzance and also to those of Biarritz and Marseilles. No visitor to Kerry in winter could fail to observe the spring-like atmosphere and the continuing growth of vegetation. The temperatures in West Connacht are similar, but Kerry owes its richer vegetation largely to the natural shelter provided in various hill-girdled valleys. There is a small diurnal range of temperature, averaging only 7°–8° in winter, and severe frosts are remarkably rare. At Cahersiveen (Valencia observatory), 7° of frost were recorded on only six occasions in thirty years: the lowest temperature was 20° F. in the winter of 1893–4. This fact does much to explain the subtropical luxuriance of vegetation, the survival of many evergreen forms and the rich development of ferns and mosses. Winter is the season of heavy rainfall: Cahersiveen has three-fifths of its mean annual rainfall of 56″ between October and March, with rain on three days out of four, but an annual average of only five days with snow.

The southwest coasts of Ireland are subject to high winds and exposure affects the vegetation almost to the point of domination. Sheltered valleys have woods of oak, birch, ash, hazel, holly and even arbutus, with many introduced trees in great luxuriance (pp. 69–70), but the rich woodlands of Killarney and Glengarriff are exceptional and blanket bog covers most of the hills and valleys of the peninsulas, with clearings on better-drained soils or even in the heart of the boglands. The Forestry Commission has extended its work in Kerry and West Cork and now has twelve holdings covering 50 square miles, of which the largest cover 7 square miles at Kilgarvan and 7 square miles at Killarney. There are other large forests at Cahersiveen, Kenmare and Kilgarvan, and also in the valleys beside Glengarriff, Co. Cork. Brosna, in the uplands of north Kerry, and Rosscarbery have forests also, but so far the Commission has acquired very little land for planting in the Dingle peninsula.

Kerry and West Cork belong in many ways to the maritime west of Ireland in their agricultural life. Fig. 68 shows how limited the agricultural area was in 1841 and it is not remarkable that in 1891 almost the whole of the peninsular areas were congested, together with the poorest parts of the uplands and lowlands in north Kerry. On a 1966 re-assessment no part of this region would rank as congested. The loss of population in the former congested districts has been heavy and continuous. But the very small farms typical of Donegal, Mayo and Galway are much less prominent in Kerry and West Cork, and medium-sized farms of 30–100 acres are the main group, with a small but important percentage of large farms. Since 1891 there has been an interesting adjustment of population to farming resources. The areas ranked as the poorest in 1891 have lost heavily in population, while the more favoured parts of the north Kerry lowlands have lost less sharply or even remained virtually stable in numbers. The southwest of Ireland does not now present the severe problems of

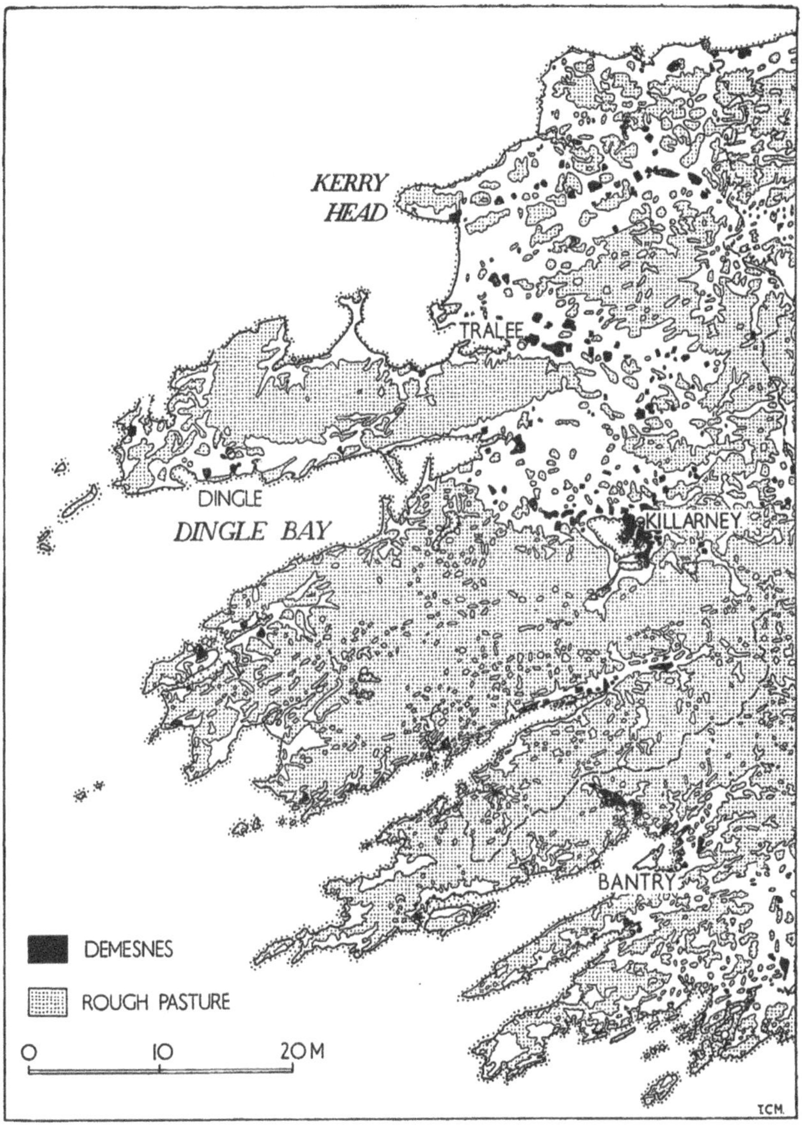

FIG. 68. DEMESNES AND ROUGH PASTURE, KERRY AND WEST CORK, *c.* 1841

When the Congested Districts Board sought land for new farms at the end of the nineteenth century, it was found that the areas available in the west were limited, and that many of the demesnes were hardly suitable for new farms as they consisted partly of woodland – as around Killarney – or were on poor land. This map shows that most of the demesnes were small.

First published in 'Pre-Famine Ireland', Manchester, 1957, p. 236 agricultural adjustment still found in Donegal, Mayo and Galway, largely because it never experienced the universal division of cultivable land into very small farms.

The proportion of land given to crops and pasture in Kerry and West Cork has declined fairly steadily since 1871, when it was 70% of the whole area as compared with 50% now: the essential fact is that one-fifth of the total area has been allowed to revert to rough pasture in less than eighty years. Similarly, 10% of the entire area was sown to crops in 1851 but only half as much by the end of the century, and arable cropping remained at this low ebb except in the wars of 1914–18 and 1939–45: in both these wars the farmers increased the land tilled by eating into their meadows and pastures rather than by reclaiming new land. The bogs of Kerry are an important source of pasture, fuel and bedding for animals; in addition, Kerry ranks third as a peat producing country, but the amount drawn from the bogs by the farmers (180,000 tons in 1964), partly for home consumption and partly for sale, has declined. Farmlands are overwhelmingly devoted to meadow and pasture throughout this region, but the proportion of improved pasture declined from 638,000 acres (nearly 1,000 square miles) in 1871 to 347,000 acres by 1945 (540 square miles) but rose to 395,100 acres (618 square miles) in 1967. The acreage of meadow has increased from 63,000 to 107,200 acres (98 and 168 square miles): the trends in West Cork have been similar. In 1967, approximately 20% of the improved land of Kerry was meadow and 6% arable, compared with 17% in the war years: the rest is pasture.

Livestock are the mainstay of farming throughout this region, and the number of cattle (in Kerry) increased from 159,000 in 1851 to 268,000 in 1911, since when there has been an increase to 316,700 (1967). In the whole area two-fifths of the cattle are milch cows and creameries are most numerous in north Kerry though they are spreading their activities by sending travelling separating stations into the remoter areas. Pig production has developed to a limited extent but was badly hit during the years 1939–45, though the number increased to some 60,000 in the nineteen-fifties and 69,100 in 1967: most of the pigs are sold at fairs for the bacon factories in Tralee, Limerick and Cork. Poultry rearing expanded steadily from the middle of the nineteenth century, and the number kept increased from 211,000 in 1851, to well over 1,000,000 by the inter-war period, though there has been a decline to 397,000 by 1967: formerly a special line was turkeys for the Christmas market. Store cattle are raised widely in the region, but most are sold before they are two years old to dealers at the fairs. Sheep farming is found primarily on the uplands of the peninsulas, and least in the north of Kerry and around Skibbereen. The number of sheep has more than doubled since 1851, and is now 173,700 (1967) in Kerry. The lot of the farm population in this region has been greatly improved during the past hundred years. The retrogression of so much pasture land to a wild condition may seem regrettable to some, but the productivity of the area as a whole is greater than before. Possible improvements include the extension of fodder crop production, the grading-up of pastures, the establishment of more creameries or branch separating stations and the provision of better transport facilities. At Castlegregory, on the north side of the Dingle peninsula, sandy soils over limestone have proved well suited to growing onions and carrots, which are marketed through a co-operative organization where the produce is carefully graded.

INDUSTRIES

Domestic industries were never strongly developed in Kerry and West Cork, though there was formerly cottage weaving of flannel in the Dingle peninsula and of tweed around Kenmare. Factories produce homespuns and socks at Kenmare; and a knitwear factory has been established at Skibbereen. Agricultural industries include creameries scattered through the countryside, bacon curing at Tralee, and milling at a number of places, including Bantry, Castleisland, Skibbereen and Tralee. In Tralee, industries associated with agriculture are most typical, but there is also a boot factory, a large enamelware factory and various small concerns supplying toys, furniture and other articles. Killarney has one old-established footwear factory, and a crane factory established by a German company, but relies for its prosperity mainly on the tourist trade and the general market facilities for the neighbourhood. Listowel now has a cutlery and knife works, but Tralee has been the town best able to attract new industries: the other towns remain small market centres sharply conscious of the competition of larger rivals such as Tralee and the relatively distant but economically powerful centres of Cork and Limerick.

POPULATION

Co. Kerry had 112,800 inhabitants in 1966, 62 to the square mile, and 294,000, 162 to the square mile, in 1841. This loss, on first sight startling, is not phenomenally high by Irish standards. But the loss continues: between 1961 and 1966, the Kenmare rural district declined by 5·5% and the Cahersiveen R.D. by 6·5%, following losses of 8·4% and 7%

respectively in the previous five years, and of 7% (Kenmare) and 5·5% (Cahersiveen) from 1951 to 1956. In the whole of Kerry the decline since 1891 to 1966 has been 37% but in West Cork 55%, from 64,000 to 29,100. Town and village populations number almost one-third of the whole in Kerry and remain fairly stationary; but in West Cork the declines in the town populations have been comparable with those in the rural population, and the towns still have approximately one-fifth of the entire population as they have had for the past seventy years. The significant facts are that even now more than seven-tenths of the population are living in dispersed habitations, that the town population is not increasing but tending to fall, especially in the case of the smaller centres, though a few in north Kerry have been revivified by such developments as the founding of creameries. The settlement pattern shows a fairly even distribution of population in most of the lowlands, with some ribbons of habitation and cultivation on hillsides or in valleys, and scattered groups of houses in the middle of the boglands. In some cases, the houses and farms lie almost exclusively along the roads through the bogs, a curious form of settlement found not only in Kerry but also in Donegal and Sligo (Figs. 80, 85). The actual density per square mile of improved land varies between 100 and 200, but is even higher in some districts, though it nowhere reaches the extremely high levels of northwest Donegal or West Connacht, for the farms in the south-west are larger and emigration has been a substantial relief to population pressure (Fig. 69).

SUBREGIONS

Kerry and West Cork may be divided into four parts, each of which consists of a number of smaller and in many cases distinctive entities (see Folder II*a*). The major divisions are: (A) the lowlands, and (B) the uplands of north and central Kerry; (C) the six peninsulas, and (D) the lowlands forming the extreme southwest of Cork. Though the whole area is part of Armorican Ireland, there is a sharp distinction between the hilly peninsulas of Devonian sandstone and the Carboniferous plateaus of north and central Kerry. The former landscape is merely the western and seaward extension of the long mountain ranges of Cork and Waterford, and the latter is continuous with similar landscapes of like geological formation in Limerick and Clare (see pp. 372–3).

A. THE LOWLAND OF NORTH AND CENTRAL KERRY

These lowlands, formed mainly of Carboniferous limestones with

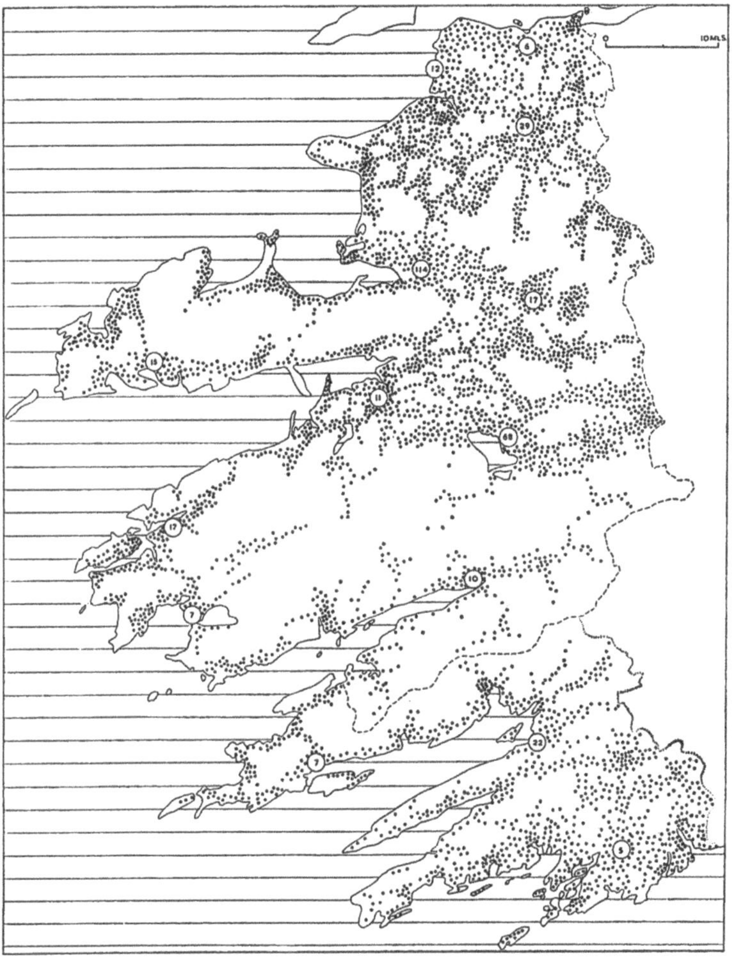

FIG. 69. DISTRIBUTION OF POPULATION, KERRY AND WEST CORK

One dot to 25 persons

In the six peninsulas settlement is mainly along the coasts though there are also farmed areas in various valleys. North Kerry shows an alternation of bogs and agricultural land and here also the desolate Carboniferous plateaus are unoccupied. The extreme southwest was very densely peopled when fishing was prosperous but has seen marked population declines during the past seventy years. Note that the Blasket islands (top left-hand inset) are no longer inhabited (p. 363).

patches of shales and flagstones forming higher ground, especially near the Shannon, are surrounded by uplands formed mainly of Coal Measures. Peat bogs have developed on the widespread alluvial river plains and also on the many other areas of poor natural drainage: comparison of Fig. 69 with a topographical map will show that settlement is sparse or non-existent along many river valleys, though there are also considerable patches of well-drained land continuously occupied and threads of settlement through the bogs. Half the farms of this lowland are of 30–100 acres, one-tenth of more than 100 acres, and the rest of less than 30 acres. The main economic staples are store and dairy cattle, pigs and poultry, and the crops are grown mainly for fodder. Grassland, in pasture and meadow, predominates: at the worst it is little different from the natural rush-filled alluvial flats, but at the best it is a rich, well-drained pasture and meadow. Most of the farmers plant some of their fields with crops intermittently, and a rotation recorded near Killarney consisted of oats on lea ground, followed in the second year by potatoes and roots, then by oats with which grass seed is planted, and in the fourth year hay, after which the field reverts once more to pasture for several years.

The density of the rural population averages 100–200 per square mile of improved land and over 80 per square mile of the entire area. Since 1891 there has been a decrease of as much as one-third in some areas, but in others the very slight declines, or even the small increases, have been ascribed to the success of some of the creameries. At its lowest level, farming in this lowland was so poor in 1891 that some districts came within the scope of the Congested Districts Board, whose inspectors reported that there were very few extra sources of income available, though there was some fishing in the Shannon and Cashen estuaries, but there is very little now. Three towns serve the lowland as its main market centres – Listowel (2,800), Tralee (12,000), and the marginally placed Killarney (6,900): there are some smaller towns such as Castleisland (1,700) and also a number of agricultural villages with a few hundred inhabitants. Listowel is a typical small market town; Tralee, the county town, is the most prosperous and industrialized place for many miles, and Killarney, made more prosperous by recent industrialization, has a number of tourist hotels and is a market town at the meeting place of uplands and lowlands.

B. THE UPLANDS OF NORTH AND CENTRAL KERRY

These uplands are made up of Coal Measures with some flagstones and shales. The Coal Measures consist of shales, sandstones and grits so highly folded, crumpled, faulted and fractured that the coal seams, if existing at all, are economically useless. Intricate as their underlying structure may be, the uplands present long featureless lines to the observer at Tralee or Killarney: they rise fairly sharply from the surrounding lowland to 600 ft., and possess considerable relatively level surfaces around 600–800 ft., 1,000 ft and 1,200 ft. Their highest crest is Knocknefaune, 1,441 ft., on the Kerry border, where the Blackwater has its source. A large number of river valleys, drained to the Laune, the Maine and the Feale, carry tongues of lowland into the very heart of the plateau. Some of these valleys are wide and open and have farmlands up to a height of 600 ft., or even in some places up to 1,000 ft., but continuous areas of settlement are found only in the more open valleys, such as the middle Feale, and elsewhere settlement is restricted to clearings on hillsides with a few farms scattered among the bogs that are so widespread over all the more level areas and continuous on the higher surface.

The poorer parts of the uplands, mainly in the east, were included in the congested districts in 1891. Most of the farmers in them had little more than 3 acres of arable land divided between potatoes and oats, with a patch of greens, and rotation took the simple form of two years' potatoes, one year's oats and a return to grassland. Mountain grazing was allocated to the farmers. At a slightly higher farming standard the rotation was ley, oats, potatoes, oats and grass. The main cash staples were bacon, butter and eggs at the poorer levels of farming, and more intensive store and dairy farming on the better farms. Since 1891 the changes have not been revolutionary, though some creameries have been established in the uplands as well as on its margins, and the population has declined by one-third so that the once-congested areas are, with one small exception, no longer congested; on the other hand, many fields and farms have reverted to rough pasture. The density of population is now around 50–60 to the square mile, but generally at least double this in the areas actually settled. Many farms are several miles away from the villages that together include about one-tenth of the whole population.

C I. THE KERRY HEAD PENINSULA

Covering only some forty square miles, this peninsula consists of a ridge of Old Red sandstone stretching from Kerry head to the Cashen estuary. At its western end, in the part that is truly peninsular, it has hills over 700 ft. high, but elsewhere it is mainly 200–300 ft. high, with many stretches of cliffs along the coast. Bogs cover some of the more level land, but most of the peninsula is occupied by farms placed along the roads, and the population density reaches 200 per square mile of cultivated land. Having lost one-quarter of its population, this area has a general density of population of over 100 to the square mile. Relatively closely settled still, it was almost entirely congested in 1891, but only one area would be included on a reassessment. The small farms in 1891 had an acre each of potatoes, oats and meadows, a patch of greens, some pasture, and access to rough grazing. Ploughs were used and seaweed was collected for manuring cropland and as a top dressing for grassland: the only form of rotation was to plant potatoes and oats in successive years on land subsequently left as pasture for several years. Some men worked seasonally as fishermen but they, like the labourers, normally had a smallholding and kept a few pigs and some poultry. Since then the changes have not been great, though there are now creameries in the area, which is economically allied to the north Kerry lowland.

C 2. THE DINGLE PENINSULA

Forty miles long and from three to ten miles wide, this peninsula rises from the Kerry lowlands into the Slieve Mish mountains, which have a central line of peaks culminating in Baurtregaum, 2,796 ft. high. The track of the Dingle railway (now abandoned) cuts through the peninsula to the west of the Slieve Mish range, crossing it at a height of 691 ft., and the ridge is continued westward in rolling moorlands around 2,000 ft. high with no point below 1,000 ft. In the east the uplands are formed of Old Red sandstone, but to the west the 'Dingle beds' appear: these consist of grits, sandstones and shales so completely folded and faulted that their age (perhaps lower Devonian) and relationship to other strata are highly controversial. The upland surface is remarkably smooth except where it is seared by a corrie glaciation that has become a valley glaciation in a few places, of which the best evidence is the magnificent series of cliffed hollows on the north side. At Dingle, the mountains swing northwards into the Brandon range, which culminates in Brandon mountain, 3,127 ft., and consists of a craggy ridge maintaining an altitude of 2,500 ft. to within a mile of the sea, finally terminating in cliffs at Brandon head, 1,200 ft. above the water. The cliffs are not sheer, for landslips have dislodged huge crags of rocks and the cliffs resemble a heap of ruins.

Lowlands stretch along each flank of the peninsula from the neighbourhood of Tralee, reaching Brandon bay on the north and the Inch sand dunes on the south: between Inch and Dingle, hills of 800–1,000 ft. reach the sea and the lowlands are restricted to a few small river valleys along the coast with various inland basins traversed by the old railway and the main road. In the west, however, there are extensive lowlands, notably around three natural harbours, Dingle, Ventry and Smerwick: these lowlands are separated from one another by hills around 1,000 ft. high. The last expression of this hilly topography is the Blasket islands, the most westerly part of Europe, a group of six small, shelterless islands with many outlying rocks. The largest island, the Great Blasket, a knife-edge four miles long and half a mile wide, rises at two points to nearly 1,000 ft. This island, the only one inhabited, was evacuated in 1953.

The density of population over the whole area averages little more than 50 to the square mile, and on the cultivated land 70–150 per square mile. The loss of population since 1891 has been 53% and as a result no part of the peninsula would be congested on a reassessment. Almost all the peninsula was congested in 1891, and the areas not then congested had a rateable valuation per head only just above the statistical limit. The arable areas on the small farms varied from 3½ to 4½ acres in equal proportions of oats, potatoes and meadow with a small patch of greens. The only rotation was to sow potatoes for one or two years, oats for a year and then grassland. Farmyard manure, sand and seaweed were used as fertilizers. One-fifth of the farmers were recorded as very poor in the eastern part of the peninsula in 1891, while few were so regarded in the west. Small farms were not then dominant in the Dingle peninsula: today only one-fifth are of less than 15 acres, but one-half are less than 30 acres. Most of the remainder are medium-sized and the farm of about 30 acres is the most characteristic unit, especially in such areas as the central part of the peninsula, both in the long valley carrying the road through Anascaul and on the gently sloping hillsides around Dingle and farther west. However, extensive areas have been allowed to revert to rough pasture, especially on the boglands and towards the upper limits of cultivation, which varies according to the type of soil and drainage, but reaches 700 ft. in places. Many small farms have been amalgamated and the general standard of housing has risen: two-storey slated houses have replaced thatched cottages now serving as out buildings. Yet poor thatched cottages surrounded by a few fields of indifferent quality are still a picturesque, if depressing, feature of the landscape.

Dingle (1,400) draws trade from the west part of the peninsula and has monthly fairs for cattle, sheep and pigs attended by buyers from Tralee, Limerick and Cork. The most westerly town in Europe, Dingle now has a co-operative creamery, but its light railway has gone though it is linked to Tralee, thirty miles away, by road services. Fishing is a traditional activity in Dingle, but the number of men engaged has sharply diminished from 600 in 1891 to a few score at present, though the catch is now frozen in a modern plant, apart from the small amount sold locally or in other parts of Ireland. Dingle also has a small boat-building yard. At present Dingle has a growing tourist industry with a variety of accommodation and facilities such as sailing and car hire. Several of the shops in the town acquire considerable profit from sales to tourists, including not only the visitors who stay for some time but also those who tour by bus or car, enjoying the fine beaches and magnificent scenery.

C 3. THE IVERAGH PENINSULA

This peninsula has great topographical diversity, due largely to glaciation and to post-glacial adjustments of drainage. Structurally the peninsula is a huge anticlinal curve of Old Red sandstone, complicated by many minor undulations in the strata, and the strike generally coincides with the direction of the longest mountain ranges and the main longitudinal valleys. In the east evidences of ice-moulding include such deep U-shaped valleys as the Laune, the Gap of Dunloe, the Caragh and the Behy, the last with five wonderful lake-filled corries: there are corries in the mountains, glacially roughened surfaces on the mountain sides, and a profusion of moraines. The whole district, lowland and upland alike, owes its great beauty of crag and lake to ice action, but there are also important remnants of the pre-glacial surface, which was apparently a hill country having the Reeks as a steeply peaked range and Mangerton, Stoompa and the Paps as the crests of a more gently moulded upland. Today the Reeks have been sharpened by corrie development, but the mountains farther east still preserve their gentle, maturely moulded curves above abysmal corries and glacial valleys similar to those of the Dingle peninsula. Five main ice-tongues coalesced at one stage to make a piedmont glacier on the lowland, and a series of crescentic moraines was laid down: at a later stage lakes were held up and terraces formed, some of which, between Headfort and Kenmare, and equally to the east of Killarney, provide singularly fertile soils. Three lakes on the lowland, Caragh, Leane and Guitane, are held up by moraines, and most of the corrie lakes scattered through the mountains are also dammed by moraines of late glacial origin (Fig. 70).

Farther west the evidence of ice sculpture in the mountains is equally clear, especially in the long range of mountains, with twenty peaks over 2,000 ft., which extends from the country between Killarney and Kenmare to its final termination in the high cliffs near Waterville; hidden away in this range are many remote and rarely visited corrie basins. To the south

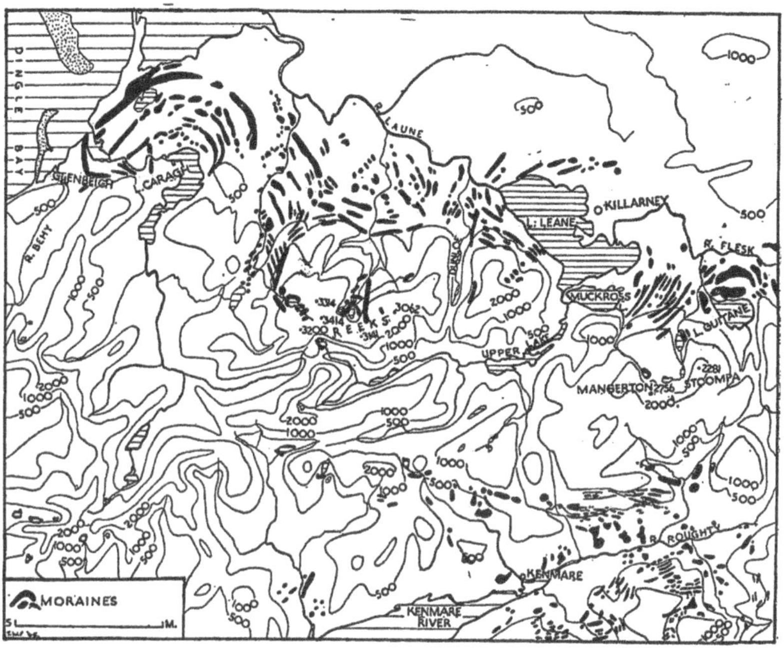

FIG. 70. RELIEF AND MORAINES, CENTRAL KERRY

The areas shown on this map include some of the most strikingly glaciated valleys in Kerry; and the various moraines associated with the local icesheet include an interesting crescentic series to the north of lough Caragh.

Compiled from data on the 1" to 1 mile Geological map Killarney and Kenmare district, drift edition 1922, and on a map by W. B. Wright in 'Memoirs of the Geological Survey of Ireland', 'The Geology of Killarney and Kenmare', 1927, p. 63

of this range lies a lesser line of hills, 1,000–1,400 ft. high, separated one from another by short transverse valleys drained to the Kenmare 'river': to the north there are isolated hills 800–2,000 ft. high surrounding lowlands drained partly by the rivers Cummeragh, Inny and Ferta, all of which flow parallel to the strike. In the west of the peninsula there is an intermingling of hills and lowlands not unlike that in the west of the Dingle peninsula, and the coastline consists of high cliffs where the hills reach the sea, alternating with featureless lowlands or drowned valleys such as the Ferta, on which Cahersiveen is placed. Valencia island, normally approached by ferry across the sound, consists of hills and lowlands similar to those of the mainland.

Little more than one-quarter of all the land in the peninsula is farmed and the density of population is now under 40 to the square mile compared with nearly 60 in 1891: the decline has been as much as 60% in some parts of the peninsula. Almost the whole peninsula was congested in 1891, but no part is congested today, though in spite of the heavy decline the density of settlement on the improved land is mainly around 150–200 per square mile. Fig. 69 shows that there are some very remotely placed communities in this area; the siting of farms and fields, influenced more by drainage than by altitude, is mainly on the lowlands or on the lower slopes of the hills, with several farms at 400–700 ft. About one-fifth of all the farms in the peninsula are of less than 15 acres, one-quarter each of 15–30 and of 30–50 acres respectively. The average farming unit is about 30 acres and, although the proportion of very small farms is much less than in West Connacht or in Donegal, the rudimentary type of farming reported in 1891 still survives on many farms. In 1891 the smallest farms generally had about 3 acres of arable land, of which half was sown to potatoes and the remainder to oats and green crops, and any form of rotation was limited to the growth of potatoes for two or three years followed by oats or green crops. There were also fields in meadow and pasture and the cattle and sheep were pastured on rough 'mountain and moor' grazing lands, held mainly in common or by several farmers jointly (in severalty). In some cases the amount of common land allocated to each farmer was in direct proportion to the valuation of his farm. Only spades were used on the smallest farms and seaweed with sand were the usual manures: to this day the farmers collect seaweed and sand, accompanied by ass and cart; even the ass with panniers is a very usual feature on the beaches. The farm's cash sales included butter, eggs, poultry, pigs, calves, sheep and wool – in general the products of mixed farming with an emphasis on dairying.

In spite of all the changes of the past sixty years, many parts of the peninsula still have some very poor farming communities. Around Caragh and Glenbeigh, only a few miles from the relatively rich Killarney lowlands, there are very closely settled areas on the moraines of the piedmont glacier or on the bogland itself; most of the farmhouses are thatched cottages. Cahersiveen (1,650), the only town, serves the extreme west: Killorglin (1,100) and Kenmare (1,100) are the largest of a few market villages. Fishing has now virtually ceased, though many farmers along the coast have boats and catch fish for home consumption and occasional sale: before the 1914–18 war there was a market for herrings in Russia and Germany and for mackerel in America, and large quantities were pickled for export, notably on Valencia island, an area of some 10 square miles with 847 people in 1966 (926 in 1961). This peninsula has now become widely known for the 'Ring of Kerry' tour followed by coaches and private motorists. On the south the Kenmare 'river', a fine ria, is surrounded in its upper reaches, notably around Kenmare, by a narrow lowland of Carboniferous limestone, partly drift-covered: the soils are rich dark loams producing large crops of swedes, mangels and potatoes. Some enterprises in Kenmare itself are worthy of mention. There is a co-operative creamery making butter and cheese, a small warehouse for homespuns, the last relic of a home knitting industry once widespread in the neighbourhood, and a quarry producing lime for building and general purposes and carrying on a large trade in a country of soils markedly deficient in lime. To these two new industries have been added, providing small paper articles and hair curlers.

C 4. THE CAHA PENINSULA

This peninsula is an anticline of Devonian sandstone flanked by slates on the coasts of the Kenmare river and Bantry bay. A long ridge, followed by the Kerry–Cork county boundary for nearly thirty miles, has no pass below 1,000 ft. along its whole length. Various lateral spurs and deep valleys, some of them with precipitous sides, run north and south from the main ridge. Low passes occur only in the extreme west, north of Castletown, and the mountainous ridge persists as a recognizable feature into Dursey island, which is separated from the mainland by a submerged pass in Dursey sound. The Carboniferous slates and grits to the north and south of the peninsula form hills rising to 600 ft., in Bear island to 800 ft. The coasts have several harbours and sounds, with islands. The peninsula was covered with ice coming from the Kenmare river area, though some of the higher summits, such as Hungry Hill, 2,251 ft., may have been ice-free. There are a few accumulations of drift on the southern side of the peninsula, but there are many striated rock surfaces with little or no soil, plentifully strewn with erratics and possessing *roches moutonnées*. Only about one-third of the total area is improved land.

The whole of the peninsula was congested in 1891 but none of it would be congested on a reassessment: the population has been halved and is now less than 60 per square mile though the settled areas are closely occupied, with as many as 150 per square mile. Nearly one-fifth of the population are in villages, such as Castletown (730), the largest of several small market centres. There are more small farms here than in the peninsulas farther north: in Castletown rural district, almost two-thirds are of less than 30 acres. The agricultural practice on the smallest farms is similar to that in the peninsulas farther north: spade cultivation is general; seaweed, sand and farmyard manure are used; the main crops are oats and potatoes, and the only rotation is to plant oats for one year, potatoes for two years, oats for a further year, and then grass. The main economic reliance is on dairy produce, with store cattle and sheep. Many small farmers depend largely on pigs and poultry. The decline in fishing has removed the main source of additional income in the Caha peninsula and helps to explain the heavy decline in population. In 1891 there were over a thousand fishermen employed in catching mackerel, which was sold fresh in spring and summer and cured in the autumn: but all this is little more than a memory now for commercial fishing is almost extinct.

C 4 (*a*). THE AREA AROUND BANTRY BAY

At the head of Bantry bay, Carboniferous slates and grits form a synclinal trough and are eroded into hills and valleys of Armorican trend: the simple correlation between hills of Devonian sandstones and valleys of Carboniferous limestones is lost here, for some strata of the Carboniferous system, such as the Coomhola grits, stand up as hills, which reach 1,797 ft. in the Shehy mountains and over 2,000 ft. to the north of Glengarriff, though most of the hills are merely 600–800 ft. high. Boulder clay is widespread but not universal in the valleys, and there are also numerous drumlins, some of which lie as separate hills with bare rock exposed on either side while others rise from a general boulder clay spread, notably around Bantry: in most cases the drumlin base is about half a mile long.* The coastal lowlands and the lower hillslopes are settled up to 700–900 ft. in places and there are 100–150 people per square mile of improved land but only 60 per square mile of total area. Bantry is a small market town (2,300), but it has now no significance for fishing. At Whiddy island a new oil port is being built to take tankers of 300,000 tons.

C 5. THE SHEEPSHEAD PENINSULA

The finger-like peninsula is an anticlinal ridge of Old Red sandstone flanked by Carboniferous slates and Coomhola grits. The main ridge, of sandstones, rises to 1,136 ft. in Seefin but is mainly below 1,000 ft. and is cut by several passes: it terminates in Sheep's head, a hilly promontory with sharp cliffs. The Carboniferous slates and the Coomhola grits form the coastal lowlands, but they too have cliffed shorelines with few inlets.

* These drumlins were noted in the *Geological Survey Memoir* (1864) as hills with green cultivated slopes quite distinct from the rocky hills (presumably *roches moutonnées*) covered with heather and gorse.

The hills are sharply glaciated by ice that came from Bantry bay, and there are many bare rock surfaces with some spreads of boulder clay in the small south-facing valleys of the peninsula. Settlement and cultivation is restricted on the north side to a narrow and discontinuous coastal fringe, but the south-facing valleys are occupied to a height of 400 ft. The whole area was congested in 1891, when the population density was 100 per square mile: it is now half this. More than half the land is mountain and moor grazing, most of it held in severalty, and the smaller farms had in 1891 only an acre each of oats, potatoes and meadow, with a plot of green vegetables. The only known rotation was to sow potatoes for two years and then oats for a year before putting the land back to grass; ploughs were used only on the larger farms, and farmyard manure, seaweed and sand were used as fertilizers. The income was and still is acquired mainly from the sale of cattle, sheep and dairy produce, but the fishing has virtually ceased.

C 6. THE MIZEN HEAD (SCHULL) PENINSULA

Lying between Dunmanus bay and Roaring Water bay, this peninsula consists of a number of isolated hills separated by lowlands; around Dunmanus bay, Carboniferous slates and grits form hills and valleys similar to those at the head of Bantry bay, but the remainder of the peninsula is of Devonian sandstone and its highest point is Mount Gabriel, 1,339 ft. high, to the north of Schull harbour. The coasts, sharply cliffed in Mizen head, are penetrated by several inlets both on the north and the south. The drift covering is thin over much of this area, but some good cultivable soils are developed on the sandstones. More than half the area is in crops and pasture, and farmlands cover most of the lowlands with patches of the hillsides up to about 600 ft.: the average farm size is now about 30 acres, and less than one-eighth of the farms are of less than 15 acres. The whole area was congested in 1891, but none of it would be congested now as the population has declined by three-fifths, from 125 to 50 per square mile. The heavy decline is due to the virtual elimination of fishing as a resource: in 1891, there were more than 500 men engaged in mackerel fishing, but few now, though at Schull there is one of the new fish-freezing plants erected with government aid. From 1961 to 1966 the population declined by 5%, having decreased by 9% from 1956 to 1961. The farming on the smallest holdings in 1891 was similar to that of areas farther north, with the liberal application of sand and seaweed in addition to farmyard manure; and part of the resource of each farm was the mountain grazing, most of it held in severalty. There are no towns in the peninsula and its marketing needs are met by four small villages, with one-fifth of the total population, or by Bantry and Skibbereen. The main staples are store cattle, sheep and dairy products.

D. THE SKIBBEREEN LOWLANDS

In this area the Ilen river cuts transversely across the general Armorican trend of the country for eight miles and gathers the waters of many longitudinal tributaries before turning sharply west at Skibbereen. Both Devonian sandstones and Carboniferous slates or grits form hills rising to 400–600 ft. near the south coast and to nearly 1,000 ft. farther inland. There is a fine coast of submergence, with several islands and some striking cliffs. The Skibbereen rural district corresponds almost exactly with this subregion: its loss of more than half in population since 1891 is due to the heavy emigration and to the marked decline of fishing along the coast. There were well over 1,000 fishermen in 1891 but only a few score now. The farming on the smallest holdings resembles that on similar areas in the southwest, but only one-seventh of all the farms are of less than 15 acres and the average holding is medium-sized, around 40–50 acres. Crops cover some one-sixth of the land and include oats, mangels, sugar-beet and even wheat, but the main purpose of the farmers is to produce store cattle and milk for the creameries. The density of population averages 60 per square mile, but 84 on the improved land. One-quarter of the whole population lives in towns and villages, of which Skibbereen (2,000) is the most important: these centres have, like the rural areas, declined in population with the decay of fishing and the heavy emigration from the neighbourhood.

CHAPTER 17

Limerick and Clare

The outlook of this region is to the west, and Limerick, its regional capital, is the largest town on the Atlantic seaboard of Ireland. America is not far distant in the imagination of its inhabitants, and the Shannon airport has brought the New World much nearer in travelling time. Though significant for passenger traffic, the airport deals with little merchandise, and the port of Limerick has only 3% of the country's overseas trade. However numerous the aeroplanes, however great the financial return given by travellers, the life of this region can be affected only to a limited extent and farming will remain its chief resource. The link with America that is most vital is the record of emigration extending over a century and more: to quote an old Clare woman on her family: 'Some are in America, but two are away over in England.' Life in Clare and Limerick centres around the farmsteads that cover its lowlands, interrupted only by a few ranges of hills and by the wonderful bare limestone plateaus beside Galway bay.

LANDFORMS

The three main structural elements in this region are a number of Armorican ranges in the east, a Carboniferous limestone lowland in the centre, and Carboniferous plateaus in the west. The Armorican ranges are anticlinal ridges comparable with those in the south of Ireland from Kerry to Kilkenny and, like them, having Devonian sandstones overlying Silurian slates and grits. Where the Devonian sandstones survive, as in Keeper hill (2,278 ft.) south of Nenagh, there are sharp hills, but most of the summits are lower, generally from 1,000 to 1,700 ft. in the Arra mountains, the Slieve Bernagh, the Slieve Aughty and the Cratloe hills. The Silurian strata have weathered into lower ground and these hills, though prominent, are hardly a barrier to communication: the Shannon is in effect a natural frontier, especially in lough Derg, which spreads to a width of ten miles between central Clare and the Nenagh district. Spenser spoke of the 'spacious Shenan spreading like a sea': but the river is bridged at Killaloe and O'Briensbridge as well as at Limerick.

In the west of Limerick and the west of Clare, upper Carboniferous strata form plateaus of limestones, sandstones and shales, of which the limestones have weathered into a karstic landscape with scarp slopes and dry valleys. Some of the limestone pavements have vertical joints that have been widened into a series of cracks varying in width from a few inches to several feet, all of them filled with soil and richly overgrown with shade-loving plants. The hills that rise above this stony desert consist partly of bare limestones and partly of slopes covered with heath plants and hazel scrub, according to the local conditions of exposure. In places there are mighty scarp edges: thus the amphitheatre around Ballyvaughan on Galway bay is bounded on three sides by very steep, terraced rock faces above a flat, grassy valley, and there are equally impressive scarps guarding a waterless lowland around Kinvarra, Co. Galway. In this karstic lowland most of the drainage is underground and a cave in Slieve Elva has been traversed for four miles: others are now being explored. The rapid runoff recorded in some places suggests a well-organized subterranean drainage: near Ballyvaughan, the fields were reported dry two hours after the rain had ceased for the first time in forty-eight hours. On the other hand, the writer observed cascades of water pouring down the west slopes of Slieve Elva after a day's rain in June 1937, and was told that there, at least, the phenomenon was frequent.

On either side of the Shannon estuary, sandstones and shales with useless Coal Measures overlie the upper Carboniferous limestones, and form featureless uplands rising in places above 1,000 ft. This dreary and monotonous landscape is continued southwards in Kerry and Cork: everywhere it has long undulating ridges with wide open valleys and few features of scenic interest. Its edge is a very definite escarpment on the west side of the Limerick lowland to Foynes on the estuary, but in Clare the scarp is less marked, though it is recognizable on the west side of the Fergus estuary. The Clare section of the 'Carboniferous plateau' consists primarily of lowland, but also has some flat-topped hills with terraced sides, one of which, Slieve Callan, rises to 1,282 ft. Sandstones, especially the local flagstones, stand out in ridges and form some of the remarkable cliffs that extend along the coastline at intervals from Loop head to the cliffs of Moher, which rise to more than 600 ft. sheer from the sea with some outlying stacks showing very sharp marine attack. This western area is partly covered by drift, largely of local materials, but also contains some limestone, though its soils are markedly acid and support a calcifuge vegetation. Throughout western Clare and Limerick, rushy grass-heaths occupy much of the lower ground, interspersed with some bogs, while the hills are peat-covered moors similar to those of north Kerry and the adjacent areas of Cork.

The limestones that extend through central Clare from Galway bay to the Fergus estuary form a number of ridges and low craggy hills, 150–300 ft. high, some of which are covered by drifts and are good farmlands. The general grain of the country is northeast to southwest, and the drainage reaches the surface in a number of rivers, which appear for a few miles and then disappear, or in some cases emerge only for a few yards in a collapsed cave. 'Turloughs', the most extraordinary feature of all, are limestone hollows in which water lies for periods depending on the amount of inflow and outflow. They vary considerably in size but may be as much as a mile across and 30 ft. deep and the largest, the Carran depression, is over two miles long from northeast to southwest and at its widest over a mile broad. This depression is entirely enclosed with no surface outlets for its drainage; it is subjected to periodical flooding and has been described as 'the best example of what is probably a true polje in the British Isles'. In many turloughs the lowest part of the depression becomes a marshy pasture, even a dry one, with a small residual lake, at times extinguished according to the fluctuations of the water table. But not all the lakes are of this type, for some have an impervious floor and persist throughout the year without such striking changes in extent. In all, the lake country extends through thirty miles from central Clare to Co. Galway, preserving throughout its southwest–northeast trend.

In the Limerick lowland, the rockfloor is of very varied composition, and a recent worker has claimed that it includes a complete Carboniferous sequence from Lower Limestone shales to Millstone Grit flags, though the higher strata, consisting of shales and Millstone grits, are preserved only in the hill-girdled basin lying to the southeast of Limerick around lough Gur. The lowland also possesses what Geikie has called 'the most compact and, for its size, one of the most varied and complete of all the Carboniferous volcanic districts of Britain',* and a further topographical element is a number of anticlinal ridges in which Devonian sandstones appear. In the main, the Limerick lowland is some 150 ft. above O.D., rising to around 350 ft. with a few minor hills, and the general appearance is a rolling hummocky landscape smoothed by glaciation; the ridges of Devonian beds, with igneous intrusions, rising to 948 ft. near Ballingarry, offer no vital barrier to communications. The two volcanic phases of the Carboniferous period have left behind a wide variety of igneous rocks, including basalts, trachytes and beds of ash, either as dykes

* Quoted by Ashby, D. F., in *Proc. Geol. Ass. Lond.*, vol. 50, 1939, p. 324.

FIG. 71. GEOLOGY OF THE LIMERICK DISTRICT

Most of the area is covered with drift, chiefly of a clayey nature, and there are wide spreads of alluvium near the Shannon, but Carboniferous limestone and igneous rocks appear at the surface in a few places.

Based on the 1″ to 1 mile Geological map Limerick district, drift edition 1906, by permission of the Minister for Finance of Ireland or interbedded with the Carboniferous strata. In places the igneous intrusions stand up as hills, particularly in the loosely defined 'basin' to the southeast of Limerick, where a basalt forms Knockseefin, 749 ft., and trachyte vent-plugs form Deuk hill and Kilteely hill, at 782 ft. and 680 ft. respectively. Not always, however, is past vulcanicity so clearly reflected in the landscape, though among several low hills due to intrusions is Carrigogunnel which, crowned by a medieval castle, rises above the sloblands around the Shannon and has a cliff face of stratified ash overlying columnar basalt: a crag 100 ft. high, it has a long tail of boulder clay to the south and provides a fine example of crag-and-tail formation. Glacial action has left a clearer imprint in the Limerick lowland than in Clare, but drifts are not present everywhere and some of the limestones and volcanic rocks appear as drift-free islands with soils of local origin. The limestone soils are in many places unretentive of water and shallow, and some of the volcanic soils are well drained but not more than 2 ft. deep. The soils derived from the drift and from glacifluvial meltwater deposits are varied and range from sands or gravels in eskers and esker-like mounds to thick clays likely to become waterlogged at any time. Intermediate types between these extremes include limestone boulder clays and red sandy soils, of which the latter are extensively developed north of the Shannon at the foot of the Cratloe hills. Bogs are relatively few, though some have developed on freshwater shell marls, which are the sites of small lakes impounded by drift deposits. A few lakes still survive, but most have long since perished by infilling; in addition, the peat they once contained has been removed by a population not too plentifully supplied with fuel.

THE SHANNON FROM LOUGH DERG TO THE SEA

The river Shannon falls into two sections that are divided at Killaloe, the bridgetown at the southern end of lough Derg. Above Killaloe, the river falls only 55 ft. in more than 100 miles, but below Killaloe, it falls by 100 ft. in a journey of twenty miles to the tidal reach at Limerick. Lough Derg, twenty-two miles long and two to ten miles broad, interposes an effective barrier between the lowlands on either side but at Killaloe it narrows into a river which is dammed up to a height of *c.* 109 ft. at Partreen weir (Fig. 72) where the water is 32 ft. deep. Embankments have been built to protect the surrounding countryside and around the weir there is a fish ladder, 600 ft long, built to preserve the Shannon fisheries. At Partreen the Ardnacrusha canal begins, and the lower Shannon is now a shadow of its former self for it merely receives the water not needed for the canal,

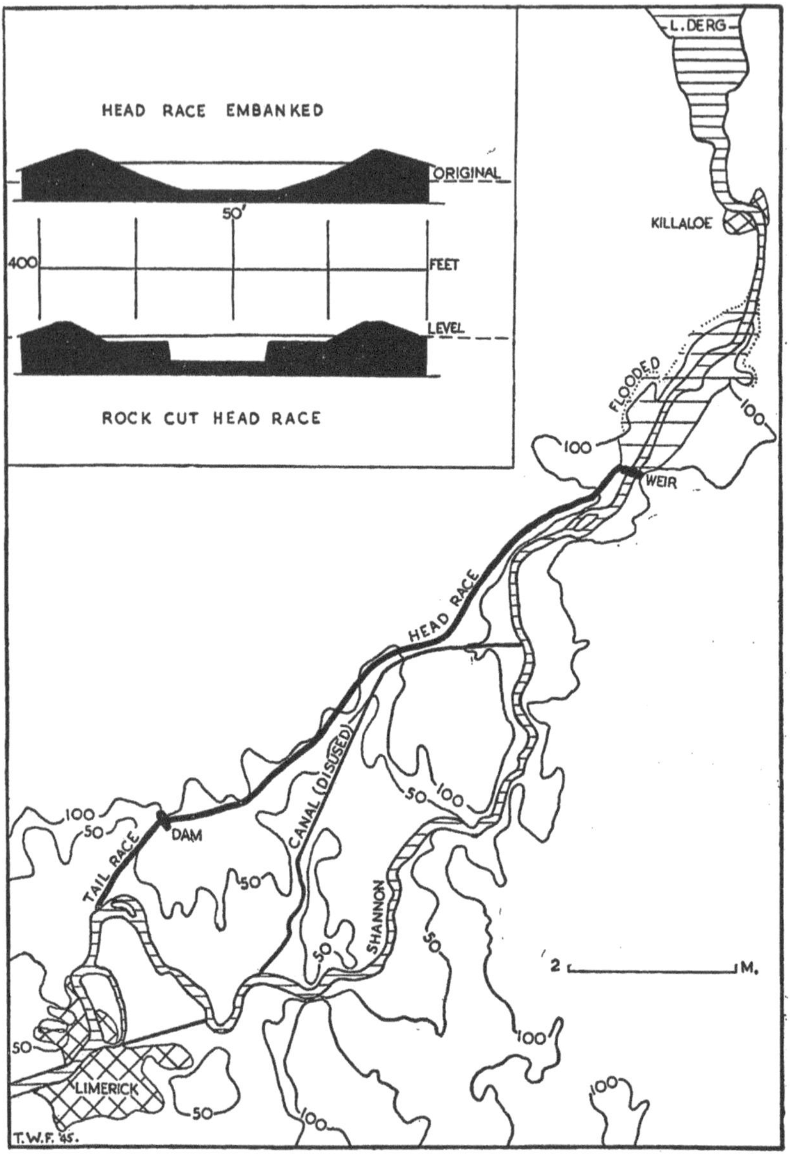

FIG. 72. THE SHANNON FROM LOUGH DERG TO LIMERICK

The changes consequent upon the construction of the Ardnacrusha power station are discussed in the text.

Based on the 6" to 1 mile maps of the area by permission of the Minister for Finance of Ireland and on 'Engineering', vol. 131, 1931, pp. 250–1 which is nearly nine miles long. The headrace above the power station is seven and a half miles long, 37 ft. deep and nearly 300 ft. wide, and the dykes are made of earth covered with a 2 ft. layer of clay and faced with layers of stone and concrete. In a few places the headrace flows through rock cuttings, and the tailrace is almost entirely rock-cut. The general effect resembles a major ship canal and the old Shannon navigation canal is now completely superseded; two modern locks carry barges past the Ardnacrusha power station, where the whole drop of 100 ft. is concentrated. This hydroelectric station was constructed between 1925 and 1929: it was hoped to make the Shannon into one storage basin by making the deep weir at Partreen, by adding other weirs on lough Ree and lough Allen, and by building embankments on either side of the river for several miles. Unfortunately the Shannon has proved to be a river particularly susceptible to low water levels in drought periods, probably due to the depletion of underground supplies.

Two miles above Limerick, which is more than fifty miles from the open sea, the Shannon becomes tidal. The estuary is a drowned river valley, fringed on either side by sand and mud banks, and by many square miles of alluvial flats composed of silt, loam and clay. These flats, which are mainly 'intake' or land reclaimed by embankments and protected by levees, require artificial drainage if they are to become good pastures. They penetrate along the various tributaries of the Shannon, such as the Maigue, which is tidal for seven miles to Adare, and the Fergus, which is tidal to Ennis. All the small rivers flowing into the Shannon are aggrading their courses and depositing large quantities of alluvium in times of flood. The Fergus estuary has a number of islands formed of Carboniferous limestone covered with boulder clay, several of which are inhabited. In this estuary a considerable amount of *Spartina townsendii* has been planted, and reclamation is proceeding actively. The Shannon west of the Fergus narrows to less than one mile between the upper Carboniferous shales and sandstone country on either side, but it widens and deepens towards the open sea between Loop head on the north and Kerry head on the south.

VEGETATION AND LAND USE

This region has a mild, humid climate comparable with that in other parts of the Atlantic seaboard, and its natural vegetation is highly varied and of unusual interest, especially in the limestone Burren area of Clare, which is a botanical museum including specimens of the alpine, the Mediterranean (Lusitanian) and the American floras. The *Dryas* and gentian (*Gentiana verna*), both alpines, grow on the limestone pavements, but the intervening cracks are thickly occupied by the maidenhair fern (*Adiantum*) of Mediterranean origin: probably these varied species, once having arrived, were able to settle and maintain themselves, as the area is naturally treeless and has never been cleared for agriculture except in small patches. It is generally believed that the drift cover has been entirely removed, though a few relics may still remain, and its former presence may be partly responsible for the mixture of calcicole and calcifuge species: there are, however, notable absentees from the flora, such as gorse, which is so widespread on the acid soils of Ireland.

The vegetation cover of the Burren is not continuous, and while some visitors comment on the singular extent of the bare limestone surface, others are impressed by the luxuriance of the vegetation: possibly the desolate wilderness of a first view blossoms as the rose on closer inspection. Certainly the vegetation cover is thicker than that of the limestone pavements in the Pennines of England, but the Burren lies near sea level and has a winter with few frosts. Tree development is impeded by strong winds, but the vegetation is able to rise above the purely xerophytic level which would be encountered in a dry climate. The ecological sequence, according to Tansley, may follow one of two lines of seral development. Pioneer mosses and lichens form a thin layer of black humus with such calcicole grasses as sheep's fescus (*Festuca ovina*) intermingled with *Dryas*, wild thyme (*Thymus serpyllum*), and other plants. This vegetation is calcicole in emphasis and provides excellent pasture where grasses are dominant. But in some places the seral development takes a different turn, and a limestone heath may develop in which heather (*Calluna vulgaris*) and crowberry (*Empetrum nigrum*) are intermingled with bracken and tufts of *Molinia*, which is more typical of a bogland flora.

No woodlands occur on the more exposed coastal sections of the Burren, but scrub appears in some sheltered hollows. This is foreshadowed near the coast in the occasional presence of hazels and hawthorns, or the Scotch rose (*Rosa spinosissima*) in the fissures of the limestone pavements, which are more characteristically the habitat of the maidenhair fern and a variety of woodland field-layer plants. Hazel-dominated scrub is seen where there is adequate shelter, though the trees are rarely more than 10 ft. high: intermingled with the hazel, there may be hawthorns, blackthorns, ivy, the Scotch rose, the spindle tree, birches, ash, holly and elm with a woodland ground flora. In some places, however, notably in one section recorded by Tansley, the hazel bushes were suffering from the effects of wind and the upper shoots were drying and withering: not far away, in an even more exposed site, the hazels were killed to the base so that a heath was formed. Only in the interior of Clare, in the district between Ennis and Corrofin, is ashwood seen; in one wood investigated, the ash was dominant, with birch and hawthorn abundant, and wych elm, hazel and even beeches, self-sown from a neighbouring plantation, also present. Much of the scrubby woodland in sheltered places away from the sea consists of hazel thickets rather than ashwoods.

The Burren vegetation in its present form shows very clearly that microclimatological factors are a crucial influence in its distribution. The limestone pavement has one climate, the intervening damp, black-soiled creeks another: a moderate degree of shelter enables scrub to develop, but only a yard away no tree can survive. The valleys around Ballyvaughan have rich grasslands, for they are aligned north-south and protected from the prevailing and sometimes overpowering westerly winds. Again, the distribution of individual species is highly varied: here *Dryas* spreads over many square yards, almost alone; there gentians are supreme, though usually in grasslands. On some almost soilless areas a few planed junipers or dwarfed yews occur: otherwise the limestones are bare until one reaches the next tiny depression with a little shelter. The Burren vegetation is of interest at any time of the year, but attains its full glory around the third or fourth week in May, when both the gentians and the *Dryas* are in full bloom.

The Burren flora is continued to some extent on the limestone country of central Clare, but nowhere else, except on the Aran islands (p. 422) is its development at all comparable with that in northwest Clare. Many reasons might be advanced in explanation, including the closer human occupation of central Clare, with long-continued and intensive grazing: in addition, woodland probably formed the main vegetation cover away from the Atlantic coast and the ground flora was never able to develop so completely: central and eastern Clare have patches of oak- or ash-dominated woodlands that appear to be survivals of the natural forest cover. The Burren flora also ceases abruptly around Lisdoonvarna, where shales of Coal Measure age produce a heavy, wet soil, frequently black in colour, and the vegetation is rushy or heathery with peat on the hills and on some of the ill-drained lowlands.

Some fifteen properties covering 50 square miles, of which 44 square miles can be planted, have been acquired for forestry: of these four, at Mountshannon, Tuamgreary, Broadford and Cratloe, are on the flanks of the Clare hills lying west of the Shannon, and another four are in the mountains lying on the east side of the Shannon. Two in the southeast of

Co. Limerick, the Kilfinane and Galtee forests, are in mountainous country; of these the Kilfinane forest covers five and the Galtee three square miles. Tulla, in Clare, a forest of some five square miles, covers upland country and near Ennis an area of three square miles, almost all of it previously scrub or rough pasture, has been planted. The Adare forest in Co. Limerick is partly former demesne woodland: this is in fact the only foresty plantation in the lowland of Co. Limerick. In Clare some old woods have been replanted at Kilrush, and a new plantation on previously open land has been made in the north of the county, at Ballyvaughan, covering 950 acres. But in west Clare there are no plantations, probably due to the natural hindrances to tree growth in so maritime a climate.

The proportion of land given to crops and pasture is 89% in Limerick, 69% in Clare and 82% in Nenagh rural district. In Clare the proportion of improved land has declined since 1891, when it was 78%, but the proportion has not been materially altered in Limerick for a century. In effect, Clare shows the same tendency to abandon land that is marked in such western counties as Kerry, Galway, Mayo, Leitrim and Donegal, but Limerick's experience is similar to that in most of the Central Lowland. Most of the rough pasture is used by the farmers, and is either attached to particular holdings or (more rarely) held in severalty. The rough pasture consists of the uplands within the region, the Burren almost in its entirety (though not the grass-covered valleys leading to Galway bay), and some areas of peat-covered or wooded lowland in northeast Clare and elsewhere. The status of the reclaimed land along the Shannon and its tributaries is varied, though much of it forms valuable pastures for at least part of the year.

In west Clare at the humbler end of the agricultural scale, farming approximates to the subsistence level characteristic of the western seaboard of Ireland, but in east and central Limerick, the farming is comparable in standard with that of the Munster dairying belt. The rateable valuation of land ranges from £2 to £3 per head in the sandstone and shale belt of west Clare to more than £6 per head in east Limerick. Some of the poorest parts of the area, which is comparable to that in north Kerry, are very little removed from the level of congested districts: in the year 1909, part of Co. Clare (all except the rural districts of Ennis, Corrofin and Meelick) was brought within the area administered by the Congested Districts Board. The average size of the farms varies from 40 acres in west and northeast Clare to 50–80 acres in Co. Limerick and around Nenagh. In other words, small farms are frequent in the west, and medium-sized farms become increasingly prevalent towards the east: in no part of the region are large farms of more than 100 acres prominent, as even in the east they number only one-fifth of the total number of holdings.

Throughout the region the proportion of land cropped is low, and in some districts virtually all the improved land is given to hay and pasture. In Clare and Limerick in 1967 25% of the improved land was meadow and only 2% in crops: wheat, grown in two wars, was almost completely abandoned between them; again, turnips and mangels, grown on a few thousand acres as feedstuffs, are of quite secondary importance compared with oats and potatoes. During the war of 1939–45 the proportion of land cropped was raised to 12–13%, almost equally distributed between oats and potatoes, but by 1955 the cropped area had declined to 6% once more. Even in 1851 less than one-quarter of the land was arable, and the past century has seen a steady decline. A writer of 1837 speaks of the small amount of tillage in Co. Limerick and adds that there were clauses in some leases to prevent the cropping of more than 1 acre in 20 or even, in extreme cases, 1 acre in 50. Many of the farmers had herds of as many as 400–600 cattle and a farmer bringing more land under the plough was regarded as a man of declining fortunes. These comments of pre-Famine days show that then, as now, wealth lies in the herds rather than in crops.

Cattle rearing has become the mainstay of farming throughout the region, except in the Burren and the adjacent limestone areas around Corrofin, where sheep rearing is much more important. Excluding the lowland sheep districts of north Clare, sheep farming is mainly restricted to the uplands and has declined steadily during the past century, so that now there are hardly a quarter of the sheep supported in the seventies, when this form of farming reached its zenith. On the other hand, the number of cattle has doubled under the stimulus of the milk-demanding creameries and the export trade in stores. Almost every farm in Limerick supplies milk to the creameries that have spread through the county since 1890 and there are numerous creameries in county Clare, based on central depots at Ennistymon and Kilrush, of which the former has fourteen and the latter fifteen branches. The store cattle trade is also important and there are good fattening pastures near the city of Limerick. Surplus store cattle are bought by dealers who take them to pastures in the Central Lowland or ship them to Britain through the port of Cork. Pig production fluctuates from year to year, but has its inevitable association with the dairy industry: the animals are sent to the well-established bacon factories of Limerick. Poultry rearing here, as generally in the Republic, has declined in recent years.

INDUSTRIES

Nowhere is the creamery more important than in Co. Limerick, which has some of the oldest in Ireland: Co. Clare also has a network of rural creameries based on Ennistymon and Kilrush. Many of the creameries have a network of branch separating stations, and the functions of some include the production of cream and cheese as well as butter: some are also depots for eggs and poultry, and a few have a mill for maize meal and flour. Generally they are small and employ hardly more than a dozen people, but they are a form of co-operative industry in the closest possible touch with the countryside. At Knocklong a factory developed on a creamery basis provides condensed milk.

In Co. Limerick the small towns are almost devoid of industries, except for a few sawmills and small factories making mineral waters, always a localized trade in Ireland. In Co. Clare, Kilrush has a flour mill dating from 1811. At Kilrush there is also a ceramics factory. At Ennis, the major town of Clare, there has been considerable industrial growth. A braid mill was established here in 1934, and various textile based industries have been established since then, including the processing of cotton yarns, the manufacture of carpets, elastics and terylene curtains. Other industries include the making of spectacle frames and a variety of concerns connected with jewellery, notably diamond cutting and the making of imitation jewellery and of jewellery boxes. Another industry is the manufacture of plastic tubing, beading and wall tiles. At Milltown Malbay there are factories for woven labels and knitwear (the latter with 100 workers) and at Killaloe cardboard boxes and hand woven tweeds are made. At Nenagh there is an aluminium works with over 200 employed and the town's other factories provide cotton yarns, metal tubes and blended spices.

At Shannon airport fifteen miles west of Limerick beside the Fergus estuary, some 4,000 workers are employed in factories established since 1959. Over 200 acres have been allocated to factories and warehouses, and the aim is to attract industries competing in export markets by inducements such as custom-free facilities, the building of standard factories and financial grants. The products include pianos built by a Dutch firm, transistor radios of which some are made by a Japanese firm, textiles and textile machinery, drilling machines for gold and diamond mining, plastics, and various engineering components, notably those for aircraft. Several hundred houses have been built with various community facilities such as shops, schools and churches. The expectation is that the resident population (2,174 in 1968) will increase to 6,000 within a few years and possibly to more later.

POPULATION

This region covers 2,500 square miles and had a population in 1966 of 229,100, of whom 110,100 (48%) live in towns and villages. Limerick is by far the largest, with 58,100, and Ennis (9,200 including its suburbs) stands out as a major town, the administrative and commercial centre for a large area. Nenagh had 4,600 people in 1966 and the other towns, much smaller, include Newcastle West (2,500) in Co. Limerick, Kilrush (2,700) and Kilkee(1,400) in the extreme southwest of Clare. There is a large number of smaller towns and market villages. In 1891, the population was 311,000 and the decline of almost one-third since is due to a loss, of at least one-third, in the rural districts, which is less than in most parts of rural Ireland: high in some areas, it was for a long time low in most of Co. Limerick due, it was said, to the stability brought by the creameries. Nevertheless, rural population decline now appears to be universal.

The rural population is spread fairly evenly through the region, and generally averages 60–100 per square mile, though in the thinly settled country around the Burren it is less than 50 to the square mile. This low density in north Clare is of recent date, as a century ago the area was closely settled: evidence of the change is given by the very large number of ruined houses to be seen. Most of the gaps in the population map (Fig. 73) are mountainous country, lowland peat bogs and heaths, or the limestone pavements of the Burren. There are also some less obvious features worthy of comment: these include the relatively heavy settlement of areas in the west of Clare and the west of Limerick, both of which have a large number of very small farms. Market towns are few in number; the Limerick lowland has a considerable number of market villages, possessing many of the functions of towns, but not towns in size or appearance: there are also a few such places in Clare. In all, one-tenth of the population lives in these market villages, which are primarily agricultural in their life: the smallest of them are little more than a cluster of farms with a few shops, churches, inns and schools.

Industrial life, apart from Limerick, is hardly more than the processing of a few agricultural staples with a few successful efforts to meet a local

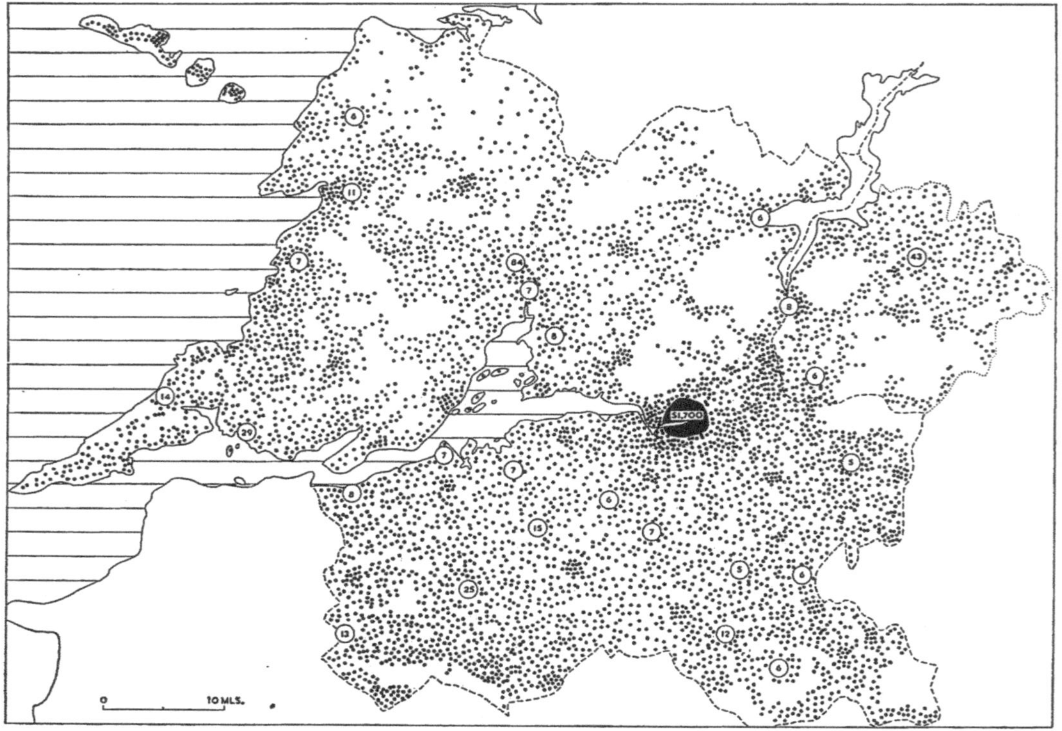

FIG. 73. DISTRIBUTION OF POPULATION

LIMERICK AND CLARE

One dot to 25 persons

Most of the unpopulated areas are uplands, though there are some bare limestone karsts near Galway bay. Towards the west, both in Clare and Limerick, there is a substantial proportion of small farms, and the density of population is rather higher than in the Limerick lowland.

The Aran islands are included on this map, but discussed on pp. 420–4.

demand for clothing which has been the basis for an expansion into the wider markets beyond the neighbourhood. Of the whole employed in Co. Limerick (excluding the city) 63% work on the farms, 14% in manufactures and trades, and the rest provide services intended largely for the farming community. In Co. Clare, the comparable percentages were 65% and 15% respectively (in 1961). There are now less than 60 fishermen though as recently as 1926 there were 400 fishermen in the Shannon estuary and along the coast of Clare. The tourist trade has not achieved the development that the natural beauties of Clare would warrant, though there are several hotels at small centres along the west coast, especially Kilkee. Lisdoonvarna, placed just south of the Burren country, is a spa with sulphur and iron springs and consists mainly of hotels that receive large numbers of visitors in the summer months. There are some interesting limestone ravines contrasting with the black-soiled country of Carboniferous shales, but the attractions for visitors extend far beyond the waters or the scenery, and the 'season' is strongly social in character. So great are the crowds that the church cannot contain the congregations, and altars are erected in the churchyard at which Masses are said by the numerous priests included among the visitors.

LIMERICK

This city is the one large distributing centre for the whole region and has more than half its entire town and village population. It is perhaps reasonable to correlate the lack of strong market towns with the long-established supremacy of Limerick as a regional capital, for it is at once a port, an industrial town, a cathedral see with two bishops, the meeting place of many roads and railways, and an active commercial town and distributing centre. The town has spread along the east side of the Shannon, with suburbs on the west. The old harbour lies between the two Shannon bridges, with quays on the main river and in the Abbey creek, but the main part of the town today is below the bridges along quays that extend for nearly a mile and lead to a floating dock with water 24 ft. deep. The port is entered by coasters or tankers with cargoes of grain, coal, oil, petrol and a few other commodities. Store cattle are not exported through Limerick as the sea journey to Britain is too long, and it is therefore preferable to send them to Dublin by special train: consequently, the exports are limited to dairy produce, with small supplementary cargoes of slates, timber and hides (Fig. 74).

Like Dublin, Waterford and Cork, Limerick is a Norse foundation at

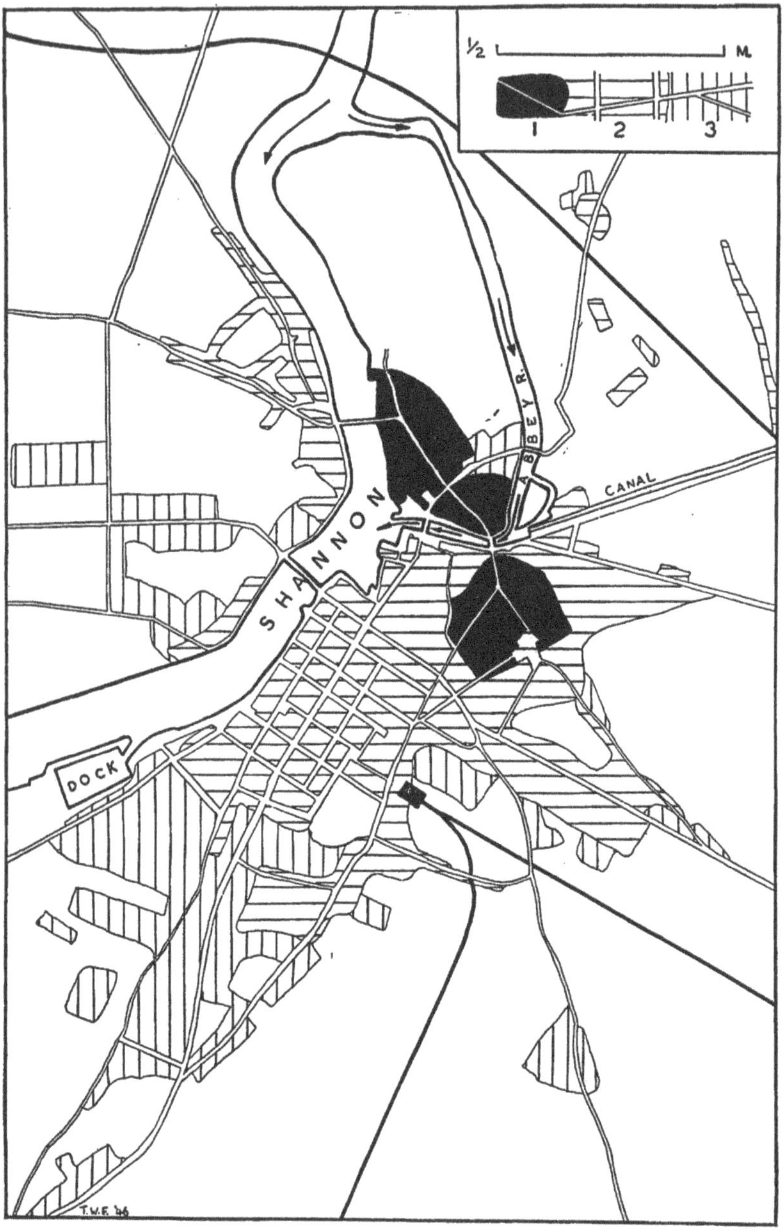

FIG. 74. THE CITY OF LIMERICK

1, Medieval walled town; 2, Expansion to 1841; 3, to 1925.

The main features of the town, discussed on pp. 385–8, include the hour-glass shape of the early walled town, the Georgian planning in a rectangular pattern, and the late nineteenth and early twentieth century spread on both sides of the river.

The outline of the medieval town is from Leask, H. G., 'The Ancient Walls of Limerick', North Munster Antiquarian Journal, Limerick, vol. 2, 1941, pp. 95–108 the lowest ford of a river. The town first grew up on what is now known as King John's island, a drift-covered Carboniferous limestone ridge 40 ft. high surrounded by the Shannon and the Abbey river. Part of this island is reclaimed land still subject to floods, but it is a defensible and often defended site with a ford leading to firm ground on the west bank of the river. Its name commemorates, as few things do except Magna Carta, King John, who came in 1210, after which a bridge was built across the Shannon. The town on the island was High Town, or the English settlement, and in time this was supplemented by 'Irish town' on the opposite bank, with the result that Limerick was like an hour-glass in shape: all of it was walled in the fifteenth century and a bridge was added to connect the two parts. In the eighteenth century the walls were allowed to decay and the town spread southwards along the firm ground east of the river into the Newton Pery district, and also on the west side. From this period of expansion, Limerick has inherited the wide streets and dignified architecture characteristic of late Georgian planning, but a considerable part of the city is congested. In 1950 the area of the city was extended by the inclusion of an additional four square miles with a population of over 9,000: the population within the city as defined in 1950 is 55,900 but another 2,200 live in contiguous suburbs (1966). Clearance of the old and generally poor housing near the centre has caused a decline of population in the inner wards. Through Shannon airport, Limerick has become one of the entrances to Europe from the New World.

The industries of Limerick are varied, and the town's factories are concerned mainly with the processing of the agricultural output drawn from the immediate neighbourhood, with the milling of imported grain, and with the satisfaction of the home demand for clothing, agricultural implements and building materials. Three large bacon factories, all of which have been established for more than a century, have sold Limerick bacon and ham widely in home and export markets: subsidiary lines include lard and sausages, and, in the case of one firm, vegetable canning. Five hundred people are employed in a factory making milk products, tinned cream and butter. Milling is in the hands of one large combine, though a few small concerns have managed to survive: home and imported grains are used to make flour and the feedstuffs so widely sold in the dairying and cattle-rearing countryside. A tannery has branches at Tralee and Cork, but the footwear industry is almost undeveloped here. Ready-made clothing and waterproofs are made by a firm which has developed an export sale. Building trades are intensively developed, and several large firms either provide materials ranging from timber (chiefly imported) to lime, or else work as contractors in the town and surrounding areas. A machinery works supplies gates and railings for the builders, and also makes some farm implements. All the varied enterprises so far discussed belong to the pre-1932 phase, though some have been expanded since that date: almost the only recent addition is a nail-making works. Outside the town, at Mungret, there is a large cement factory. In the main Limerick is an industrial centre for the processing of agricultural staples and the supply of clothes and building materials in the neighbourhood: in all this, the industrial activity shows a definite geographical influence.

SUBREGIONS

Limerick is a regional capital and therefore given separate treatment. Each of the five subregions distinguished has some distinct quality in its landforms: each has some distinct agricultural qualities. This must remain a crucial factor in the consideration of such an area, for fishing, the tourist trade, quarrying and (except for Limerick) industry are all of quite minor importance. The welfare of the people rests largely in developments such as those of the creameries, which bring direct profit to the farmers, for agriculture is the region's main resource.

1. THE LIMESTONE COUNTRY OF NORTH CLARE

In the Burren, very restricted areas are available for settlement, but the valleys opening to Galway bay and a few small coastal lowlands have farms given to sheep rearing, with some store cattle raising. To the east lies the turlough area, with its fluctuating lakes and drift-covered limestones separating bare limestone areas: here, too, sheep rearing is more important than cattle production. The subregion as a whole is thinly settled, and the farms are mainly medium-sized or large: the decline in population since 1891 has been more than one-third. Ennis is the major town, with Corrofin and Ballyvaughan as market villages.

2. THE SANDSTONE AND SHALE COUNTRY OF WEST CLARE

In the upper Carboniferous plateau area, sandstones, especially flags, stand out prominently, but most of the area is lowland, drift-covered, and having some acid and ill-drained soils with bogs. But apart from the higher parts of the ridges, a few coastal heathlands in the Loop head peninsula, and the boglands, there is close settlement with many small farms, though the average farm size is 40 acres. The major creamery is at Ennistymon, and the main emphasis of the farming is on the rearing of young cattle and on milk production, with some pigs and poultry. The two largest centres are Kilkee and Kilrush: Ennistymon (1,100) and Milltown Malbay (700) are smaller market centres. The loss of population since 1891 has been approximately two-fifths.

3. WEST LIMERICK

Structurally similar to the west of Clare, and separated from it only by the Shannon estuary, this subregion has a well-marked scarp edge that divides it clearly from the Limerick lowland. Its valleys are closely settled and medium-sized farms predominate. Much of the land is poor, with rushy pastures, but the creameries so dominate the economic life that half the cattle are milch cows, and sheep are almost entirely absent. The surplus young cattle find their way to richer pastures through fairs at such centres as Abbeyfeale (1,300) and Glin (700): the nearest town is Newcastle West. In population it has lost by one-third since 1891, but the density is still approximately 100 per square mile of improved land.

4. THE LIMERICK LOWLAND

This region lies at the western end of the Golden Vale, and is linked with a somewhat similar area in Tipperary: to the north lies the Shannon and the Silvermines mountains, to the south the Galtees and the Ballyhouras. As a lowland, it is relieved from monotony by low volcanic hills and Armorican ridges. Milk production for creameries is more important than anywhere else though there is cattle fattening near Limerick. The population per square mile of cultivated land is 70–120 per square mile, and for fifty years from 1891 there was only a slight decline (10%) in the population of this lowland. In spite of some change since 1941, the area remains dominantly agricultural, industrialized only to a slight extent through the creameries, and having its markets either in Limerick or in a number of small market towns and villages.

5. THE UPLANDS AND LOWLANDS OF THE LOWER SHANNON

Sharp landscape contrasts exist here, as the Armorican mountain ranges are surrounded by lowlands and trenched by corridor valleys. The uplands have some sheep farming, but stock raising is more important everywhere for the creameries have not the power they possess in region 4. Nenagh is a market town with small industries, and there are several market villages, of which the largest is Killaloe (800), the see of a diocese covering a large part of Clare. As a whole the loss of population has been very heavy, in places more than half since 1891, and the present population density has fallen to 40–50 per square mile of total area or 60–90 per square mile of improved land.

CHAPTER 18

East Connacht

A difference of farming pattern, together with poorer soils, more heathlands and bogs, and a much damper climate, marks Connacht off from Leinster, and in spite of all the improvements made during the past seventy years the Shannon still separates the 'peasantry' of the west from the farmers of 50–100 acres in the east. Through Connacht, from the Shannon to the Atlantic coast, the number of people living on each square mile of improved farmland steadily increases, the towns, with a few conspicuous exceptions, become mere villages and industrial enterprises almost disappear. The land, in some places at least, cannot support the population except in a state of near-poverty or even actual want, and the continual high emigration is a clear adjustment to such circumstances. Even between 1961 and 1966, there was a population loss of 7%, and there are still no signs that the outward movement is arrested. Since 1891, the decline has been about 60%, and as a result many small farms have been united into larger holdings. Some two-thirds of the region is improved agricultural land, and the remainder consists either of bogs or of bare limestone lowlands. The drift cover is extremely thin or even non-existent around Galway bay and on the long low ridge dividing the Shannon from its tributary, the Suck, but considerably thicker in much of the region and piled into drumlins towards the north. A series of eskers and gravel mounds crosses the region from east to west and carry many historic roads across bogs and marshes. The towns exist mainly as market centres for an agricultural community of comparatively humble status, though a few have some minor industrial developments, and Galway city is now, as for centuries, a major regional capital of the west.

To the west of lough Mask and lough Corrib, the hillfoot forms a structural and human divide between the peninsular upland of west Galway and the lowlands of east Galway, south Mayo and south Roscommon. On the south, the lowlands merge into the turlough area of central Clare, which lies between the Burren hills and the desolate Slieve Aughty upland. The Shannon bogs and 'callows' are a distinct barrier on the east and a striking change of landscape is seen along the margins of the drumlin belt, which maintains its identity across the whole of Ireland. As a whole this East Connacht region consists of the areas drained, partly by streams with underground courses, to the Shannon, Suck and Clare rivers; and the wide spreading interfluves, 200–400 ft. high near the Shannon, fall gently towards Galway bay, where they reach the coast in peninsulas and islands with sandbanks and reefs offshore.

LANDFORMS AND LAND USE

Most of this lowland is composed of Carboniferous beds, chiefly limestones but with some sandstones that appear at the surface without exerting any marked effect on the landforms. The whole of this lowland belongs to the Armorican structural province of Ireland, and its western boundary is the line of junction with the Caledonian west. Devonian sandstones and earlier rocks break through the general Carboniferous surface in places: several of the inliers fail to break the lowland level but two appear as hills in Slieve Aughty, an upland reaching 1,200 ft., and in Slievebawn, a ridge that extends for twelve miles with a north-northeast trend in east Roscommon and a maximum elevation of 864 ft. There is a gradual rise from sea level in Galway bay to 400 ft. between the wide river valleys, with a few hills of 500 ft.

The term the 'land of bare limestones' is perhaps an over-emphatic description of even the least drift-covered section of this lowland, but the covering is deep only on the margins of the drumlin belt. The drumlins, which are varied in form and may be as much as 70 ft. high and half a mile long, occur to the east of lough Mask and around lough Kilglass, immediately west of the Slievebawn. As a rule the southern limit of the belt of drumlins is a fairly clear line, but it has never been precisely mapped. The thinness and even the absence of the drift cover over much of the region is a puzzling feature which has been ascribed to the paucity of waste near the centre of an icesheet located either in Sligo and Leitrim or in Galway and Mayo. The drift cover outside the drumlin belt is rarely more than a few feet thick, whereas it is more than 100 ft. thick in many districts east of the Shannon. Its virtual absence around Galway bay and on the long interfluve running through the heart of Roscommon between the Shannon and the Suck may be due partly to post-glacial denudation. Some areas are karstic and have exposures of bare limestones: in such districts cultivation and settlement is mainly restricted to patches of drift. This is especially true in the waterless country that extends from central Clare (p. 373) to Galway bay and northwards to Tuam.

Turloughs are hollows filled with lakes at times and completely dry at others: the periodic filling shows the fluctuations of the underground drainage system in a climate with abundant but often curiously distributed rainfall from season to season. In this belt, with its intermittent drift cover, very few rivers manage to maintain a permanent course (Fig. 75) and the surface has a large number of sinkholes, some of which expose the underground drainage. It is reasonable to regard the turlough as the more splendid relation of the sinkhole, as both are due to the subsidence of limestone strata. Such large permanent lakes as lough Corrib and lough Mask are almost certainly expressions of the same vulnerability of limestone to what is essentially chemical weathering or solution, though their great depth, which reaches 191 ft. in lough Mask and 152 ft. in lough Corrib, suggests that glacial scouring has played some part in their evolution. A sinkhole or smaller hole (locally known as a 'sluggy') may be nothing more than a tiny depression too small to appear on almost any map: those shown in Fig. 75 are only a fraction of the total number. The normal form is a bowl-shaped hollow, varying in diameter from a few feet to many yards and having, in most cases, a dry base filled with stones and loose clay. Sinkholes occur through the whole of East Connacht, notably around Castlereagh, Co. Roscommon, and Williamstown, Co. Galway, and there are individual turloughs in the same areas; but the turlough and sinkhole country is most continuously developed in the part of the lowland stretching from central Clare into central Galway.

The drifts are varied and consist mainly of local materials, though there are numerous erratics of granites and schists derived from the Caledonian area. The bogs are most widespread in the east, particularly around the Shannon and the Suck, where they cover several square miles in some places. The drier limestone areas are bog-free, but bogs occur in numerous depressions and on drift-covered level areas even in the east. The so-called 'eskers' provide a further element of variety in a landscape that to the superficial observer is merely monotonous: they show a variety of form similar to that seen east of the Shannon (pp. 272–3). Many 'eskers' have no clearcut direction, and are in fact mounds of sand and gravel spread out in a very complicated pattern, yet there appears to be a north–south series joining an east–west series and some maintain their identity for many miles; notable examples are the Athlone esker to the south of lough Ree, and the Clonmacnoise esker, which is continued across the Shannon and the Suck (Fig. 48, p. 284). It is possible that some 'eskers' are kame-like marginal accumulations comparable with the great Salpausselka ridges of Finland, or again that others are successive deltas laid down during the withdrawal of an icesheet and therefore possessing an alignment at right angles to the

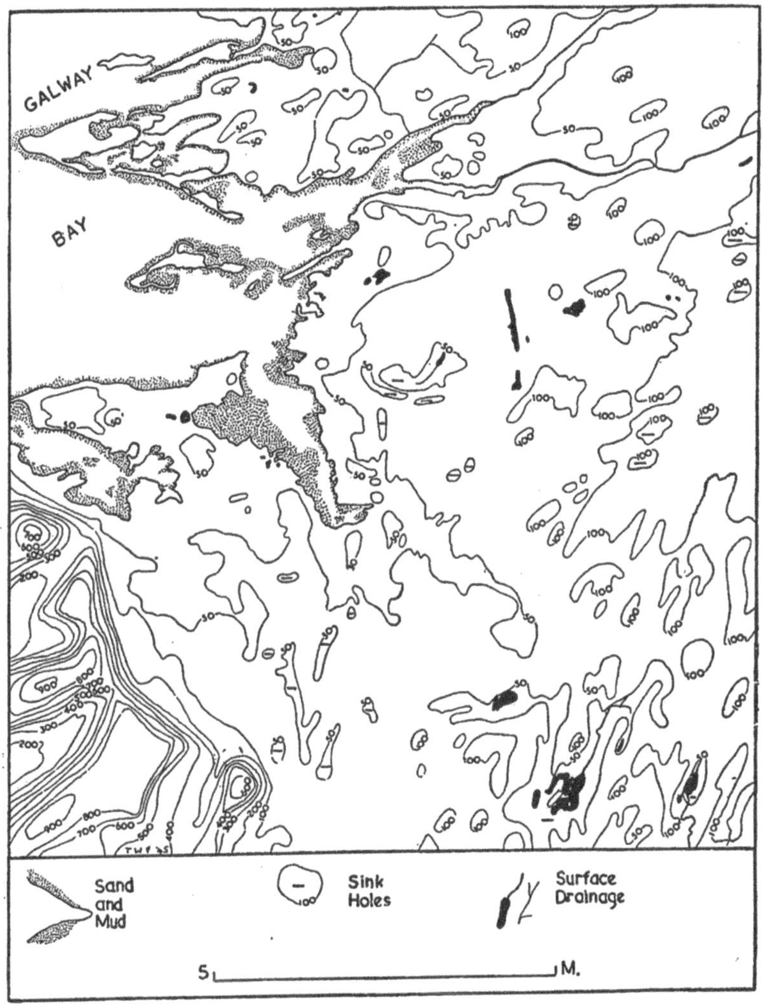

FIG. 75. THE KARST AREA, GALWAY BAY

Very little surface drainage exists in this area: note especially the escarpment forming the edge of the Burren area in Co. Clare.

Based on the 1" to 1 mile maps, sheets 105, 106, 114, 115, by permission of the Minister for Finance of Ireland marginal series. If this be so, the analogy with Finnish conditions is close: at present it is only possible to state that there is general evidence of a withdrawal of the icesheet towards the north and west.

The natural vegetation of the region ranges from heather-dominated moorland on the poorest and most stony soils to patches of natural or sub-spontaneous woodland on the limestones and on the drifts so liberally represented. While much of the limestone area has never borne trees, there are some patches of scrub and even of woodland. The scrub consists in places almost entirely of hazel, but it is found mixed with hawthorn and other scrubs and may form an almost impenetrable thicket of the *maquis* type. Ash woods also occur though oak is also very prevalent and in places is more frequent than ash: it is probable that locally the leaching out of the limestone content of the soil has favoured an oak rather than an ash type of woodland. Near Galway bay, for example, in a well-developed limestone area, thickets of hazel and occasional woods are characteristic, and the woods show the shearing of trees from ground level on the west side to a normal height on the east. Some of the islands in lough Mask are covered with ash, oak, birch, holly and alder on drift-covered Carboniferous limestones; the woods are wind-planed and the tallest trees at the eastern ends of the islands are only 30 ft. high. Similar areas in lough Carra are wooded, with birch, buckthorn, ash, willow, juniper and various other trees in smaller numbers. Pines, though not native, are spreading sub-spontaneously in many of these woods.

Apart from scattered patches of scrub and a few untouched natural woodlands in such remote places as islands in the lakes, trees are mainly found in the numerous demesnes which break the dreary monotony of an open limestone country with grey-walled fields. The Forestry Commission has taken over some 70 square miles for planting, largely in old demesnes of which some, such as those at Ballygar, Gort, Loughrea and Woodford, each cover more than four square miles, though most are much smaller. In all the Commission's fifteen holdings cover less than 3% of the region. East Connacht shares a general lack of forests with other parts of Ireland, but the acquisition of land is not easy, for many of the limestone crags or turf bogs are used either for rough grazing or, in the case of bogs, for home supplies of fuel. The proportion of this region given to crops and pasture varies from three-fifths around Galway bay, to more than four-fifths in the neighbourhood of Roscommon. The land used only for rough grazing includes Slieve Aughty, some parts of the peat bogs and a few of the stoniest tracts of the limestones around Galway bay and in Roscommon. But some of the limestones have great agricultural value and form excellent sheep pastures; parts of the Shannon peat bogs are more truly marshy 'callows', suitable for cattle feeding and perhaps best regarded as water meadows. Over most of the lowland the area of improved land has been considerably extended during the past hundred years, largely by the reclamation of boglands after they have been cleared for fuel.

Agriculture is by far the most important source of employment, as in the rural districts at least three-quarters of the entire 'employed' population or over nine-tenths of those in the 'producers and repairers' group work on the land. Few labourers are employed for commerce is rudimentary and farming mainly at a humble level. The cash resources are low and the holdings are small, though a definite specialization has been evolved in the rearing of sheep and young cattle, many of which are destined to be sent eastwards across the Shannon, some through the famous fairs of Ballinasloe. This is almost the only lowland part of Ireland in which sheep outnumber cattle and are the main cash staple of many farmers. Their association with dry limestone pastures is natural enough, and it is perhaps not surprising to find approximately 1,000 sheep per 1,000 acres of crops and pastures in the almost driftless belt that extends from north Clare through Galway into south Mayo around Ballinrobe.

Farms of less than 30 acres predominate in northeast Galway and in the adjacent parts of Roscommon and Mayo, all of which included considerable areas regarded as congested districts in 1891. In spite of the very heavy declines of population since then, small farms are still supreme in the landscape and the density of population is over 100 per square mile of improved land. The rough pastures are intensively used and many farms cover little more than 20 acres, intermingled with larger farms and demesnes. Some of the holders migrate to Great Britain as seasonal labourers year by year, or earn money as labourers on the larger farms, or on roadwork, or peat-cutting, or even in the towns. The reports of the Congested Districts Board showed that most of the small farms were little removed from the subsistence level: as a rule 3–4 acres were divided equally between oats and potatoes, with a small patch of meadow, and the remainder of the farm was pasture. The inducement to migrate was very strongly felt by those in a position to leave: in some districts 'all the young people migrate to America before they are twenty-five'. Except that Great Britain is now the outlet for emigrants, the position remains much the same and the area still shows distinct signs of population pressure in spite of long-continued emigration. In south Galway the small farms are less numerous than farther north and the pressure on the land is less acute. Very few areas came within the statistical limits laid down for a congested district in 1891, though there are still islands of poverty in what is hardly a land of plenty. However, medium-sized farms, especially those of 30–50 acres, are the most usual type, and the number of farms of more than 100 acres is everywhere significantly small. The density of the rural population is now 50–100 per square mile of improved land, and the rate of decline since 1891 has been more than 50% in some areas.

Farming is strongly pastoral in emphasis with at least two-thirds of the improved land in grass, and most of the remainder in meadow though some 10–15% of the farmland is arable in some areas. Most of the cropped land is in oats and potatoes, intended for use as foodstuffs and feedstuffs, though efforts have been made to encourage the growth of wheat and sugar-beet, the latter for the factory at Tuam. During the war the acreage of wheat increased to 27,000, but in 1967 it was 2,700 (in Co. Galway), and the sugar-beet acreage diminished from 12,000 in 1945 to 4,500 in 1955, 7,800 in 1961 and 4,300 in 1967. Most of the acreage of these crops is in east Galway: little land is given to them in Mayo and Roscommon. There are no creameries in this region and the cows are kept mainly for butter production, though a good deal of milk is fed to calves, which are kept for a year or more and then sent forward to Westmeath and Meath for fattening. In addition, many farms in the area act as intermediate feeding grounds between the poor pastures of the west and the rich pastures of the east. Cattle are most numerous in central Roscommon and sheep in Galway on the driftless limestones, though there are local variations of wide range. Although specialization on sheep and cattle has raised the farmers above the subsistence level, many problems remain. Large numbers of the farms are too small to be economic, and even if the creameries penetrate into the area, many of the farmers will find it difficult to supply milk daily as they have only three or four cows: creameries are generally more successful in areas dominated by medium-sized farms. As yet, pig keeping is unimportant, and poultry keeping has greatly diminished.

INDUSTRIES

Galway, which in 1966 had 26,300 people, is the only important industrial centre of this region. There is little sign of any processing factories for local agricultural produce, as the various mills depend almost entirely on imported grains. Woollen mills, some with shops in the town, provide goods of local design in excellent taste. There are large buildings and con-

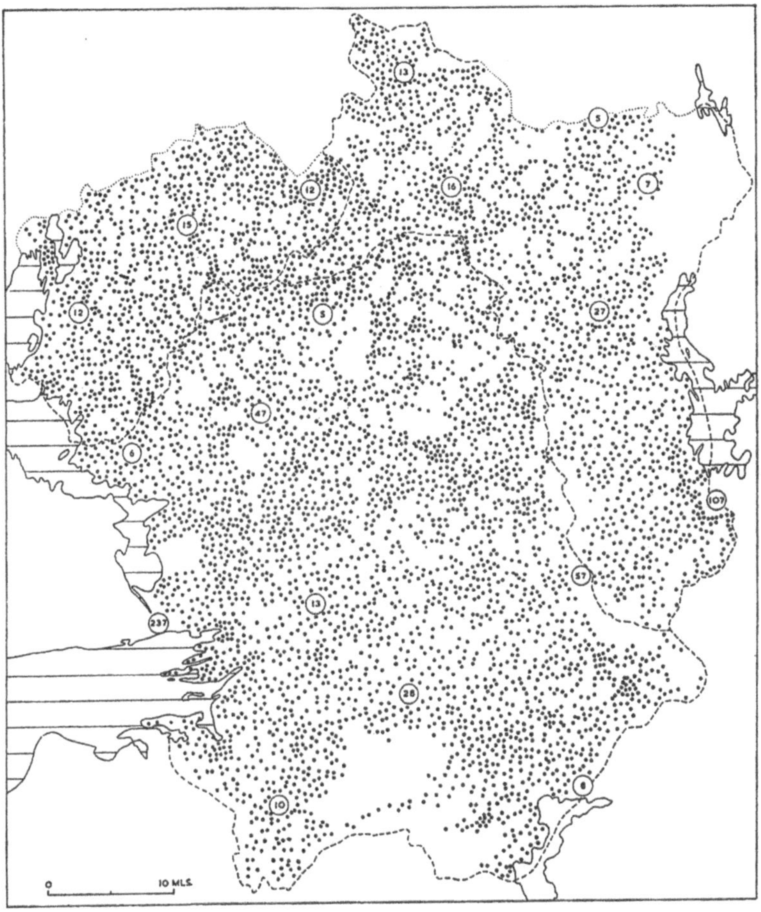

FIG. 76. POPULATION DISTRIBUTION, EAST CONNACHT

One dot to 25 persons

This map shows, perhaps more perfectly than any other in this book, the interruption of farmlands and settlement by bogs and, on the Shannon–Suck, interfluve, or around Galway bay, limestone karsts. Towns are widely spaced and, generally, small; the largest are Galway, in some senses the regional capital of Connacht, and Ballinasloe.

tracting firms with sawmilling, furniture works and several engineering enterprises, some of which serve a wide market. Galway's status as a minor regional capital is reflected partly in its printing works, and in the success of chemical and fertilizer factories. The town has shared in the advance of the consumer industries, including a hat factory and a successful

O

textile printing works. The mineral water industry is well represented, but there is no brewery. A bone china works employs over a hundred operatives. There are also concerns making various souvenirs for the tourist trade, and providing goods made from Connemara marble. Although Galway has never been conspicuously successful as a fishing centre, a plant has been opened for the freezing and processing of fish in an effort to encourage fishing off the west coast.

The remaining towns of the region have very few industries and are little more than market rural centres. Tuam has one of the four new sugar-beet factories in a building of strikingly modern design, and also has sawmills and a mineral water factory. During the 1950s, the cotton industry was introduced to this region with spinning at Loughrea and a processing plant at Portumna, beside the Shannon. Ballinasloe, a town long famous for its autumn fairs, has a mill providing what is advertised as 'protein meal' and a footwear factory. At Gort, a small town in the extreme south, there is a grain mill and a sawmill. In south Mayo the most notable development is a bacon factory at Claremorris employing some 150 people: in the town there are also some building and contracting firms and a factory making laminated springs. Ballinrobe has a small knitwear factory, and Ballyhaunis a furniture factory but in general the industrial development is slight. The industrial position of Castlebar, Westport and Ballina is discussed on p. 413. In Co. Roscommon there is also little industrial growth though there are bacon factories at Ballaghaderreen and at Ruskey, on the Shannon, and mineral waters are manufactured at Ballaghaderreen. In Roscommon there is a mosaics factory which has grown gradually from its small beginnings in 1947 and a small leather and plastics factory established in 1961.

POPULATION

Essentially rural in character, this region of 2,700 square miles has almost two-thirds of its population living outside all towns and villages. The dot map shows the many gaps in the settlement pattern, most of which are uninhabited peat bogs: the largest single break is the Slieve Aughty upland, south of Loughrea. Fig. 76 also demonstrates that there are many places where a ribbon pattern of settlement reflects the road system, especially around Ballinrobe and in the area between the towns of Roscommon and Castlereagh. Several roads follow the tongues of cultivable land on eskers and moraines through bog-covered lowlands. A close study of Fig. 76 will show that the general northwest-southeast line of the bogs is not without its influence on the distribution of settlement. In addition, some of the driftless country is practically devoid of habitations, especially around Galway bay and, to a lesser extent, in some parts of Co. Roscommon.

The population of this region was 303,000 in 1891, 233,000 in 1936 and 191,000 in 1966, and has declined by 37%, but the towns and villages increased from 43,000 to 61,600 within the same period, largely through the growth of Galway. The decrease in the rural population is therefore 50% since 1891 and the density per square mile outside the towns is 48, compared with 96 in 1891. There were 130 people per square mile of improved land in 1891, and 65 in 1966, but this general figure hides the fact that the number of people living on the land is still comparatively high, especially in such pockets of close settlement as the south of Mayo and the adjoining areas in Galway and Roscommon. The reports of the Congested Districts Board drew attention to these pockets of over-close settlement and said that in them 'the land is incapable of maintaining its population . . . the majority of the able-bodied men are compelled to spend from one-third to one-half of their lives as migratory labourers, absent from their homes, and unable to attend to the cultivation of their land'.* At the poorest levels this picture is substantially true today, but the congestion has been alleviated in various ways, some of which were foreseen and even set in motion by the Congested Districts Board, though seasonal migration still remains and the region is therefore supporting fewer people than it appears to maintain.

Too many people lived in this region in 1891 under the then existing form of agricultural organization, and emigration has been part of an essential adjustment that has meant a general rise in the standard of living. The same desirable results might perhaps have been achieved through rural industrialization combined with co-operation, but these have not been a feature of the area. Improvements suggested in the eighteen-nineties included the promotion of better systems of agriculture with superior breeds of livestock and poultry, the provision of artificial manures, and the construction of accommodation roads to remove the isolation from which some farmers suffered acutely. All these have been achieved to some extent and, coupled with them, housing schemes have done much to improve the standard of life. But perhaps the most important development of all has been the acquisition of new lands for settlement and the reorganization of many holdings formerly dissipated into a large number of strips. Most of the demesne lands have been taken

* Base Line Reports of the Congested Districts Board, 1891: this quotation is from a 'General Report referring to the inland districts of Mayo, Galway, Roscommon and Sligo'.

over for new tenants and the enlargement of the holdings has gone on to the present day. In both 1926 and 1936, there were about 36,000 farmers in this region, but while 67% of them had less than 30 acres each in 1926, only 60% of them had such small farms in 1936: in 1946 there were 30,500 farmers of whom 57% had less than 30 acres. Between 1926 and 1936 the number of farmers having 30–50 acres increased from 21 to 26% of the whole, and by 1946 to 29%: the number having 50–100 acres increased from 8% in 1926 to 10% in 1936 and 11% in 1946. Large farms of over 100 acres have been estimated to cover only 4% of this area. More recent statistics for rural districts are not available but the same trends continue. In the whole of Co. Galway there were 23,800 farms in 1936 and 20,400 in 1961: the proportion of farms of less than 15 acres fell from 27% to 18% and of 15–30 acre farms from 32% to 30%, but farms of 30–50 acres increased from 26% to 32%.

The comparative stability of many of the smaller towns since 1891 is a fairly usual characteristic of rural Ireland. This region, which has a low buying power among its rural community, had 15% of its population in towns in 1891 and 32% in 1966. Only in Galway, in effect a regional capital, is there marked growth: in addition to its industries and wide trading contacts, it has a university college and is the centre of a diocese. Galway increased in population from 13,800 in 1891 to 18,300 in 1936 and to 26,300 in 1966: its suburb, Salthill, has gradually become a seaside resort with a large number of hotels but the old fishing village, the Claddagh, is now completely submerged under a new housing scheme. The main fishing is for eels and salmon in the Corrib. In 1966, only five other towns had more than 1,500 people: Ballinasloe (5,830), Tuam (4,740), Loughrea (3,000), Roscommon (2,730) and Castlereagh (1,600). These few towns do not provide a complete marketing framework for so large a region, but there are also a number of smaller places, hardly towns in the normal sense of the town, which act as trading centres: of these the largest have 1,000–1,500 and the smallest some 200 people.

SUBREGIONS

East Connacht is almost entirely a lowland and its demarcation into subregions depends largely on the nature of the local soils and drainage which influence the farming pattern and the distribution of population. The presence or absence of drift and the extent of the boglands in each part of the region are factors of primary significance, but the division of the land into small farms of modest economic status has been almost equally important in the shaping of the landscape as it appears today. Even more, the agricultural landscape of tomorrow will probably differ only in detail from that of the present, though the farms are expanding rather than contracting, housing is improving, and the standard of living is at least not declining. Five subregions are delimited here; reference to Folder I will show their essential geomorphological basis.

I. THE LIMESTONE COUNTRY OF CENTRAL GALWAY

The belt of country that extends from the neighbourhood of Gort to Galway bay and thence northwards to Tuam and Ballinrobe consists of Carboniferous limestones which are either driftless or else have merely a thin covering. Few rivers manage to maintain a course on the surface except for the Robe, the Clare and two small streams flowing to Galway bay, but there are numerous sinkholes and some larger fissures, including collapsed caverns (Fig. 75). The extensive but intermittent 'turlough' lakes that characterize central Clare are found only around Gort within this region, though there are many small depressions of comparable formation. Bogs occur only around lough Corrib, but the cultivated area is limited to three-fifths of the whole as there are stretches of bare limestones that are covered with a vegetation reminiscent of maquis or garigue. This subregion therefore has some features in common with the adjacent Burren district of northwest Clare (pp. 377–9) and is also comparable to central Clare (pp. 373,379) and Roscommon (pp. 391–2). It rises gently from a fragmented coast with many peninsulas and islands (Fig. 75) to a ridge 200–500 ft. high. This ridge forms a low water-parting between the Clare and the Suck drainage basins to the east of Athenry, and is clearly limited on the south by the Burren hills and Slieve Aughty. On the north there is no clear boundary, but rather a gradual transition into drift-country with drumlins around lough Mask.

Agriculturally this region is distinguished by its lowland sheep farming, which is more fully developed here than anywhere else in Ireland. In both Tuam and Gort rural districts there are four times as many sheep as cattle, and in all the flocks number 300,000, or more than one-quarter of the total for the entire province of Connacht. The farms are mainly in the 30–50 acre group, with a substantial number of smaller farms and a few larger ones; only in a few places is there an economic decline to the level of a congested district. The density of population is now 50–75 per square mile and nearly 100 per square mile of improved land. Galway is the major market centre and Tuam, a cathedral city with a Roman Catholic archbishop and a Church of Ireland bishop, provides the northern part of this subregion with trading contacts, but otherwise the only 'towns' are such places as Ballinrobe (1,240), Gort (1,000) and Athenry (1,330) and a few villages such as Kinvarra (300) on Galway bay.

2. THE SLIEVE AUGHTY

These uplands, which extend over some 250 square miles, are of Armorican structure and expose Old Red sandstones at the surface with occasional inliers of Silurian strata. They consist of long monotonous ridges rising to 1,000–1,200 ft., and are covered by bogs. The Slieve Aughty are by-passed by almost all travellers and offer little scenic attraction; their isolation is accentuated by lough Derg, a water barrier of considerable significance, on the east. Settlement is restricted to the valleys and some lower hillslopes: the farmed lands are generally surrounded by rough pastures to an even greater extent than Fig. 76 would suggest. A substantial part of this upland area was congested in 1891 and in spite of a heavy loss of population, part of it has hardly risen above the congested level since. The farms generally have about 3 acres of meadow and a share in the mountain grazing, which is either individually owned or grazed in common by two or more holders. Home-made butter, eggs, pigs, calves and sheep are the main sources of income, and as neither migratory labour nor local employment has ever been common in the district the emigration rate has been particularly heavy. The main problem of the farmer in such an area as this is access to markets: many farms are as much as fifteen miles distant from Loughrea or Gort and therefore depend on the intermittent fairs and on the inconsiderable commercial facilities of Woodford, a village of 240 people.

3. THE PORTUMNA–BALLINASLOE LOWLAND

This subregion is in many ways a typical piece of the Irish Central Lowland, as it consists of drift-covered rolling country, floored by Carboniferous shales and limestones. Mainly 200–400 ft. high, it has patches of bogland, especially on the divide between the Clare and Suck rivers and near the rivers, but it is most remarkable for its 'eskers', which extend across it in sinuous lines that are almost invariably followed by roads. These dry ridges are of particular importance where they cross bogs, especially those developed around the Shannon and the Suck. An esker carries the road from Loughrea to the important Shannon crossing at Banagher; from Banagher it is eight miles upstream to Shannonbridge and thirteen miles downstream to Portumna. Similarly Ballinasloe lies on an esker at one of the few places in the Suck valley where there is firm ground on each side. The isolating influence of the Shannon is due not only to its infrequent bridges and its wide bogs but also to its flooding in most seasons.

In no sense, however, is this lowland a congested district; although the farms are generally smaller than those in Tipperary and Offaly, the counties on the opposite side of the Shannon, they generally average 50–60 acres. Rough pasture, chiefly bogs, accounts for one-quarter of the total land. The main emphasis is on sheep and cattle rearing: while sheep are generally two or three times as numerous as cattle, the area lacks the very definite specialization on sheep which has developed around Galway bay. The rural density per square mile is now around 50–60, following a decrease of approximately one-third since 1891, and the area has two market towns, Loughrea and Ballinasloe. But these towns are eighteen miles distant from one another and the area depends for its marketing on various smaller centres such as Portumna (910), an attractively kept small town at the head of lough Derg.

4. THE SHANNON–SUCK INTERFLUVE, ROSCOMMON

The interfluve between the Shannon and the Suck consists of an almost driftless expanse of Carboniferous limestones at an altitude of 200–400 ft. though to the north of lough Ree, Slievebawn, a ridge composed of Devonian and Silurian strata, rises to 864 ft. A few Armorican ridges of similar rocks appear in this interfluvial area, but fail to break its generally smooth rolling line (see Folder I). Extensive bogs are developed around the Suck and the Shannon and also in various depressions near the town of Roscommon, but the area is most notable for its limestone pastures, which vary from karstic fields of little or no agricultural value to rich grasslands that support large numbers of cattle and sheep.

Sheep are in a distinct majority here, especially towards the south, where they are three times as numerous as cattle (in Athlone II rural district), but farther north there is some of the best cattle-fattening country west of the Shannon, between Strokestown and Boyle. In general terms, this Roscommon subregion has a farming pattern that includes both sheep and cattle production, with the latter prevailing towards the north. Pastures and meadows cover nearly nine-tenths of the improved agricultural land and cropping supplies only a limited quantity of foodstuffs and feedstuffs. In some ways Roscommon, like Longford, is an ante-chamber of the Meath–Westmeath cattle fattening and grazing area, as its economic organization rests upon the satisfaction of demands from areas farther east. The farming pattern here is fairly typical of those areas west of the Shannon that are not congested but yet are dominated by holdings of small extent. The majority of the farmers have less than 30 acres, but medium-sized holdings are also very numerous and there are a few large farms: the average is about 40 acres. As there is little industry in the area, and its people depend almost entirely on the land, the town and village population is less than one-fifth of the whole. Emigration has been heavy and since 1891 the population has declined by more than one-third, so that the density is now 50–75 to the square mile.

5. THE CLAREMORRIS–CASTLEREAGH AREA

Southeast Mayo, west Roscommon and northeast Co. Galway are included in this subregion, which consists of densely settled bog-strewn country very little above the status of a congested district in 1891, and still dominated by small farms with an average size of 25–30 acres. Here the Carboniferous rockfloor is almost entirely drift-covered, with a series of eskers running north–south at right angles to the east–west series; the land slopes gently from an altitude of 200–400 ft. towards the Suck valley in the east and the Clare valley in the west. Patches of bog are frequent but rarely extensive and some three-quarters of the total area is improved farmland. The population is almost entirely dependent on agriculture and its farms average 25–30 acres: on the farms as much as 15% of the land is given to crops, chiefly to the fodder staples, oats and potatoes, but cattle are more important than sheep and the sale of young cattle is linked with the home butter production and with pig and poultry raising. In 1891 the density of population was 150–200 per square mile, but this has been reduced to 75–100 as a result of particularly heavy emigration: in some parts the loss has been 50% or more since 1891. The market centres are all small and include Castlereagh (1,600), Ballaghaderreen (1,260), Ballyhaunis (1,150) and Claremorris (1,480), each of which gathers up the trade of as much as 100 square miles. These places provide interesting examples of places which, though towns in function, are little more than villages in size owing to the restricted purchasing power of the areas they serve. Emigration continues actively from this area: for example in the five years 1956–61, the population of Co. Roscommon declined by 7·1% and there was a decline of 5% in the following five years to 1966. The bacon factories at Claremorris and Ballaghaderreen are almost the only signs of industrialization: the subregion, like others, still has more people than it can comfortably support under the present economic organization. The solution of its problems may lie partly in the local establishment of agricultural processing and possibly other industries, in the development of co-operative buying and selling, and in the enlargement of the holdings. All have begun already on a very small scale, but, some would say, on so small a scale that their effect is as yet imperceptible.

CHAPTER 19

West Connacht

West Connacht is a hard environment of mountains and bog-covered lowlands, peninsulas and islands, with few areas of fertile soil. One may walk for as much as twenty miles across lowland country having no sign of settlement, but houses cluster thickly in the restricted areas occupied by small farms. The Congested Districts Board became responsible for the welfare of the vast majority of the inhabitants in 1891, and their investigations covered both the densely settled seaboard and the thinly occupied interior, which has severe problems of isolation through lack of transport. The reports showed that the evil genius of poverty was rampant and that there was very little land available for resettlement in the region or in such immediately adjacent areas as East Connacht, which still has its own problems of excessively small farming units. 'It was', says Micks, 'the deplorable conditions of the people in Connemara, Achill and Erris that produced the general feeling of sympathy that made the Congested Districts Board popular.' * The small farms still remain though improved in many cases by rearrangement into groups of small fields instead of scattered plots of infinitesimal size, and also by fencing the arable land with high walls to keep out animals. Most of all, the housing has been revolutionized and many of the hovel-like cottages are now relegated to the status of farm buildings, while the families live in new, if less picturesque, homes.

In spite of all the changes, this region still remains relatively humble in its economic position. Most of the market centres on the west coast are little more than villages, though there are some prosperous towns, such as Ballina in the Moy valley or Westport in the Clew bay lowland. Galway, the regional capital of a large part of Connacht, lies on the eastern fringe of this region. The dominance of agriculture is greater here than in any other part of Ireland for there is very little indigenous home industry, and although fishing is a supplementary resource of many people living on or near the coast, all efforts to make it commercially successful have failed. The scenic attractions of West Connacht are considerable, but here, as elsewhere, the employment given by the tourist industry is small. It is difficult

* Micks, W. L., *An Account of the Constitution, Administration and Dissolution of the Congested Districts Board*, 1925, p. 138.

to imagine any development that can provide a substantial measure of employment in the region, and a greater resource than any of the above has been seasonal migration, year by year, and emigration of a more permanent character: the seasonal migrants' savings still balance family budgets as they have done for more than a century, and the postal orders of the sons and daughters in dispersion are constantly useful.

West Connacht is an environment of great beauty, with towering mountains rising above wide valleys and lowlands, fine rivers and lakes, sea cliffs several hundred feet high, peninsulas guarding long inlets that are almost fiordic in character, and numerous islands both near the coast and some distance from it in Atlantic waters. Surrounded by the sea on three sides, this region may be defined on the southeast by lough Corrib and lough Mask, and in the northeast by a line drawn through the lowlands near Castlebar to the Moy estuary near Ballina. The Aran islands are also included in this chapter, as their main link to the world is through Galway. These islands are composed of bare limestones partly covered with 'made land' and have an Irish-speaking population possessing many traditional customs and crafts. A major problem is transport though a steamboat runs frequently throughout the year to a pier beside deep water at Kilronan on Inishmore. There are no piers on the smaller islands and all passengers, stock and merchandise are conveyed by curraghs.

LANDFORMS AND LAND USE

Mountains cover only a small part of West Connacht but form a backcloth in every lowland landscape. The three main structural elements are first, a highly metamorphosed rockfloor of pre-Cambrian date which forms lowlands and, where quartzites occur, conical hills; second, Ordovician and Silurian sedimentary rocks of highly varied types, best seen in the mountains around Killary harbour; and third, Carboniferous strata, mainly limestones and sandstones, preserved around Clew bay and in north Mayo. If a rise of 250 ft. in sea level took place, the present valleys and cols would be filled up and there would be a number of hilly islands: the valley sides are steep and in places precipitous, especially around Killary harbour, where Silurian strata stand out in a series of bluffs above the lonely shores of the 10-fathom deep fiord.

The pre-Cambrian floor is formed of gneisses, schists, Galway granites and, around Recess, the serpentines known as Connemara marble: its form is complicated and probably more than one phase of pre-Cambrian metamorphism was involved. The Galway granites, of pre-Cambrian date, form gently rolling country rising to a maximum altitude of 1,198 ft. between Galway and Recess. Far more impressive are the conical hills of pre-Cambrian quartzites, most of them over 2,000 ft. high in the Maumturk mountains, the Twelve Bens, Achill island, the Nephin Beg range and the isolated peak of Nephin itself. In all there are more than thirty quartzite peaks over 2,000 ft., which stand above the lowlands as sentinels of which it has been said, 'Waste thou wast, waste thou art, waste thou shalt remain'.

Ordovician and Silurian rocks outcrop from the shores of lough Mask to the Atlantic and form the mountainous country on either side of Killary harbour. A wide variety of strata is represented, for the rocks range from extremely coarse conglomerates to the finest sands; and metamorphism has produced slates and mica-schists, some of which are highly resistant to erosion. Mweelrea, 2,688 ft., is the highest point in West Connacht, and there are several other peaks in the extensive upland around Killary harbour. This upland may be a Devonian peneplane, as it is covered in places by Carboniferous rocks: preserving a smooth plateau-like surface, it has been cut by fault movements into a series of blocks separated by deep and generally wide valleys which became avenues for ice-tongues in the Quaternary period. Probably at one time Carboniferous strata covered most of this region, but long-continued denudation has exposed such underlying features as the long, broken ridge, Caledonian in origin, which extends from a point near Westport past Castlebar into the Ox mountains. In north Mayo, Carboniferous beds are exposed in the Ballina lowland, and sandstones west of the town are covered by a dreary expanse of bog at 200–400 ft., with hills 600–800 ft. high near Killala bay and some fine cliffs on the coast. There are extensive areas of Carboniferous limestone around Clew bay and also around Castlebar, but a drift cover prevails and areas of karstic quality are few.

Glaciation has left its mark heavily in West Connacht, in at least two phases. The mountains have numerous corries and cliffed valleys with morainic deposits, and rolling lowlands with numerous erratics and extensive blanket bogs are characteristic of the areas round the mountains. Near Pontoon, lough Conn, for example there are granites forming a mammilated surface reminiscent of Labrador or Finland. Much of the blanket bog lies on a bare rock surface, and there can be no doubt that the rock area exposed has been steadily increased by the removal of peat. The best areas for farming are patches of well-drained drifts, such as the drumlins, moraines and eskers of the Ballina lowland. Around Newport and Westport hundreds of drumlins cover the lowland, separated from one another by peat bogs, marshes and lakes, with a few driftless areas showing a karstic quality. Drumlins form an archipelago in Clew bay where the islands are already cliffed on the west side by wave attack. There are also drumlins around lough Conn and lough Mask though in this area the spreads of drift are less clearly moulded into the drumlin form than in Clew bay. All the drumlins run parallel to the striae, and are related to movement from an icesheet centred over the Twelve Bens and the Maumturk mountains north and northwest to Clew bay and south towards Galway bay. This alignment of the drumlins affects the whole pattern of roads, fields and farms and is therefore a significant feature of the regional geography.

The coasts of West Connacht, as varied as the land behind them, are merely the present margins of a land recently subjected to marine submergence. There are drowned valleys with long muddy estuaries or long winding channels, peat bogs partly covered by the sea, drumlin islands now subject to vigorous wave attack, and truncated hills with cliffs several hundred feet high. The sand dunes developed along some parts of the low-lying coasts are prone to move inland under the stimulus of heavy and constant westerly winds, though many have been stabilized by planting marram grass. As a whole the waterways are unattractive to navigation as there are numerous submerged rock belts and sandbanks, few good natural harbours and many inlets with banks of sand and mud. The Atlantic swell is heavy for most of the year and storms are violent: in addition inlets such as Killary harbour have severe local winds dangerous to shipping. All these factors help to explain why West Connacht possesses, apart from Galway, only two ports, both of little significance, Westport and Ballina, and also why fishing on a commercial scale has never been successful. But the island people sail Atlantic waters in their curraghs, made of laths covered with tarred canvas, which are said to be better suited to the rough seas than wooden boats as their very lightness gives them buoyancy.

Two-thirds of West Connacht has never been cultivated and preserves a vegetation cover little altered from the natural state, especially where peat is still accumulating. But turf cutting alters the natural drainage and begins the drying-out of bogs and the removal of the thin rind of peat, known as 'scraws', exposes the solid rock: at Carraroe, for example, the ice-polished and striated granite is now exposed and on Inishbofin the area of bare rock is annually increased. In contrast to the bogs, the stony mountains of quartzites and gneisses have large areas of soilless rock or scree with very little vegetation on their steep slopes. A few native woodlands survive on hillsides: near lough Conn there are large areas of oak and oak-birch

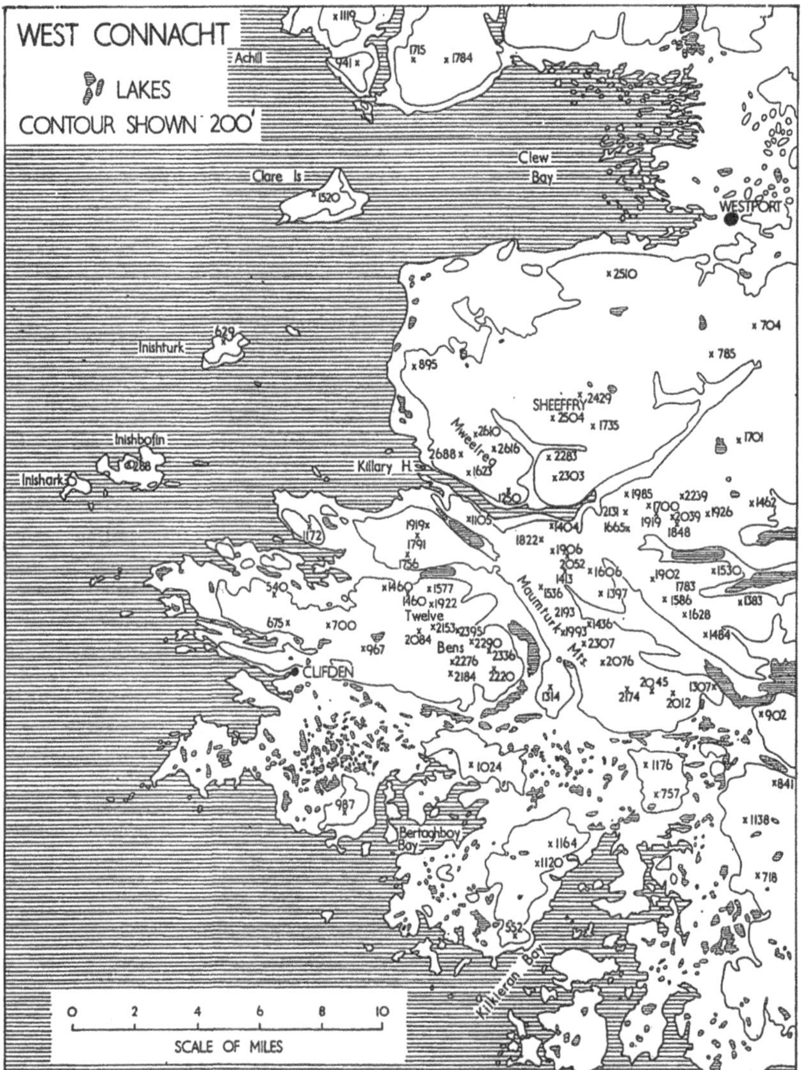

FIG. 77. WEST CONNACHT, FROM CLEW BAY SOUTHWARDS

Clew bay has a number of islands formed of drumlins; south of this bay is Croagh-patrick (2,510 ft.). There are numerous peaks over 2,000 ft. north of Killary harbour and several more to the south, notably in the Twelve Bens and the Maumturk mountains. But most of Connemara is a lowland, with bogs and lakes of great variety.

woods with rowans, holly, ash, hazel and a rich ground vegetation. But most of the region has no woodland above the status of scrub, which is found in sheltered places as far west as the islands lying off the coast: thus, the east side of Clare island has scrub 4–8 ft. high with hazel, birch, holly, rowan, oak, willow and a ground flora that includes bog-myrtle and honeysuckle. The first enterprises of the Forestry Commission were the plantations at Ross, between Galway and Oughterard, and at Foxford, both well away from the Atlantic, but in recent years so much more land has been acquired that the Commission now has 90 square miles, of which two-thirds is plantable. Much of the forest is on drained bog in sheltered valleys such as those of Glenamoy and Ballycastle in north Mayo, those around the Nephin mountains or near Croagh patrick. But there are also lowland forest sites such as Ballynahinch, seven miles east of Clifden: significantly no forests are planted on the Atlantic shore. The new forests will enhance the fine scenery.

Blanket bog, the climax vegetation over the greater part of the area, is interrupted over many square miles of lowland only by a multitude of lakes with brown peaty water which is strongly acid (pH 4–5). In Iar Connacht, some 300 square miles of bogs is enclosed by roads that run from Galway through Oughterard to Maam Cross, then south to Cashla bay and east to Galway again: this bog is apparently on a driftless granite rising in broad undulations from sea level to 1,000 ft. The detailed plant distribution, however, shows interesting local transitions from *Sphagnum* and other plants favouring a saturated habitat to *Calluna* in better-drained areas, though *Molinia* is conspicuous everywhere, and cotton-grass moors extensively developed. Similar bogs occur over almost all the more level parts of Connemara to the Atlantic coast, in the valleys stretching to Killary harbour and Clew bay, and in north Mayo, where more than 400 square miles is uninterrupted bog, treeless, trackless and extremely isolated. The continuity of the bog landscape is unique even by Irish standards, for the bogs in Donegal or Kerry or in the Central Lowland, though of considerable area in the aggregate, are broken into a number of patches.

On and near the coasts the most striking effect of climatic exposure is the virtual elimination of tree growth or its restriction to sadly planed scrub. Equally significant features are the bog developments and the survival of some unusual immigrant species which found a climatically favourable environment that was also treeless: these include three Mediterranean heaths and the *Saxifraga spathularis*, none of which is found in Britain. This saxifrage is hardy and is found clinging to bare rocks at all altitudes, though it may be near its northern limit in Mayo as it is much less luxuriantly developed in Donegal. The three heaths are *Erica mediterranea*, *E. mackii* and *Dabeocia cantabrica* (St Dabeoc's heath): the first, or Mediterranean heath, is the most widespread and covers large areas south of Clew bay and a wide belt of country in north Mayo, including Achill and the Mullet. The two other heaths, *Erica mackii* and *Dabeocia*, have a more restricted range: but the latter extends through west Galway and also occurs in patches through Mayo from Cong to Achill. A native of south-west France, Spain and the Azores, *Dabeocia*, like *Saxifraga spathularis*, is not a tender plant for it climbs to 1,900 ft. on the Twelve Bens and survives on exposed heaths to the edge of the Atlantic. *Plantago* sward is perhaps the most complete expression of a climate dominated by seawinds; it is found on the exposed coasts of Inishark and Inishbofin, on the west of Clare island, and on Achill island up to 300–400 ft.

In the Clew bay lowland and the Moy valley some three-fifths of the total area is farmland, but elsewhere hardly one-fifth has been reclaimed and cultivated: even so, farming is the essential means of sustenance for almost all the population, whatever their supplementary resources may be. At least half the unoccupied area is lowland peat bog rather than mountain country, and the present limits of settlement are the frontiers that human enterprise has reached in its struggle against intractable soils. The peat bogs, which are used for pasture, bedding materials for animals, and fuel for the home or for sale, are by no means negative areas in the farm economy. However, one feature of great interest in the land use of this region is the extent to which the natural vegetation remains little altered from its primeval state of bogs harbouring rare plants, or scrubs of bushes rising to the stature of trees only in a few favoured localities.

Agriculturally, West Connacht is comparable with west Donegal; it has the same preponderance of small farms, similar high densities of population on cultivated land, and a general routine of subsistence farming. The holdings average 10–30 acres; they are smallest along the coast where the population per square mile of improved land is most dense. Thus, in the Mullet peninsula, part of the Belmullet rural district in the northwest corner of county Mayo, most of the farms have about 12 acres of improved land of which a considerable proportion, in places half to two-thirds, is used for crops, chiefly oats and potatoes. In such areas it is still possible to see relics of the former hamlets and small patches sown to various crops, but the Land Commission had reorganized the arrangement of fields and houses so that they are normally in strip fields with a new house on each farm. Not everywhere has this been achieved: in parts of Connemara there are innumerable small fields surrounded by stone walls made from the boulders dragged out of the soil. The farms are no longer regarded as economic and it is one of the features of such western areas that the Land Commission has been making, in the first half of the twentieth century, peasant holdings that were regarded as providing an adequate economic existence for a family in the late nineteenth century but are no longer so regarded. The holders must have other resources to provide what is now thought to be even a minimum standard. Even so the Belmullet area was remarkably stable in population up to the nineteen-fifties, as the farming income was supplemented by the seasonal migration of agricultural workers to England, by the profits of lobster fishing, by local labouring and money from emigrants.

In fact many of the men spend a large part of their time working elsewhere and the women emigrate in large numbers. There are few landless fishermen: in 1961, only 258 were returned for Galway and Mayo counties, compared with 240 in 1951. Seasonal migration was well established long before the Famine and the Congested Districts Board reports included such statements as (several hundred) 'labourers leave this district annually for Great Britain: the time of departure varies according to the size of the holdings and the poverty of the families, some starting in March and others in May and June . . . they save on an average £8 each in ordinary years and return home in October or November'.* The profits are now much greater and the men of Mayo scatter through the more prosperous farming districts of Great Britain year by year. The division between a farm labourer and an actual farmer is often difficult to draw in such areas as this, for the labourer in most cases possesses a certain amount of land and merely hires out his services locally to the few holders of large farms. In addition both those who are returned as 'labourers' in the Census classification and many of the humbler farmers work on the roads for parts of the year, especially the winter months. Other resources of the men on the land include peat-cutting and fishing, the former for sale in the towns and on some of the islands, and the latter for home consumption or intermittent sale.

INDUSTRIES

Home industries are a traditional feature of life in Co. Galway and to a lesser degree in Co. Mayo. Wool from the sheep shearing is spun and knitted into socks or woven into flannel or rough tweeds, in some places

* *Congested Districts Board, Base Line Report*, Rath Hill district, Co. Mayo (east of Blacksod bay): similar remarks are made for other districts.

for home use but in others also for sale. In 1891 some families derived as much profit from the women's home industries in winter as they acquired from the men's summer migrations. These industries have not died out, as in Kerry, nor have they provided the foundation of factory enterprise, as in Donegal, but 'bawneen' tweeds and 'Connemara' socks still find their way into the town shops. The flannels and tweeds are generally worn undyed and possess an off-white colour, but some of the fabrics are dyed in colours of varying brightness, some of which are made from lichens or other local materials. Efforts by the Congested Districts Board to develop industries were only partly successful, but distinct success has been scored by the woollen mill at Foxford, where rugs and blankets are made. Achill island has a knitting industry at Doagh, and a recent development is fish curing, with shark fishing and the preparation of shark oil. There are also fish-freezing plants at Galway and at Carna.

The three market towns on the fringes of West Connacht live mainly by trading with the surrounding rural areas but also have a number of industries. Ballina (6,100), a small port, has mills producing flour and a variety of animal feedstuffs, works producing concrete, and factories providing milk products, fertilizers, mineral waters, and goods from reclaimed rubber. In Castlebar (5,630) the main industries are a bacon factory and the making of felt hoods for hats. Westport (2,930) has a variety of industries which include sewing cotton, shirts, footwear and leather laces, hosiery, grain milling and brush making. In all these towns and in some smaller places there are builders and contractors. As elsewhere in Ireland this list includes some factories directly serving the farmers by processing pigs or milling feedstuffs, or by providing fertilizers and building materials, but it also includes a few works showing no geographical influence at all, though providing useful employment in towns threatened with slow decay.

POPULATION

The region covers 2,160 square miles and in 1966 had a population of 95,700, 44 to the square mile, as compared with 169,000, 78 to the square mile, in 1891: the decline by 43% is not high by Irish standards. The town and village population has increased from 19,000 in 1891 to nearly 25,760 in 1966, chiefly through the expansion of Ballina and Castlebar. Market facilities are provided also by smaller places such as Clifden (990) and Belmullet (700), both of which are remote 'towns' established early in the nineteenth century. Dispersed settlement is general, except on the Aran islands (pp. 420–4), and Fig. 78 shows that the coastal fringes of

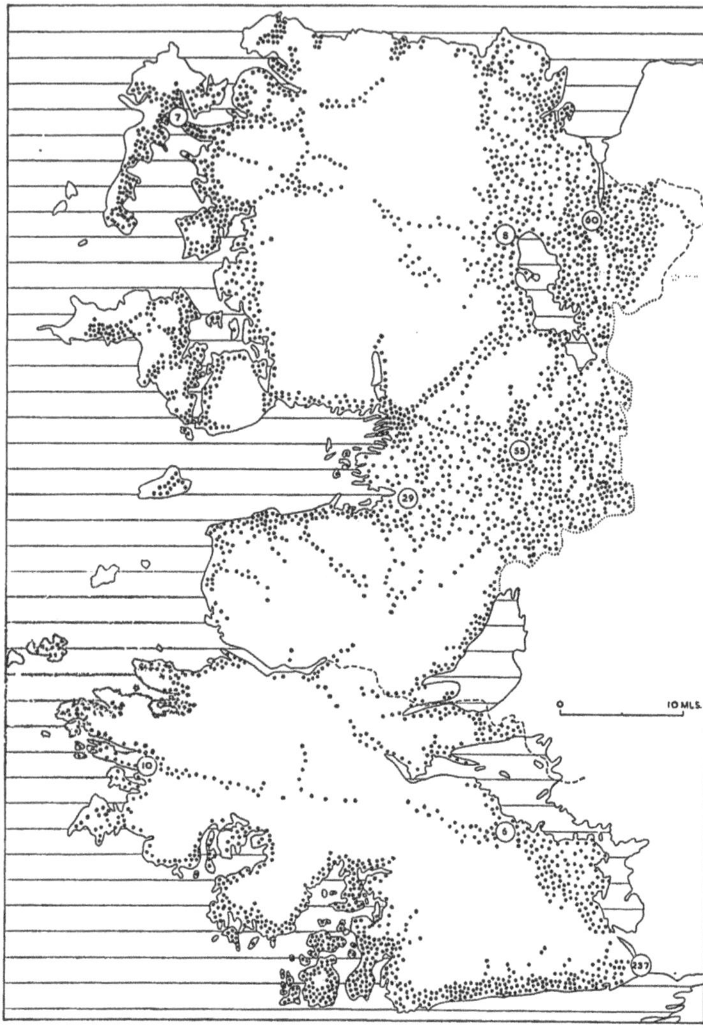

FIG. 78. POPULATION DISTRIBUTION, WEST CONNACHT

One dot to 25 persons

Except in the Clew bay lowland, the Ballina lowland and around lough Corrib and lough Mask, most of the population live near the coast.

Galway bay and of the Atlantic shores are very closely occupied, but only for a depth of one mile or so from the shore. There is a similar type of settlement along the western shores of lough Corrib and lough Mask, but few farms occur in the interior valleys. Finally, the lowlands extending from Clew bay to the Moy estuary show a continuity of occupation that is typical of the Irish Central Lowland, though there are numerous gaps where peat bogs or isolated hills occur.

The pattern of rural settlement is so closely woven that the term 'dispersed' cannot be applied to it without some qualification. Here, as in western Donegal, the farmhouses are so near one to another that many roads resemble a village street lined by houses. Another feature shown by Fig. 78 is that, except in the Clew bay–Moy lowlands, almost all the population lives near the sea. The ribbon of small-farm settlements west of Galway, the closely farmed coastlands of Connemara, the lines of houses and fields on either side of Clew bay illustrate this point; the shores are uninhabited only in Killary harbour and in places where the coast is formed of cliffs, as in north Mayo or Achill, or of sand dunes, as on the west of the Mullet. Inland, the limit of settlement is rarely altitudinal; more generally it is a line of demarcation set by the farmers themselves between their fields and their rough pastures, which generally consist of bog-covered lowland rather than hills. In all the circumstances it is perhaps surprising that almost three-tenths of the area has been made into improved land for farming.

SUBREGIONS

The landscape contrasts that exist in West Connacht are considerable: quartzite peaks rise above dreary boglands, smooth-topped mountains hide steep-sided trough-like valleys, drumlins interpenetrate with marshes and bogs, and the coastal waters are studded by islands and peninsulas. In such an environment there are many microcosms, each with its own individuality and problems: every island, every peninsula, every remote group of farmsteads is a small human region in some way distinct from its neighbours. A broader sweep also reveals contrast. Connemara has a mass of tiny lakes and rocky heaths which is matched only in the Outer Hebrides within the British Isles, but the Moy lowland is an actively farmed and closely settled country with numerous drumlins, in some ways similar to the whole drumlin belt that stretches across Ireland to the vicinity of Dundalk. There are therefore both small and great contrasts within the region: less mountainous than Donegal or Kerry, West Connacht shares with them a variety of scene. It may be divided into six subregions, which will now be considered.

I. THE LOWLANDS OF SOUTHWEST GALWAY

Two regional names occur in this subregion, Iar Connacht in the east and Connemara in the west. Neither has any precise definition, but Iar Connacht is the country beside Galway bay which has a straight east–west coast with no islands: the name Connemara is applied to the district west of Cashla bay, which consists of many peninsulas and islands, most of them inhabited. These lowlands are bog-covered expanses of the pre-Cambrian floor, and away from the coast there is an almost deserted area, rarely penetrated by travellers and almost devoid of any sign of human occupation. Almost the entire population lives either near lough Corrib on the limestone country or in a coastal belt about one mile wide from Galway westwards. The islands are as densely settled as the mainland but the interior, though low-lying, is desolate and uninhabited, except for a few farms along the road between Oughterard and Clifden. The poverty of the farmers in this area seventy to eighty years ago staggered most observers, and not least the inspectors of the Congested Districts Board, who reported that in some places half or even three-quarters of the population were desperately poor. Their estimate of poverty was necessarily relative and subjective, but its prevalence was undoubted. One inspector analysed the amount of work involved in manuring an acre of land: some stable manure mixed with heather, sedges and weeds was used, and in addition seaweed was brought from the shore, where it had previously been collected and dried. For 1 acre, 36 horse or donkey loads, or 200 'women's basket loads' were needed; the donkey, with its baskets, would make the journey six times a day and six days of good weather, rarely consecutive, would be needed. Under such conditions it is not remarkable that the arable acreage on a farm rarely exceeds 2–3 acres divided between oats and potatoes: equally significant is the concentration on the seaboard. Cattle and sheep are kept on the farm pastures and on the heathlands. Fishing and turf-cutting still provide, as in 1891, extra sources of income and emigrants' money still comes into the area. Modern government welfare schemes include tomato-growing in greenhouses, supplied from a central depot: to travellers along the coast road this Gaeltacht scheme may be a puzzling feature. An even more modern addition is the peat-fired electricity station at Screeb. Clifden is the main market centre in the west and Galway in the east, but much of the area's commerce is transacted at roadside shops or monthly fairs.

2. THE MAYO–GALWAY UPLANDS

These uplands, covering more than 400 square miles, have summits 2,000–2,500 ft. high: the sharp quartzite peaks of the Twelve Bens and the Maumturk mountains contrast with the smooth summits of the Ordovician and Silurian strata that form the Partry mountains, Mweelrea and the other hills north of Killary harbour. The combined effects of faulting, river action and glaciation have produced a network of deep valleys, one of which, submerged, forms Killary harbour. Only along the coast is there continuous settlement: the lowlands elsewhere are almost unoccupied. The farming is similar to that of the coastlands in the southwest of Galway, with 2–3 acres on each holding tilled for oats and potatoes. Cattle and sheep are pastured both on the farms and on the surrounding mountain grazings. Clare island had 167 inhabitants in 1966 (205 in 1961) and Inishbofin 247 (248 in 1961): Inishark had 92 in 1966 and 108 in 1961. All these inhabited islands have farmers living partly on fishing with pensions and emigrants' remittances. Men go away each year, or for periods of a few months or even years, as seasonal migrants; there is some home weaving and knitting in the east among the communities of Joyce's country and the Partry mountains. In addition many households are supported largely by emigrants' remittances. Though the population has declined almost everywhere by one-quarter to one-half since 1891, one division, Owenadorraun, at the northern entrance to Killary harbour, almost doubled its numbers from 1891 to 1936, when it had 136 inhabitants; later there was a decline to 102 in 1956 and 92 in 1966. Under a settlement scheme land was reclaimed, fences built and new houses erected, but at considerable expense and only for tenants willing to accept a humble manner of life: as a rule, emigration offers an easier remedy for economy stringency to the inhabitants. This subregion has some of the most isolated communities in Ireland and possesses an air of remoteness rarely equalled elsewhere.

3. THE CLEW BAY LOWLAND

This lowland may be regarded as a single subregion though it includes three distinct geomorphological units. The first is made up of Carboniferous limestones and sandstones, covered with a swarm of drumlins, which form the country around Westport and Newport with the numerous islands at the head of Clew bay. Secondly, a low drumlin-covered ridge, 200–400 ft. high, made up of Silurian and metamorphic rocks, crosses the lowland from Westport to Castlebar; it belongs to the Caledonian structural line that is continued through Pontoon (lough Conn) to the Ox mountains. Third, Carboniferous limestones, also drumlin-covered, are developed between Castlebar and lough Mask. Although these three separate morphological units may be distinguished, this sub-region is primarily a drumlin belt with a mixture of good well-drained soils on the slopes and desolate boglands, marshes or lakes in the hollows. The surface drainage is so immature that there are a large number of lakes in the hollows, one of which, Castlebar lough, is three miles long. Some of the land between the drumlins is entirely driftless and streams such as the Aille sink into the ground to reappear some miles away.

Agriculturally, this lowland is one of small farms averaging 20–25 acres most of which are worked by the holders and some family helpers: there are very few labourers. The farming is mixed, with cattle, sheep, pigs and poultry as the main sources of income: little more than one-tenth of the improved land is cropped and approximately one-third of the area is rough grazing, either in the form of peat or bare limestones. The towns of Westport and Castlebar are market centres in origin, but also possess a few small industries: their trading activities extend over wide areas in the hilly country to the north and south. The density of population is still almost 100 per square mile of improved land, in spite of a decline in numbers by more than one-third since 1891. Though more richly endowed than the congested districts and having some fertile soils, the Clew bay lowland is a small-farm landscape typical of the greater part of Connacht.

4. THE COASTS AND PENINSULAS OF WEST MAYO

Mayo's western fringe is a complex series of peninsulas and islands, with closely settled coastal lowlands everywhere. In Achill quartzite peaks rise to 2,200 ft. and in Corraun Achill to 1,784 ft., but elsewhere lowland prevails on the rockfloor of metamorphosed sediments. Settlement and agriculture are almost restricted to a maritime belt and only a tiny fraction of the population live more than two miles from the shore. Apart from mountains and the Mullet sand dunes, the uninhabited areas consist of boggy heaths of vast extent. The farms here are almost the smallest in Ireland and have about 4 acres of arable land divided between potatoes and oats, pastures of similar extent and access to rough grazing, generally held in common. The chief produce sold consists of young cattle, sheep, pigs and eggs.

Fishing is hampered by the lack of any sales, but turf-cutting is remunerative in some places. Domestic industries are deeply rooted in this area, but the industrial developments that have arisen from it are slight though not unsuccessful. Migratory labourers leave many parts of this area, conspicuously Achill, for three months or longer. Standards of living have steadily improved in this western area, and there are now bus and lorry services to Belmullet and Achill, which had a railway connexion to Westport from 1895 to 1940. In 1937, Dr R. Lloyd Praeger wrote of Achill,

> . . . in 1898 . . . the population, far larger than the meagre agriculture and spasmodic fishing could support, was maintained by American money sent home by absent sons and daughters. Housing conditions were still very bad: Dooagh was a huddled cluster of poor thatched cabins . . . but now things are changed for with government aid all the old hovels are gone and a quite extensive village has arisen of houses not, as in so many places in Ireland, of design quite alien to the people and the environment. There is a little more tillage, due to continued peat-cutting and drainage.*

5. INTERIOR MAYO

Almost uninhabited, this subregion lies between the coastlands of west Mayo and the Ballina lowland. A few farms are scattered in cleared patches among the boglands, and small communities live in valleys along the north coast, but otherwise an area of nearly 400 square miles is unoccupied. Most of this area is a lowland which is thickly covered with blanket bog interspersed with numerous lakes. It is crossed only by roads from Castlebar and Ballina to Belmullet. The bogs rest on Carboniferous sandstones in the northeast and on the pre-Cambrian floor to the west. The Nephin Beg range consists of ice-sharpened peaks of quartzite, 1,250–2,400 ft. high and reaching 2,646 ft. in the isolated Nephin, two miles west of lough Conn. The lower and more rounded hills developed on the pre-Cambrian rocks reach the north coast in a series of cliffs, some of which are several hundred feet high.

6. THE BALLINA LOWLAND

This subregion lies between the Ox mountains and the desolate lowlands discussed above: it is connected to the lowlands around Castlebar and Swineford by the Foxford gap. It is floored by Carboniferous limestones that are covered by glacial deposits, mainly moraines north of Ballina and drumlins to the south. Good pastures and arable fields exist on the various glacial deposits, but the amount of bog on level ground increases towards the south, and in some places the fertile meadows surrounding the Moy appear as a ribbon of green surrounded by the purple-brown bogs. The farms are generally small with an average size of 25–30 acres, but several

* Praeger, R. L., *The Way that I Went*, Dublin, 1937, pp. 192–3.

demesnes have been parcelled out and added to the farmed area, partly in an effort to resettle some of the poorest farmers. The density of population is around 80 per square mile and there has been a decline of approximately one-third since 1891. The poorest parts of the area on the flank of the Ox mountains were congested in 1891 and are still at a humble economic level.

Ballina, at the first crossing of the Moy, is a small port used mainly by tramp steamers, which bring in some of the coal, timber and grain for north Mayo and part of Co. Sligo, but only about one boat a week now uses the port of Ballina. The town also has a railway terminus, but its bus connexions now bring more people into the town; there are daily buses to Dublin, Sligo, Belmullet, Killala and various other centres. Railway lorries also deliver goods through the countryside and the isolation of some communities has been lessened by these services. Ballina, with its wide hinterland, is a relatively prosperous town and increased in population from 4,800 in 1891 to its present 6,100. The wide range of small industries is noted on p. 413. There is an export trade in the famous Moy salmon and eel fishing survives on the river; the fish are trapped at weirs, kept alive in large floating tanks in the river, and then shipped to London and Liverpool during the winter months. The commercial side of fishing is better developed here than in many parts of Ireland; in addition, Ballina is visited by many anglers for its salmon and trout fishing. The town is generally said to be more prosperous than many of its kind, yet it lives by trade rather than industry and has no new large-scale enterprise brought by the recent government programme. Its commercial strength lies largely in its widespread contacts and it has no rival of any consequence in an area covering at least 300 square miles, with a population of some 40,000. A small regional capital, it has a weekly press, five banks, hospitals, a Roman Catholic cathedral and several hotels, but its prosperity contrasts strongly with the poverty of most of its hinterland, excluding the richer areas of the Moy valley. By contrast, Killala, an older place, is now a decayed village of some 340 people with a famous round tower and a Church of Ireland cathedral. The harbour has silted up and now is used only by fishing boats, but there are many warehouses built to store goods in transit. The village now has empty houses, with deserted shops and stores.

THE ARAN ISLANDS

The Aran islands stretch across Galway bay, forming a barrier running roughly northwest–southeast; there are three main islands – Inishmore, the largest, Inishmaan and Inisheer. All three are closely settled in relation

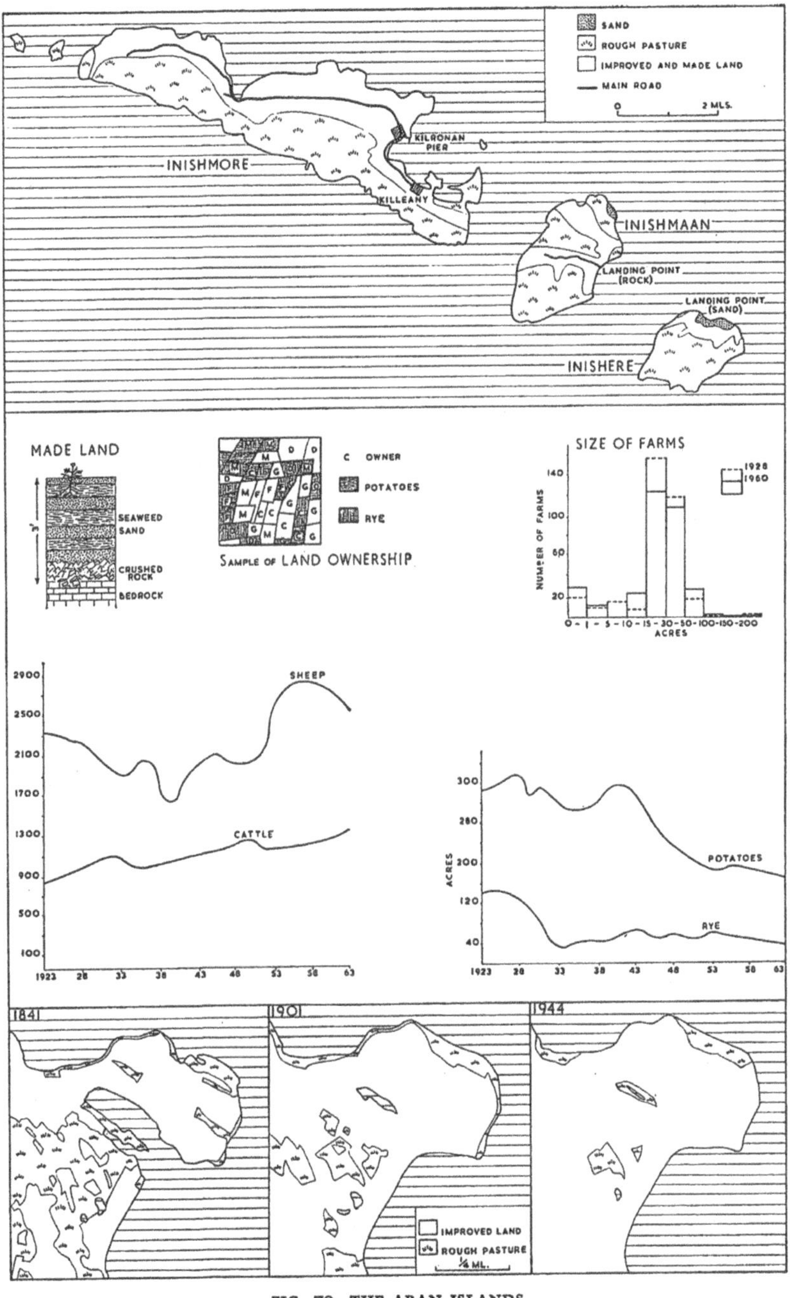

FIG. 79. THE ARAN ISLANDS

The area of rough pasture in 1841 and 1901 for part of Inishmore (northeast of Kilronan) are taken from the 6″ to 1 mile maps; the 1944 observations were made by Joan M. Ward, B.A. Note the soil profile in 'made land' and the ownership of fields by five men.

to their resources, for they consist of bare Carboniferous limestones, laid down in nearly horizontal beds and appearing at the surface as a series of grey terraces having limestone pavements comparable with those of the Burren in Co. Clare; between the terraces there are beds of shale, now fertilized by sand and seaweed. The stony appearance is accentuated by a number of erratics, largely of granites, brought by an icesheet from Connemara, by the numerous archaeological remains of prehistoric and early Christian times, such as large forts, and finally by the stone walls, 5–6 ft. high, built around the tiny fields. Stone is abundant everywhere, even in the memorials built by a people at once pious and filially minded to departed members of their families. The cliffs on the southwest side reach 200 ft., but the land slopes gently towards the inhabited northeast, and the human outlook is towards Galway and Connemara rather than to the desolate Burren coast which physically it resembles so closely. As in Clare, a large part of the flora is confined to the joints in the limestone pavement, which have been widened by solution and have a thin vegetable soil at the bottom. The maidenhair fern (*Adiantum* sp.) with leaves as much as 2 ft. long is frequent in cracks, and gentians (*Gentiana verna*) are also common: Aran, like the Burren, shows an intermingling of northern and southern types. There are no trees on the islands, but shrubs are found in rock crevices among the low-growing herbs, where they manage to resist the onslaughts of high winds and grazing animals. The shrubs include the spindle tree, blackthorn, guelder rose, hawthorn, oak and hazel.

Unpromising as this environment may appear, it supports an Irish-speaking community of 1,615 people in an area of eighteen square miles, of which only one-third is cultivated. In all the circumstances it is remarkable that any land is cultivated at all, but the islanders have made their 'gardens' (fields) by piling layers of sand and seaweed on the top of crushed rock (Fig. 79): the fields are laid out in lazy beds and in some parts of the islands the cultivated area has been steadily extended. Fig. 79 (lower part) shows an area near Kilronan now almost covered with arable fields, not necessarily sown each year. The farmers' chief crop is potatoes intended for foodstuffs, which grow well in the limey soil of these islands, and may fetch good prices as an early crop in some seasons. Oats are grown but rye is the traditional cereal, used for fodder and – to a diminishing extent – for thatching. The main form of rotation is an alternation of rye and potatoes, though land is left as pasture for several years after it has been used for crops. Everywhere the holdings are small and dissipated into a number of separate fields. Fig. 79 shows a patch of land covering approximately 70 acres which is divided into nearly forty fields owned by five separate farmers, none of whom live on their land. Cattle and sheep are pastured on the farms and on any waste patches that will support them including much of the limestone country on the Atlantic coast. Few pigs are kept now and poultry are reared only for domestic use. The cattle are taken to Galway for sale and beef is hardly ever consumed, though mutton and pork appear on the table at intervals.

The cash resources of the islanders are slender and fluctuating. They are almost self-sufficient in their economic life, as they grow or catch their own food, make their own clothes and thatch their own houses. The food consists mainly of potatoes, with eggs, bread, butter, tea and fish caught by lines suspended over the cliffs or by hand lines or small nets from the curraghs. These boats are made locally of a lath framework, either of driftwood or bought timber, covered with tarred canvas, which is usually old flour sacks. Curraghs are used not only for fishing, but also for transport to the mainland or to take people, cattle and surplus potatoes to the steamer. The boats spend four years on the open sea and four more in coastal waters, and are generally seaworthy: when not in use, they are carried on to the shores and rested upside down on flat stones. The men of Aran fish the waters around the islands, but their cash profits are now negligible, as the export trade to Britain and the United States, which flourished during the 1914–18 war, was lost in 1920. But lobster fishing brings some profit.

Clothing is made of wool spun from sheep fleeces which is then knitted or woven into garments of distinctive design. The men wear homespun trousers, sombrely dyed, a knitted sweater and a waistcoat, sometimes sleeveless, also of homespun tweed. This tweed has recently acquired some popularity in Dublin and is sold by the islanders to tourists. The traditional type of shoe is a 'pampootie', a kind of moccasin made from cowhide and bound together with strips of the same material. Another local craft is basket-making from willows or coiled straw rope. Even furniture is made on the islands: in some of the poorer cottages only furniture made from driftwood is seen. Unlike some rural communities, the people of Aran are house-proud and will use financial windfalls for improving their houses or building new ones: many were built after the 1914–18 war and a number of two-storey slated dwellings have been erected with government subsidies since then. One past industry, which had a brief revival due to war scarcities, is kelp burning. Seaweed was collected from the rocks or collected from the shore and carried beyond high water mark and laid out for drying. When dry, it was laid out in stacks 6 ft. high, 6 ft. wide and 8 ft. long, each of which weighed about four tons and yielded one ton of kelp by burning during a specified period of July. The price paid for kelp varied according to the iodine content: based on the gathering of seaweed it was an occupation favoured mainly when fishing was unprofitable.

An aura of romance is frequently cast over the life of the Aran islands and not without reason, for its people still possess a degree of self-sufficiency that is rare in the modern world. As a result, some sixty years ago,

> much of the intelligence and charm of these people is due to the absence of any division of labour, and to the correspondingly wide development of each individual, whose varied knowledge and skill necessitates a considerable activity of mind. Each man can speak two languages. He is a skilled fisherman, and can manage a curragh with extraordinary nerve and dexterity. He can farm simply, burn kelp, cut out pampooties, mend nets, build and thatch a house, and make a cradle or a coffin.*

At present one major resource of the islanders, who need at least some money to buy fuel from the mainland as they have none on the island except cowdung, is cattle, which are shipped by steamer to Galway. There is also a profitable trade in potatoes, including some early varieties grown successfully in the mild climate; this trade, however, has fluctuated considerably in recent years. But fishing has failed, kelp burning has ceased, the selling of homespuns may suffer through changing fashions and the tourist industry is precarious and in any case not invariably helpful. On the other hand, there has recently been a considerable increase in the number of visitors, many of whom only make the round trip by boat from Galway to the three islands. Recent experiments in economic reorientation include the rearing of turkeys and battery chickens, and the production of flower bulbs for the Irish market. From the nineteen-twenties large quantities of cement have been imported, both for house building and for the water troughs to conserve supplies for dry periods. Although there is no suggestion that the islands should be evacuated there has been considerable emigration in the past and will be more in the future.

* Synge, J. M., *The Aran Islands*, 1910, pp. 156–7.

Material on the modern economic geography, including the data on stock and crops has been acquired from the unpublished thesis of Mr M. B. Gahagan, B.A. (1964).

CHAPTER 20

The Drumlin-Drift Belt

This closely settled belt of country, a hundred miles long and forty wide, extends from Dundalk to Sligo bay: primarily lowland, it is covered with drumlins and ridges of drift, though in Sligo and Leitrim Carboniferous strata survive in residual hills reaching 2,000 ft., which form the only striking gaps in the settlement pattern. On the south the drumlin-drift belt is defined by a line drawn from the Clew bay lowland through Mayo into north Roscommon and thence into Leitrim, Cavan and Louth. On the north, however, it is much more difficult to discern a regional boundary on a morphological basis, for the drumlins of the Erne valley are continued into south Donegal and those of Monaghan and north Louth belong to the great drumlin swarm that extends through the heart of Ulster to Strangford lough and the suburbs of Belfast. But if there is no clear morphological boundary, the Border between the Republic and Northern Ireland provides an effective human and economic divide. Here the drumlin-drift belt is divided into two sections: (*a*) the *western*, which includes the Carboniferous plateaus of Leitrim and Cavan, with the lowlands around the Ox mountains; and (*b*) the *eastern*, which consists of the hilly country of Ordovician and Silurian rocks in Cavan and Monaghan, and the low-lying Carboniferous-floored areas between them.

A. *THE WESTERN HILLS AND LOWLANDS*

The western part of the drumlin-drift belt, from Sligo bay to the upper Erne valley, covers nearly 1,900 square miles, seven-tenths of which is cultivated land. Much poorer than the eastern part, this region includes some areas under the care of the Congested Districts Board in 1891 and in 1966 had a population of 117,740: its agricultural lands are closely settled and dominated by small farms. The non-cultivated land consists partly of uplands, varied in structure but rising generally to *c.* 2,000 ft., and also of the peat bogs which lie between the drift-covered ridges and drumlins.

LANDFORMS AND LAND USE

The uplands of this region include the Ox mountains, a Caledonian fold with pre-Cambrian metamorphic rocks intruded by granite, the Curlews, an Armorican fold with Old Red sandstones and Silurian beds, and also a number of impressive plateaus formed of Carboniferous strata. Lowlands stretch around these various hills and penetrate them in constricted and frequently steep-walled valleys: the uneven rockfloor has been plastered by a variety of drifts with the result that level surfaces are few and the lowland usually possesses an undulating, even a hummocky character. A number of lakes diversify the scene still further, especially in the upper Shannon valley and that of its tributary, the Boyle: in addition, there are many peat bogs which are, however, generally of small extent. The Ox mountains have rounded summits from 600 to almost 1,800 ft. high west of the Collooney gap, and maintain their identity as a number of isolated hills on the shores of lough Gill to Manorhamilton at the foot of Benbo, 1,367 ft. high. The maximum width of the Ox mountains is seven miles, but they form a clear population divide between the closely settled lowland on either side: the only important break is the Collooney gap, which carries the road and railway to Sligo. The mountains are a desolate expanse of spongy bog, which has a depth of 2 ft. on the numerous massive erratics that occur on the high ground. An Armorican range, the Curlews, reaches 867 ft. near Boyle and is continued to the east and west in low hills 500–700 ft. high. Carboniferous plateaus 1,000–2,000 ft. high are formed of a variety of sandstones, shales and limestones, with some Coal Measures. Almost everywhere, the summits are flat and the valley sides steep, though sometimes cliffed due to the outcrop of resistant beds. A few valleys, notably Glencar (near Sligo), show the effects of many landslips and have long sheets of talus beneath the crags. This area was near the centre of the icesheet at one phase and its deep wide valleys show clearly the effects of ice moulding.

Relics of the natural vegetation include extensive bogs and patches of natural woodland. Bog covers most of the Carboniferous plateaus and is especially well developed on the Ox mountains. On the top of Carrowkeel, a hill overlooking lough Arrow, it occurs in hollows within a limestone pavement. The ground flora of the bogs varies considerably: heathery moors are frequent on the hills and indicate good natural drainage, as for example on Carrowkeel, where rainfall percolates rapidly into an underground drainage system. The accumulation of bogs is due partly to the high rainfalls, which are probably at least 60″ per annum in almost all the uplands. In the lowlands the curious arrangements of the drifts has meant that many hollows have become natural sites for bogs and lakes: in south Leitrim, for example, the whole scene is a mixture of parallel tracts of farmland and of bog or lake, and in Mayo peat bogs form a sea of waste between islands of cultivated land.

Patches of natural woodland survive around lough Gill and include among their constituents the arbutus which, though undoubtedly native, is separated by 160 miles from its next station in Co. Kerry. The arbutus occurs here on sites similar to those around Killarney, such as rocks and rough ground close to lough Gill, and on islands in the eastern half of the lake; it is frequently associated with the yew. The arbutus survives on limestones, sandstones and metamorphic rocks alike and cannot, therefore, be classed as either calcicole or calcifuge. Two types of oak forest, which have respectively calcicole and calcifuge affinities, have been recognized around Sligo: in the calcicole type, oak occurs with ash and hazel and a rich ground flora that includes anemones, primroses and bluebells; in the calcifuge type, oak occurs with birch, holly and hawthorn and the ground flora is much poorer with bilberry (*Vaccinium myrtillis*), sometimes bluebell, and wood sorrel: there are many grades between the two extremes of rich and sparse oak forest. Many of the sheltered glades have a remarkably rich growth of ferns and mosses, and garden plants such as *Montbretia* are naturalized in some places. The luxuriance of growth recalls that of Kerry under generally similar but slightly warmer conditions: peat bogs grow rapidly on glacial erratics, walls have an abundance of plant life and even the thatched roofs of neglected or abandoned cottages are quickly overgrown.

Rough pasture covers approximately one-third of this region, and consists mainly of bogs; more than half is attached to farms and forms a useful supplement to the enclosed fields. The extent of rough pasture has varied considerably during the past century, especially in Co. Leitrim, where it was 28% of the area in 1851, only 22% in 1891, but 26% in 1967: this loss of ground to rough pasture is a feature common to all the upland areas of Ireland. So far some 65 square miles has been acquired for afforestation, and several of the plantations cover old woodlands, including some treasured relics of the native vegetation or areas immediately beside them. In Collooney, near the gap carrying the road to Sligo, some 4,700 acres have been bought for forests, but almost one-third of this area is bare rock or heathland that will not support trees. Some of the woods around lough Gill have been incorporated in a forestry estate of 1,700 acres. The largest new plantations are in the Ox mountains and in the

Carboniferous uplands, including Manorhamilton, Drumkeeran, Ballyfarnon, Glenfarne and Swanlinbar. The general aspect of the landscape is open, with farmlands and bogs stretching to the hills limiting the lowlands on every side.

Agricultural land classed as 'improved' covers seven-tenths of the region and forms the main resource of the inhabitants. Everywhere the small farm is dominant, the farming pastoral in emphasis and the labour force small, but there is little employment available in non-agricultural occupations, except in the few towns. At the lowest financial levels the rural population are smallholders living on poor and small farms but at the higher financial levels they are farmers of 40–50 acres with a reasonably prosperous existence: nearly two-thirds of the whole farmed area is divided into holdings of 15 to 50 acres. Economically the farmers depend mainly on cattle. Crops grown on hardly one-tenth of the improved land are mainly oats and potatoes set in carefully manured fields or parts of fields. Spade cultivation, which is fairly general here, restricts the arable land to a few acres and means that the cattle depend chiefly on the pastures rather than on any feedstuffs grown on the farm. The acreage of meadow has, however, gradually increased and now covers as much as one-fifth of some holdings. The cattle farming has two main objectives: milk production and the sale of young beasts at fairs. The dairy farming of this region is less industrialized than that of the Munster creamery belt, although one-quarter to one-half of the cattle are milch cows and over half the cows in Co. Leitrim and nearly three-quarters of those in Co. Sligo are suppliers of milk to the creameries or separating stations. But there are many farms too remotely placed for milk to be taken to a creamery profitably and outside the creamery network the farmers depend mainly on the supply of young beasts at fairs.

Dairying is a foundation for the store cattle trade and many of the young beasts are sent to the 'plains of Boyle', limestone grasslands with nourishing pastures to the east of Boyle; from this area they proceed through the Central Lowland to their ultimate destination. The number of cattle kept varies from 24 to 27 per 100 acres of crops and pasture, but few of these are more than two years old. Sheep are much less numerous than cattle except on the Ox mountains and on some of the upland limestone pastures, but until recent years very large numbers of poultry were kept: very few pigs are kept. A farmer of 30 acres might possess 10 cattle, of which 4 would be milch cows and the rest young beasts for sale, 30 hens, and in some places, a few sheep and pigs. Cash returns are small, but few farmers at this level depend solely on their holdings, for they either migrate seasonally, or vicariously through the sons, or receive periodical remittances from relatives abroad, or work intermittently in the neighbourhood of their homes, as carters, road menders or on larger farms. Most of the people are obviously poor, and emigration is proceeding at so rapid a rate that many areas lost one-fifth or even more of their population between 1936 and 1966.

POPULATION

The total population is 117,740 (1966), 63 to the square mile, compared with 254,000, 137 to the square mile, in 1891, a decrease of 54%. As in other parts of Ireland this loss was merely spread from one area to another: in Co. Sligo, for example, comparison of the figures for district electoral divisions in 1891 and 1936 showed that some had decreased by as much

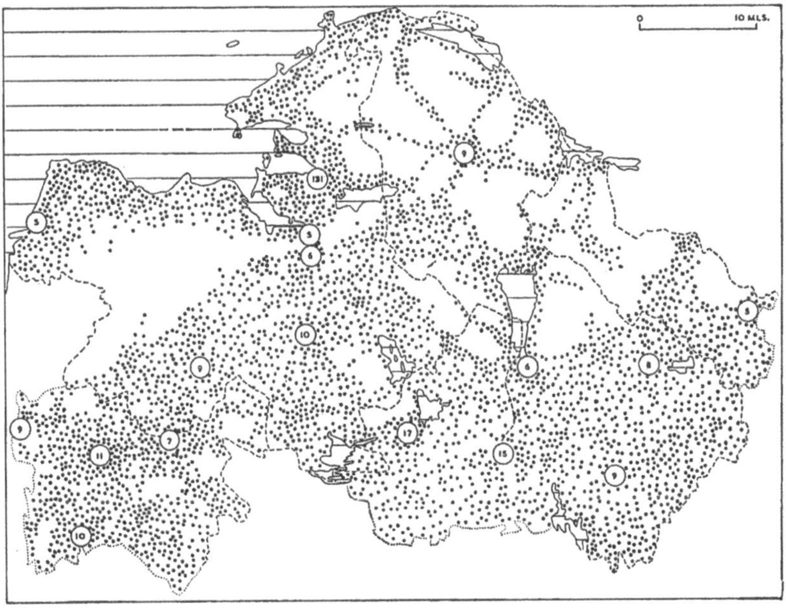

FIG. 80. POPULATION DISTRIBUTION, WEST PART OF THE DRUMLIN-DRIFT BELT

One dot to 25 persons

The settlement is comparatively close, though significant blank areas include the Carboniferous uplands and stretches of lowland bog. There is some measure of similarity between this map and Fig. 76 (p. 397), which shows the population distribution in East Connacht.

as one-half and others by as little as one-tenth. Recently, however, decline has been heavy and almost universal: from 1956–61 Co. Leitrim's population declined by 9·7% and from 1961–6 by 8·7%, and all save three of its seventy-eight district electoral divisions showed a decrease and all the twenty-one divisions of Swinford rural district declined, with an average of −10·5% (1961–6). But the town and village population, 32,450 in 1966, has changed little in recent years and now numbers more than one-quarter of the whole compared with only 11% in 1891. Over most of the lowlands the settlement pattern is fairly even and interrupted only by stretches of bog (Fig. 80). In some places the farmhouses line the roads that pierce the reclaimed bogs, conspicuously in the north-east and northwest of Co. Sligo, and in many cases the houses are placed on the roads around the drumlins, over which the cultivated fields extend continuously. While the density of population is much lower than that of west Donegal or west Connacht, it is higher than that of the east of Ireland, where the larger farms support fewer people than a similar extent of land divided into smaller holdings.

The towns are mainly market centres serving large areas of country with a low buying power, and there is little sign of rural industry, except for creameries with branch separating stations and a few small mills. In Co. Leitrim, Carrick-on-Shannon has a curious range of factories, providing electrical accessories, marquees, jewellery with souvenirs, and toys. At Manorhamilton (890) there is a knitwear and glove factory and at Drumshanbo (540) a jam factory. The town of Sligo (13,400), one of the minor regional capitals of Ireland, is a small port and has a strong wholesale and retail trade. Its factories produce bacon, flour, frozen foods, clothing and drawing instruments. There is a long-established mill at Ballysadare. At Collooney the joinery and fibre glass industries have been introduced and at Tobercurry (940) there is a lock and hardware factory and also a precision engineering and tool works. Boyle (1,800), in the north of Co. Roscommon, has a grain mill. There are few signs of growth in Sligo, and it seems likely to remain merely one of the larger market towns rather than an industrial focus, though it is also an administrative centre, a rail terminus, and an important road junction, with buses travelling through Mayo as well as the adjacent counties of Leitrim and Donegal.

SUBREGIONS

This region falls into four subregions, each of which is a distinct morphological unit and in some ways a human entity also. The Ox mountains, with their clear Caledonian southwest–northeast line, divide the drift-covered lowlands into two parts, linked only by a few deep gashes to which the name 'alt' is given. The Carboniferous uplands stand up as plateaus above the wide valleys drained by tributaries of the Erne, the upper Shannon and the Bonet river: smooth summits, sandstone and limestone scarps, drift-strewn valleys and lakes in glacier-scoured troughs, all combine in a hilly landscape with a population farming smallholdings in remote glens. Everywhere agricultural activity dominates the scene, villages are few and small, and towns insignificant, apart from Sligo and Boyle; for this is a country of simple farming, almost untouched by industry and economically at the stage of primary production from the land.

1. THE SLIGO–LEITRIM COASTLANDS

A continuous belt of lowland extends from the mouth of the Moy estuary to the mouth of the Erne, with a coast mainly composed of low cliffs of Carboniferous limestone interrupted by sand dunes, notably on Sligo bay. The coast is broken by three estuaries, which allow tidal waters to reach Sligo, Ballysadare and Drumcliff, but there is little fishing though Sligo salmon is widely famed. Generally two to five miles wide, the lowland is a closely farmed, densely settled drift landscape with many drumlins over a rockfloor of Carboniferous beds. The farms average 20–30 acres, but many are much smaller and the houses are very closely spaced along the main roads and almost equally frequent along the subsidiary roads. Almost entirely congested in 1891, this lowland has since seen a large reduction of population and many of the small farms have been united: for example at Grange near the foot of Ben Bulben, the Congested Districts Board made a series of strip farms with a road lined by farmhouses at its upper limits. The holdings, all of which slope down to a common salt marsh grazing on the seashore, were originally one field wide, but over one-third of them have since disappeared by successive amalgamations. Though extreme congestion has vanished from this lowland, it is still primarily one of small farms having little cropping and depending mainly on cattle, sheep and poultry, with access to peat bog and hill grazing on the Ox mountains and the Carboniferous plateaus. Some farmers cut turf for sale, especially on the large bog flanking the Ox mountains, and many work seasonally on the roads. Along the coast there are a few hotels, especially at Rosses Point, which has a famous golf course on the sand dunes, and at Strandhill.

2. THE OX MOUNTAINS

This Caledonian ridge is an uninhabited moorland cutting across the region from the Moy valley to Benbo mountain, near Manorhamilton. Monotonously even in outline, it is composed of varied metamorphic and igneous rocks, and is bog-covered almost everywhere; in addition, a bog covers approximately twenty square miles of drift-covered Carboniferous limestones on the north of the ridge, interrupted only by narrow ribbons of farmland along the valleys. Towards the Collooney gap, quartzites appear in low ice-moulded hills 500–900 ft. high and the ridge overlooking lough Gill is broken by a number of 'alts', or deep gashes. These hills are either bare rock, heather moor, or rough natural woodland, some of which have been replaced by neat plantations.

3. THE CARBONIFEROUS PLATEAUS AND VALLEYS

Two worlds are here: deserted, generally bog-covered uplands 1,500–2,000 ft. high, and valleys, mostly a mile or more wide, with small farms and numerous roads. Where the valley sides are less steep than in Glencar or Glenade farms climb to an altitude of 600 ft. or more, but for the most part the plateau provides only sheep pastures or remotely placed reserves of peat. The Coal Measures preserved here include workable seams in the Arigna district, east of lough Allen. In spite of the heavy declines of population, generally by as much as one-half since 1891, the holdings remain small, with an average of 15–30 acres, and the population density is 100–150 per square mile of improved land. On each farm a few acres are tilled with the spade, and sown with oats and potatoes, but the economic staples are farmers' butter, young cattle, sheep, poultry and eggs. The monthly fairs are of considerable importance for stock sales in this area and the ordinary commercial life depends very largely on periodical visits to the market villages which, though small, perform the functions of towns.

The Arigna Coalfield. This coalfield stands second in production and employing capacity to the Castlecomer field, and its exploitation was accelerated by wartime fuel shortages. In the past most of the coal was used for domestic fuel, and much of it taken to Arigna station, the terminus of a branch line from Ballinamore. In the middle of the nineteenth century, local coal was used for an ironworks at Drumahaire. The coalfield is remotely placed beside lough Allen, and in such a locality rare visitors are surprised to see miners and remains of past industrial activity. The seams outcrop on the flanks of hills beside the lake, and are so narrow and of such limited economic value that in the past their working was restricted to periods when imported coal was scarce. The coal is now used for power generation.

4. THE DRUMLIN-DRIFT LOWLANDS

These lowlands stretch from the upper Moy valley across the Shannon to the upper Erne and are bounded on the north by the Ox mountains and the Carboniferous plateaus, but on the south merge gradually into the lowlands of Central Ireland, from which they are distinguished by their numerous drumlins and their exceptionally close settlement. Most of this lowland is limestone-floored, and its drumlins are fertile, well-drained mounds sown mainly to arable crops with cattle pastures on bogs in the depression between them. The drumlin landscape may be seen in a well-developed form around the Curlew mountains, part of a ridge of Old Red sandstones and igneous rocks 700–800 ft. high that intersects the lowland around lough Key and lough Gara. It may also be seen near Keishcorran and Carrowkeel, two limestone hills 1,000 ft. high to the west of lough Arrow. The drumlins farther east around the Shannon and towards the Erne have so disorganized the drainage system that a complex of peat bogs, marshes and lakes separates them one from another. But more detailed survey work is needed to distinguish gravel drumlins from those with rock cores, or again from sheets of drift plastering the many low ridges of solid rock that floor this lowland.

As a whole, the lowland is a drift-covered expanse, mainly 200–400 ft. high, with a complex interpenetration of cultivable land and rough pastures or lakes. The proportion of rough pasture decreases from west to east: in Swinford rural district (Mayo) it covers 30%, in Mohill, Co. Leitrim, 15%. This contrast is closely reflected in the population distribution map but Swinford still (1966) has 111 to the square mile of improved land as compared with 72 in Mohill. Both rural districts have lost about two-fifths of their population since 1891 and are losing it still, but Swinford, with an average farm size of 20 acres, remains nearer to the level of a congested district than Mohill, with a mean farm size of 28 acres. Everywhere the loss of population has been heavy, and everywhere small farms run by the holder with some family worker dominate the scene. This subregion, however, can only maintain so dense a population by agriculture on a small farm basis with low cash incomes derived from the sale of milk or butter, young beasts, poultry and eggs.

Rural industries are almost unknown, but the creameries so far established provide a useful economic outlet that removes poverty from some farms. The towns and villages in this lowland have only one-tenth of the population: the largest of them is Boyle, in the centre of a good pastoral district, and the others are mere villages, such as Swinford (1,120) and Kiltamagh (900) in Co. Mayo; Tobercurry (940) and Ballymote (920) in Co. Sligo; Carrick-on-Shannon (1,640), Ballinamore (740) and Mohill (870) in Co. Leitrim. Each of these 'towns' has its own market area and is separated by several miles from its nearest neighbour, but, though indispensable, they must inevitably remain small centres of ordinary shopping enlivened by periodical fairs. The fairs also take place at various small villages, most of which have 100–300 inhabitants, and at other chosen points in the countryside which are dignified by long-standing tradition.

B. *THE EASTERN HILLS AND LOWLANDS*

The eastern part of the drumlin-drift belt includes the whole of Co. Monaghan, all of Cavan except the northwest, the northern part of Co. Louth and the Carlingford peninsula. The county boundaries which form part of the Border are regional limits only in a political and economic sense for they show a supreme disregard of physical features: thus, the Cavan–Fermanagh boundary crosses the upper Erne valley lowland, and the Monaghan boundary winds through the wide Finn corridor-valley, climbs to the top of the Slieve Beagh at 1,200 ft. and then descends to cross the Blackwater valley. In effect the county boundaries divide the Erne valley, a natural corridor, between the two political units, and also make a frontier in the middle of the closely settled Blackwater valley half-way between Monaghan and Armagh. The boundary between the counties of Armagh on the one hand and Monaghan and Louth on the other also passes through a densely peopled agricultural landscape; having penetrated within four miles of Dundalk, it swings round the base of the Carlingford peninsula and runs through Carlingford lough where, during the 1939–45 war, the inhabitants of Warrenpoint looked from their darkened homes to the lights of Omeath across the water.

Generally hilly, this region consists almost entirely of improved farmland: it differs from the western area discussed on pp. 424–34 in various ways. In the first place, it is of much greater agricultural productivity, for its soils bear better crops and the land is assessed at approximately twice the valuation per acre of areas farther west. More land is cropped and the agricultural scene is similar to that across the Border where the farms, though small, are very actively cultivated and support a considerable number of cattle in proportion to their area. A second difference between the eastern and western parts of the drumlin-drift belt lies in the proportion of the rural and town populations: in the west only 28% live in towns and villages but in the east 43%. The market towns are larger, more industrialized, and in most cases increasing at least slightly in population. Dundalk, the largest centre, is a more active town than Sligo, its equivalent as a minor regional capital and port in the west. Arising out of this more equitable balance between towns and farmed countrysides, one may note a third difference: this eastern region shows traces of an industrial tradition spread through the whole area, though now concentrated in the towns. Though flax processing and milling are now extinct it is still possible to recognize at least some traces of that interpenetration of farming and rural industries that is associated with northeast Ireland or, in effect, the historic province of Ulster.

LANDFORMS AND LAND USE

Ordovician and Silurian beds, comparable with those forming the Southern Uplands of Scotland, form irregular hills from 400 to 900 ft. high in a band of country approximately twenty miles wide between the Dundalk lowland and the Monaghan corridor. This hilly country, with its east–north-east trend, which is continued through Armagh and Down and finally appears on the south side of Belfast lough (pp. 17–18), is of irregular outline due to the uneven nature of the rockfloor. The drift cover has partly concealed the underlying sharp features, and provides soils varied from light and stony sands to heavy clays. The post-glacial drainage is complicated, with a number of small lakes and peat flats. Around Dundalk, Silurian rocks form a lowland, but in the Carlingford peninsula igneous rocks, intruded in Tertiary times, form rugged mountains.

The lowland that stretches from the upper Erne through Monaghan to the Blackwater valley is floored by Carboniferous limestones covered by drifts, mainly in the form of drumlins. In central Cavan the Erne spreads out into lough Oughter, a lake which is an intricate series of channels separating drumlins one from another: several smaller lakes lie between drumlins, but some of them are mere ponds and in many places there is merely a peat flat. Within the lowland corridor, the drumlins are orientated quite independently of the rockfloor but their trend, southwest around Clones, south around Monaghan, and southeast around Carrickmacross, bears a direct relation to the glacial striae. In some instances the drumlins coalesce into one another, forming a landscape of numerous little hills with highly varied conditions of soil and drainage. It is not certain, however, that all the so-called drumlins consist entirely of drift for some may possess a rock core: nor is it possible to make a clear distinction between actual drumlins and ground moraine plastered on an uneven rockfloor, such as that existing in the Ordovician and Silurian belt.

There are few relics of the natural vegetation in this region, as almost the entire area has been absorbed by farmlands. Cultivated land occupies virtually all the Ordovician and Silurian hilly country, and also extends up the Carlingford mountains to an altitude of 700 ft. on suitable slopes, though in these mountains many of the steeper slopes are uncultivated and some of the level ground is also left as rough pasture on account of its waterlogged condition: it is the moderate slopes that are favoured, because they possess good natural drainage. The Carlingford peninsula has the largest forestry plantations within the region; already they cover over four square miles. There is also a large forest area, of nearly five square miles, at Scotstown, Co. Monaghan. Another large forestry estate, at Cootehill, consisted of four square miles of wooded demesne, virtually all of which is now covered with plantations. Smaller demesnes have been acquired and at least partially planted at Monaghan, Belturbet, Killashandra, Bailieborough and Kingscourt. There are other demesnes that might be used for forestry plantations, but this region will never possess large forests. Typically, it is one of small farms interrupted only by a few patches of bogland and by very occasional woods.

Farms of 30–40 acres are the main element in the agricultural landscape, and most of them include a few fertile fields given to crop production. Mixed farming prevails and the main economic staples are young cattle, milk, eggs, pigs and poultry. There is a general average of some 30 cattle per 100 acres of improved land: in Co. Cavan and Co. Monaghan 33% of all the cattle are milch cows. The specialization on dairying is seen in the prevalence of creameries, for in Co. Cavan over four-fifths and in Co. Monaghan seven-tenths of the cows provide milk for the creameries: even so, the concentration on dairying is less complete than in Munster for the trade in young stores is important and favoured by many farmers: the young stores are sent to grazing and fattening pastures, chiefly in Co. Meath and its appendage, Co. Westmeath. Sheep rearing is prominent in the Carlingford peninsula which has the only upland pastures of any considerable extent in the area.

Arable farming is fused with livestock production, especially towards the east. Interesting transitions, or agricultural gradients, may be observed: in the west, nine-tenths of the land is in pasture or meadow, but in the east four-fifths. There is also an increasing dependence on arable feedstuffs as one moves eastwards; very gradually, but none the less clearly, the transition to the intensive small-farm economy characteristic of Northern Ireland is seen. The main arable crops are oats and potatoes, but barley and turnips are grown very infrequently except in the east around Dundalk and Carrickmacross. Some farmers have a surplus of oats or potatoes, and in a few cases a considerable part of the annual income is derived from their sale: oats are ground locally in various mills scattered through the countryside. Flax growing was formerly widespread in this area and through the countryside one encounters the retting troughs in streams and the small scutching mills once worked by water power. Little flax has been grown since the 1939–45 war. And the total cropped average has declined to less than 5% of the improved land in Cavan and less than 10% in Co. Monaghan. The obstacles to arable farming lie in the prevalence of small farms, whose holders do not generally keep much labour-saving machinery, and also in the nature of the drumlins. Even a single drumlin may possess varied soils, which may be well drained and coarse-textured near the top, clayey and waterlogged towards its foot: if the slope of a drumlin is steep, cultivation may be so difficult that the land is best used as pasture and meadow.

INDUSTRIES

This region shares with the remainder of Ulster a tradition of home industry, chiefly in linen making. The farmers of Co. Monaghan grew their own flax, prepared it, and made it into linen until the early part of the nineteenth century, when factories acquired the trade. The small farms provided food as well as flax, but most of the income came from the domestic industry and the fields were in some cases neglected. But the old scutching mills have now gone with the virtual extinction of flax growing. Most of the old rural mills have gone also, though there are survivors in the villages, including Ballyconnell (520) and Killashandra (408). Creameries are scattered through the countryside in villages and provide, here as elsewhere, a direct contact with the Irish and export markets. Villages are relatively numerous in the region, and a few of them have some industrial life such as Emyvale, Co. Monaghan (270), which has footwear and furniture factories, and Shercock (290) which has a factory whose products are advertised as blankets, tapestry and tie silk. The Triassic clays at Kingscourt (840) have been used for the brick and tile industry and here too cement preparation has been developed.

Most factories are in the towns, some of which are small. A few have a surprisingly large variety of industries such as Belturbet (1,100), which has shoe making, two factories providing chemicals such as adhesives, a firm producing artesian well and other boring equipment, and a mineral water plant. Comparably, Bailieborough (1,170) has an engineering works, an iron foundry and enamelling works, another making bolts and nuts, a footwear factory, and the allied industries of sawmilling and furniture making. Ballybay (1,100) has knitwear and tanning industries, and at Carrickmacross (2,100) there is a well-established footwear works and factories providing clay goods and grain driers, the last opened in 1963. Castleblayney (2,100) has factories providing footwear, girls' clothing and tubular steel furniture with a large milk products factory combined with milling, seed preparation and exporting and agricultural engineering, as lough Egish in the countryside. In Cavan (4,200) there are bacon, mineral water and tractor rebuilding factories, and at Clones (2,040) there are knitwear and tile factories. A variety of textiles are produced at Cootehill (1,500), and in Monaghan (4,020) the industries include furniture, bacon, milling and clothing.

Dundalk (21,700), a minor regional capital, a port, and a thriving commercial and industrial town, gives employment far beyond its boundaries. As a port is is used mainly by coal boats and most of the raw materials come by road and rail. Dundalk stands half-way between Dublin and Belfast though its significance as a railway town has been reduced by the closing of the lines to Enniskillen and to Greenore (p. 248). There is an ironworks in the town, but many of the manufactures are allied to the distributive trades; among the products are beer, mineral waters, furniture, motor bodies and sawn timber. Agricultural demands are met by factories for producing fertilizers, for making egg cases and even sheep branding fluid. Agricultural production is seen in mills producing meals and wheat foods and also in bacon curing. There is a large tobacco factory and the textile industry is represented by a linen mill and works making a variety of cotton and linen goods. But the largest factories, some of which have 1,000 employees, are those making boots and shoes, ranging from the heaviest types worn by farmers to the smartest ladies' wear. There are, in all, several thousand factory workers in Dundalk, but it remains essentially a market town in appearance and lacks any definite factory quarter. In size, and in its industrialization, it is comparable with Drogheda (p. 276), but Dundalk's port and river hide themselves discreetly from the town whereas the quay-lined Boyne dominates Drogheda. At Ardee (2,920) in Co. Louth there is a timber industry with chair making and a cotton spinning and doubling enterprise.

POPULATION

Fig. 81 shows that the population is distributed fairly evenly over almost the whole area, for there are few unoccupied patches except for the mountains of the Carlingford peninsula. Close as the settlement may appear both on the ground and on the map for 1961, given below, the population outside towns and villages has declined by more than half

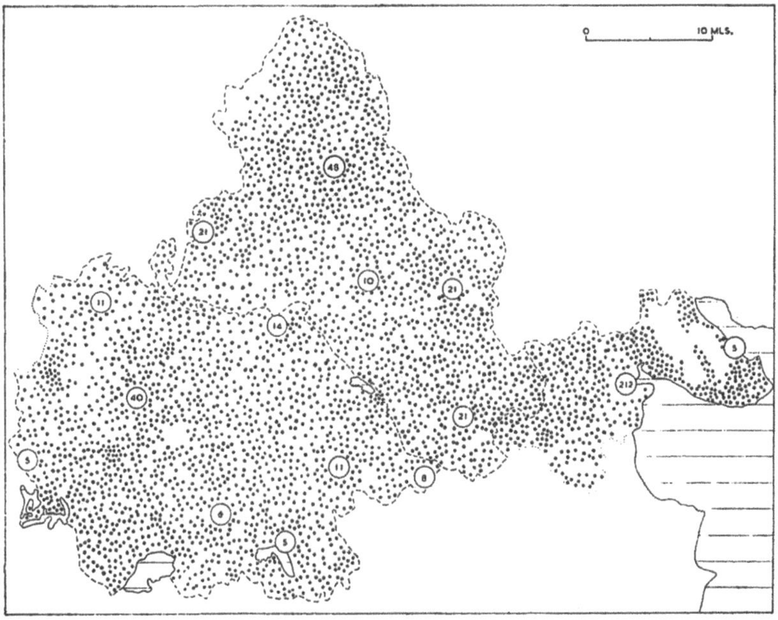

FIG. 81. POPULATION DISTRIBUTION, EAST PART OF DRUMLIN-DRIFT BELT

One dot to 25 persons

This small-farm country has a remarkably close and even settlement pattern, very similar to that across the Border in Northern Ireland (see Fig. 87, p. 481).

since 1891, from 180,000 to 70,010. This is remarkably heavy even by Irish standards, and indicates a decline in density per square mile from 150 to 58 per square mile, or 183 to 71 of improved land. The town population has increased from 39,000 in 1891 to 52,200 in 1966: virtually all the towns have declined in population though Dundalk expanded from 12,400 in 1891 to 21,700 in 1966. The long-standing industrial tradition of this area accounts in part for the general maintenance of the town populations which, as a group, now include 41% of the whole as against 18%

in 1891. In summary, the decline from 219,000 in 1891 to 122,200, a loss of 44%, is due to heavy rural emigration and as a result the overall density is now 101 compared with 180 in 1891.

SUBREGIONS

Five subregions may be distinguished in the eastern part of the drumlin-drift belt: the upper Erne lowland, the corridor lowland of Monaghan, the drift-covered hilly country of Ordovician and Silurian strata, the lowland around Dundalk and, most distinctive of all, the Carlingford peninsula. The essential unity of the area lies in its farming pattern, in its drift-derived soils, and in its mixture and interpenetration of pastoral and arable farming. Its many agricultural transitions, discussed above, modify its uniformity without destroying its unity. Finally, all its five subregions, though fairly clearly defined on the south, are artificially limited by the Border on the north.

1. THE UPPER ERNE VALLEY

The Erne, which rises in the hilly country of Co. Cavan, spreads into a series of 'lakes' which are an intricate maze of channels between cultivated peninsulas and islands. Almost the whole of the Erne lowland is covered with drumlins and drift ridges, which are spread over a Carboniferous rockfloor around lough Oughter and a rockfloor of Ordovician and Silurian strata between lough Oughter and the southern shores of lough Gowna. Many of the peat flats in hollows are developed on the site of former lakes or fens. The lowland, mainly around 200–400 ft. high around the Erne, rises as low hills to 850 ft. near lough Gowna. The countryside is divided into small farms, averaging 32 acres, and over two-thirds of the land is in pasture, with meadow and crop land, chiefly oats and potatoes, sharing the remainder. A few demesnes with woodlands diversify the scene, but less than one-tenth of the area is bog, and the main products of the farms are young cattle and milk for the creameries. The largest market centre is Cavan (4,200).

2. THE MONAGHAN CORRIDOR

This corridor from the Erne valley to the lough Neagh basin is used by roads, the Clones to Armagh railway and the Ulster Canal. On the north-west the Slieve Beagh, of upper Carboniferous strata, rises to 1,255 ft.; on the southeast the Ordovician strata forms a hilly edge rising above the corridor, which is mainly 200–400 ft. high. An intricate mass of drumlins covers Carboniferous limestones and provides highly varied soil conditions and a drainage pattern almost as complicated as that of the Erne lowland. The indeterminate divide between the Erne and the Bann drainage systems runs through the centre of the lowland, which has numerous small lakes and peat flats. There are, however, no peat bogs of large extent to interrupt the agricultural scene, which is dominated by small farms averaging 30 acres. Mixed farming prevails with milk for creameries and young cattle as the main products, though arable farming becomes more prominent towards the east. Clones (2,040) had its market area bisected by the Border, but it remains a small commercial and industrial town. Monaghan (4,020) was little affected by the political changes and is a good trading town with modern factories. The railway and canal are closed.

3. THE HILL COUNTRY OF CAVAN AND MONAGHAN (Fig. 82)

The hilly country between the Dundalk lowland and the middle Erne valley is plastered with drifts, chiefly in drumlin form. The underlying rocks are mainly Silurian slates and grits, but a basin floored by beds of Carboniferous and Triassic age occurs around Carrickmacross, Ardee and Kells. In this basin the upper Carboniferous strata stand out as hills 400–600 ft. high, but the limestones and the lower Triassics have been worn down to 200–250 ft.; to the west, the Silurians rise in sweeping lines to more than 1,000 ft. with a general level of around 500–600 ft. Everywhere lakes or peat bogs exist in the hollows and cultivated fields cover the drumlins and hillsides. Most of the farms are small and consist only of a few fields, with a total extent of around 30 acres. As much as one-fifth of the land is arable, and almost one-fifth of the grassland becomes hay. The potato harvest is a source of income on most farms and both cattle rearing for sale and for milk to the creameries are sources of profit; but flax growing and scutching is now extinct. In spite of a heavy decline in population, the density per square mile is nearly 100, a figure fairly typical of an area dominated by small farms. Carrickmacross (2,100), Castleblayney (2,100) and Cootehill (1,500) are the largest of the numerous market centres, most of which are mere villages scattered through the area at distances of about ten miles from one another.

4. THE DUNDALK LOWLAND

The present position and recent expansion of Dundalk have been discussed on p. 438: it is a town of sufficient size to show a markedly uneven edge, partly through its planless spread along various roads from the compact Georgian nucleus. The countryside around the town consists of

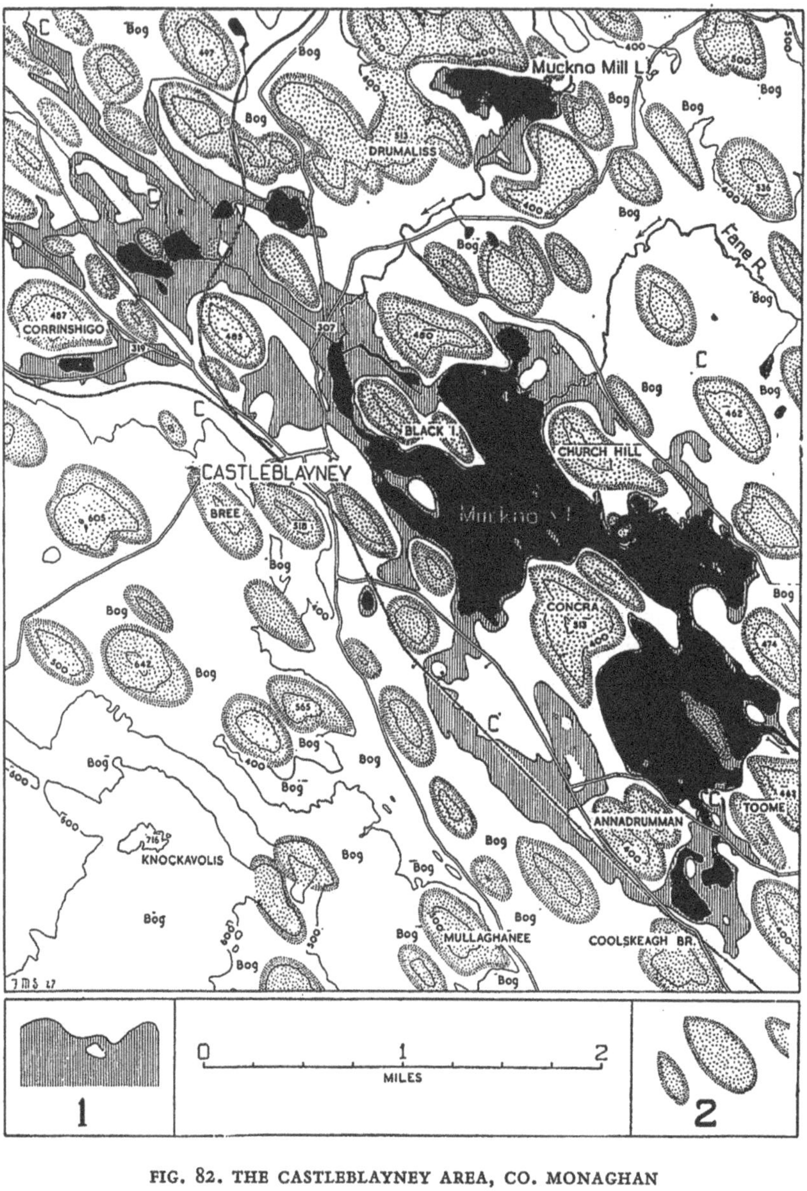

FIG. 82. THE CASTLEBLAYNEY AREA, CO. MONAGHAN

1, Area of former lake bed; 2, Drumlins; 3, Crag and tail formation.

Like many other lakes in the drumlin-drift belt, lake Muckno has been substantially reduced by sedimentation: the area of the former lake bed was 3 square miles but it is now merely 1·4 square miles. The drumlins are aligned from northwest to southeast: in some cases a number are joined to form an impacted series.

Map drawn by F.M. Synge, and originally published in 'Irish Geogr.', vol. 1, No. 4, 1947, p. 107 drift-covered Silurian strata, reaching an altitude of 200 ft. and diversified by drumlins rather than dominated by them. In this lowland the farms are mainly medium-sized, with cattle grazing and arable cropping, largely for potatoes and oats: the small-farm economy of Ulster and the large-farming grazing country of Meath meet in this lowland. Around Ardee the population density in the rural areas diminishes, due to the increasing prevalence of the large farms characteristic of the eastern section of the Central Lowland (cf. Fig. 47, pp. 280–1 and Fig. 81, p. 439.

5. THE CARLINGFORD PENINSULA

This peninsula takes its name from the historic town of Carlingford, which clusters around an ancient castle and abbey. It is made up of Silurian and Carboniferous strata through which basic lavas, chiefly gabbros and acid granites, have been intruded (Fig. 83). The ringdyke of gabbros stands up as a remarkable double line of crests with Slieve Foye (1,935 ft.), but the granites form lower hills such as Clermont Carn (1,674 ft.) or else are eroded into a basin between the gabbro hills. These hills have a large extent of bare rock stripped of its soil by the earlier Irish glaciation and by heavy erosion ever since: no peat bogs have accumulated on the craggy slopes. The smoother granite hills, on the other hand, have a deep peat-cover, actively cut in recent years. These negative areas of settlement are used only for timber plantations or for sheep grazing: below them the boulder clay of the second glaciation stretches from the coastal lowland into the valleys as a cover at least 50 ft. thick, in which the present rivers are deeply trenched: there are also numerous traces of meltwater channels, especially on the south side of the peninsula.

While the limit of cultivation has receded on the hillsides and many traces of abandoned farms may be seen, the combination of fertile soils, moderate rainfall (Greenore average 32″) and the accessibility of home and overseas markets has induced comparatively intensive farming. Almost two-fifths of the land is arable and large quantities of potatoes, oats, barley and turnips are grown. The economic staples include lambs, young cattle, poultry and butter, and the general picture is one of mixed farming similar to that of the Cavan–Monaghan country. The farms, however, are too small to provide more than a simple living for the population and most of the houses dotted through the landscape are little more than cottages, many of them thatched. Each farmhouse is surrounded by a few fertile fields, many of which are bounded by wide stone walls made of the innumerable stones thrown up in ploughing.

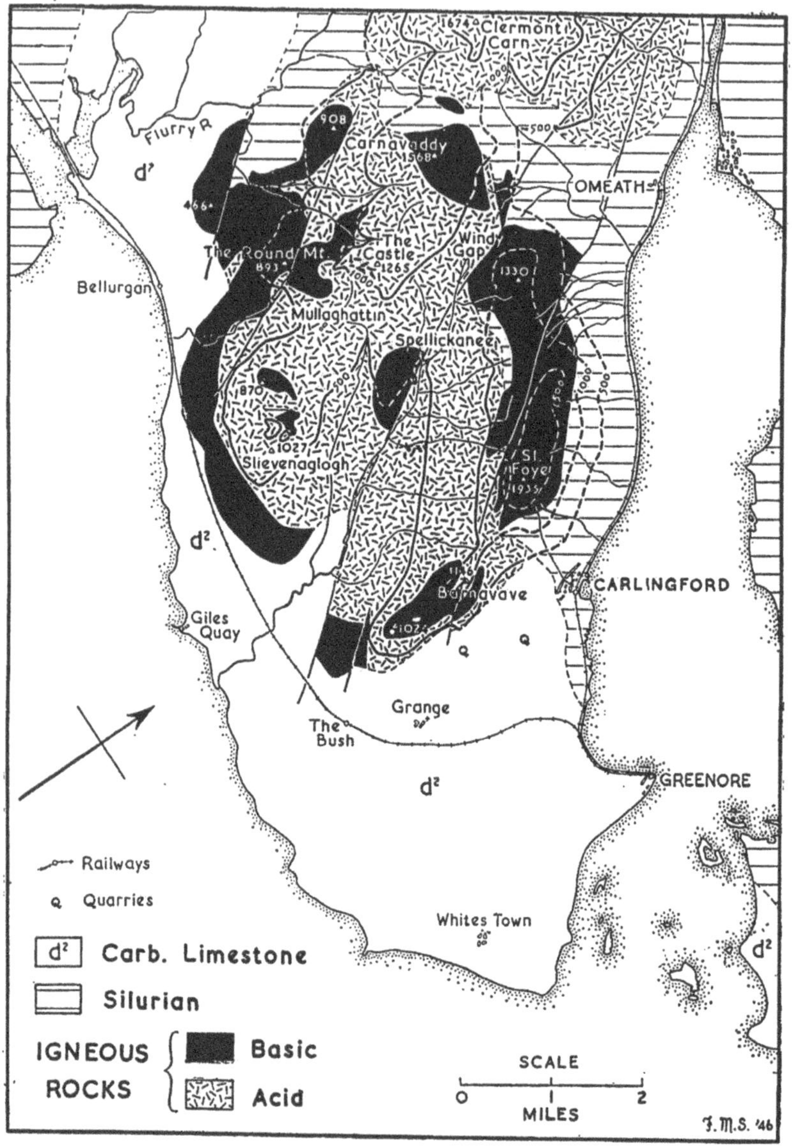

FIG. 83. GEOLOGY OF THE CARLINGFORD PENINSULA

Carboniferous limestones form a lowland, the Silurian beds the lower hillsides facing Carlingford lough, and the igneous rocks the sharp and jagged hills which give the peninsula its mountainous appearance. The railway is now closed.

Map drawn by F. M. Synge and based on data in Richey, J. E., and Thomas, H. H., 'Quart. J. Geol. Soc. Lond.', vol. 88, 1932

The whole peninsula is part of the market area of Dundalk. The villages include Carlingford (550), Omeath (260) and Greenore (140), all of which have some catering for tourists.

Surprisingly Greenore, having lost its railway in 1957 (p. 248), has recently revived as a port. In 1960 a container traffic service to Preston was inaugurated and achieved quick success but in 1963 a rival service from Preston to Dublin began to operate. The company based on Greenore acquired road transport vehicles and a new ship and by 1966 provided a service three times weekly to Preston and twice to Sharpness. The imports are now materials and manufactured goods and the exports are of both agricultural and industrial origin from Northern Ireland as well as the Republic. Already successful in shipping cattle to continental European ports and to Alexandria the company managing the port have begun to send cattle to Glasgow. A site of 175 acres has been allocated to industry and a subsidiary of a Swiss plastics firm, with 30 employees, makes sheetings, eggpacks and film sent to Northern Ireland by road and to Britain by sea. It is at least a beginning and the prospects for Greenore appear to be bright as its advantages include an absence of the congestion sometimes experienced in Dublin and Belfast, the availability of casual labour from the local farmers and efficient road transport.*

* The information in this paragraph is gathered from Gillmor, D.A., 'The revival of Greenore, a minor seaport', *Irish Geogr.*, vol. 5, no. 4, 1967, pp. 330–3.

CHAPTER 21

Donegal

The political separation of Donegal from Co. Londonderry is a factor of major geographical significance, for the richest areas of Donegal lie in the Foyle basin, where its natural regional capital, Londonderry, is also situated: yet the Foyle basin is divided between two states and Londonderry has become a frontier town. The east of Donegal is a fertile agricultural lowland, but the west is a hard environmental school in which the population wrings a meagre return from poor land and can only maintain even a simple standard of living by seeking extra sources of income through seasonal or permanent emigration, home industries, intermittent labouring or factory employment. The western seaboard was penetrated by light railways under government schemes, but the last of these was closed on December 31, 1959, and as the former lines of the Great Northern Railway to Bundoran and Ballyshannon were closed in 1958, the county now has no railways at all (see pp. 244–5, 250). The land communications focus on Derry, Strabane and Enniskillen, and most of the traffic from other parts of the Republic comes into Donegal through one of these towns.

LANDFORMS AND LAND USE

The rocks of Donegal consist primarily of schists, quartzites and granites; and the influence of the Caledonian folding is clearly seen: the northeast-southwest line in the mountains and valleys has in places an almost diagrammatic emphasis. Quartzites, weathered into their not unusual conical form, make sharp lines of hills in the Errigal–Muckish range, the Mulmosog–Aghla hills, the Knockalla mountains with the Urris hills (separated by lough Swilly), and the Slieve Snacht in Inishowen (Fig. 84). The Caledonian trend is almost as striking in the granite hills on either side of the Gweebara depression. The Dalradian rocks forming Donegal were already invaded and metamorphosed by igneous masses when they were rearranged by the Caledonian folding: structurally similar to the Scottish Highlands and to the country forming West Connacht, they have been steadily denuded through most of geological time, though there have also

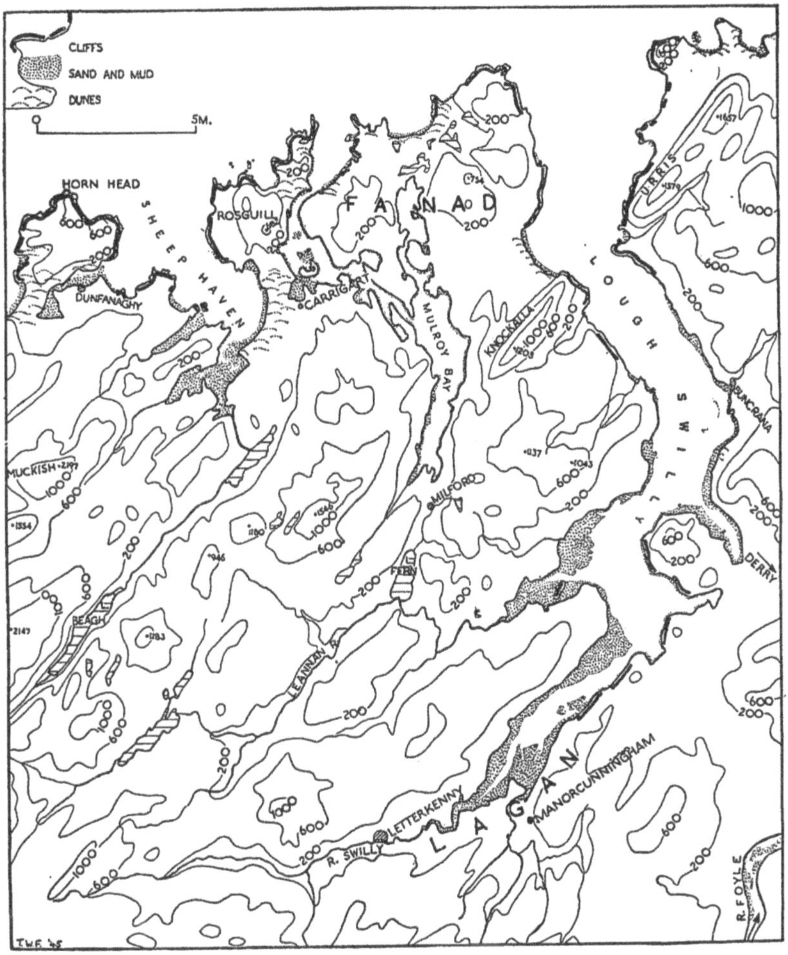

FIG. 84. NORTH DONEGAL, PHYSICAL FEATURES

The Caledonian line is clearly revealed in the topography of this area, and three long inlets, Sheep Haven, Mulroy bay and lough Swilly, add to the diversity of the landscape.

been a few marine transgressions, one of which is recalled by the presence of Carboniferous strata at 1,900 ft. on Slieve League.

Long-continued denudation of the Caledonian mountains in Donegal produced an erosion surface, possibly Devonian in date, at *c.* 800–1,000 ft., which has yet to be thoroughly explored. It is widely represented, however, and many of the peaks in the mountain heartland rise from it, notably in the Errigal–Muckish range. Faulting and igneous activity, seen in many dykes, probably accentuated the Caledonian line in Tertiary times: in addition, the erosion surface around 200 ft. may have developed during the Tertiary era. This surface is seen in many districts such as the lowland around Donegal bay, the extreme western end of the Slieve League peninsula, the Rosses, and a narrow coastal strip around the Bloody Foreland. Hills of quartzite rise sharply from this surface, and probably represent monadnocks similar to those in the heart of the mountains. The 200-ft. surface falls gently towards the present coast and also spreads into the valleys of various rivers that now have incised courses and are cutting down to the present sea level. It truncates rocks of varying resistance to erosion, including Carboniferous strata, schists and gneisses.

Donegal shows the strong structural imprint of the Caledonian folding because erosion, though so long-continued, never succeeded in masking completely the influence of rock-structure or in reducing the area to a common base level. The drainage pattern shows the interplay of rivers following the Caledonian trend with those at right angles to it. Those rivers flowing northeast–southwest, or the reverse, probably follow subsequent lines; the consequent direction, northwest to southeast, is also strongly represented. Complicated as the drainage pattern may appear, it is in the main rectangular, due to adjustment to Caledonian structures.

The severe glaciation to which Donegal was subjected so modified the landscape that almost all the detailed features show its influence. At its fullest extent the ice may have covered the whole mountain area, sending out tongues to the surrounding lowland fringes, now partly submerged. The valleys were widened, deepened, and filled in places with morainic material: some are narrow U-shaped troughs, especially in the granite area, while others are wide and open, especially in the schist country. The drifts on the moorlands have now a thick peat cover, through which large erratics appear in places. On the lowlands there are sharp contrasts: the lowland peneplane around Donegal bay is covered with drumlins as far west as Killybegs, and smaller drumlin 'swarms' occur around Ardara and also around Milford; drift of various types is widely distributed and forms some rich soils in eastern Donegal. In the west, however, glacial erosion, rather than deposition, is more apparent as large areas, notably in the Rosses, have bare rock surfaces strewn with erratics or very thinly covered with drift. Heather moors cover much of this drift, but a considerable area is reclaimed and farmed, with black-soiled damp fields from which stones have been removed and piled into walls or laid in heaps here and there. Many hollows are filled with small lakes which, in parts of the Rosses, are almost as numerous as those of the peat bogs of Co. Mayo.

Western Donegal is virtually treeless, due to many factors which include the persistence of moderate winds, chiefly westerly in origin, with occasional gales. On exposed coasts, including those of Horn Head, flecks of spray are blown inland even on fine days so a halophytic vegetation may develop even at a height of 500 ft. on cliffs: this 'spray-level' vegetation is a curiosity rather than a widespread feature. High winds prevent tree growth and exposed trees and hedges show the killing-off of shoots common to the rest of Europe's Atlantic fringe. Most of the climatic conditions of the west coast are suited to bog development, which depends largely on an excess of ground water. West Donegal has a heavy rainfall, infrequent sunshine and a scarcity of drying winds, both of which result in deficient evaporation. The bog accumulation is most marked on level or gently sloping areas such as the discontinuous peneplane surfaces about 200 ft. high, and on uplands 800–1,000 ft. high. In effect, the present land use of Donegal shows very clearly the combined effects of climate and slope.

Towards the east of Donegal, the degree of saturation becomes less acute, though much depends on local conditions of exposure. The sheltered inlets of the north coast have a certain amount of semi-natural woodland, now reduced by clearing for fuel and structural timber: in the neighbourhood of Dunfanaghy, Creeslough and Carrickart, and along the shores of Mulroy bay, ash, alder, hazel and oak appear in that order of relative frequency, with willows and birch, as the most abundant constituents of the natural woodlands that survive along the sides of sheltered creeks and valleys. Around Mulroy bay, much of the rocky land near the shore and on the islands has semi-natural woodlands, chiefly oak and birch with holly, bracken, heather and mosses as the field layer. In places gorse (*Ulex europaeus*) makes a continuous scrubland similar to that seen in many parts of western Donegal. Generally woodlands and scrublands in a semi-natural state have survived only in fairly steep and rocky hillsides. The Forestry Commission has seventeen estates in Donegal, many of them in the mountains, such as those of Meeninroy and Gweedore: others such as Ards and Mulroy are beside the sheltered ramifying loughs in the north of the county. In Donegal, as in West Connacht, some tree planting has been done on drained peat surfaces.

The proportion of land under crops and pasture has declined steadily and continuously since 1851, and the intensification of agriculture during two world wars did not arrest the decline: 58% of the area was improved land in 1851, but only 39% in 1967, though there has been little change for several years. The area sown to crops has declined from 16% of the whole county in 1851 to a mere 6% in 1967, and the decrease both in improved land and in that used for crops is obviously related to the decrease of population by 57% from 255,000 in 1851 to 108,550 in 1966. Every stage of land-degeneration may be seen, from fields recently invaded by bracken, heather or gorse to long-abandoned holdings where the former cultivation is suggested only by remains of walls or a few stones which were once part of a house. But though the area cropped is small in relation to the whole of Donegal, it is large in proportion to the area of improved land. In the extreme northwest, as much as half the improved land is under arable cultivation, chiefly of oats and potatoes: this feature shows a subsistence economy rather than cash-crop farming. The farmer depends partly on rough grazing for his stock and uses his few fields to grow oats for his animals and potatoes for his family. In east Donegal, one-quarter to one-third of the improved farmland is tilled, largely for foodstuffs and feedstuffs, though most of the farms here are above the subsistence level. In south Donegal, however, only one-tenth of the improved land is cropped. Oats are the main 'corn' crop of Co. Donegal and potatoes are found on almost every farm: turnips are grown in small patches for feedstuffs. Barley and rye (once a familiar crop in the west of Ireland) are now comparatively rare; flax growing fluctuates markedly – for example from 2,500 acres in 1953 to practically none at all – and mangels and sugar-beet, unsuited to the climate, are hardly grown at all.

In west Donegal the bogs and moors are an important resource for the community. They provide pasture and bedding materials for cattle and sheep, and are an invaluable source of fuel. The struggle to maintain the farmlands is desperate in some places, as heather, bracken, gorse and hawthorns are ever ready to creep into the pastures from the moorlands and hedgerows and many of the fields are already so closely covered with *Juncus* that they slip very easily into a semi-wild condition. In human terms, old age, ill-health or the migration of workers often means in time the almost imperceptible but yet certain reversion of the land to a wild or semi-wild state. As a rule, the farmers cling tenaciously to the arable land, which is sown mainly to oats and potatoes in western Donegal, and manured by sand, shells and seaweed brought from the shore; artificial manures are also used. The perpetual battle with saturated and acid soils goes on in the little fields surrounded by high walls built to keep out animals. Towards the east, the farmlands become progressively richer, with larger fields, more machinery and fewer very small farms. One difference between west and east in Donegal is that the amount of labour necessary to raise 1 acre of oats and potatoes in the west is vastly greater than that necessary in the east: it follows, therefore, that the farms are smaller on the poorer lands of the west, and the density of population in relation to the cultivated area is remarkably high.

Cattle are the mainstay of Donegal farming and their numbers have remained fairly even during the past hundred years: there are normally between 140,000 and 170,000, of which approximately one-third are milch cows. Only one-fifth of these are suppliers of creameries and most of the farm butter is consumed at home or sold locally. Cattle are sold mainly when less than two years old to dealers from the eastern counties, and Donegal shares very fully in the internal cattle trade by which calves and young cows make their first journey to pastures farther east in Ireland and their second to slaughter either in Dublin or in Great Britain. Few pigs are kept: in 1967 there were only 18,000, though there are schemes of encouraging pig rearing in the Gaeltacht areas. In Donegal, as in rural Ireland generally, the loss of the export trade in eggs and poultry has been unfortunate: the Congested Districts Board reports showed that many of the poorest people in the west would have been almost destitute but for the sale of eggs and the former success of poultry raising affected all types of farms profitably. At the end of the nineteenth century the number of poultry was approaching a million, which was exceeded in this century, but by 1967 there were only 489,000. Sheep rearing has been subject to sharp fluctuations: in 1851 there were 82,000 sheep but the number increased rapidly to 214,000 in 1871. The number declined during both world wars and in 1967 was 253,000, most of which are in the uplands.

Between 1851 and 1945 there was a marked decline in the area given to improved pastures, from 456,000 acres to 242,000 acres in 1945 but increased to 327,000 acres by 1967. The area of meadow was 31,000 acres in 1851 but increased to 57,000 by 1945, eventually diminishing to 45,700 acres by 1967. Crop land covered 197,000 acres in 1851 and 126,000 acres in 1945 but by 1967 only 69,500 acres, about one-third of its extent in 1851 and little more than half that at the end of the 1939–45 war. The decline in cropping is universal in the west of Ireland, and associated with a decrease in manpower: nevertheless, compared with other western areas, Donegal is remarkably well farmed. But it is not farmed to its maximum capacity, partly because the financial interest of many families lies in migratory labour, local odd jobs, fishing, peat cutting for sale or under government schemes, and local industries. The intensification of agriculture would mean a considerable expansion in the growth of fodder crops, including hay, and the improved drainage of pastures. More labour would be needed, but here the possibilities of future development rest with the potential emigrant as surely as they rest with the agricultural reformer.

INDUSTRIES

Donegal has an industrial development of considerable social value with its roots in the widespread domestic production reported by the Congested Districts Board. In 1891 there was home sewing (chiefly of shirts), weaving, knitting and 'sprigging' or hand embroidery. Various charitable attempts to establish 'sewing rooms' for girls formed a link in the chain of development from home to factory industry; many of the 'factories' still have outworkers and some are little more than receiving depots for homemade goods. Home workers in the nineteenth century were shamelessly exploited; they tramped many miles to collect wool, shirts or linen from depots and then worked ceaselessly for a few shillings a week. Each of their activities has led to a definite type of industry, and 2,100 people were employed in textile work in 1961, over two-thirds of them women. And this industry remains prosperous.

Home knitting is now definitely subsidiary to the factory production, but its present location reflects its origins. In east Donegal the industrial development began in association with Londonderry, but since the Border was established, factories have been established in various small towns and villages of Donegal. There are large knitwear factories at Buncrana, Glenties (both with over 200 workers) and another at Letterkenny. At Carrickart in the north a 'home industries' factory provides a variety of woollen goods. In the southwest there is home tweed weaving, partly changed into a workshop industry as under such conditions control of design and craftsmanship is easier. At Kilcar, in the extreme southwest there is a factory with 150 indoor and 100 outdoor workers, and there are several collecting depots for tweeds in Ardara, and factories also at Downings and Clonmany in north Donegal. A large factory established in 1859 at Donegal town, produces tweeds and clothing and has 250 workers. Even larger is an old-established factory at Convoy, producing worsted, serges, tweeds, coatings, blankets and rugs. There are 350 workers and every effort is made to increase the export trade, especially with North America. The home shirt industry has been replaced by factories, with Buncrana as the main centre. Other factories exist at

Muff, just inside the Border, at Lifford, at Castlefinn, at Ballybofey in the Finn valley, at Carndonagh and Moville in Inishowen, and in Donegal town: several of these have been established since 1922. At Killybegs there is the only surviving factory in the British Isles producing hand-knotted carpets: the firm has 75 workers.

The embroidery industry has preserved its tradition of fine handwork against all machine-made imitations, and at Glenties one firm has 4 'inside' and 200 'outside workers'. Mountcharles in south Donegal is a centre for the craft. Other local industries include glove making and home knitting in the west. There are numerous small enterprises serving local needs, but an interesting specialization is the oatmeal production of east Donegal, which is established in small towns such as Letterkenny and Buncrana, or villages such as Moville, Castlefinn, Ballindrait, St Johnston, Ramelton and Milford (with a large bakery). All these mills make other products such as wheat flour, maize meal, bran, linseed cake. Killybegs now has a considerable fishing industry, and its facilities include an ice factory, fish processing and fish meal plants and a boat building and repairing yard.

POPULATION

Co. Donegal had 108,550 people in 1966, 58 to the square mile. The population has declined by 63% since 1841 and by 42% since 1891: this decrease is less severe than in many other primarily agricultural counties of Ireland. The explanation of this relatively low rate of emigration lies in many factors such as the opportunities for seasonal migration within the county itself from west to east and also to Great Britain (primarily to Scotland), the development of domestic and factory industry, fishing, and, it is commonly said, the resource and enterprise of the people themselves. But from 1961 to 1966, there was a population loss of 4·7% and emigration is well established, for at least one-quarter of a million people have left since the Famine. The loss has been spread through the whole country with no little clear correlation between the relative wealth or poverty of the different districts and the rate of migration, for the richer east has lost as severely as the poorer west, on a long term assessment. But a closer investigation of the Census shows that many of the remoter areas are now losing heavily in population. Even from 1961 to 1966, fourteen out of the twenty-seven divisions forming the Glenties rural district lost at least 10% of their population and there were similar losses elsewhere, mainly in the remoter areas. In 1891, 89 of the 149 district electoral divisions of Co. Donegal were 'congested': in 1936, 49, in 1951,

33, in 1961, 17 and in 1966, 13. Although the west of Donegal appears to be over-populated on any reasonable statistical basis, the inhabitants cling to their homes, aided by the profits of industry, seasonal migration, fishing, turf cutting, remittances from affluent relatives abroad and government assistance.

The density of population in relation to the area of crops and pasture shows an extraordinarily high concentration in the west and northwest, which is comparable with that of west Mayo (Fig. 78). In many of the limited areas farmed and settled there are as many as 400 people per square mile of improved land. This density of population seems fantastically high, but even so it is approximately half the density of a century ago. It could not be maintained without the resources already mentioned and the widespread use of 'mountain pastures' (moorlands and bogs at any altitude) for grazing sheep and cattle. In Dunfanaghy rural district over half of the farms are of less than 15 acres: some of the small farms are so close together that an area of nominally dispersed population strongly resembles one continuous settlement, especially where most of the farms line the roads. Similar conditions occur in various valleys, though in many cases the settlement is less continuous and is concentrated in certain areas with favourable soil drainage (Fig. 85).

In the drumlin country of south Donegal settlement is much more widespread than in the Congested Districts of the western coastlands, but at the same time less intense in relation to the area of improved farmlands. The farms are larger on the average than in the west: in the Donegal rural district about one-quarter of the farms are of 1–15 acres and one-third of 15–30 acres: the position in the Ballyshannon rural district is much the same. Most of the remaining farms are of 30–50 acres, with a small sprinkling of larger farms. Towns and villages have one-quarter of the total population in Donegal rural district: these include local market centres which have small industrial developments, such as Donegal, Mountcharles and Killybegs. Ballyshannon (2,230), at the mouth of the Erne, has similar functions, but Bundoran (1,420) is one of the few places in the Republic to derive its income mainly from holiday traffic.

The east of Donegal resembles the south of the county rather than the west in its distribution of population. It has a few very closely settled areas, notably in the northern peninsulas, some parts of which are not unlike the Congested Districts of the west, but generally the density of population is 100–200 per square mile of occupied land, spread fairly evenly over the lowlands and the mountain valleys. The Finn valley is typical of these: it has a number of market centres with a few small

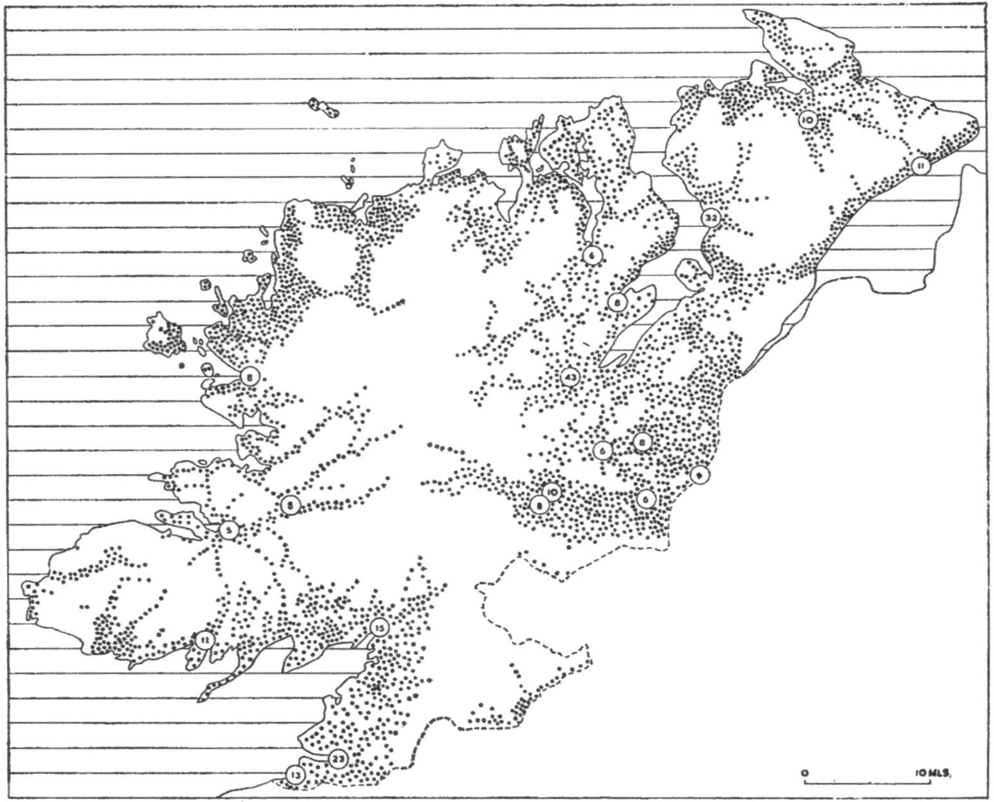

FIG. 85. POPULATION DISTRIBUTION, DONEGAL

One dot to 25 persons

On the west coasts and on some of the islands there are remarkably high densities; in the south the settlement is mainly near the sea, though there is also some occupation of the mountain valleys. The east has larger farms than the west and the population distribution may usefully be compared with that of areas east of the Foyle.

industries, such as Lifford (925), Castlefinn (600), Stranorlar (910) and Ballybofey (1,030). These two last-named places are adjacent. Letterkenny (4,530) and Buncrana (3,100) on lough Swilly are also market and industrial centres: the towns and villages have about one-quarter of the total population. The high densities of population are in part a reflection of the prevalence of the small farm, even in the east of Donegal. This richer east receives a number of servants and labourers from the west, some of them seasonally: even so, it has lost one-third of its population since 1891.

SUBREGIONS

Donegal may be divided into eight subregions, with many transitional areas between them.

1. THE DRUMLIN BELT AROUND DONEGAL BAY

This subregion stretches from the south of the Erne to Fintragh bay, west of Killybegs. Carboniferous rocks consisting of limestones, shales and sandstones form a band of low-lying country five to seven miles broad; and the long finger-like peninsula ending in St John's point is the last outpost of Carboniferous rocks in this part of Ireland except for a small exposure in Fintragh bay. This subregion, structurally part of the Irish Central Lowland, derives its unity from the presence of drifts: there are remarkable drumlin swarms between Ballyshannon and Donegal, where the landscape is closely settled and industriously farmed, but to the south of Ballyshannon the drumlins are interspersed with patches of bare limestone. West of Donegal town, drumlins and farmlands prevail in the valleys, but the ridges between them have windswept peat bogs with only intermittent agricultural settlement. On the westward journey, every valley seems less fertile than the one before it; trees become fewer and the soils begin to possess the black colour of land wrested from the bogs. East of St John's point, the drumlins overlie limestone; west of it, they cover schists and gradually give place to *roches moutonnées*. The farming is mainly cattle rearing and dairying, some sheep raising, with oats and potatoes as the main crops. Most of this area was congested in 1891, but would not be on a reassessment, as the loss of population has been over one-third. The local home industries, embroidery and knitting, have led to industrial developments in Donegal and Mountcharles. Fishing was important in the latter years of the nineteenth century but has declined since: the St John's peninsula, Inver and Killybegs were the main centres. But Killybegs is now a major fishing centre, with much-improved harbour facilities, fish-processing plants and a boat-building enterprise. Here too the tourist trade is expanding.

2. THE HILL BASTION OF THE LOWLAND

Around the south Donegal lowland there stretches a moorland and hill area almost without inhabitants, though some farms exist on the valley sides which penetrate it and separate one line of hills from the next. Few farms are found above 600 ft. and most of the level or gently sloping areas above the valleys are desolate peat bogs, crossed only by lonely roads: the only pass is the Barnesmore gap at 550 ft. The hills are formed mainly of schists and rise in places to 1,500 ft., but the Blue Stack mountains overlooking lough Eske are a rocky granite range with summits over 2,000 ft. Hidden away to the north of lough Erne is a small lowland area of drumlins covering Carboniferous rocks, chiefly famed for the village of Pettigo, half in the Republic and half in Northern Ireland, the centre for excursions to lough Derg, where pilgrims perform austere penitential exercises.

3. THE SLIEVE LEAGUE PENINSULA

This is best known for the precipices, rather than cliffs, that fall 1,900 ft. from the top of Slieve League, and over 1,000 ft. from the top of Slieve Tooey, into the sea. Quartzites form these hills and several others, such as Mulmosog to the south of Ardara. The remainder of the peninsula consists of mica-schists, moulded into lowlands and ridges, of which the most striking is Crownarad, to the west of Killybegs and Fintragh bay. Drift deposits are covered with peat except in the limited areas that are cleared and settled.

The settlement, though limited to a small part of the peninsula, is surprisingly dense, especially in the valleys drained to Tawny bay, which include the Glendarragh, a U-shaped glaciated valley to the west of the Crownarad ridge, the Ballaghdoo river valley, the Glen river valley through Carrick, and Glengesh, a valley west of Mulmosog. Other settled lowlands include the Glencolumbkille area and the peneplaned area around Malinmore and Malinbeg in the extreme southwest. All these areas have 150–200 persons per square mile of improved land and therefore show that peculiarly high concentration of population which marks many districts in the west of Ireland. This southwestern peninsula, one of the poorest districts in the country in 1891, would still rank in part as congested on a 1966 reassessment in spite of a population loss of as much as 50%. In 1891 'American money' was the most important item in most family budgets, though today remittances from emigrants in Britain are a main resource of many households. The tradition of home industries has remained, and some men still work as handloom weavers though there is a tweed factory at Kilcar. A recent development is the cultivation of vegetables for processing, including celery, cabbages, beans and carrots, notably around Kilcar and Glencolumbkille.

The least favourable areas for settlement are the extensive peat bogs developed on the level patches of land around 200 ft., 400 ft. and 800 ft., all of which appear to be fragments of peneplane surfaces. There are, however, some cultivated areas in the valleys cut deeply into these level surfaces, for example in Glengesh, which has farms on the valley floor and on the gentler hillslopes where the valley opens out northwards towards Ardara. These farms consist of a few fields divided mainly between meadow and pastures, with some sown to oats and potatoes; in some cases these crops are found only on plots within the fields. Most of the fields are divided by stone walls, but there are a few hedges of hawthorns and hazels though windbreaks are rarely seen, yet elderberry bushes have been known to resist all storms and winds. Here, as elsewhere in Donegal, the peat bogs provide pasture for animals as well as fuel.

4. THE COASTAL LOWLANDS AND ISLANDS OF WEST DONEGAL

These lowlands and islands between the hillfoot at Ardara and the Gweedore river are built mainly of granite, though quartzites form isolated hills at Maghery, near Dungloe and on Aranmore. The coastline is one of submergence in middle youth, with many islands, some of them linked together by spits, and the bays are being filled with detritus. Along the coast, cliffs alternate with low shores and stretches of sand dunes. The lowland as a whole is seared with rocky hills and smaller outcrops, many of them ice-moulded, and strewn with lakes among peat bogs. The fields have black bogland soils, fertilized with sand and seaweed, and the stones lifted from the fields have been piled into heaps or made into walls.

The lowland may be divided into two sections:

(*a*) *From the hillfoot south of Ardara to the Gweebara river*. These lowlands have intermittent agricultural settlement with a density of 150–250 per square mile of improved land. There are mountain valleys with farmlands on hillslopes around Ardara and Glenties, two villages which are route and market centres with tweed and hand-knitting industries. In 1891 three-quarters of the population in this lowland lived on holdings valued at less than £4, each of which consisted of about 6 acres divided equally between meadow and the two main arable crops of the west, oats and potatoes.

Seasonal migration was hardly known in this area as there was a local demand for labour on large estates. Sheep and cattle were pastured on the extensive 'mountain' (i.e. bogland) pastures. About two-fifths of the annual income came from the knitting of the women and children . . . 'they all knit, down to the tiny children of five or six; they knit in the fields, they knit in the houses, they knit as they walk along the roads'.* The men, however, did not knit and in fact had a comparatively easy time while at home, for some did little beyond setting the crops and taking in the harvest. Since 1891 this area has lost over one-third of its population, so that only part of it would be congested on a reassessment. But the type of life remains fundamentally similar to that of 1891, even though the factory development is more apparent and the returns from seasonal labour are vastly increased. And knitting still survives.

(*b*) *The lowlands from the Gweebara river to the Gweedore river.* This area consists of peat bogs with innumerable lakes in hollows and areas of discontinuous settled land in between. There are also some farmlands in river valleys, such as the Gweebara river. The density per square mile of improved land is generally over 150 and reaches 300 in places, notably along the coast where sand and seaweed are available. The entire area was congested in 1891 and would still be congested on a reassessment: the poor farmland areas to the south of Dungloe have lost from 25% to 50% of their population since 1891, but those to the north of Dungloe had actual increases to 1936, a most unusual phenomena in Ireland: since then there has been a decline. In parts of the Rosses it is difficult to believe that any population at all could be supported, but among the stony lands one may see a couple of fields with black soil, a cottage (generally with only two rooms), and the pile of turf gathered for the maintenance of the constant fire.

The Rosses were very closely studied by W. L. Micks, whose report for the Congested Districts Board was a model of its kind.† Only 3% of the holdings were valued at more than £4, and 84% were rated at less than £2, yet only 4% of the families were described as 'very poor'. The holdings consisted of about 2½ acres, of which 1 acre or a little more was devoted to potatoes, ½ acre to 1 acre to oats, and 1 rood each to green crops and meadow. A form of seasonal migration to mountain grazing (transhumance) was practised and in some cases there was a second holding on higher ground. Less well endowed in natural resources than almost

* Congested Districts Board, Base Line Reports 1891: report on the Glenties district.

† This report on the Rosses is given in Micks, W. L., *An Account of the Constitution, Administration and Dissolution of the Congested Districts Board*, Dublin, 1925, pp. 241–58.

any other part of Donegal, the inhabitants could not possibly exist on the produce of their holdings alone. Mr Micks gives an example . . .

> in Keadue there are exactly 100 families, while the valuation of the townland is £103 4s. The population is 505, so that the valuation per head of the population amounts to four shillings and one penny! And yet, owing to the wages earned in migratory labour, the people of Keadue are in ordinary years well fed according to the standards of the district, and also comfortably housed and clad.

The women cared for the home and farm during the summer, knitting and sewing at all possible times, while almost all the able-bodied men, girls and even children went away as migratory labourers. The men went to Scotland primarily for agricultural work, but also to industrial occupations. The girls and children went to farms in east Donegal, Londonderry and Tyrone, especially to the Lagan. Micks suggests that some men came back with as much as £20 saved, but in any case more than half the annual cash income of £43 came from the migratory labour (say £10 from Scotland and £6 from the Lagan) and the knitting and sewing of the women (say £7 10s.). Knitting and migratory labour now give large returns.

The success of the local co-operative movement has done much to maintain the comparative prosperity of this area. There are now factories for knitted goods at Dungloe and Kinclassagh, one of which has nearly 100 workers, and farm produce is marketed through the Templecrone Co-operative Society. The movement was in origin an attempt to remove the evil of the 'gombeen man' who, calling at the cottages, gave the minimum price for eggs and butter and charged the maximum price for sugar and tea. In most cases his commercial activities were conducted solely on a barter basis. In 1940 the railway from Strabane to Burtonport was closed, but there are useful bus and lorry services. Burtonport is a centre for boat traffic to various islands, of which the largest is Aranmore, with almost 300 persons per square mile of cultivated land. The remainder of the island consists of heather moor, heavily grazed. The population decline is heavy: Aranmore island had 948 people in 1961 but 847 in 1966. Lobster fishing for the export market brings some profit and there is now a marinated herring factory at Burtonport.

5. THE LOWLAND FRINGE OF NORTHWEST AND NORTH DONEGAL

These lowlands consist of land sloping from about 200 ft. to the present sea level, with coastlines of sand dunes, battered rocks or low cliffs. Hills rise steeply from the 200-ft. surface and the underlying structure is of great complexity, with granites, diorites, quartzites (which form a hill over 1,000 ft. high at the Bloody Foreland), and many types of metamorphic rocks, including limestones, seared by dykes. Off the coast there are a number of inhabited islands, including Tory, seven miles out. Horn head is a magnificent tombolo linked to the mainland by sand dunes, which climb its southern slopes to a height of over 200 ft.: on the north side there are cliffs nearly 600 ft. high.

Settlement in this area is, with few exceptions, limited to the better drained areas less than 300 ft. high. To the south of the Bloody Foreland there is almost continuous agricultural settlement along a winding road where the farms are so small and the cottages so close together that the resemblance to a village street is striking. Tracks lead off to other settlements and the density is about 400 per square mile of occupied land. Potatoes and oats grow in black-soiled fields surrounded by walls as much as 5 ft. high, and there are patches of meadowland and green vegetables. On the coast there are sand dunes and boglands, inland there are mountain pastures. This area was congested in 1891 and would be congested now though the population has declined by one-fifth since 1891. 87% of the holdings in this area were valued at less than £2 in 1891 and it was estimated that two-fifths of the whole population found employment annually as seasonal labourers. The men went mainly to Scotland but also to the Lagan: women and children went exclusively to the Lagan. The people were reported to be more industrious and energetic than in many areas of Donegal. Recently a trading estate of 200 acres has been established at Bunbeg, and already two British firms have factories producing woollen textiles and carpets. Close by, there is a Gaelic theatre in a fine modern building. On the whole the signs of social change are considerable, for new houses have been built under government schemes and the roads improved. Tory island, which in 1961 had 240 inhabitants, covers little more than one square mile in area and is made up of bog-covered granites rising to nearly 200 ft., and quartzites in the east forming ground nearly 300 ft. high, with striking cliffs. The population lives by fishing and farming their starved fields, aided by doles, emigrants' remittances and seasonal migration. In spite of the isolation, the decline of population here has been slight.

East of the Bloody Foreland peat bog covers the gently sloping coastal lowland, but farmlands reappear at Meenaclady, and around Gortahork and Falcarragh there are signs of that increasing richness towards the east of the county which is a feature of northern as well as southern Donegal. The farms are mainly of 5–8 acres, of which more than half is divided between oats and potatoes, the former predominant, and the remainder is grazing. The density of population on the farmlands is high, and in places reaches 300 per square mile. Near Gortahork there is the headquarters of the local tomato-growing scheme, from which plants are sent to greenhouses on farms in the district. With heat it is possible for the farmer to make good profits. Seasonal migration is still widespread though home weaving and fishing have declined almost to extinction. The Dunfanaghy area round the market village had a considerable demand for local labour on estates, one of which, on the south-facing slopes of the Horn head peninsula, has since been divided by the Land Commission, and stands out as a rich fertile area with a well-drained red soil in marked contrast to the black-soiled fields of the smaller farms in the higher parts of the peninsula. Between Falcarragh and Dunfanaghy there are some excellent farms, and the fairs at Falcarragh (470), an expanding and prosperous village, gather up a great deal of stock. Dunfanaghy (360), a village with government offices, hotels, a bank and shops, is on the main tourist road through north Donegal: these places, though so small, discharge the functions of towns for the large areas they serve.

6. THE MOUNTAIN HEARTLAND OF DONEGAL

The southern part of the Donegal mountains has been treated above as the bastion of the south Donegal lowlands (subregion 2, p. 457). The remainder of the mountain area interpenetrates closely with the coastal lowlands of the north and is cut by a number of deep glaciated valleys with some agricultural settlement. In this area the peneplane around 700–800 ft. is extensively developed, mainly as desolate bog-covered moorland though there are some bare rock surfaces: from it rise lines of hills, which in the north include three ranges, the Muckish–Errigal line of seven quartzite peaks, 1,500–2,500 ft. high, the Derryveagh rounded granite hills of similar altitude and the Glendowan mountains, 1,000–1,800 ft. In this northern part of the subregion, the Caledonian line is shown with particular emphasis: to the south there are lower hills of granites and schists and a prominent quartzite peak, Aghla mountain near Glenties, almost 2,000 ft. high, but the Caledonian trend is less clearly emphasized.

Settlement is restricted to the valleys and is continuous only along the Finn, the Swilly, the Leannan and their tributaries. Some small and isolated agricultural communities exist in remote parts of the mountain mass, where the farmers each have a few acres of arable land used partly for oats and potatoes, and share some 'mountain' or common pasture for sheep: the other stock includes a couple of cows and some poultry. Thinly peopled in 1891, the mountain areas have lost about half their numbers by emigration: in the Errigal valley power stations, run respectively on water and peat, have been constructed to supply the remotest parts of Donegal with electricity (p. 205).

7. THE NORTHERN PENINSULAS

These northern peninsulas are deeply penetrated by arms of the sea, Sheep Haven, Mulroy bay, lough Swilly, and lough Foyle. The peninsulas increase in grandeur and size from west to east; Mulroy bay is a delicate miniature of lough Swilly which in its turn is much less spacious than lough Foyle. The uplands of central Donegal have their structural continuation in these peninsulas, which have lines of hills of Caledonian trend in quartzites and schists. The populated areas are coastal lowlands and hillslopes, continuous along lough Foyle but discontinuous elsewhere, and some lowlands and valleys which penetrate the mountains. The general picture is one of extreme fragmentation in topography and settlement.

(*a*) *Rosguill.* Rosguill is a tombolo consisting of hills of granite and quartzites rising to nearly 700 ft. The coast is rocky, with cliffs and islets. Small farms are scattered along the few roads and there are isolated holdings on the hillsides; the area, congested in 1891, would still be congested today. Rosapenna, at the southern end of Rosguill, is a small tourist centre offering among its attractions a magnificent golf course on the sand dunes separating Mulroy bay from Sheep Haven. Carrickart, at the southern end of the sand dunes, is a much commercialized village selling Donegal tweeds and rugs. South of Carrickart there are some settled agricultural lands on patches of lowland and in valleys around the quartzite hills that dominate Mulroy bay. The density of population on the settled land is high, generally over 200 to the square mile, in spite of a loss of more than one-third since 1891. There are small farms with extensive areas of rough grazing. At Downings (180), on a site facing Sheep Haven, there is a tweed mill and centre for hand weaving.

(*b*) *The Fanad–Knockalla Peninsula.* The peninsula between Mulroy bay and lough Swilly falls into two parts, the Fanad and the Knockalla range. The Fanad, a hilly area of great topographical variety almost bisected by a tributary arm of Mulroy bay, is a clearly marked entity possessing its own regional name. Few acres are of greater structural complexity: granites and quartzites form low but jagged hills, most of which are less than 700 ft. high, diorites and felstones have been intruded, and there are bands of metamorphic limestone which are recognizable from the replacement of heather by gorse. Bare rock surfaces and *roches moutonnées* moulded by ice are widespread, but drumlins occur near Mulroy bay. An extensive area of sand dunes holds up shallow lakes. The result is a country of extraordinary diversity with frequent changes of scenery, heather-covered hills, boglands, sea, small lakes, but few areas of good agricultural land. The population lives mainly by farming small holdings having black peaty soils manured by sand and seaweed. Most of the houses are scattered along the roads, with a number of hamlets but no villages, and some of the farms are very isolated. The density per square mile of cultivated area is over 200, though the Fanad has lost one-third of its population since 1891.

Knockalla is a tangled hill mass formed of mica-schists, quartzites and various intruded igneous rocks, with a general altitude of 500–1,200 ft. Settlement is restricted to the lower hillsides and to the lowlands, which are of considerable extent on the lough Swilly side, especially around Rathmullen (440), a market village with some facilities for tourists. Carrowkeel, on the west, is a village in a small amphitheatre of agricultural land. The farmlands among the hills have been wrested from the prevailing peat bogs, but in the lower parts of the valleys and along the lough Swilly coast there are some fertile soils. As a general rule, the upland farms are small and the lowland farms larger. The density of population is about 150–200 per square mile of cultivated land on the lower slopes, and over 200 in the upper parts of the valleys; the area has lost more than one-third of its population since 1891.

(*c*) *Inishowen*. Inishowen, or Owen's island, is a mountainous peninsula lying to the north of the Burnfoot depression, which is eight miles long, and consists of an old seabed elevated in geologically very recent times. Schists form the greater part of the Inishowen peninsula and are moulded into hills rising to nearly 1,600 ft. from a moorland about 800 ft. high, which is broken by various valleys filled with glacial and glacifluvial deposits. Two lines of quartzite hills dominate the peninsula, the Slieve Snacht range between Buncrana and Carndonagh, and the Urris hills in the northwest: each of these ranges shows the Caledonian trend. There are also quartzite hills in the Malin area rising to 900 ft., and in Dunaff head at the entrance to lough Swilly. The settled lowlands are discontinuous except around Carndonagh in the north, but many of the hillsides and interior valleys are covered with farmlands.

The population density is 150–200 per square mile of cultivated land towards the north. Although this area was congested in 1891, and part of it would still be congested today, it has not the poverty of many districts farther west. The smallest farms consisted of about 7 acres as against 4 acres in the Slieve League area or the Rosses. The sources of extra income were numerous: labour was in demand locally, in the Lagan or in Scotland, fishing was practised and the domestic occupations included weaving, spinning and sewing for shirt factories in Londonderry. The population has declined by more than one-third since 1891. The towns of the peninsula have some industrial enterprises, mostly established since the Border was created. Buncrana is basically a small market centre with one long street, and a market centre with a mill, still surviving, as its industry but it also has a large hosiery factory, shirt factories and even a tobacco and snuff factory. There are various service trades including building and printing. At Carndonagh (1,060) there is a works providing children's clothing, and Moville (1,060) has a long-established and still functioning grain mill and a shirt factory. At Carndonagh, alcohol is made from potatoes and from imported molasses, some of which are brought to Buncrana harbour. In spite of industrialization, Buncrana still retains its tourist traffic, and Moville has visitors during a short holiday season.

8. THE LOWLANDS OF EAST DONEGAL

These lowlands extend from Mulroy bay to lough Swilly and lough Foyle and consist mainly of mica-schist country overlaid with glacial deposits in various forms such as drumlins around Mulroy bay, boulder clays, gravels and sands farther east. Alluvium, generally ill-drained natural pasture, occurs chiefly along the Foyle and its tributaries, and there are 'intakes' of reclaimed land along lough Swilly and in the Foyle valley: but level areas are comparatively infrequent and the lowland scene is diversified by isolated hills 600–900 ft. high, some of which are formed of quartzite, metamorphic limestones or igneous intrusions. The lowlands are not of uniform fertility: the area around Mulroy bay has some good lands and some poor lands, but the large farms are invariably found on the good lands. The Lagan, the area to which so many seasonal labourers go annually from the poorer districts of Donegal, lies around Letterkenny and Manorcunningham (Fig. 84). The term 'Lagan' is sometimes applied also to the Burnfoot depression. West of the Foyle, moorlands on the small hills break the continuity of settlement only to a slight extent, and the farmlands reach 600 ft. in places. The density of population on the improved land is between 100 and 200 to the square mile. In

Stranorlar rural district, one-eighth of the farms are of more than 100 acres and the remainder are divided fairly equally between medium and small-sized farms. The position around Letterkenny is similar and comparisons with neighbouring regions exhibit the usual paradox of Irish farming – the poorer the land, the smaller the farm

Mixed farming is practised, with oats, potatoes, some wheat and formerly flax as the chief crops, along with cattle for meat, dairy produce and sheep on the higher ground. As a relatively rich agricultural region, east Donegal is served by market villages almost reaching the status of small market towns, such as Stranorlar, Ballybofey and Raphoe. Some of these as well as other centres, such as Convoy, have small industries, chiefly textiles, which employ several hundred people. 32% of the population in the Stranorlar rural districts are in towns and villages compared with 19% in Dunfanaghy and 25% in Glenties. Although east Donegal absorbs labour from other parts of the country, it has lost one-quarter of its population since 1891. Three observations may be made here. Firstly, a prosperous rural area may lose population through the desire of farmers' sons and daughters to enter professions or occupations in towns and cities elsewhere in Ireland, in Great Britain, or in America. Secondly, the poorer farms in such an area are likely to be taken over by richer farms as subsidiary holdings. Thirdly, it may seem strange that these richer eastern lowlands should show a population decline almost as marked as that of the congested districts of the Rosses and the north-western coastlands, but the secret lies largely with the experience and wishes of individuals.

CHAPTER 22

Northern Ireland

Northern Ireland is not Ulster but only two-thirds of it, nor is it a natural region but rather an area of varied landscapes, hills and valleys, uplands and lowlands, farmlands and peat bogs, rivers with numerous islands broadening out into lakes, coasts of fine cliffs alternating with sand dunes. For several generations recognized to be the most industrialized part of Ireland, it is not predominantly an industrial area, for its manufacturing activities are located primarily in Belfast and district, or in the city of Londonderry, with many works in smaller towns. Like the Republic, it lacks any coalfield to act as a magnet for mining and industrial enterprise and its prosperity depends on the import of raw materials and the export of finished products. What is unique in Irish experience is the growth of Belfast from the middle of the nineteenth century into a vast and spreading town, having its economic life based both on the shipbuilding of the Lagan estuary and the historic linen industry that developed first in the houses of the people, then on water-powered sites and finally in great factories. Yet the heart of Ulster lies in its countryside in these six counties just as in the three others, Donegal, Cavan and Monaghan, that now lie across the Border.

When the six counties were separated from the remaining twenty-six, it was not intended that the Border should retain its present position, and the British Government set up a commission to make a readjustment based on social, economic and geographical factors. The Report of the Commission was ready in 1925 but never published or implemented: therefore, the Border has itself acquired sufficient importance in the life of the region to be, in human terms, a major geographical and economic divide, which one is obliged to accept as a limit of some human significance. On the east Carlingford lough divides the two political units of Ireland: this fine inlet, with hills on either side, carries a shipping route to Newry, a town that was a great canal port in its day. Easy ways are open through the passes between the Carlingford mountains and Slieve Gullion, to Dundalk and the Central Lowland; and Co. Armagh meets Monaghan in what is generally a farmed lowland, lacking any effective natural barriers such as major rivers or hills, apart perhaps from Slieve Gullion. The border runs through the drumlin belt in a complicated fashion between Armagh and Monaghan, follows the river Blackwater for several miles, and then twists round through an agricultural area to include most of the Erne basin, but neither the headwaters nor the mouth. Both the control of the river for flood prevention (p. 471) and its use for a hydroelectric scheme (p. 205) has involved diplomatic arrangement between Belfast and Dublin. And the boundary with Donegal, though winding through uplands between Pettigo and Strabane, follows the Foyle for several miles, winds round the outer suburbs of Londonderry and reaches the sea through lough Foyle.

What has happened is that a series of county boundaries have been elevated to frontier status. No part of Northern Ireland is inaccessible, for lowland routes penetrate everywhere. From Belfast, easy passage-ways lead beside the basaltic scarp to the sheltered glens of Antrim, to the drumlin-covered lowlands of east Down and the Ards peninsula, or through the Lagan valley to the Bann, which is crossed at Portadown. From the lough Neagh area ways open to the Erne valley with its spreading lakes, to the middle and lower Bann with their bogs and settled farmlands, to the hilly country of Tyrone, and southwards to the drumlin-drift belt and the central lowland. Physically the lough Neagh lowland is apparently designed by nature to draw travellers from many sources by road and rail to its focus in Portadown or to the ancient and dignified ecclesiastical capital Armagh. But Belfast is now a focus of more economic and administrative importance than either of these and therefore the Ulster Basin is merely a passage-way having no regional capital even though it is densely peopled, intensively farmed, and of some industrial importance. When the Bann is crossed, there is a lowland route to lough Foyle followed by the railway, an easy climb through hilly country to the Roe valley and thence to Londonderry, or an inviting passage to the Strule basin and the Erne valley. Historically the west of Ulster consisted of the basin of the Foyle and the Strule, together with various smaller river valleys approached by passes through cols in the hills that surround the basin. Londonderry town is its natural focus and the Foyle its great artery of commerce by river, road and railway, but the political boundary of 1921 has separated Donegal from the remainder of Ulster and insulated Londonderry from part of its natural hinterland.

Half the governmental province is formed by the three counties of Antrim, Down and Armagh, that is the area east of the Blackwater and the lower Bann. These two rivers, which became diocesan and county boundaries several hundred years ago, are almost as difficult to cross as the Shannon, and still separate the closely settled east of Ulster from the less densely settled west. In these three eastern counties, the imprint of settlement by immigrant Scots and English is most clear: here considerable areas have a distinctly Protestant way of life and outlook, though the descendants of the Catholic Irish or Scots from the western isles predominate in such areas as the Glens of Antrim, south Down and south Armagh, and form a substantial minority almost everywhere else. The east of Ulster has been transformed from a land of poverty to one of wealth in the last four centuries by the labour of its inhabitants, and two phases of its history are crucial, the plantation and immigration of the seventeenth century and the industrialization of the nineteenth. It is a country of careful farming and industrial towns, strongly individualistic, and differs so markedly from the rest of Ireland that its 'regional consciousness' is entirely beyond question. But in the three western counties the Protestant and Catholic communities are complexly intermingled, for the imprint of the Plantation has been far less sharp, though it is clearly apparent in Londonderry and district, where the tradition of the settlement of 1610, largely by people of English origin, is carefully preserved.

LANDFORMS AND LAND USE

The interpenetration of hills and lowlands is the salient feature of the Ulster landscape, which includes a variety of structural elements. Ancient schists and gneisses appear at the surface in the extreme northeast of Antrim, and over a much larger area west of the Bann, where they form a definite mountain chain, with several peaks over 2,000 ft. high in the Sperrins, and a number of lesser peaks, many of them over 1,000 ft., elsewhere. This ancient floor is structurally analogous to the Scottish Highlands and is part of the Caledonian province of the northwest and north of Ireland (pp. 14–17). Covering this ancient rockfloor there are various Palaeozoic sediments, which include the Devonian strata seen in the hilly country to the northeast of Enniskillen, and a wide range of Carboniferous beds, of which the newer members form plateaus to the southwest of the two Erne loughs and in Slieve Beagh, to the north of the upper lake. Carboniferous beds are also exposed on the coast to the east of Ballycastle.

North of the Lagan valley the most remarkable topographical feature is the basalt plateau, whose scarped edges are followed by the Antrim coast road and are equally well seen at the western extremity of the basalt outcrop, parallel with lough Foyle. This great basaltic lavaflow stamps a large part of northeast Ireland: both in the west and in the east it forms a plateau, but in the centre it underlies the Bann valley as a basin-like floor. The basalt scarps have numerous landslips, but for many miles the black basalts lie directly on white Cretaceous chalks, so well seen from the Antrim coast road: in places too it is possible to discern, beneath the chalks, soft Liassic clays, and red Triassic marls and sandstones. The Cretaceous chalks normally form only a thin ribbon and lack even a distinctive calcicole flora, though they have been a useful source of lime: on the high ground between Murlough bay and Torr head, there is an area recalling the chalk downlands of England, at approximately 900 ft.; this down-like area is immediately adjacent to a typical piece of basalt country covered with a peat soil and flora which is advancing steadily on the chalk with a raised and rounded edge about 1 ft. high. Elsewhere, drifts and glacifluvial gravels mask both the thin exposures of Liassic clays and the Triassic clays, marls and sandstones that floor the Lagan valley. The Triassic clays have been used for brick-making near Belfast, a city of red brick, and the sandstones, too soft for building materials, form gathering grounds for waters tapped by wells for the whiskey-distilling and mineral-water manufacturers of the city.

South of the Lagan–Bann corridor, Ordovician and Silurian rocks, with granitic intrusions, form a hilly lowland that extends from the coast through Down and Armagh into Co. Monaghan. This belt consists generally of low hills which, however, rise up to 1,200 ft. in Co. Armagh. The main uplands, on the other hand, are of igneous origin and include the Slieve Croob, a granite mass reaching 1,735 ft. on which the Lagan rises, the Mourne mountains, a sharply peaked mountain range with several summits over 2,000 ft. and the Slieve Gullion ringdyke, which is adjacent to the similar ringdyke in the Carlingford mountains. These uplands all stand up decisively above the surrounding lowlands: they are of very diverse history, for the Slieve Croob and the associated Newry granites are of early Devonian age, and the Mournes are of Tertiary formation. Consequently the Mournes show a much more youthful type of topography, with its sharpness accentuated by a local glaciation that left a number of cirques in the heart of the mountains, especially on the east and north-facing slopes. Another remarkable feature of the countryside is its hundreds of drumlins, spread over areas below 1,000 ft. and especially numerous on the lower ground in Armagh and north Down: in the south, around the Mourne mountains, they become less distinct and grade into crag-and-tail formations or even thin plasterings of drift on an irregular rockfloor. Within a single square mile of this Palaeozoic country there may be numerous drumlins, a heathery crag with a boulder clay tail, a peat bog and a small lake; wide views are rare and the scenery seems to some dreary and monotonous in its constricted horizons, but to others fascinating in its constantly changing vistas.

All the river valleys of Northern Ireland have been affected in various ways by glaciation, both by the plastering of drifts and by the deposition of glacifluvial sands or gravels associated with the far greater rivers and lakes that were fed by meltwaters. The 'glacial' lough Neagh was, according to Charlesworth, considerably larger than the present lake, which occupies a structural basin dating from pre-glacial times and is therefore, it is claimed, the oldest lake-basin in the British Isles as well as the largest. The Lagan was filled by a large lake impounded by ice near the coast, and smaller lakes were formed in all the eastern valleys of the Antrim plateau and on the north coast in the valleys near Ballycastle: a remarkable series of deltas in the Carey valley, east of the town, at 680 ft., 510 ft., 430 ft. and 340 ft., with others at lower levels, indicate stages in the lake's history. One of the most striking routeways through the area, from Portadown to Newry and Carlingford lough, is probably a meltwater channel: it carries the main road between the two towns. Carlingford lough, a fiordic inlet, was filled by a powerful glacier which has left some massive erratics weighing several tons.

Over the greater part of its course, the Erne spreads into a maze of waterways called lakes for convenience: thus the two Erne loughs are separated only by the river crossing at Enniskillen and the upper Erne lough is divided only by a few miles of complex waterways from lough Gowna, in Co. Cavan. All these lakes have numerous islands and peninsulas and in effect form one vast sheet of water between various drumlins and low hills: around them is an extensive network of bog and alluvium in which a number of 'loughlets' exist. There is every indication that the growth of vegetation, clearly perceptible around the shores of existing lakes, has gradually reduced them to their present size; until recently (p. 40) a 10 ft. rise in the water level could enlarge upper lough Erne to twice its normal size. This lake has an area of twenty-nine square miles, of which 5·3 square miles consist of islands, and the shore is 100 miles long. The Erne has a normal level of *c.* 150 ft. in the two loughs, but at Belleek its concentrated waters, which drain a catchment basin of 1,689 square miles, descend in a course marked by rapids, now harnessed for a power scheme, to Ballyshannon, three miles away. Lofty and thickly wooded banks stand on either side of the river.

Non-agricultural land is restricted to the uplands of the Antrim plateau, the Mourne mountains and hills such as Slieve Gullion in the east, and to the Sperrins and certain lesser hills west of the Bann: there are some interesting fens around lough Neagh (p. 66). The intensive occupation of this region has restricted both forest and bog to those areas where farming is either impossible or barely remunerative, and many of the smaller lowland bogs have been entirely removed by a population in need of fuel and unable to afford imported coal, though bogs still survive in parts of the Bann valley and on the ridge between the Bann and the Main. Peat covers most of the Antrim plateau, and is well developed also on Slieve Croob and the Mourne mountains, though in the last-named drainage channels are eroding the peat so rapidly that already it has disappeared from the principal mountains where the 'summit heaths' (or 'grasslands') are developed on a soil consisting of 'black earth' and granite gravel. At the same time, peat is accumulating in certain places, such as a col near Slieve Donard; while there is every reason to suppose that the present rate of erosion is greater than that of peat-accumulation on sloping ground, areas of a level character may still see the development of peat. Farther west, beyond the Bann, the moist climate with an annual rainfall of 35″–40″ on the lowlands provides environmental conditions favourable to bog growth both on the flatter uplands and in the areas of impeded drainage in the Erne valley and elsewhere. The Sperrin mountains, for example, have large areas of bog in which the schistose rocks rarely appear at the surface, and the basaltic hills normally have a peat cover also. In each case, forestry plantations have been established successfully on many of the hillsides, even on peat, and the advocates of tree planting claim that this is a suitable form of land use, as much of the hill country is too wet for sheep.

Intensive agricultural occupation in the east has reduced the woodland almost to extinction, but the carpets of bluebells and primroses seen in spring on the slopes of the basalt escarpment, the Glens of Antrim and elsewhere, are evidence of the field layer of vanished woodlands. In the Mourne mountains, where there are several old woods and some new plantations, *Rhododendron ponticum*, originally introduced as undergrowth in a wood near Hilltown, has spread wild over many acres of stony hillside. The existing woods consist of groves of trees on some of the better-drained bogs, a few thickets of hawthorn, blackthorn, whin and blackberry on some of the poorer clay soils together with the plantations of the demesnes and of the government's afforestation programme. Many of the demesne woods are on good soils and some excellent timber has been produced. Only 11 square miles have been officially acquired in Co. Armagh, in many ways the most closely farmed county of Ireland: Co.

Antrim has over 30 square miles scheduled for planting but although there are a few old woods of considerable extent, such as those around Knocklayd hill, near Ballycastle, the country remains comparatively treeless, partly because the hillslopes in the glens have such a high grazing value that it has not been possible to acquire any large block of land at an economic price. In Co. Down the extension of forests is much more likely, as already the Ministry owns over 16 square miles and there are a number of demesnes in the area east of Belfast around Bangor and Newtownards. On the lower slopes of the Mourne mountains, the main forests are at Rostrevor, beside Carlingford lough and at Newcastle, where beech of fine quality has been produced and conifers such as larch and silver fir have been equally successful in the valleys leading to the Slieve Donard summit. But it seems improbable that the hillslopes and valleys of the Mourne mountains will ever possess the forested character that marks the Wicklow mountains and various other ranges in the south, as most of the granite soils are infertile and support only a thin covering of peat. In 1955 the Tollymore Forest Park was opened: in 1967 a second Forest Park was opened at Gortin Glen Forest, near Omagh.

On the numerous islands of the two Erne loughs, and also around some of the lake shores, there are vestiges of a forest which is either natural or semi-natural: some of the islands have dense mixed woods in which oak, ash, alder, birch, aspen, hazel, hawthorn, holly and other familiar deciduous trees occur. The woods on the bare limestone ridges to the west of the lake have sessile oaks, birch, hazel or ash locally dominant, and in spite of continuous cutting these deciduous woodlands remain one of the main scenic features of the lough Erne country. Planted woodlands are a notable feature of Co. Derry, which now has almost 3% of its area in government plantations: over 100 square miles is to be used in Tyrone and Fermanagh. The major woodlands in the lowlands are associated with various demesnes, such as Florencecourt, Co. Fermanagh, to the south of lower lough Macnean, where the original Irish yew (*Taxus fastigiata*), now widespread in cultivation, was found around 1767 on limestone.

In 1966 three-fifths of the farms in Northern Ireland were of less than 30 acres, 20% of 30–50 acres and 17% of 50–100 acres. The fields vary in size from *c.* 2 acres to *c.* 6 acres on the larger farms and in some landscapes the actual space taken up by hedges and ditches is a comparatively high proportion of the total area. On some of the higher ground towards the upper limits of cultivation the hedges spread into the fields as thicket of hawthorn, holly, gorse and blackthorns, but in general the farmers give careful attention to their land, though the proportion sown to crops is now as low as during the inter-war period. Slightly more than half of the province is in crops and pasture, and of this 30% was in crops in 1939, well over half during the war years, but only 16% in 1966. This does not include hay, cut on 30% of the improved land in 1939, but on rather less during the war years and 26% in 1966. Until 1960, potatoes and oats were the chief crops of Northern Ireland, but in 1961 the acreage of barley increased markedly due to the introduction of new varieties, so that the percentage of the arable area given to crops (excluding hay) was oats 25%, barley 53%, potatoes 18% and other tillage 4%. Wheat was grown on 3,000 acres in 1966, compared with 5,000 acres in 1961 and has therefore ceased to be a crop of significance. Flax, grown on 120,000 acres at the end of the war diminished almost to extinction by 1959: it is no longer recorded in official statistics. The agricultural life has three main qualities: first, it is based on the output of small farms in which crops cover a few vital fields; second, the general aim is the rearing of stock for meat and milk, and third, there is an historical link with the linen industry. Oats, potatoes and barley are used for feedstuffs and sold; potatoes are sold in the towns and there is an export of seed potatoes, especially from the Ards peninsula, under conditions of rigid official inspection for quality. Flax cultivation has a strong traditional association with the Lagan and Bann valleys together with the area around Strangford lough, and its decline to extinction is due both to the increase of imports and to the reluctance of farmers to grow a crop demanding much severe hand-labour. All through Ireland mixed farming is characteristic and changes from year to year: as shown above, an historic crop, flax, disappears and barley, little grown until 1961, has become dominant.

East of the Bann, the stock on the farms is normally the main source of cash income and the number of animals kept is high in proportion to the area farmed. In most rural districts there is an average of 25–35 cattle per 100 acres of farmland, of which one-quarter to one-half are milch cows, up to 40 pigs, 200–600 poultry and a varying number of sheep. The last are most numerous on and around the various uplands, especially the Antrim plateau, the Mournes and Slieve Croob, and least numerous in the Lagan and Bann valleys. The general tendency is for the number of cattle, sheep and pigs to remain fairly steady but the egg and poultry trades have become more profitable through efficient marketing in Great Britain. Poultry keeping was at its height in the immediate post-war years when the British demand was insatiable. Lambs, which are cheaper to rear than calves, commonly begin life on the upland farms of Down and Antrim and are fattened on the surrounding lowlands. Similarly, calves are moved from the higher mountain farms or the small lowland farms to larger holdings for fattening: as explained on p. 194, many of the cattle pass through several farms before they reach the abattoir or the docks. As the supplies of meat for the Belfast market cannot always be met from home supplies, partly because many farmers concentrate on the export market, there is normally an import of cattle from the Republic. Near Belfast, both in the Lagan valley and around Comber, vegetable growing is prominent; here also the dairy cattle are more numerous than elsewhere, and commonly number at least half the total head of cattle.

Nowhere is the possible specialization of a mixed type of farming more fully seen than in Co. Armagh, to the south of lough Neagh, where most of the holdings average less than 20 acres. Many of the smallholders in the Montiaghs area near the lake have land consisting entirely of reclaimed peat bog, on which they grow sufficient potatoes, beans, peas and cabbage for the year, as well as soft fruits, especially strawberries, raspberries or red currants and rhubarb. Many of these holders are also farm labourers for a few months of the year, chiefly in the spring and at the harvest, and some draw unemployment benefit during the winter. Other farmers having slightly larger holdings that include both bog-soil and clay drumlins grow fruit, potatoes and oats on the reclaimed bogs and graze cattle on the drumlins. Some farmers devote nearly all their land to cattle grazing, and others concentrate almost exclusively on apple growing, which is more fully developed here than anywhere else in Ireland. Again, on a number of farms, apple growing and cattle grazing are combined, with the result that in years of poor apple yields the farmers have at least the resource of the cattle. Further, there are farms near Portadown concentrating exclusively on poultry, or on dairy cattle, or on cattle and sheep together. This area, too, has poultry, often in very large numbers, on nearly every farm; it shows perhaps, a wider range of specialization than is usual in Ireland, but at the same time illustrates a surprising combination of resources among a hard-working and enterprising people favourably placed near town markets.

West of the Bann, the landscape is essentially agricultural and the patches of mountain or lowland bog or of woodland appear as little more than an occasional interruption of the farmlands. For the most part the drumlins of the Erne lowlands and the varied morainic or glacifluvial deposits of the valleys elsewhere provide soils that are reasonably well drained and in places of good quality. There is an even spread of settlement (Fig. 86) almost everywhere in the lowlands and also in the mountain valleys to a general level of *c.* 800 ft. The farms are mainly of small size: in all the three counties of Fermanagh, Derry and Tyrone, those of less than 30 acres predominate, though there are a number of medium-sized holdings of 30–100 acres and a few of more than 100 acres: the last-named includes some large farms on the lowlands, whose economy is directed to feeding cattle for the town markets or export through Londonderry; some of the large hill farms specialize in sheep rearing and young cattle. The fields are generally small and the landscape shows the fragmentation that is characteristic of the whole Ulster province from the drumlin-drift belt northwards.

Production on the farms is mixed and a substantial part of the acreage is arable in the counties of Derry and Tyrone: in Fermanagh, however, there is less cropping and the pastoral side of farming is dominant. There is an average of 25–30 cattle per 100 acres of improved land, less than half this number of pigs, but very large stocks of poultry, especially in the Bann valley. Sheep are numerically most important on some of the better-drained upland pastures around Limavady, Strabane and Castlederg rather than in the Sperrin mountains, but only in rare cases are considerable numbers of sheep found in the lowlands. Derry and Tyrone have mixed farming, with one-fifth or more of the land in crops, barley, oats, potatoes and turnips, but hardly any wheat as the heavy rainfalls and the cool summers are unfavourable to its growth and ripening. Some seed potatoes are grown by the more enterprising farmers, but grass seed is a crop of declining importance. Meadow hay is grown on a substantial part of the improved land and in many districts little more than half the farmland is in pasture, a condition only equalled in Ireland in Co. Antrim and parts of the Lagan valley. Co. Fermanagh, more thoroughly pastoral in its farming activity, has in most years a mere 10% of the land in arable crops, chiefly roots, oats and potatoes; but at least one-quarter of the land is in hay which, with the pastures, provides the main sustenance of the cattle. But it will be noted that Derry and Tyrone support as many cattle in proportion to the farmed area as the more pastoral Fermanagh.

INDUSTRIES

Belfast and its neighbourhood is the one highly industrialized area in Ireland, and is associated more than any other area in the British Isles with the linen industry. The survival of this industry in Northern Ireland is due to many causes, not the least of which has been the enterprise of its inhabitants throughout the nineteenth century. Flax growing was introduced into Ireland during the eighteenth century and was soon found in every Irish county, though by the middle of the nineteenth century, when the first agricultural statistics appeared, it was already concentrated mainly in the old Ulster province. Time was when observant travellers could mark their arrival in Ulster by the sight of the fields of flax, the retting troughs in the streams and the scutching mills dotted through the countryside, but flax is no longer grown and retting troughs have become little pools for paddling children. In the early part of the nineteenth century cotton manufacture seemed likely to dominate the industrial life of the Lagan valley, but this phase passed relatively quickly (pp. 103, 142), though there is still a strong trade in textile finishing, not only for linen woven in Northern Ireland but also for cotton and rayon goods received from Britain. The shirt-making industry, located primarily in Londonderry and district but also in Belfast, receives cotton and other materials from Britain as well as linen from home sources: a shirt may be made from cotton spun and woven in Lancashire, taken to Londonderry and then returned to a warehouse in Manchester for sale.

Linen spinning is concentrated mainly in Belfast, and weaving is located mainly in Belfast and its environs. The demand for linen is now less than one-third what it was before 1914, but the introduction of synthetic fibres, particularly rayon, to the mills has maintained at least part of the production and employment. There are large plants, in all employing more than 10,000 workers, for making synthetic fibres in Northern Ireland: ever since 1932 government assistance has been available for firms entering the Six Counties, including capital grants and factory space that can be rented. Since 1959, as never before, more people in Northern Ireland have been employed in the engineering industries (54,000) than in textiles (47,000 in 1961). The closing of the largest and perhaps most famous flax mill in Belfast seemed an event almost of tragedy. But much still survives, and the linen industry still exports two-thirds of its goods to countries over a large part of the world: there is still a demand for goods of high quality made from fully bleached cloth for shirtings, sheets, damasks and cambrics, and also on the sale of unbleached materials such as canvas, duck and drills for army, sports and other clothing. Most of the major towns owe their rise partly to the building of one or more mills engaged in one or both of the preparatory phases of spinning or weaving: these towns include Lisburn, Lurgan, Dromore, Banbridge, Portadown and Newry on the south side of the Lagan valley and Larne, Ballymena, Antrim and Ballymoney on the north side. Farther west, spinning and weaving helped in the development of Dungannon, Cookstown and a number of industrial villages, including Sion Mills (1,600), established in 1835 by a family anxious both to give employment and to manufacture linen. Traditionally the industry depended partly on home-grown flax, scutched in various small mills worked only seasonally, but now abandoned: almost forgotten too is the practice of bleaching linen strips in the open air.

Shirt and collar manufacture has been established in Londonderry ever since the middle of the nineteenth century and the large mills, in all about thirty in number, are a prominent feature of the town. In origin this industry was domestic, and associated with home sewing: by 1914 some 7,000 were employed in the factories and some 10,000 in the surrounding districts, including the Inishowen peninsula and other parts of Donegal, and the rural areas of Co. Derry as far south as Strabane. The domestic sewing is practically extinct, having been killed partly by its own small returns to the worker, by the increasing accessibility of the factories from rural areas by motorbus and bicycle, and by the creation of the Border: as noted on p. 453 various factories have been founded in Co. Donegal for the trade of the Republic. At present Londonderry has some 6,000 employees in its shirt factories, more than half the total of 11,000 for Northern Ireland. There are also shirt factories at Strabane and at Coleraine, where they have replaced spinning and weaving mills. Another industry allied to home crafts is a lingerie factory at Strabane and hand embroidery works at Dungiven and Enniskillen.

In the Lagan valley, the second major industrial complex centres around shipbuilding, which depends on the import of coal and (formerly) ores from the west of Scotland and Cumberland. The Belfast shipyards stand third to the greater Clydeside and Tyne–Wear–Tees areas, and have seen the building of numerous cargo and passenger liners. The docks and shipyards are built on reclaimed land in the Lagan estuary and are served by dredged channels 21 ft. deep, which are able to receive all save the largest liners afloat. Belfast shipbuilding depends chiefly on male labour and is therefore the complement of the linen industry, which employs more women than men. These two industries together are the basis of the economic life of Belfast, and both have various ancillary trades, among which rope and cable making is prominent. One large firm in the city employs nearly 4,000 people and produces ropes, linens and twines ranging from rope-cables to string. In addition, various firms at Belfast, Portadown, Newry, Larne and Ballymena make machinery, part of which is absorbed by the shipyards and the linen trade. The aircraft industry has also developed.

It is agreed generally that a wider range of industrial enterprise is necessary in Northern Ireland, which has an unemployment rate higher than that usual in Britain (p. 118). Several new industries have been added, of varied types, for example in the suburbs of Belfast. There are well-established domestic and export industries, such as distilleries at Bushmills and Coleraine, a large tobacco factory in Belfast, bacon-curing works at Belfast and Portadown and mills at a large number of places scattered through the countryside. There are furniture factories in Portadown and some neighbouring villages, and a carpet works in Portadown has been conspicuously successful. Personal enterprise has been largely responsible for such developments, for there are no home resources of coal adequate for industrial use: local foresight with perhaps an element of luck gathered the linen trade into the northeast from the whole of Ireland and added other manufactures of agricultural origin. Hopes of a vast expansion of agricultural industries have long been held and the making of milk products has been conspicuously successful. At present the main hope lies in the building of a 'regional city' based on Lurgan and Portadown to serve partly as an area of industrial growth.

POPULATION

The general distribution of population in Northern Ireland is discussed on pp. 141–3: here it is proposed to say something on its regional distribution. The three eastern counties, including Belfast, have over a million people, with over one-quarter of the whole Irish population in less than one-tenth of its area. The overall density here is 400 to the square mile, a figure which, though less than that of England, Holland or Belgium, is higher than in any other part of Ireland except the county of Dublin. There are comparatively high rural densities of population in these counties, though they are markedly less than in the years before the Famine when, for example, Co. Armagh had 232,400 people, of whom only 40,500 were in towns and villages, and the density averaged 511 to the square mile. In 1961, Co. Armagh had 117,600 people, of whom almost half are in towns and villages. Both in Antrim and Down, two-thirds of the population is in the rural districts, but these include some areas suburban to Belfast (p. 145). The three western counties of Northern Ireland, Londonderry, Tyrone and Fermanagh, had in 1961 a population of 351,000, 137 to the square mile, of whom over two-thirds lived in towns and villages. In Northern Ireland, the boundary between industrial and agricultural activity is difficult to draw as several thousand people live in

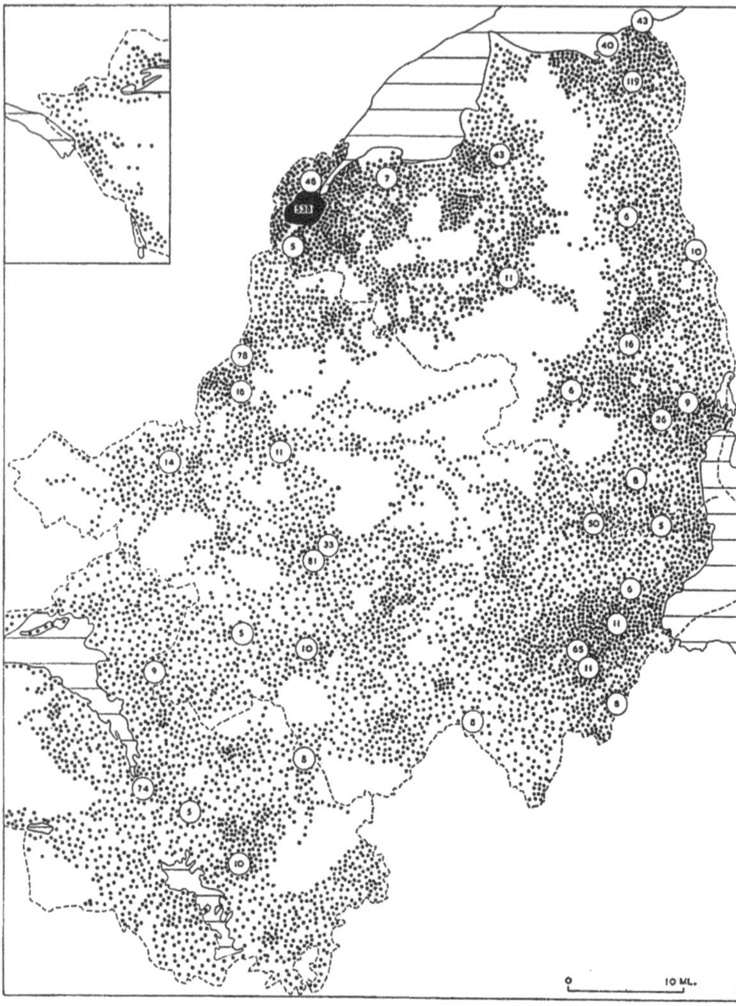

FIG. 86. POPULATION DISTRIBUTION, WEST OF THE BANN

One dot to 25 persons

The general pattern is of settlement in lowlands, and in the valleys of the Sperrins and other uplands. The densities of population are higher in Londonderry and Tyrone than in Fermanagh. The figures for towns are for 1961.

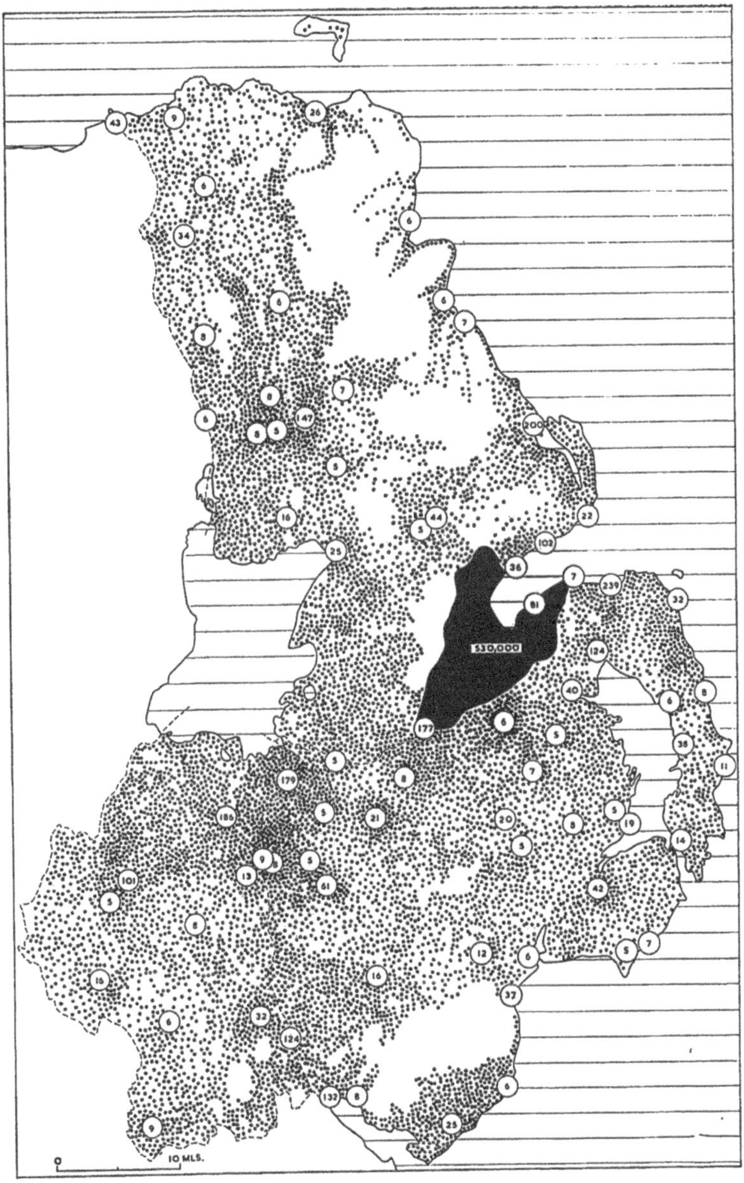

FIG. 87. POPULATION DISTRIBUTION EAST OF THE BANN

One dot to 25 persons

Apart from the uninhabited islands and some lowland bogs, the density of population is high almost everywhere, especially in the Lagan valley and the lowland to the south of lough Neagh.

industrial villages, each with a few hundred people, or go to work from rural homes and small farms, in some cases combining their employment at the mills with smallholding. In some ways the complexion of the economic life is comparable with that of Belgium, where workers are drawn into neighbouring towns with such facility that industry enters deeply into the social fabric of some countrysides. In Ireland labour is easily attracted from any rural area if it is needed, for the cash returns of town employment seem considerable to a community of small farmers who rarely allow for the fact that wages do not include food and that on farms incomes may be low but the family subsistence is assured.

In the three eastern counties, the population was 951,000 in 1841, and only 821,000 in 1891, but there was an increase to 1,030,100 in 1951, and 1,074,600 in 1961, mostly due to the expansion of Belfast. From 1891 to 1961 the rural population declined from 462,100 to 412,200, but of this some 75,400 is regarded as really belonging to Belfast: in fact the rural population has declined by approximately one-third. Meanwhile the town population has doubled: Belfast accounts for almost all of this, though there are substantial increases in some holiday centres, such as Bangor and Portrush, and some market towns with industries, such as Larne, Ballymena and Portadown. But most of the towns have changed little in population during the past fifty years and the natural tendency of industry has been to gravitate to Belfast with its immeasurable advantages of good transport and marketing facilities, and above all its port. Figs. 87 and 22 (p. 133) both show that the Lagan valley, from Belfast southwestwards to Portadown and Armagh, is one of the few densely populated parts of Ireland, and also illustrate the spread of the city into the adjacent countryside. Other densely settled areas include the slopes of the Mourne mountains (a small farm district), the neighbourhood of Ballymena, and the district around Newry and Bessbrook: the two last have several industrial villages. The gaps are mainly uncultivated bogs together with some hills in south Armagh, the Mournes, the Slieve Croob and the Antrim plateau. Only in a few areas are there less than 100 people to the square mile: generally the density is somewhere between 100 and 200 per square mile. Where it is above 200 to the square mile, the concentration is usually due to the admixture of industry or residential development with farming, though the concentration on the seaward side of the Mourne mountains is almost entirely agricultural as the profit from fishing and holiday trades is not sufficient to affect the population density to any marked degree.

The towns fall into three main groups: first, market and industrial;

second, residential; and third, holiday resorts. Some towns combine more than one of these functions and are, for example, both market centres and holiday resorts, or residential and holiday towns. Larne (16,300), the terminus of the short sea-crossing from Stranraer, has an aluminium works and is a market centre: it has doubled in population within the past sixty years. Newtownards (13,100), Lisburn (17,700), Lurgan (17,800) and Portadown (18,600) are all industrial satellites of Belfast with a considerable retail trade. More remote from Belfast but comparable in its general features with the towns just mentioned, is Newry (12,500), now declining in population. To the north lies strongly Scots Ballymena (14,700), centrally placed in the Main valley. Many of the smaller towns such as Banbridge (6,100) and Dromore (2,100) in Co. Down, Ballymoney (3,400) in Co. Antrim, Tanderagee (1,300) and Keady (1,600) show little sign of expanding in population, for the steadily increasing efficiency of transport militates against their trade and they are generally less likely to attract new factories than the larger centres, though industrial planning may arrest decay in some cases.

Residential towns include some Belfast suburbs such as Holywood (8,100) and Bangor (23,900), which has increased fivefold in the past sixty years and may be compared, in position and functions, to Bray, near Dublin. Armagh (10,000) is most famed as the major cathedral city of Ireland and has a strongly residential atmosphere: it lies off the main railway routes as its less lovely neighbour, Portadown, became the important railway centre of the lowland. Holiday resorts are spread along the coasts at intervals of a few miles from Portrush to Warrenpoint and include many villages with houses let for a few months in the summer. Portrush (4,300) receives thousands of visitors: Ballycastle (2,600), a town with its centre one mile from the sea, has added holiday catering to its market functions and a number of villages along the Antrim coast road, such as Cushendun, Cushendall and Glenarm, also receive numerous tourists. Whitehead, at the northern entrance to Belfast lough, is a centre for day visitors, who also frequent Bangor and various other places on the more residential south side of the lough. On the Ards peninsula there are various holiday centres of which the largest is Donaghadee (3,200): farther south the main tourist centres are Newcastle (3,700) and Warrenpoint (3,200) at each end of the Mourne mountains. The natural attractions of the East Ulster coasts are so great, especially around the Antrim plateau and the Mournes, that the continued growth of its tourist traffic, fed primarily from Belfast but also from Great Britain, seems virtually certain.

Fermanagh, Tyrone and Derry had, in 1961, 242,100 of their 350,850 people, 69%, in rural areas. A century of population decline has reduced the population by half in Co. Derry, by three-fifths in Co. Tyrone and by two-thirds in Co. Fermanagh. These decreases are not untypical of the Irish countryside and are similar to those in the non-industrial areas farther east. Of the towns, a number have grown slightly since 1937 but Londonderry has grown considerably since 1891, from 33,200 to 50,100 in 1951 and 53,740 in 1961. Coleraine (11,900) in the Bann valley, Omagh (8,100), Strabane (7,800) and Enniskillen (7,400) have grown recently. Some of the larger market towns possessing good communications and a measure of industrialism have captured much of the trade of the smaller towns: it is said, for example, that Londonderry has acquired part at least of the trade of Strabane fifteen miles away. Similarly Coleraine is in a relatively favourable economic position, as it is the strongest market town for many miles, and has some shipping. In 1968 courses began at the new University of Ulster, at Coleraine, in time to have 7,000 students. Portstewart (3,950), on the coast, is a holiday and residential centre of a rather quieter character than Portrush. Many of the smaller towns and villages are significant trading centres for the countryside. Fig. 88 is reproduced partly for its historic interest, as it was used as political propaganda: it shows that the Border cut through the market areas of Londonderry, Strabane and Clones. The surprisingly large market area of Enniskillen will be noted, but although before 1921 the town was frequented at times by people from Leitrim and Cavan, many of the farmers used smaller towns and villages closer to their homes.

Londonderry city stands out as the one regional capital in the west of Ulster, having, before 1921, a port hinterland that covered the whole of Donegal, a wide area of predominantly agricultural country stretching from Co. Monaghan to east Mayo and north Galway through the counties of Cavan and Longford, and also including the whole of Leitrim and Sligo. The Bann basin was served mainly from Belfast, which as a commercial port had a range of influence stretching far beyond the present six-county Northern Ireland. Londonderry and Belfast are at once competitive and complementary as ports, with the latter in a much stronger position than the former. Industrially the specialization in shirt manufacture gives Londonderry a measure of individuality in its economic life which compensates to some extent for its weakened nodality. During the war of 1939–45, Londonderry was a strongly guarded naval centre, as lough Foyle was used as an Admiralty anchorage instead of lough Swilly (in the Republic from 1921). In addition, substantial improvements

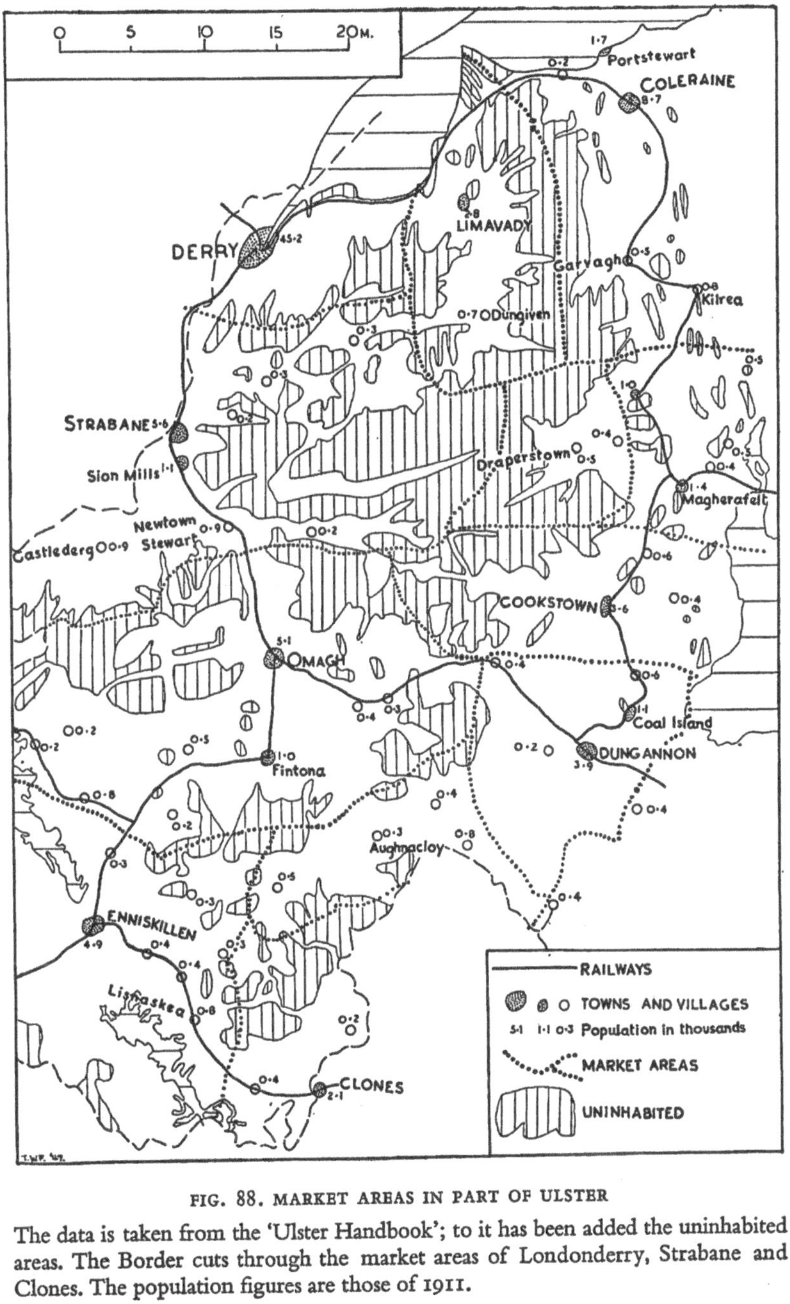

FIG. 88. MARKET AREAS IN PART OF ULSTER

The data is taken from the 'Ulster Handbook'; to it has been added the uninhabited areas. The Border cuts through the market areas of Londonderry, Strabane and Clones. The population figures are those of 1911.

have been made in the harbour facilities and the town therefore faces the last part of the twentieth century with certain recently acquired advantages to balance its frontier problems. Its very distance from Belfast, over seventy miles by rail or road, combined with the lack of any nearer competing town of comparable size, assures it of some economic strength.

SUBREGIONS

Physically the heart of the Ulster province is the area around the lough Neagh basin; economically the nodality of Belfast is so firmly established by road and rail communications that it is a feature of major importance in its regional geography. In this chapter, Northern Ireland is divided into twelve subregions, of which the first six cover the three eastern counties (E1–E6) and the second six the three western counties (W1–W6). Every subregion could be further divided into a multitude of smaller units, for the glaciation has left its mark firmly on the landscape: the soils, mainly derived from drifts, differ widely within quite small areas and the drainage is similarly varied. Even one square mile of farmland normally has fields with a wide range of fertility owing to differences in soil, aspect and drainage, and the smallness of the farm units accentuates the diversity of the landscape. Indeed Northern Ireland as a whole shows changes of scene as frequent as those which characterize the drumlin-drift belt described in Chapter 20.

E1. THE MOURNE MOUNTAINS

The Mourne range is thirteen miles long by five miles broad, aligned from northeast to southwest, and formed of granites intruded during Tertiary times into Silurian grits and slates. More than a dozen summits exceed 2,000 ft. and the highest, Slieve Donard, 2,796 ft., stands at the northeastern end of the range. The valleys are deep and the moorland country above them is smoothly moulded and open, though a corrie glaciation of late date has left a number of cliffed basins, some of which are lake-floored. On the east, between Newcastle and Kilkeel, the hills rise steeply from the surrounding lowland, but to the southeast, southwest and north there are various foothills formed of metamorphosed Silurian sediments. The deep valleys have been both accentuated and softened by glaciation: typically U-shaped in, for example, the Silent Valley, they have thick ground moraine accumulations on their floors. Borings for the dam constructed in the Silent Valley penetrated 172 ft. of ground moraine, which consisted of granitic sands, gravels, very fine silts and huge boulders up to 15 ft. long of the local granite. The lowlands around the mountains consist of Silurian grits and slates thickly plastered with ground moraine.

Agriculturally the Mournes are little more than an open sheep-run, though forestry plantations are now being developed on some of the lower slopes, especially around Newcastle. There is little settlement above 600 ft., but the coastal lowland is as closely settled and actively farmed as any area of Ulster and normally has at least one-third of its area in crops, chiefly barley, oats and potatoes, with large numbers of cattle, poultry and sheep. Kilkeel (2,300) is the main market centre. The agriculture on the north and west sides of the mountains is similar, as sheep, cattle and poultry are all prominent and the amount of land under crops is relatively considerable. Around the mountains the population has declined by 30% since 1891, but the density per square mile is still 150 near Rathfriland and 200–300 between Newcastle and Kilkeel, even though there are a few non-agricultural resources except some catering for summer visitors. Kilkeel has several boats fishing for whiting in winter and lobsters in summer. Trawlers also fish for herring in the Irish sea from August to October and off the south of Ireland later.

E2. THE SOUTHWEST (CO. ARMAGH)

From the western flanks of the Mourne mountains to the river Blackwater the country is diversified by numerous hills, covered by great swarms of drumlins, and lacks any closely marked regional limits. Structurally it consists mainly of 'Newry axis' country, of Ordovician and Silurian slates and grits, penetrated by a magnificent ringdyke with Slieve Gullion as its centre: this feature is of Tertiary age and stands immediately beside the ringdyke forming the Carlingford peninsula (pp. 443–5). The Newry axis country is generally hilly, and rises up in smooth lines to 1,200 ft. near Newtown Hamilton; its edge runs parallel to and south of the main road from Monaghan to Armagh and Portadown, where the Carboniferous, Permian and later rocks appear in the lowlands, though Tertiary basalts cover a considerable area around Portadown. In short, all the four main structural elements of the northeast, the Caledonian floor of the Newry axis country, the sedimentary beds of the lowlands, the Tertiary igneous hills and the basaltic cover, are represented in this subregion, yet the extraordinary wealth of drumlins gives it its main character as a landscape. Towards lough Neagh, however, glacial gravels and sands are spread out in large sheets to provide a fertile soil. The clearly defined valley running from Portadown through Newry to Carlingford lough cuts as a ribbon-like gash through the hill country: this ill-drained valley is followed by a canal, no longer used for commercial traffic, connecting Carlingford lough and the Bann at Portadown. The Newry Ship Canal carries three services to Liverpool every week and some general cargo with oil. But there are now no railway services.

This subregion is marked by an intensity of agricultural production on a small-farm basis, and all save a small fraction of the country is farmed, for the limit of settlement reaches 1,000 ft. and even higher in places. The small farms have a mixed economy in which as much as one-quarter of the acreage may be sown to crops, a substantial part of the remainder is meadow, and the rest pasture. Barley, oats and potatoes are the main crops, with some grass seed: the animals include sheep, cattle, pigs and a high proportion of poultry. Creameries were strongly established in south Armagh but are now superseded by milk-collecting depots: dairying is merely one of several economic resources. It is, to some extent, also industrial, having a tradition of textile industries at Newry and its neighbour Bessbrook, with Keady and Portadown. Many of the smaller towns and villages have a somewhat desolate appearance, as their trade has moved to the major centres: while the larger towns have remained stable in population or even increased, the country population has declined by 30–40% since 1891. Even so, the density per square mile is still around 120–150 rising to 300–500 per square mile around Portadown and Armagh through the fusion of intensive agriculture, including fruit growing near lough Neagh, and industry, both in the towns and in a number of villages. This high density of population suggests the transition to the Lagan valley.

Through this subregion runs the boundary between the dominantly Catholic and Protestant communities, but it traces no clear line. Portadown, for example, is three-quarters Protestant, Armagh city three-fifths Catholic and Newry even more strongly Catholic. If, as is sometimes maintained, religious allegiance is a possible basis for a revised border between the two political units, the actual line would be hard to draw, though it would probably run from the northern end of the Mourne mountains to a point somewhere in the neighbourhood of Armagh city.

E3. THE ARDS AND MID-DOWN

On either side of Strangford lough there stretches a countryside of low hills reaching up to 600 ft., with numerous drumlins, which are seen around the lough in various stages of submergence. This countryside is floored by the Silurian slates and grits which appear occasionally at the surface, especially along the coasts: it has one small upland, Slieve Croob, a granitic intrusion of Caledonian structure and possibly Devonian age. Slieve Croob is a treeless moorland, cultivated to an altitude varying from 600 to 1,000 ft., with a clear 'hill-pasture zone' forming a girdle of rough grassland partly covered by gorse, holly and stunted blackthorn around the mountain. Elsewhere the agricultural lowland prevails: laid out in small fields, it has little rough pasture except in the marshy depressions between the drumlins, which at one time held peat but are now almost entirely cleared for fuel. Small farms predominate though there is a sprinkling of medium-sized and large holdings. The proportion of land in crops is generally at least one-quarter, and on the Ards peninsula there is some cultivation of wheat as well as an important sale of seed potatoes and grass seeds. But the main purpose of farming in the Ards peninsula is to rear lambs, sheep and cattle to be fattened for the market or fed for the export store trade. There are small market centres at Ballynahinch (2,000), Crossgar (840) and Saintfield (700), all long known for cattle sales: to a great extent the farmers buy young stock from the Mourne mountains or south Armagh and then feed it for some months before its final sale. In this respect their function is similar to that of the 'graziers' in Co. Meath, but with the essential difference that the stock are fed on the small and medium-sized farms of a strongly arable character rather than on 'ranches'. Poultry rearing is also a vital resource on most farms and pig production has steadily increased. Ardglass and Portavogie are fishing centres.

Most of the towns are small, with the exception of Newtownards, an industrial appendage of Belfast. Downpatrick (4,200) has the cathedral of the Down diocese and a reputed grave of St Patrick. Along the coast of the Ards peninsula there are several holiday resorts and in the south, at the entrance to Strangford lough, is the small town of Portaferry (1,400). The density of population ranges from around 100 to nearly 200 per square mile, and is that of an agricultural community living mainly on small farms without any marked industrial development. Since 1891, the population has declined by approximately one-quarter.

E4. THE BELFAST CORRIDOR

The entrance to Belfast from the sea runs through the lough for nearly twelve miles before passing through the deep artificial waterways that lead to the harbour. Some four square miles in the centre of the city is reclaimed land, occupied partly by docks and warehouses and partly by houses. To the west of the city lie the basaltic scarps of Divis and Cave hill, to the east the drumlin-covered hills formed of Ordovician and Silurian slates and grits; between these two clearly marked limits is the Lagan valley, floored by Triassic sandstones and marls covered with drifts, lacustrine sands and gravels of glacifluvial origin and, near the coast, estuarine silts. The development of Belfast is intimately linked with its emergence as a great port. It also possesses a natural nodality that has been steadily accentuated by the building of modern trunk roads and railways: on the northwest a low pass leads from the lough into the fertile valley of the Six Mile water and on to the town of Antrim; to the east there is the Dundonald gap, an old east–west fault valley in the Ordovician and Silurian hill-country, having Triassic sandstones comparable with those of the Lagan valley. All round the lough shores there is the lowland fringe occupied by seaside suburbs, and to the south and west there is easy access to the middle Bann valley and the country around lough Neagh.

A charter was granted to Belfast in 1613, and it became a garrison and market town having both English and Scottish settlers. In the early part of the nineteenth century Belfast was still a compact town on the west side of the Lagan, having a number of Regency buildings of which some still survive. As the city spread, the more fortunate among the population settled to the north along the Antrim road or to the south along the Malone ridge: the poorer classes settled on the west side, where many of the linen mills were established: at Ballymacarrett, already an industrial suburb by 1840, engineering and chemical works were established. These, with the vast shipbuilding yards on the Lagan, dominate the east side and the central shopping, commercial and professional quarters are found on the west side of the river: here too are the main railway stations, neither of them inconveniently far from the city centre. Belfast, unlike Dublin, has few tenements but a multitude of streets having small family houses, many of which fell to the ground during the air raids. But Belfast, like Dublin, has spread by the building of Corporation estates and other houses. The expansion is most marked on the northeast and the south, as the west side expansion was limited by the overshadowing hills and by the industrial character of the neighbourhood: in 1958, the Belfast rural district was abolished, and part of it, on the northeast shore of the lough, became the urban district of Newtownabbey, which had 37,500 inhabitants in 1961.

Belfast city has spread from its original nucleus over the lower slopes of the basaltic escarpment, around the lough shores, especially on the south side towards Bangor, into the Dundonald gap, and along the Malone (sands) ridge which stands between the Lagan and the little river Blackstaff, a marshy area largely avoided by settlement and therefore isolating the south suburbs of the Malone ridge from the smoky western industrial districts of Belfast. The continued growth of the city may eventually engulf such independent centres as the ancient town of Carrickfergus, the modern seaside resort of Bangor, and the linen spinning and market town of Lisburn, for all these, and several others even farther from the city such as Donaghadee on the Ards coast, are already its dormitory towns. It is possible to recognize a Belfast 'conurbation' extending from Lisburn to central Belfast and then bifurcating into two long fingers on either side of the lough to Carrickfergus and Bangor respectively, a south–north distance of twenty miles; the maximum east–west extent of settlement, from Legoniel to Dundonald village, is hardly ten miles.

The Lagan valley has at once an industrial, a residential and an agricultural character, and to the south of the somewhat congested city these three elements in its life are closely intermingled. Several of the bleach mills and weaving factories are in rural surroundings, some of them on the canalized Lagan which, only a short distance from the centre of the city, flows through a pleasantly wooded and secluded glen cut in the glacifluvial sands of which the Malone ridge is also made. The beech woods of this valley, and those of the numerous demesnes in the vicinity, are almost the only woodlands in the lowland, though the country has a well-wooded appearance due to the hedgerow trees, with occasional shelter-belts and small groves planted for ornamental purposes.

Farming is favoured by the combination of light, generally sandy soils, a moderate rainfall (Lisburn mean annual 34″) and the accessibility of the Belfast market. More than half the cattle are dairy cows, and some of the rest stores received from other parts of Northern Ireland for fattening or export from Belfast. Normally eight- or nine-tenths of the farmland is in grass, and the main arable crops are oats and potatoes. Market gardening is a specialization found in the neighbourhood of Lisburn and also to the east of Belfast in the Dundonald gap. Towards lough Neagh, however, the more typical Ulster farming based on arable cropping combined with cattle grazing is found with a certain amount of horticulture. The density of population is high, mainly over 400 to the square mile, but the rural areas unaffected by industry or by the spread of Belfast have declined in population by 10–20% since 1891.

E5. THE ANTRIM UPLAND

The greater part of this upland is formed of basalts, though part of the ancient crystalline floor and various Palaeozoic sedimentary rocks, including some Carboniferous Coal Measures of little value, occur in the extreme northeast. Beneath the wall-like basalt outcrops the Cretaceous, Liassic and Triassic beds appear, and many of the soils around them are derived from a variety of components, including glacifluvial sands and gravels. The plateau surface, around 1,000–1,500 ft., with a few rounded hills such as Trostan, 1,817 ft., is a dreary peat bog redeemed by the glens and the narrow coastal fringe: in these sheep are the major resource, though as much as one-quarter of the land is tilled, for barley, oats, potatoes and hay. Although the population per square mile in the inhabited areas is around 100–150 to the square mile, the decline of population since 1891 has been as much as 40% around Ballycastle, a neighbourhood with a Roman Catholic majority both in the town and the glens. Farther south towards Belfast there is an agricultural transition towards dairying and cattle rearing, though the plateau between Larne and Carrickfergus is mainly a sheepwalk. From Larne southwards, the coastal fringes have a steadily increasing density of population, which reaches 300–400 to the square mile: even on Island Magee, a peninsula flanking Larne lough, the density of population, almost wholly agricultural, is around 150 per square mile. The town of Larne is a packet-station, market and industrial centre, and also a railway terminus and the gateway to the Antrim coast, a famed tourist route; it stands at the seaward end of the wide gap leading to Ballyclare and Antrim.

E6. THE MAIN AND BANN VALLEYS

The Bann rises in the Mourne mountains and flows across the lowlands around Portadown to enter lough Neagh which, with an area of 155 square miles, is the largest lake in the British Isles. Formed in Tertiary times, probably Eocene, it lies, everywhere except on the north, over the 'lough Neagh clays', the last and not least puzzling of the sedimentary formations in Ireland (p. 25). Several other streams flow into the lake, including the Blackwater from the southwest, which traverses a limestone valley from Monaghan, the Six Mile Water from central Antrim, numerous streams from the west, and finally the important river Main, which flows parallel to the Bann itself, but from north to south. The Bann leaves the lake as a considerable river and flows through the collapsed centre of the basaltic plateau to the Atlantic. There is easy communication between the Bann and the Lagan valleys, but both streams are county boundaries over part of their courses: the Bann is still in many ways a barrier as there are extensive bogs around it and the bridges are placed at intervals of eight to ten miles. The main northward route by road and rail avoids the Bann valley and runs through the Main from Antrim to Ballymena to reach the Bann at Coleraine, Co. Derry. Therefore this subregion may be thought of as including the western flanks of the Antrim upland, the Main valley, and the low ridge forming the interfluve between this routeway and the Bann itself.

As a whole, this subregion presents a great range of agricultural types, some industrial development, and landscapes that owe their variety partly to the influence of glaciation and post-glacial developments. Around the southern shores of lough Neagh there are fen peats in many stages of reclamation, with an agricultural life focused partly on cattle grazing but also on fruit production: patches of uncleared bog are intermingled with intensively farmed and densely peopled enclaves: small fruits and vegetables are grown on the easily worked peat soils, and the most important cash crop is strawberries, with raspberries, blackcurrants, rhubarb, carrots, leeks, parsley, spring onions and celery. East of the lake, in central Antrim, the farmlands have arable cropping, mainly oats and potatoes, intensive hay production and cattle both for milk supply and for meat: the population density is around 100 per square mile and the only town, Antrim (1,450) is a small centre. To the north of lough Neagh, the peat bogs are interspersed with farmed drumlins. The Main valley, on the other hand, is far more closely settled, and the town of Ballymena is the market centre of a wide area that includes the valleys cut in the western part of the Antrim upland. Twenty miles separate Ballymena from Ballymoney, the nearest market town to the north; eleven miles to the south is Antrim. Ballymena is the centre of a mixed farming district in which almost half the improved land is divided between crops and hay, and cattle are reared both for the store trade and for dairying: sheep production is also important, especially near the uplands, and there is also a considerable number of pigs and poultry. Ballymena has linen factories and the footwear industry is developed here. The Main and Bann valleys were formerly the chief flax growing area of Ireland, as the crop was well suited to the rich basaltic soils.

W1. THE WEST FLANK OF THE BANN VALLEY

Between the hill country of Derry and Tyrone on the west and the river Bann with lough Neagh on the east, there is a lowland some ten miles wide and fifty miles long, closely settled, actively farmed and of some commercial significance. The structure of this subregion is complicated: in the north, basalts floor the lowlands, but to the south, around Maghera, sedimentary rocks of Carboniferous and later ages appear at the surface. These include some Coal Measures around Coalisland, some Triassic and

Cretaceous beds, and the lough Neagh clays near the lake. It is, however, the glacial and glacifluvial deposits that form the parent material of most of the soils for, as Charlesworth has shown, there are swarms of drumlins, spreads of outwash gravel and sands, and, on the flanks of the uplands, lacustrine deposits laid down in ice-dammed lakes. There are no soil profiles available, but the soils of the farmed country are generally reported to be reasonably fertile and well drained, though some low-lying areas are covered with bogs, notably near the Bann, where they form appreciable gaps in the settlement pattern (Fig. 86).

Agriculture has a strongly arable complexion in this subregion and one-quarter to one-third of the improved land is cropped with barley, oats and potatoes, with as much as one-eighth to one-quarter under meadow hay; little more than half is permanent grass. Barley has replaced flax on the farms here, and there are three other main cash crops – oats, potatoes and grass seed. The cattle include both stores and milch cows, and creameries formerly had some measure of success in the area; pigs, poultry and, near the uplands, sheep are also sources of income. Small farms of less than 30 acres predominate, and the rural densities of population, some 150 to the square mile of improved land, remain high in spite of a decline of 30% since 1891. The towns are market centres approximately ten miles one from another, all of some antiquity and partly revivified during the seventeenth-century plantations and settlement, but with the clearer imprint of eighteenth-century planning and nineteenth-century industrialization. Of these towns, the most northerly, and successful, is Coleraine, four miles from the mouth of the Bann, a port, market and industrial centre of 11,900 people, an increase of 11% since 1951. The other 'towns' include Kilrea (950), Garvagh (550), both now entirely subsidiary to Coleraine, Maghera (1,600), Magherafelt (2,460), Cookstown (5,000), Stewartstown (620) and Dungannon (6,500): of these places Maghera, Magherafelt, Cookstown and Dungannon have increased considerably in population during and since the 1939–45 war. Cookstown, laid out around 1750 by Mr Stewart, the local landowner of the day, consists of a street one mile and a half long: Maghera also has one main street with subsidiary roads, but Dungannon, on a limestone hill, has another familiar Georgian plan, a square with four principal streets.

W2. THE DERRY–TYRONE UPLANDS

Two structural features appear in these uplands: first, basalt plateaus, almost flat-topped, with scarped edges overlooking the Roe valley; and second, the smoothly rounded Sperrin hills, which include several summits over 2,000 ft. and are made up of ancient schists and grits with various intrusions. The basalt plateaus are similar to those of Antrim and like them are covered with peat, except where some forests are now being planted: considerable areas are above 1,000 ft. and in Benbradagh, near Dungiven, there is a summit of 1,535 ft., with 1,825 ft. in Mullaghmore. The Sperrin mountains are merely the main range in a wide hill belt which extends from Slieve Gallion, 1,735 ft. (north of Cookstown), to the Foyle and westwards into Donegal. With the basalt plateaus, they form an effective divide between the lower Bann valley and the Roe valley. No railway crosses these uplands, though roads climb through passes at 750–900 ft. between the Bann valley on the east and the Roe and Foyle valleys to the west.

W 3. THE CLOGHER VALLEY

This wide open valley lies between Devonian uplands with summits around 1,000 ft. on the north, and a Carboniferous plateau, the Slieve Beagh, reaching 1,255 ft. on the south: it is floored by Carboniferous rocks overlaid by drifts and is an essentially agricultural area hardly touched by industry. The Clogher valley light railway, which ran through the heart of the area, was closed in 1941 (p. 244). The farming is mainly pastoral and normally at least three-fifths of the improved land is in permanent pasture, with meadow on most of the remainder: the transition from the type of farming with arable crops figuring prominently to the more purely grassland type of farming is apparent in a traverse of the valley from east to west. Cattle are kept for stores and milk and poultry-keeping is of particular importance. Most of the farms are small, of less than 30 acres, and the density of population is almost 120 per square mile of improved land. There are no towns and the market centres are villages such as Aughnacloy (800), Ballygawley (430), Augher (220) and Fivemiletown (780). In effect, this subregion is one of the many Irish backwaters, entirely agricultural and not unprosperous but off all the main routes.

W 4. THE ERNE VALLEY

Two lakes, upper and lower lough Erne, occupy the lower part of this valley, and Enniskillen, its one important town, stands at the crossing place between them. Not surprisingly the main roads avoided the centre of the valley for the lakes spread outward across the marginal bogs in times of flood until the recent drainage scheme was carried out. Drumlins on the valley floor are aligned from northwest to southeast and make a multitude of islands in the lakes or peninsulas on their margins: the valley is overlooked on the west by Carboniferous plateaus, some of which have scarped edges, but on the east side there are easy passage-ways into the Clogher valley and the Foyle-Strule basin. On the whole, this is an agricultural area, producing store cattle, having a considerable interest in dairying and a great development of poultry keeping; in some years less than one-tenth of its improved land is under crops, though at least one-quarter is meadow. The farms are mainly small with an intermingling of medium-sized and large holdings, and the density of population, 70–80 per square mile of improved land, is akin rather to that of the pastoral Central Lowland of Ireland than that of the more arable areas of Ulster. The rural areas have declined by 25% during the past sixty years. Enniskillen is the main centre. Lisnaskea (980) and Irvinestown (930) which are respectively ten and twelve miles from Enniskillen, are the only other 'towns', though the area around upper lough Erne has its natural market focus in Clones, Co. Monaghan. Viewed on a political map this part of Northern Ireland, itself highly mixed in political sympathies, is surrounded on three sides by the Republic, which has both the mouth and the upper reaches of the Erne river. The drainage scheme is noted on page 40.

W 5. THE FOYLE–STRULE VALLEYS

The Strule and the Foyle together form the main north–south stream of this large basin and receive a number of important tributaries which join the river at right angles on each side: two of the major left-bank tributaries, the Finn and the Deele, are in Co. Donegal and join the river on the nine-mile stretch that is the boundary between Northern Ireland and the Republic. But the basin is a well-defined unit, having one main north–south arterial route along the main river, joined by several tributary routes, and its communications are focused on a number of market towns of which Omagh, Strabane and Londonderry are the chief. In spite of the Border, Strabane and Londonderry are still the main centres for communications for Co. Donegal as well as for the valleys of the uplands to the east.

Structurally the basin is formed of ancient metamorphic rocks from lough Foyle to a point a couple of miles north of Omagh and its valleys are separated by hills from 1,200 to 1,800 ft. high, of which those near Strabane continue the line of the Sperrins while others, such as Bessy Bell, 1,387 ft., near Newtownstewart, stand up in isolation above the lowland. Around Omagh the lowland widens out into a basin, floored by Devonian and Carboniferous beds and comparable perhaps with parts of the Central Lowland of Scotland. Everywhere there are drifts and some glacifluvial deposits, particularly in the valleys of the Sperrins, which ice held up marginal lakes: in general the soils are of considerable fertility and land is cultivated to an altitude of *c.* 800 ft. The farming is mixed with at least half the land in permanent pasture and the remainder used chiefly for hay in the south towards the Erne valley, but for crops, barley, oats and potatoes, farther north. Financial returns are given largely by the stock reared both for the store trade and for fattening and sold mainly at the monthly fairs, together with milk production, partly for the canning factory at Omagh. Sheep are kept in large numbers only near the mountains.

Small farm units of 30 acres or less are the most usual type, with an admixture of medium-sized and large farms. The rural density of population per square mile of improved land is 110–120 in the south of the basin around Omagh, 150 in the central area, including Strabane and Castlederg and higher still near Londonderry. The market centres include Omagh (8,100) and Strabane (7,800) and several smaller places that are mere villages. There is a ring of minor market centres approximately ten miles from Omagh, including Sixmilecross (250), Fintona (1,300), Dromore (500) and Newtownstewart (1,300), all of which have lost much of their trade to the larger centres. Strabane has shirt factories and a lingerie works but like Omagh it is mainly a market town, though it has lost part of its trade to Lifford, a village in Donegal just across the river. Castlederg (900) is a minor market-centre ten miles from Strabane and Sion Mills (1,600) is an industrial village.

W 6. THE ROE AND FAUGHAN VALLEYS

Between lough Foyle and the west edge of the basalt plateau lies a hilly area, with crests around 800–1,300 ft. and valleys in between them: the two main rivers are the Roe and the Faughan. The area is one of ancient metamorphic rocks, though there are Carboniferous sandstones around lough Foyle and in the Roe valley and Triassic beds also appear in the Roe valley with a thin band of Cretaceous rocks immediately below the basalt scarp. The coastal plain consists of raised beach and alluvium, with spreads of glacifluvial sands and gravels and there are also some 'intakes' or slob-lands on the lake shores, formed by sedimentation behind walls or embankments made when the railway was built. The entrance to lough Foyle at Magilligan point is marked by a long belt of sand dunes. Both along the coasts and inland, many of the soils are well-drained sandy drifts and the farms generally have at least one-third of their land under crops, with another tenth in hay and the remainder as permanent pasture. Cattle are kept primarily as stores, though there is some fattening for the meat market of Londonderry and a few farmers specialize in dairying, or sheep, which are far more numerous than cattle on the hills. Almost half the area is rough grazing, much of which is well suited to sheep. The density of population in the rural area is around 110 per square mile of improved land, and the main town market is naturally Londonderry, though Limavady (4,300) and Dungiven (600), each seventeen miles away from Londonderry, are two ancient plantation towns and market centres in the Roe valley.

The City of Londonderry. Derry is a town of great antiquity, chosen as the centre of the London Companies' plantation in the seventeenth century, as it was a defensible and often defended site on a hilltop over 100 ft. higher than the Foyle, here 300 yards wide. This hill, a rock-knoll of schistose sedimentary rock covered with boulder clay, is surrounded by the Mary Blues Burn, an old channel that is still below the level of high water at spring tides, which are kept back by retaining walls. The nucleus of the town is the old walled city with its cathedral built in 1633 on the site of an earlier church, and its stately eighteenth-century buildings: this is still the main shopping, commercial and to some extent also residential part of the town, and the walls are well preserved. In the mid-nineteenth century, Londonderry had already spread beyond this nucleus along the ridge and also across the Mary Blues Burn to the rising ground beyond, and over the river to the suburb of Waterside. Derry was still, however, a town of merely 10,000 people and grew gradually during the second half of the nineteenth century to 40,000 in 1901 and 53,700 in 1961. The buildings spread across the rising ground on each side and Waterside became an industrial and residential suburb: docks and warehouses lined the river bank and four separate railway stations were built, two on each side of the river. The lack of planning is only too apparent in the city away from the original nucleus, especially along the riverside. The working-class suburbs occupy the lower ground with the stately old houses in the centre and the attractive modern suburbs spread out on high ground above them. The industrial strength of Londonderry is mainly in shirt-making, though some of the smaller factories have been closed recently. Its port was greatly improved during the war of 1939–45 when it was used as a naval base: since the war it has acquired chemical and food industries with oil refining. The older quays on the Foyle are no longer well kept and Londonderry profits comparatively little from its nearness to the Atlantic as the commercial strength of Belfast is incomparably greater. Nevertheless, it remains the regional capital of the northwest. Even so, in 1965 Coleraine was named as the site of the new university.

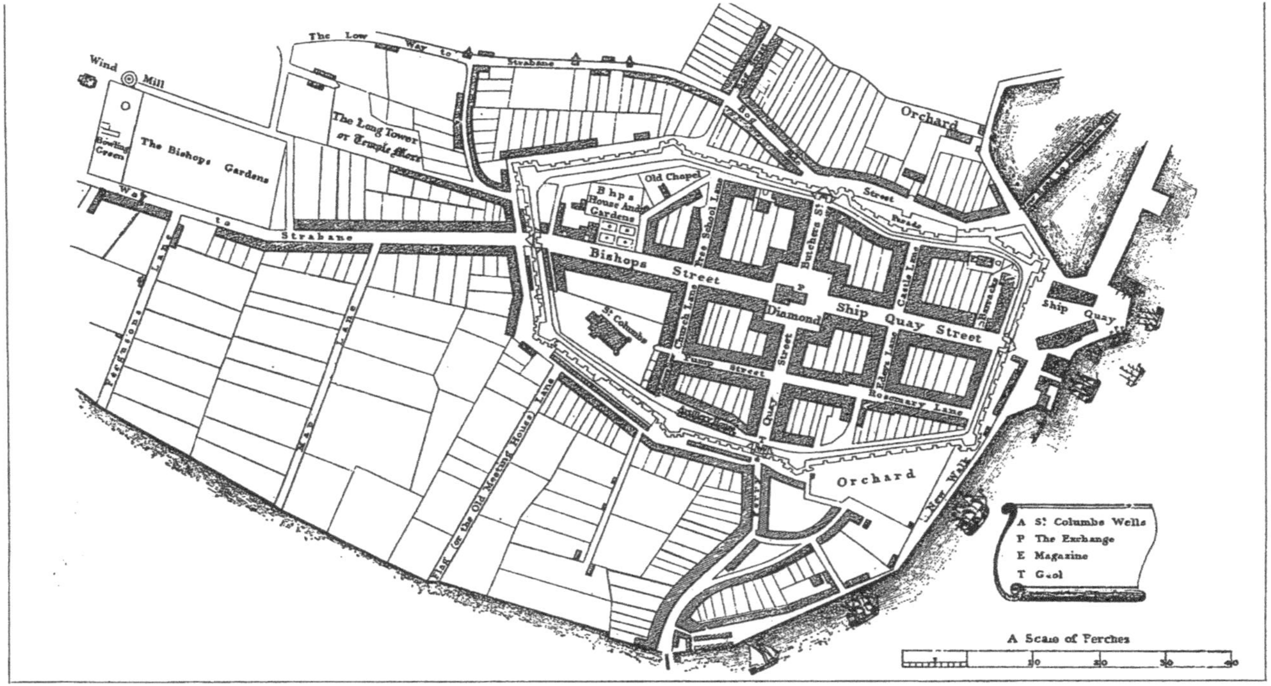

FIG. 89. LONDONDERRY IN 1788

The city within the walls could no longer contain the inhabitants, whose houses already spread in ribbons along various roads. The low ground (Bog Side) later became densely occupied with poor property: much land has since been reclaimed beside the river. But the main part of the town, within the walls, remains much the same and the Diamond is an attractive feature.

Reproduced from the 'Ordnance Survey Memoir 1837' *(see p. 539, section 4)*

CHAPTER 23

A Changing Island

It has been the purpose of this book to examine carefully the Ireland of the present day and, to a lesser degree, of the past. By deliberate intention the microscope has been turned on the countrysides of this land, rich and poor, wooded or treeless, mountainous or lowland. There is a diversity of economic standards among the agricultural population who range from the relatively prosperous farmers of 100–200 acres in the richer areas of Leinster or Munster to the cottage-farmers of 10 acres or less in the west, particularly in the Connacht province. Ulster has some of the most industrious farmers of the country, but it too is a province in which many of the farms are small and the standard of living extremely simple. Year by year the young people leave the rural areas, and steadily the rural population declines as it has done without interruption since 1845: there is no sign that the adjustments of the past century are complete yet.

Population decline makes Ireland a land unique in Europe – perhaps in the world. Hysterical speakers comment on 'the drain of our best life', 'the lifeblood of a nation drawn away', 'the flight from the land': less hysterical but romantically-inclined people see in 'peasant' ways an 'irrational' way of living that has immense but undefined satisfactions alleged to be of a spiritual character. Holy poverty, it seems, is embraced by all and sundry as they revel night by night in the heritage of an ancient culture and toil day by day on their fields among the stony wastes of Connemara. In fact the past century has seen the steady union of farms into larger holdings, the replacement of many – but by no means all – of the poorest houses by new ones, more comfortable but pitifully ugly, and a reduction of the rural population by half or even two-thirds in most countrysides.

Year by year, cottages inhabited by old people are abandoned after their death for no one is willing to farm so little land unless there is some other source of income available. A man working on the road, labouring on a farm, cutting peat on a bog, or having seasonal employment in a forestry plantation, will perhaps take such a cottage as a home with a useful patch of ground on which he may keep a cow, a few sheep, some poultry and grow his potatoes and oats; otherwise the smallholding has had its day. The farms isolated by great distance from neighbours, or on islands, are also abandoned. When the author first saw Beginish, an island near Valencia, Co. Kerry, in 1941, six families lived there: only three remained in 1952, though there were still six inhabitants (five males) in 1966. And by 1952, the inhabitants of the Blasket islands, to the west of the Dingle peninsula, were eager to settle on the mainland as the people of other western islands had done before them. There is a recession of settlement in all the harder environments of Ireland, perhaps due to restlessness inspired by education, the wireless, television, contact with tourists or – possibly most vital of all – accounts of the better living available elsewhere. The poorest people of the land are realists.

In spite of the great changes of the past hundred years, there is a remarkable range of population density in the rural areas of Ireland. It might have been expected that the poorest and most densely settled countrysides would have lost population at a higher rate than the richer and less densely settled areas: but it has not been so. There is a range of rural population density from less than 50 to more than 400 per square mile: there is *not* an even lowland density of population in Ireland, nor has there been for more than one hundred years, and probably much longer. One of the most interesting features of Ireland is that north of a line from Dundalk to Galway the farms are in general smaller and the rural density of population higher, than to the south of this line. In other words, Connacht and Ulster are more densely peopled, and farmed in smaller units, than Munster and Leinster. Why should this be? It has been so for a hundred years and almost certainly for a much longer period: but the reasons for this known fact of agrarian life probably lie deep in an historical past not so far revealed by historians. May it be that the Norman imprint on Leinster and Munster, combined with the plantations of parts of Co. Cork and the former King's and Queen's counties, was strong enough to prevent too minute a division of the land; and that in Ulster land was subdivided into minute holdings because its people lived partly by home industry in the eighteenth and early nineteenth century, whereas Connacht was the refuge of the dispossessed in Cromwellian and other wars? And why has Kerry larger farms than West Connacht or Donegal? No easy answer can be given to these problems: but an answer may eventually be found by some research worker in history, or possibly in historical geography.

Towns are not native to Ireland. Norsemen, Normans, Elizabethans and Stuart planters established almost all those that now exist, and others that could not now be called towns. Scattered through the country at distances of 8–16 miles there are quiet market centres, of which some have industries and some have not; in the northeast the linen industry has so strengthened the towns that they are larger than those of the Republic. Belfast is a phenomenon unique in Ireland – a market town which grew into a major industrial city during the nineteenth century as linen, shipbuilding, rope-making and other industries succeeded, primarily with imported raw materials but stimulated by enterprising families of great commercial foresight. Dublin has its industries also but it is the Edinburgh of Ireland – a commercial, administrative, residential and cultural centre rather than Ireland's Glasgow. Although re-industrialization in the Republic had as one of its aims a revival of country towns by scattering industries through the land, many of the new industries have been established in Dublin and its immediate neighbourhood: the city has grown rapidly in population and the housing estates of the Corporation and of the speculative builder have spread alarmingly through the adjacent countryside. Over half the population now live in towns – a higher proportion than ever before. And, incredible as it may seem to some visitors, the towns have been greatly improved during the past hundred years. The wretched houses they had once have been swept away so that little trace now remains, and streets have been rebuilt though fortunately the Georgian houses possessed by some towns have survived. Beside each country town there is a housing estate but within the average town there are still wretched houses, some of which are undeniably slums. And in many towns there is a regrettably large number of ruined cottages and outbuildings still standing. Although many country towns now have far fewer people than they had a century ago, they have a higher proportion of their inhabitants in remunerative employment. At the present time, almost every country town wants an industry and once a factory is established workers will be drawn from the adjacent countryside from a radius of surprising extent.

By deliberate choice, much of this book has dealt with population densities, for these provide a key to the appearance of the countryside. Far to the west, farms are so small and houses are so close together within the limited cultivated area that roads resemble village streets: in the extreme east, where farms are large, pastoral, and run with few labourers, farmhouses are dotted sparsely in the scene and labourers' cottages are placed here and there. The central group consists of farms of medium size, so unusual a hundred years ago, so prevalent now; areas dominated by farms of this size have neither the congestion of the west nor the sparse human occupation of the east. And there is an interesting relationship between the farmed countryside and the market towns: in the west, beyond the Shannon, they are generally small and comparatively poor, in the east larger and richer. As market centres they reflect very clearly the comparative buying and selling power of rural populations though some have now profited by industrialization and their smart new factory may appear as incongruous as the new housing estate. Many of the towns have a square, a market place or a 'diamond', and houses along roads branching from this centre, but some consist only of one long street. The new houses, if built under the county council, line a series of planned roads; if built privately, they are just anywhere. Planners will shake their heads: but some of the towns are hopeful places, with a few enterprising families who have brought them prosperity and even a modicum of culture. Of others it is kinder to say nothing at all.

Emigration has entered deeply into the fabric of Irish life. 'So many of the best go', one observer said: and they do. It seems that they will go even if work is available locally for there are complaints of a scarcity of workers on the land and too rapid a turnover of labour in some of the new factories. Probably the appeal lies in the wish to see new places, to have a wider range of experience, to join some friend or relation in Britain or America. So long as labour is needed elsewhere, so long will Irish workers be available: and by this means the Irish communities abroad (this includes Britain) will be replenished and increased. Many Irish people regret it and many hoped that in time emigration might cease: there are signs that the population of the country as a whole is now virtually stable, as the increase in the towns is approximately equal to the loss from the rural areas and the outward movement from the country is numerically similar to the natural increase (relatively high) by the excess of births over deaths. In some recent years, emigration has been less than the natural increase.

At various points in this book it has been stated that the standard of living has risen considerably in recent times. But there can never be an Ireland of eight million people again unless agriculture is intensified beyond all present dreams and industry is expanded at a rapid rate. Far from employing more people, farming employs less and industry has its problems, especially in the northeast where the linen trade has not the certainty of demand that it once knew and shipbuilding is from its very nature deeply sensitive to world demand. Northern Ireland's need is for a wider range of industry and its government hopes to attract even more of the light industries characteristic of these times. Coal, economic minerals and other raw materials are available in negligible quantities though the processing of agricultural produce offers some hope of industrial expansion. So far, protection of industry in the Republic has made it possible to give people work in factories, though if both parts of Ireland enter the Common

Market tariff protection will gradually diminish. But so far the industrial policy has been reasonably successful: others tried to develop works in earlier times and many failed but this time the government has seen that there were few failures. And above all, they provided electricity almost everywhere from the water-power of rivers as well as from imported coal or oil and – most interesting of all – from peat. Industrialization is regarded as the potential solution to the emigration problem.

A climate damper than that of Britain, the prevalence of lowland bogs as well as mountain peat wastes, an array of lovely but infertile mountains and of ill-drained lowland soils, all restrict the Irish farmer. How often he 'saves the harvest' from torrential rain; how limited the success of the schemes for preventing the flooding of the rivers has so far proved! But many of the farms could be made to grow more grass and better crops given a better organization of drainage by the farmers, and a more sedulous eradication of weeds and a closer attention to fertilization. Education could help to improve farming methods but the courses most favoured in the technical schools are those opening the way to the towns rather than those in rural science. At least the Irish farmer raising stock and growing crops is producing food, which is generally sure of a sale in the nearest and best export market – Great Britain, though not always at the price the Irish farmers think to be reasonable. But the demand of the customer may change. For a long time eggs and poultry were major exports from the Republic to Britain, but they are not bought now though bacon, butter and cheese remain valuable items in the export trade.

It is as well to realize that Ireland's resources are not unlimited: a century ago Sir Robert Kane hoped that all the population in need of land could be settled on cleared bog country or on mountain sides. Some areas of bog have been cleared and settled, but most mountainous areas have lost land from cultivation. The advocates of afforestation wish to take over many valleys that appear derelict and have already taken some but in places their interest conflicts with those of sheep farmers who need the mountains as rough pasture. There are many adverse factors, conspicuously in Connacht, Donegal, and to a lesser extent in Kerry and West Cork, where so large a proportion of the countryside is bog or marsh. To the west of the Shannon, bog alternates with ground moraine, having some soils rewarding to effort. The richest areas lie in the south and east, notably in the great valleys of Munster, the warm fields of Wexford, the lush pastures of Meath, the fertile mixed soils of the spreading lowlands around Belfast, lough Neagh and the Bann valley.

Much of the regional diversity of Ireland has as a basis the varying quality of the glacial drifts and glacifluvial deposits. These range in thickness to 200 ft. and more, covering – so far as one can discover – a lowland surface that at some stage was reduced to a peneplane. Nowhere have drifts shaped the landscapes more clearly than in the drumlin belt that stretches right across the country and covers one-eighth of the area. But the general line of the country has two clear lines – one from northeast to southwest, and one from east to west – the first Caledonian and the second Armorican. There are plateaus also of Carboniferous rocks comparable to the Pennines of England though without their rich coalfields, and of Tertiary basalt in the northeast. And on the margins of this island there is an enchanting variety of coasts but the inhabitants have gone to sea less than any other people of western Europe. To be the gateway of the New World has not been the role of Ireland in European history: although millions of her sons and daughters are in America, her closest present links are with Great Britain and so with Europe.

Bibliography

The works noted for each chapter include the general sources used, together with any detailed studies available. Much is owed to the geological and geomorphological work published in Ireland, some of it by British and continental writers. There is an increasing number of geographical studies each year, published in *Irish Geography* and in a variety of other journals. In Chapter 5, Historical Geography, reference has been made to individual paragraphs by a series of numbered references: elsewhere the sources are indicated by a general statement except when detailed location of an individual source has seemed desirable. For several journals, the standard abbreviations given in the *World List of Scientific Periodicals* have been used, but titles of other journals have been given in full.

The abbreviations are listed here:

Advanc. Sci.	*The Advancement of Science*, London
Econ. Geogr.	*Economic Geography*, Worcester, Mass.
Geogr. J.	*Geographical Journal*, London
Geogr. Rev.	*Geographical Review*, New York
Int. Geol. Congr. Brussels (1922)	International Geological Congress (also London, 1948)
Irish For.	*Irish Forestry*, Dublin
Irish Geogr.	*Irish Geography*, Dublin
J. Ecol.	*Journal of Ecology*, London
Quart. J. Geol. Soc. Lond.	*Quarterly Journal of the Geological Society*, London
Proc. Geol. Ass. Lond.	*Proceedings of the Geological Association*, London
Proc. R. Irish Acad.	*Proceedings of the Royal Irish Academy*, Dublin
Sci. Proc. R. Dublin Soc.	*Scientific Proceedings of the Royal Dublin Society*
Scot. Geogr. Mag.	*Scottish Geographical Magazine*, Edinburgh

CHAPTER 1 (*General Survey*)

Pioneer works include Howarth, O. J. R., *A Geography of Ireland*, Oxford, 1911, and Fitzgerald, W., *Historical Geography of Early Ireland*, London, 1925; there are chapters on Ireland in the standard works dealing with the British Isles, especially Mackinder, H. J., *Britain and the British Seas*, 1902; Herbertson, A. J., and Howarth, O. J. R., *Oxford Survey of the British Empire*, vol. I, 1914; Stamp, L. D., and Beaver, S. H., *The British Isles*, London, 1954; Demangeon, A., *Les Isles Britanniques*, Paris, 1927, and translation by E. D. Laborde, London, 1939. A particular debt is owed to the writers of three works – Cole, G. A. J., *Ireland the Outpost*, 1919; Cole, G. A. J., and Hallissy, T., *Handbook of the Geology of Ireland*, London, 1924; and Praeger, R. L., *The Botanist in Ireland*, Dublin, 1934. With the latter should be noted the same author's popular but informative *The Way that I Went*, London and Dublin, 1937. Ireland receives careful treatment in Solch, J., *Die landschaften der Britischen Inseln*, vol. 2, Vienna, 1952, pp. 1110–1280.

CHAPTER 2 (*Structure and Landforms*)

The literature available on the physical geography of Ireland is extensive but scattered, and this bibliography may be supplemented by those in the works listed.

1. *General Works. The Memoirs of the Geological Survey*, 1858–91, are still an important source. Reprints and extensions of the originals include a series on economic minerals, listed in Bishopp, D. W., *A Short Review of Irish Mineral Resources*, Dublin, 1943, and *Memoirs* dealing with the country around Dublin, 1903; Belfast, 1904; Cork, 1905; Limerick, 1906; Londonderry, 1908; Killarney and Kenmare, 1927. In these special memoirs drifts are considered in some detail and some reference is made to soils and agriculture. Kilroe, J. R., *Soil Geology of Ireland*, 1907, contains a large number of sections and diagrams, and is remarkable for its attempt to link both the solid and the drift geology with climate, agriculture, water supply and afforestation. A preliminary account of present pedological work is given in Gardiner, M. J., 'The National Soil Survey', *Irish Geogr.* vol. 4, no. 6, 1963, pp. 442–53. Cole, G. A. J., and Hallissy, T., *Handbook of the Geology of Ireland*, London, 1924, gives an excellent review of the historical geology and has many pertinent comments on morphology. Newer substantial works are Charlesworth, J. K., *The Geology of Ireland*, Edinburgh, 1953; *Historical Geology of Ireland*, Edinburgh, 1963. For a simpler treatment see Nevill, W. E., *Geology and Ireland*, Dublin, 1963. Much of the recent work is summarized in the handbooks issued for

the British Association meetings held in Ireland, *Belfast in its Regional Setting*, Belfast, 1952, and *A View of Ireland*, Dublin, 1957.

2. *Regional Works.* Two important general collections of papers with bibliographies were written for field meetings of the Geologists' Association: Charlesworth, J. K., and others, 'The geology of Northeast Ireland', *Proc. Geol. Ass. Lond.*, vol. 46, part 4, 1935, pp. 441–86 and 493–502; Smyth, L. B., and others, 'The geology of southeast Ireland, together with parts of Limerick, Clare and Galway', ibid., vol. 50, part 3, 1939, pp. 287–351: a supplement to the last paper is 'The geology of eastern Ireland', *Int. Geol. Congr.*, London, 1948. Local studies of a geomorphological character are now relatively numerous: for the Leinster chain, see Farrington, A., 'The topographical features of the granite-schist junction in the Leinster chain', *Proc. R. Irish Acad.*, vol. 37, B9, 1927, pp. 181–92, and 'The pre-glacial topography of the Liffey basin', ibid., vol. 38, B9, 1929, pp. 148–70: see also Haughton, J. P., 'Low-level surfaces of erosion in the Wicklow mountains', *Irish Geogr.*, vol. 1, no. 2, 1945, pp. 34–8. Two papers on erosion surfaces with one on the coast of the northeast are summarized in Whittow, J. B., Davies, G. L., Stephens, N., 'Physiographic evolution in Ireland', *Advanc. Sci.*, vol. 14, 1958, pp. 381–91. Old and not yet solved problems are opened also in Davies, G. L., 'The age and origin of the Leinster mountain chain: a study of the evolution of south-eastern Ireland from the Upper Palaeozoic to the later Tertiary', *Proc. R. Irish Acad.*, vol. 61, B5, 1960, pp. 79–107. See also Davies, G. L., 'Cyclic surfaces in the Roundwood basin, County Wicklow', *Irish Geogr.*, vol. 5, no. 3, 1966, pp. 150–60. A paper of unusual historic interest is Jukes, J. B., 'On the formation of some river valleys in the south of Ireland', *Quart. J. Geol. Soc. Lond.*, vol. 18, 1862, pp. 378–403: the same theme is developed in Miller, A. A., 'River development in southern Ireland', *Proc. R. Irish Acad.*, vol. 45, B14, 1939, pp. 321–54. Arising from this work, there are two interesting papers: Farrington, A., 'The South Ireland Peneplane', *Irish Geogr.*, vol. 2, 1953, pp. 211–17, and Miller, A. A., 'The origin of the South Ireland Peneplane', ibid., vol. 3, 1955, pp. 79–88. A detailed study of the area south of Dungarvan is Orme, A. R., 'Planation surfaces in the Drum Hills, County Waterford, and their wider implications', *Irish Geogr.*, vol. 5, no. 1, 1964, pp. 48–72. Recognition of surfaces is also the theme of Proudfoot, V. B., 'Erosion surfaces in the Mourne mountains', ibid., vol. 3, 1954, pp. 26–35. An important short paper is Cole, G. A. J., 'Some features of the Armorican (Hercynian) folding in southern Ireland', *Int. Geol. Congr. Brussels*, 1922, pp. 423–33. Among many geological papers, mention should be made of Kilroe, J. R.,

'The Silurian and metamorphic rocks of Mayo and north Galway', *Proc. R. Irish Acad.*, vol. 26, B10, 1907, pp. 129–60: Wager, L. R., 'The geology of the Roundstone district, Co. Galway', ibid., vol. 41, B5, 1932, pp. 46–72; Richey, J. E., and Thomas H. H., 'The Tertiary ring complex of Slieve Gullion (Ireland)', *Quart. J. Geol. Soc. Lond.*, vol. 88, 1932, pp. 776–849: and for the Burren of Co. Clare, Douglas, J. A., 'The Carboniferous limestones of County Clare', ibid., vol. 65, 1909, pp. 538–586. Many other geological papers have points of geomorphological interest but mention can be made only of a few: Oswald, D. H., 'The Carboniferous rocks between the Ox mountains and Donegal bay', ibid., vol. 109, 1953, pp. 167–83; Dawson-Grove, G. E., 'Analysis of minor structures near Ardmore, Co. Waterford, Eire', ibid., vol. 3, 1955, pp. 1–21: Hodson, F., 'The beds above the Carboniferous limestone in north-west County Clare, Eire', ibid., vol. 109, 1953, pp. 259–79; Nevill, W. E., 'The Millstone Grits and lower Coal Measures of the Leinster coalfield', *Proc. R. Irish. Acad.*, vol. 58, B1, 1956, pp. 1–16: Capewell, J. G., 'The stratigraphy and structure of the country around Sneem, Co. Kerry', ibid., vol. 58, B8, 1957, pp. 167–83. The official Report of the Drainage Commission 1938–40, Dublin, gives useful information on the Shannon and the Barrow. The literature on caves has grown considerably in recent years, and includes several papers in *Cave Science*, in the *Proceedings of the Spelaeological Society* and in *Irish Geography*: four studies may be mentioned – Coleman, J. C., and Dunnington, A. J., 'The Pollnagollum cave, County Clare', *Proc. R. Irish Acad.*, vol. 50, B5, 1944, pp. 105–32; Coleman, J. C., 'The Aille river and cave, Co. Mayo', *Irish Geogr.*, vol. 2, no. 2, 1950, pp. 58–60 (excellent bibliography); Coleman, J. C., 'Caves near Carrickmacross, Co. Monaghan', ibid., vol. 2, no. 4, 1952, pp. 180–3; Coleman, J. C., 'Caves in the Cong area of Galway and Mayo', ibid., vol. 3, no. 2, 1955, pp. 94–106. See also Williams, P. W., 'An initial estimate of the speed of limestone solution in county Clare', ibid., vol. 4, no. 6, 1963, pp. 432–41. In 1964 the Irish Speleological Association was founded and their journal, *Irish Speleology*, is now available (enquiries to the Editor, c/o Department of Geography, Trinity College, Dublin 2). On coasts, with a correlation between Wales and Ireland, see Orme, A. R., 'Abandoned and composite seacliffs in Britain and Ireland', *Irish Geogr.*, vol. 4, no. 4, 1962, pp. 279–91.

3. *Glaciation*. General treatments are available in Wright, W. B., *The Quaternary Ice Age*, London, 1937 (2nd edition); in Movius, H. J., *The Irish Stone Age*, Cambridge, 1942; Charlesworth, J. K., *The Quaternary era*, London, 1957: several of the papers by Charlesworth and Farrington also

comment on the correlation of Irish icesheet and mountain glaciations with one another and with those elsewhere. A paper of great historic interest is Ball, J., 'Notice of the former existence of small glaciers in the county of Kerry', *Journal of the Geological Society of Dublin*, vol. 4, 1848–50, Dublin, 1851, pp. 151–4. Another valued but venerable paper is Kinahan, G. H., and Close, M. H., 'The general glaciation of Iar-Connaught', Dublin, 1872. The eskers and drifts of the Central Lowland are mapped and described in Sollas, W. J., 'A map to show the distribution of eskers in Ireland', *Sci. Proc. R. Dublin Soc.*, vol. 5, series 2, 1896, pp. 785–822: an interesting hypothesis on esker formation is given in Flint, R. F., 'The origin of the Irish "eskers",' *Geogr. Rev.*, New York, vol. 20, 1930, pp. 615–30. See also Farrington, A., 'Unglaciated areas in southern Ireland', *Irish Geogr.*, vol. 1, no. 4, 1947, pp. 89–97. Linton, D. L., 'On unglaciated areas in Scandinavia and Great Britain', ibid., vol. 2, no. 1, 1949, pp. 25–33, and later discussion, ibid., vol. 2, no. 2, 1950, pp. 77–9. For a statement criticizing the view that parts of Kerry were unglaciated, see King, Cuchlaine A. M., and Gage, M., 'The extent of glaciation in part of West Kerry', ibid., vol. 4, no. 3, 1961, pp. 202–8. Some of the views of King and Gage are contested in Lewis, C. A., 'The glaciation of the Behy valley, county Kerry', ibid., vol. 5, no. 4, 1967, pp. 293–301, and cognate problems are discussed in Williams, M., 'Glacial breaches and sub-glacial channels in south-western Ireland', ibid., vol. 5, no. 1, 1964, pp. 83–95. On the relations between moraines in Ireland and Great Britain, see Synge, F. M., 'Retreat stages of the last ice-sheet in the British Isles', ibid., vol. 2, no. 4, 1952, pp. 168–71. A summary treatment is given in Synge, F. M., and Stephens, N., 'The Quaternary period in Ireland – an assessment, 1960', ibid., vol. 4, no. 2, 1960, pp. 121–30. The issue of *Irish Geography*, vol. 4, no. 5, 1963, pp. 311–76, is devoted almost entirely to articles on glaciation collected as a tribute to Dr. A. Farrington.

Regional studies of Irish glaciation include Dwerryhouse, A. R., 'The glaciation of northeast Ireland', *Quart. J. Geol. Soc. Lond.*, vol. 79, 1923, pp. 352–433; Charlesworth, J. K., 'The glacial geology of the northwest of Ireland', *Proc. R. Irish Acad.*, vol. 36, B12, 1924, pp. 174–314; 'The glacial retreat from central and southern Ireland', *Quart. J. Geol. Soc. Lond.*, vol. 84, 1928, pp. 293–344; 'The glacial geology of north Mayo and west Sligo', *Proc. R. Irish Acad.*, vol. 38, B6, 1928, pp. 100–15; 'The glacial retreat in Iar Connacht', ibid., vol. 39, B2, 1929, pp. 95–106; 'A map of the glacier lakes and the local glaciers of the Wicklow hills', ibid., vol. 44, B3, 1937, pp. 29–36; 'Some observations on the glaciation of northeast Ireland', ibid., vol. 45, B4, 1939, pp. 255–95; Orme, A. R., 'Drumlins and the Weichsel glaciation of Connemara', *Irish Geogr.*, vol. 5, no. 4, 1967, pp. 262–74. Important researches are discussed in the following papers by Farrington, A., 'The glaciation of the Wicklow mountains', *Proc. R. Irish Acad.*, vol. 47, B7, 1934, pp. 173–209; 'The glaciation of the Bantry bay district', *Sci. Proc. R. Dublin Soc.*, vol. 2 (new series), no. 37, 1936, pp. 345–61; 'The granite drift near Brittas, on the border between County Dublin and County Wicklow', *Proc. R. Irish Acad.*, vol. 47, B12, 1942, pp. 279–91; 'The glacial drifts of the district around Enniskerry, County Wicklow', ibid., vol. 50, B6, 1944, pp. 133–57; 'Glacial drifts of Slieve Bernagh', *Irish Geogr.*, vol. 2, no. 4, 1952, pp. 153–5; 'The Lee basin, Part 1: Glaciation', *Proc. R. Irish Acad.*, vol. 60, B3, 1959, pp. 135–166. For a detailed study of an area to the north-west of Dublin see Synge, F. M., 'The Glacial deposits round Trim, Co. Meath', ibid., vol. 53, B10, 1950, pp. 99–110. Other recent work includes Farrington, A., 'A note on the correlation of the Cork–Kerry glaciations with those of the rest of Ireland', *Irish Geogr.*, vol. 3, no. 1, 1954, pp. 47–53; Proudfoot, V. B., 'The glaciation of the Dingle peninsula', ibid., vol. 3, no. 1, 1954, pp. 36–8; Martin, S., 'Raised beaches and their relation to glacial drifts on the east coast of Ireland', ibid., vol. 3, no. 2, 1955, pp. 87–93; Stephens, N., 'Some observations on the "interglacial" platform and the early post-glacial raised beach on the east coast of Ireland', *Proc. R. Irish Acad.*, vol. 58, B6, 1957, pp. 129–49; Bryant, R. H., 'The 'pre-glacial' raised beach in south-west Ireland', *Irish Geogr.*, vol. 5, no. 3, 1966, pp. 188–203; McMillan, Nora F., 'Quaternary deposits around lough Foyle, North Ireland', *Proc. R. Irish Acad.*, vol. 58, B9, 1957, pp. 185–205; Dury, G. H., 'A glacially breached watershed in Donegal', *Irish Geogr.*, vol. 3, no. 4, 1957, pp. 171–80, and 'Glacial morphology of the Blue Stack area, Donegal', ibid., vol. 3, no. 5, 1958, pp. 242–53; Stephens, N., and Synge, F. M., 'A quaternary section at Sutton, Co. Dublin', *Proc. R. Irish Acad.*, vol. 59, B2, 1958, pp. 19–27; Orme, A. R., 'Quaternary changes of sea-level in Ireland', *Institute of British Geographers, Transactions*, no. 39, 1966, pp. 127–40. The southeast of Ireland has already proved to be an area of remarkable glacial interest: it is further discussed in Synge, F. M., 'Some problems concerned with the glacial succession in south-east Ireland', *Irish Geogr.*, vol. 5, no. 1, 1964, pp. 73–82. Combining an interest in coasts and glaciation, three papers deal with the west of Ireland: Guilcher, A. and King, C. A. M., 'Spits, tombolos and tidal marshes in Connemara and West Kerry', *Proc. R. Irish Acad.*, vol. 61, B17, 1961, pp. 283–338; Guilcher, A., 'Drumlin and spit structures in the Kenmare river, south-west Ireland', *Irish Geogr.*, vol. 5, no. 2, 1965, pp. 7–19;

King, C. A. M., 'Some observations on the beaches of the west coast of county Donegal', ibid., vol. 5, no. 2, 1965, pp. 40–50.

For a discussion of some Irish topographical terms, including 'drumlin' and 'esker', see Quin, E. G., and Freeman, T. W., in *Irish Geogr.*, vol. 1, no. 4, 1947, pp. 85–9, and vol. 1, no. 5, 1948, pp. 151–5.

Some papers of local interest are listed under the regional chapters, notably those of Dr Marjorie Sweeting, on County Clare (p. 532).

CHAPTER 3 (*Climate and Weather*)

The most important statistical sources are the Book of Normals, Averages of Temperature for the British Isles for periods ending 1930, Averages of Bright Sunshine for the British Isles for periods ending 1930, all published by the Air Ministry, London, and the 'Climatological Atlas of the British Isles', H.M.S.O., 1952. The rainfall statistics are contained in the annual issues of *British Rainfall*, Air Ministry, London, annual to 1939. Books dealing with the climate of the British Isles include Bilham, E. G., *The Climate of the British Isles*, London, 1938, and, for rainfall, Salter, M. de C. S., *The Rainfall of the British Isles*, London, 1921. In addition there are interesting comments on climate in Tansley, A. G., *The British Isles and their Vegetation*, Cambridge, 1939; Scully, R. W., *Flora of Co. Kerry*, Dublin, 1916, and Praeger, R. L., *The Botanist in Ireland*, Dublin, 1934. The data on the supply of water to the Shannon hydroelectric station have been drawn from the annual reports of the Electricity Supply Board, Dublin.

CHAPTER 4 (*Vegetation and Soils*)

Two works are indispensable: Praeger, R. L., *The Botanist in Ireland*, Dublin, 1934, and Tansley, A. G., *The British Islands and their Vegetation*, Cambridge, 1939. In earlier editions of this book, some attention was given to the Irish flora: this has been replaced by a section on soils, pp. 61–3. But as users of this book may wish to study floristic distributions, some of the sources are given here. For the relation of the flora of Ireland to those of other countries, see Praeger, R. L., 'Recent views bearing on the problem of the Irish flora and fauna', *Proc. R. Irish Acad.*, vol. 41, B8, 1932, pp. 125–45; Praeger, R. L., 'Relations of the flora and fauna of Ireland to those of other countries', *Proceedings of the Linnean Society of London*, session 151, part 3, 1938–9, pp. 192–213. The view that no Irish plant survived the Ice Age is expressed in Charlesworth, J. K., 'Some geological observations on the origin of the Irish flora and fauna', *Proc. R. Irish Acad.*, vol. 39, B16, 1930, pp. 358–90: on this, see Farrington, A., 'Unglaciated areas in southern Ireland', *Irish Geogr.*, vol. 1, no. 4, 1947, pp. 89–97, and Jessen, K., 'Studies in late Quaternary deposits and history of Ireland', *Proc. R. Irish Acad.*, vol. 52, B6, 1949, pp. 85–290: a summary is given in *Nature*, vol. 165, 1950, p. 800. There are sections on vegetation in the recent British Association handbooks (Belfast, 1952; Dublin, 1957).

The strawberry tree is treated in Sealy, J. R., 'Arbutus unedo', *J. Ecol.*, vol. 27, 1949, pp. 365–88. Some interesting studies of particular plants are given in the admirable book, Scully, R. W., *Flora of Co. Kerry*, Dublin, 1916; for the northeast, see Stewart, S. A., and Corry, T. H., *A Flora of North East Ireland*, 2nd edition, 1938, Belfast, revised by R. L. Praeger and W. R. Megaw. A flora specially written for use in Ireland is Webb, D. A., *An Irish Flora*, Dundalk, 1943. Local studies include the following: Pethybridge, G. H., and Praeger, R. L., 'The vegetation of the district lying south of Dublin', *Proc. R. Irish Acad.*, vol. 25, B6, 1905, pp. 124–80; White, J. M., 'Recolonization after peat-cutting', ibid., vol. 39, B21, 1930, pp. 453–76; Duff, M., 'The ecology of the Moss Lane region, lough Neagh', ibid., vol. 39, B22, 1930, pp. 477–96; White, J. M., 'The fens of North Armagh', ibid., vol. 40, B15, 1932, pp. 233–83; Armstrong, J. L., Calvert, J., and Ingold, C. T., 'The ecology of the Mourne mountains', ibid., vol. 39, B20, 1930, pp. 440–52; Webb, D. A., 'The vegetation of Carrowkeel, a limestone hill in north-west Ireland', *J. Ecol.*, vol. 35, nos. 1 and 2, 1947, pp. 105–29; O'Reilly, H., and Pantin, S., 'Some observations on the salt marsh formation in Co. Dublin', *Proc. R. Irish Acad.*, vol. 58, B5, 1956–7, pp. 89–128; Webb, D. A., and Glanville, E. V., 'The vegetation and flora of some islands in the Connemara lakes', ibid., vol. 62, B5, 1962, pp. 31–54; Webb, D. A., 'Noteworthy plants of the Burren', ibid., vol. 62, B9, 1962, pp. 117–34.

Information on the location and extent of forestry plantations is given in the Ulster Year Book, and in the Republic's *Report of the Minister for Lands on Forestry*, published at intervals. There are useful sections on forests and woods in various works including Praeger, R. L., op. cit. and Tansley, A. G., op. cit. Two historical papers of great interest are Forbes, A. C., 'Some legendary and historical references to Irish woods and their significance', *Proc. R. Irish Acad.*, vol. 41, B3, 1932, pp. 15–36; Forbes, A. C., 'Tree planting in Ireland during four centuries', ibid., vol. 41, C6, 1933, pp. 168–99; also of value are Forbes, A. C., 'Some early economic and other development in Eire: and their effect on Forestry conditions', *Irish For.*, vol. 1, 1943, pp. 6–15; 'The Forestry revival in Eire', ibid., vol. 4, 1947, pp. 11–26. A useful summary is given in McEvoy, T., 'Irish native woodlands: their present condition', ibid., vol. 1, no. 2, 1944, pp. 27–35. See also Freeman, T. W., 'Forestry and Land Use

Survey', ibid., vol. 7, 1950, pp. 24–31. Other papers include Pimlott, J., 'The history of afforestation in Northern Ireland', *Advanc. Sci.*, vol. 9, 1952, pp. 297–303, and McEvoy, T., 'Forestry in Ireland', ibid., vol. 14, 1958, pp. 307–16: both of these are informative. The lack of woodland in the eighteenth century is discussed in Young, A., *Tour in Ireland, 1776–1779*, ed. A. W. Hutton, London, 1892; the planting and disafforestation of the Georgian era is treated in Maxwell, C. E., *Country and Town in Ireland under the Georges*, London, 1940, esp. pp. 106–7. The Congested Districts Board hoped to establish several forests, but see Micks, W. L., *An Account of the Constitution, Administration and Dissolution of the Congested Districts Board for Ireland from 1891* to *1923*, Dublin, 1925, esp. pp. 26–7. A recent book (not dated but apparently 1966) is Fitzpatrick, H. M. (ed.), *The forests of Ireland*, Dublin (Society of Irish Foresters).

Soil study is now active in Ireland. In the Republic the first bulletin of the National Soil Survey was Ryan, P. and Gardiner, M. J., *Soils of County Wexford*, Agricultural Institute, Dublin 1964, and later studies are discussed in Finch, T. and Synge, F. M., 'The drifts and soils of West Clare and the adjoining parts of counties Kerry and Limerick', *Irish Geogr.*, vol. 5, no. 3, 1966, pp. 161–72 and Conry M. J., 'The soils of county Carlow', ibid., vol. 5, no. 3, 1966, pp. 204–13. See also Ryan, P. and Walsh, T., 'Character study of soils developed on Weichsel glacial deposits in Ireland', *Proc. R. Irish Acad.*, vol. 64, B24, 1964–6, pp. 465–507. Material from Northern Ireland is treated in Symons, L. (ed.), *Land Use in Northern Ireland*, London 1963.

A useful paper on peat is Dwyer, D. J., 'The peat bogs of the Irish Republic: a problem in land use', *Geogr. J.*, vol. 128, 1962, pp. 184–93. There is also a note of value in Johnson, J. H., 'The commercial use of peat in Northern Ireland', ibid., vol. 125, 1959, pp. 398–400.

CHAPTER 5 (*Historical*)

Many interesting papers have been written on historical geography, including Andrews, J. H., 'Eighteenth-century "port forelands" in the Irish Sea', *Irish Geogr.*, vol. 3, no. 2, 1955, pp. 114–17, and 'Notes on the historical geography of the Irish iron industry', ibid., vol. 3, no. 3, 1956, pp. 139–49; also 'Road planning in Ireland before the railway age', ibid., vol. 5, no. 1, 1964, pp. 17–41; Emery, F. V., 'Irish geography in the seventeenth century', ibid., vol. 3, no. 5, 1958, pp. 263–76; Johnson, J. H., 'The Irish tithe composition applotment books as a geographical source', ibid., vol. 3, no. 5, 1958, pp. 254–62; Jones Hughes, T., 'East Leinster in the mid-nineteenth century', ibid., vol. 3, no. 5, 1958, pp. 227–41. The historical literature is abundant: the general history most likely to be accessible to readers of this volume is Curtis, E., *A History of Ireland*, London, 1936. Three works on economic history are O'Brien, G., *The Economic History of Ireland in the Seventeenth Century*, Dublin, 1919; *The Economic History of Ireland in the Eighteenth Century*, Dublin, 1918; and *The Economic History of Ireland from the Union to the Famine*, London, 1921. An excellent short history is Beckett, J. C., *A Short History of Ireland*, London, 1952. To celebrate the fiftieth anniversary of the 1916 rising, twenty-one university historians gave television talks published as Moody, T. W., and Martin, F. X. (eds.), *The Course of Irish history*, Cork 1967.

1. Evans, E. E., *Irish Heritage*, Dundalk, 1942; *The Mourne Country*, Dundalk, 1951; *Irish Folk-Ways*, London, 1957. In addition much of the material relating to traditional crafts is published in *The Ulster Journal of Archaeology* and *Ulster Folklife*, Belfast.

2. Gaeltacht Commission (*Comisiun na Gaeltachta*), *Report*, Dublin, 1926, p. 36.

3. Fox, C., *The Personality of Britain*, Cardiff, 1932, 1933, 1943; for an interesting critical judgement, see Fleure, H. J., in *Geogr. Rev.*, vol. 36, 1946, p. 340, in which it is suggested that the term 'Atlantic Britain' should be used for those lowlands on the west side of Great Britain which received many early settlers. Obviously this idea is of some importance in relation to Ireland.

4. The literature of the prehistoric periods in Ireland is far too comprehensive to be detailed here. The main journals with archaeological papers are *Ulster Folklife*, the *Ulster Journal of Archaeology*, the *Journal of the Royal Society of Antiquaries*, Dublin, and the *Proceedings of the Royal Irish Academy*. In addition, there are various regional journals.

5. Collingwood, R. G., and Myres, J. N. L., *Roman Britain and the English Settlements*, Oxford, 1936, esp. p. 117. On the Irish settlements in the west of Great Britain, see pp. 278 ff.

6. The main source on the Celtic Church is Kenney, J. F., *Sources for the Early History of Ireland*, Part 1, Ecclesiastical, New York, 1929, which has an admirable general introductory chapter.

7. The travels of St Brendan are discussed in articles by Babcock, W. H., *Geogr. Rev.*, vol. 8, 1919, pp. 37–46; *Scot. Geogr. Mag.*, vol. 31, 1915, pp. 411–22.

8. Kenney, op. cit. in note 6, p. 5.

9. A valuable summary of the nature of the *tuaths* is given in MacNeill, E., *Early Irish Laws and Institutions*, Dublin, 1935, pp. 91–100.

10. The High Kingship and the various divisions of Ireland in pre-

Norman times are discussed in Curtis, E., *A History of Ireland*, London, 1936, and *A History of Mediaeval Ireland*, London, 1923 and 1938.

11. On this phase, see Kenney, op. cit. in note 6; Curtis, op. cit. in note 10, including an appendix in *Mediaeval Ireland* on the later history of the Ostmen in Dublin, Waterford, Limerick and Wexford.

12. See Curtis, E., *History of Mediaeval Ireland*, pp. 12–18, for church reform. This book also has an appendix on towns, pp. 406–18; an interesting short work on towns is Webb, J. J., *Municipal Government in Ireland*, Dublin, 1918. Counties emerged gradually in Ireland from the twelfth century onwards; on the earlier phases, see Otway-Ruthven, J., 'Anglo-Irish shire government in the thirteenth century', *Irish Historical Studies*, vol. 5, 1946, pp. 1–28. The classic work on the Normans is Orpen, G. H., *Ireland Under the Normans, 1169–1333*, 4 vols., 1911–20.

13. On the varied limits of the Pale, and its relation to the 'land of peace', see Curtis, E., *History of Mediaeval Ireland*, esp. p. 353: an important source on the 1515 limits is given in Maxwell, C. E., *Irish History from Contemporary Sources, 1509–1610*, London, 1923, pp. 81–2.

14. Curtis, E., *History of Mediaeval Ireland*, Appendix III, pp. 408–16 (in 1938 edition), gives an account of the varied fortunes of the towns and shows that the revival of the Irish language, which began in the fourteenth century, 'reached the towns last of all'.

15. The material on Ireland under the Tudors and Stuarts is drawn from the original sources in Maxwell, op. cit. in note 13: Leix and Offaly, pp. 227–41; Munster, pp. 241–54; Ulster, pp. 254–307; towns, pp. 354–371; trade and ports, pp. 372–5: see also the general introduction. For the ecclesiastical developments, see Phillips, W. A. (ed.), *History of the Church of Ireland*, Oxford, 1933–4, and Edwards, R. W. D., *Church and State in Tudor Ireland*, London, 1935.

16. An historical work, with a geographical introduction, is Moody, T. W., *The Londonderry Plantation*, Belfast, 1938; see also Moody, T. W., 'The treatment of the native population under the scheme for the plantation in Ulster', *Irish Historical Studies*, vol. 1, 1938, pp. 59–63.

17. For the work of Sir William Petty, see *Political Anatomy of Ireland, 1672*, ed. C. H. Hull, Cambridge, 1899; *Treatise of Ireland, 1687*, ed. C. H. Hull, Cambridge, 1899.

18. This section is based mainly on Young, A., *Tour in Ireland, 1776–1779*, ed. A. W. Hutton, London, 1892. Two works on the social and economic life of the eighteenth century are Maxwell, C. E., *Dublin under the Georges*, London, 1936; *Country and Town in Ireland under the Georges*, London, 1940.

19. Lewis, S., *Topographical Dictionary of Ireland*, London, 1837, vol. 2, p. 263.

20. Connell, K. H., 'The Population of Ireland in the Eighteenth Century', *Economic History Review*, vol. 16, 1946, pp. 111–24; Connell, K. H., 'The colonization of waste land in Ireland, 1780–1845', *Economic History Review*, 2nd series, vol. 3, 1950, pp. 48–71; Connell, K. H., *The Population of Ireland 1750–1845*, Oxford, 1950; Freeman, T. W., *Pre-Famine Ireland*, Manchester, 1957.

21. Edwards, R. D., and Williams, T. D. (editors), *The Great Famine: Studies in Irish History*, Dublin, 1956.

22. Maps of the dioceses are given in the Catholic Encyclopaedia, and in the *Report of Her Majesty's Commissioners on the Revenues and Condition of the Established Church (Ireland)*, with appendix, 1868: see also the annual Catholic Directory and the Irish Church Directory. There are various references to particular dioceses in Phillips, W. A., *History of the Church of Ireland, 1933–4*, and numerous studies by Lawlor, H. J., in the *Irish Church Quarterly* and in the *Louth Archaeological Journal*. Also of particular interest is a lecture by H. J. Lawlor on 'The Irish Reformation in the twelfth century', in *Ad Clerum*, Dublin, 1918, pp. 136–77. See also Day, J. G. F., and Patton, H. E., *The Cathedrals of the Church of Ireland*, London, 1932.

23. Mackinder, H. J., *Britain and the British Seas*, p. 210.

Other work on the historical geography of Ireland includes Andrews, J. H., 'Eighteenth century "port forelands" in the Irish sea', *Irish Geogr.*, vol. 3, no. 2, 1955, pp. 114–17; 'Notes on the historical geography of the Irish iron industry', ibid., vol. 3, no. 3, 1956, pp. 139–49; 'Ireland in maps: a bibliographical postscript', ibid., vol. 4, no. 4, 1962, pp. 234–43. An article in historical geography of special population interest, dealing with the year 1834, is Johnson, J. H., 'Population movements in county Derry during a pre-Famine year', *Proc. R. Irish Acad.*, vol. 60, C3, 1959, pp. 141–62. An earlier period is discussed in Emery, F. V., 'Irish geography in the seventeenth century', *Irish Geogr.*, vol. 3, no. 5, 1958, pp. 263–76.

CHAPTER 6 (*Population*)

The main statistical sources are the reports of the *Census of Ireland* to 1911. the *Census of Northern Ireland*, 1926, 1937, 1951, 1961, 1966, the *Census of Eire*, 1926, 1936, 1946, 1951, 1956, 1961, 1966. Census reports are arranged by counties in Northern Ireland, and in the Republic by subjects, such as the population of each administrative area, occupations, religion and birthplace, language: all the information is summarized in an excellent General Volume which appears, however, several years after the Census, The *Ulster Year Book*, 1925, 1929, 1932, 1935, 1947, 1950, 1953, 1956, 1957–9, 1960–2, 1963–5, 1966–8 also has general information. The whole of Ireland is included in Hunt, A. J. (ed.), 'Population maps of the British Isles 1961', *Institute of British Geographers, Transactions*, no. 43, 1968.

Historical works include Young, A., *Travels in Ireland*, ed. Hutton, 1892, which has a number of interesting comments on the time at which, apparently, the marked rise of population was beginning. Conditions in the early nineteenth century are treated in many books and papers, of which two are of special interest: Adams, W. F., *Ireland and Irish Emigration to the New World*, New Haven, 1932, and Hansen, M. L., *The Atlantic Migration*, 1607–1860, London, 1940. The various pre-Census estimates are discussed in the *Saorstat Eireann* General Report of the *Census* of 1926, Dublin, 1934, and the evidence has recently been reassessed with new material in Connell, K. H., 'The population of Ireland in the eighteenth century', *Economic History Review*, vol. 16, 1946, pp. 111–24, and in *The Population of Ireland 1750–1845*, Oxford, 1950. The circumstances of the times and the use of the potato as a dietetic staple were the subject of a Finlay Memorial Lecture, published as Salaman, R. N., 'The influence of the potato on the course of Irish History', Dublin, 1943, and in Dr Salaman's fine book *The History and Social Influence of the Potato*, Cambridge, 1949.

Emigration has been discussed by many writers in terms of its effects at home and abroad. For Ireland itself, see O'Brien, W. P., *The Great Famine in Ireland and a retrospect of the fifty years, 1845–1895*, London, 1896, and Carruthers, W. A., *Emigration from the British Isles*, London, 1929: the article by C. H. Oldham on 'Town and country life in Ireland', in *Oxford Survey of the British Empire*, ed. A. J. Herbertson and O. J. R. Howarth, London, 1914, vol. 1, chap. 17, is an interesting period piece. The works by Micks, W. L., Arensberg, C. M., and Kimball, S. T., and the Gaeltacht Commission (see bibliography for Chapter 7) all comment on the social effects of emigration. Contemporary material is available in the *Irish Statistical Bulletin* (formerly the *Irish Trade Journal and Statistical Bulletin*), published quarterly in Dublin, the annual *Statistical Abstract of Ireland*, and future historians will find interesting material in the newspapers, including the *Irish Review and Annual*, published at the beginning of the year by the *Irish Times*, Dublin. Regional studies by the author of this book are printed in the *Journal of the Statistical and Social Inquiry Society of Ireland*, Dublin, vol. 16, 1939, pp. 89–104; 'Migration movements and the distribution of population in Eire' (abstract in *Geogr. Rev.*, vol. 29, 1939, pp. 507–8); 'The changing distribution of population in County Donegal', vol. 16, 1940–1 pp. 31–46; 'The changing distribution of population in Kerry and West Cork', vol. 16, 1941–2, pp. 28–43; 'The changing distribution of population in County Mayo', vol. 17, 1942–3, pp. 85–106; 'Population distribution in County Sligo', vol. 17, 1943–4, pp. 254–68. General papers by the present author are 'The congested districts of western Ireland', *Geogr. Rev.*, vol. 33, 1943, pp. 1–14, and 'Emigration and Rural Ireland', *Journal of the Statistical and Social Inquiry Society of Ireland*, vol. 17, 1944–5, pp. 404–22. The density of population in relation to agriculture is discussed in 'Farming in Irish Life', *Geogr. J.*, vol. 110, 1947, pp. 38–56. An article dealing with a period of heavy emigration in Johnson, J. H., 'Population changes in Ireland, 1951–1961', ibid., vol. 129, 1963, pp. 167–74. Population problems are also discussed in numerous papers in the *Journal of the Statistical and Social Inquiry Society of Ireland* since it first appeared in 1847: one of particular importance is Geary, R. C., 'Irish population prospects from the viewpoint of reproduction rates', vol. 16, 1940–1, pp. 91–122. In 1955 the Reports of the Commission on Emigration and other population problems (1948–54) was published by the Dublin Stationery Office. One historical aspect of Irish seasonal migration is discussed in Johnson, J. H., 'Harvest migration from nineteenth-century Ireland', *Institute of British Geographers, Transactions*, 41, 1967, pp. 97–112.

All these papers bear on the question of the effects in Ireland: few writers deal with the Irish abroad, but two of special interest are Handley, J. E., *The Irish in Scotland, 1798–1845*, Cork, 1943; Handley, J. E., *The Irish in Modern Scotland*, Cork, 1947 (the latter marred by polemic tendencies). The fortunes of emigrants are studied in Jackson, J. A., *The Irish in Britain*, London, 1963. Two recent papers deal with the Irish abroad: Lawton, R., 'Irish Immigration to England and Wales in the mid-nineteenth century', *Irish Geogr.*, vol. 4, no. 1, 1959, pp. 35–54, and Rose, A. J., 'Irish migration to Australia in the twentieth century', ibid., pp. 79–84. Many of the works listed in the bibliographies for Chapters 5, 7 and 8 also comment on population: the population issue is so vital in Ireland that it is considered in an immense variety of books.

CHAPTER 7 (*Aspects of Social Geography*)

The materials used are drawn from numerous sources, including historical works and those on population, especially Census reports. Social factors are discussed in Arensberg, C., *The Irish Countryman*, London, 1937, and Arensberg, C., and Kimball, S. T., *Family and Community in Ireland*, Cambridge, Mass., 1940. The survival of many old customs and practices in Irish life is the main theme of Evans, E. E., *Irish Heritage*, Dundalk, 1941, and later works listed under note 1, Chapter 5; see also 'Survivals of the Irish openfield system', *Geography*, London, vol. 24, 1939, pp. 24–36. There is a considerable amount of information on Northern Ireland in Mogey, J. M., *Rural Life in Northern Ireland*, Oxford, 1947. Regional studies include Freeman, T. W., 'The Congested Districts of Western Ireland', *Geogr. Rev.*, vol. 33, 1943, pp. 1–14; 'The Graiguenamanagh basin' (Co. Kilkenny), *Irish Geogr.*, vol. 1, no. 3, 1946, pp. 61–77; 'The Castleblayney district, Co. Monaghan', ibid., vol. 1, no. 4, 1947, pp. 106–26; 'Tullamore and its environs, Co. Offaly', ibid., vol. 1, no. 5, 1948, pp. 133–50; 'The town and district of Carrick-on-Shannon, Co. Leitrim', ibid., vol. 2, no. 1, 1949, pp. 1–19; 'The town and district of Roscommon', ibid., vol. 2, no. 2, 1950, pp. 64–76; 'Buncrana and its environs', vol. 2, no. 3, 1951, pp. 93–105; Haughton, J. P., 'Kinsale – an essay in regional geography', ibid., vol. 2, no. 3, 1951, pp. 111–23; Haughton, J. P., 'Killaloe and district', ibid., vol. 2, no. 4, 1952, pp. 141–52; Freeman, T. W., 'The Irish country town', ibid., vol. 3, no. 1, 1954, pp. 5–14; 'North Inishowen, Co. Donegal', ibid., vol. 3, no. 2, 1955, pp. 59–68; Haughton, J. P., 'The live stock fair in relation to Irish country towns', ibid., vol. 3, no. 2, pp. 107–113; Haughton, J. P., 'The town of Skibbereen', ibid., vol. 3, no. 4, 1957, pp. 181–91; Freeman, T. W., 'Galway – the key to West Connacht', ibid., vol. 3, no. 4, 1957, pp. 194–205; Davies, G. L., 'The town and coalfield of Ballycastle, Co. Antrim', ibid., vol. 3, no. 4, pp. 206–15; Vaughan, T. D., 'Dromore: an urban study in county Down', ibid., vol. 4, no. 2, 1960, pp. 131–7. The papers available on Inishbofin are noted under Chapter 19.

On rural settlement there are several papers, including Macourt, D., 'Infield and outfield in Ireland', *Economic History Review*, series 2, vol. 7, 1954–5, pp. 369–76; 'The outshoot house-type and its distribution in County Londonderry', *Ulster Folklife*, vol. 2, 1956, pp. 27–34; 'The maps of the Brownlow Estate and the study of the rural landscape in North Armagh', *Ulster Journal of Archaeology*, vol. 20, 1957, pp. 114–22; 'Surviving openfield in County Londonderry', *Ulster Folklife*, vol. 4, 1958, pp. 19–28; Johnson, J. H., 'Studies of Irish rural settlement', *Geogr. Rev.*, vol. 48, 1958, pp. 554–66. See also Buchanan, R. H., 'Rural changes in an Irish townland, 1890–1955', *Advanc. Sci.*, vol. 14, 1958, pp. 291–300, which is a detailed study of a townland on the southeast coast of Co. Down. A paper showing that archaeological methods can be usefully employed to show former houses and other buildings, on lines similar to those dealing with deserted villages in Britain, is Buchanan, R. H., Proudfoot, V. B., Johnson, J. H., 'Excavations at Murphystown, Co. Down', *Ulster Journal of Archaeology*, vol. 21, 1958, pp. 115–26. The vast changes in the settlement pattern of one part of rural Ireland, no doubt of many more, is illustrated in Johnson, J. H., 'The disappearance of clachans from county Derry in the nineteenth century', *Irish Geogr.*, vol. 4, no. 6, 1963, pp. 404–14. A substantial addition to the literature is Flatres, P., *Géographie rural de quatre contrées Celtiques, Irlande, Galles, Cornwall et Man*, Rennes, 1957.

Religion. The Census volumes give full statistical information: data have also been drawn from the *Catholic Yearbook*, the *Church of Ireland Directory*, and the *Methodist Handbook*, all published annually in Dublin. The distribution of religions around the Northern Ireland boundary receives full treatment in North-eastern Boundary Bureau, *Handbook of the Ulster Question*, Dublin, 1923, a work of considerable interest as an indication of the views of the time. For Belfast, see Jones, Emrys, 'The distribution and segregation of Roman Catholics in Belfast', *Sociological Review*, Keele, vol. 4, no. 2, 1956, pp. 167–89. This receives further discussion in the same author's *Social Geography of Belfast*, London, 1960.

Political aspects. Always a subject of polemic, these have received little serious attention by geographers and there is still no detailed study of the Border though the present author suggested this as a subject to a research student in 1938. To some extent profitable lines of general inquiry are suggested by the *Handbook of the Ulster Question* mentioned above. The fascinating speculations on the possible revision of the Border are raised in Andrews, J. H., 'The "Morning Post" line', *Irish Geogr.*, vol. 4, no. 2, 1960, pp. 99–106. A modern work is Heslinga, M. W., *The Irish Border as a Cultural Divide*, Assen, 1962: this book will carry a varying amount of conviction from one reader to another. See an interesting review by J. H. Andrews in *Irish Geogr.*, vol. 6, no. 6, 1963, pp. 455–7. Barritt, D. P., and Carter, C. F., *The Northern Ireland Problem: a study in group relations*, London, 1962, is a valuable study of the outlook of various groups, but both authors hardly seem to realize that most people are not as tolerant and charitable as themselves. A temperate and analytical paper is Johnson, J. H., 'The political distinctiveness of Northern Ireland', *Geogr. Rev.*, vol. 52, 1962, pp. 78–91.

Language. Census reports are the main source: see also *Comisiun na Gaeltachta* (Report of the Gaeltacht Commission), Dublin, 1926. See also *The restoration of the Irish language*, 1965: this report was published by the government commission on the Irish language. It recommended measures to retain and develop the use of Irish whenever possible as a feature of national policy. Two economic and planning studies of Irish-speaking areas have been written by Symes, D. G., 'The population resources of the Connemara Gaeltacht', 1965, and 'The population resources of the Gaeltacht', 1965: these form nos. 1 and 2 of *University of Hull Publications, Miscellaneous Series.*

CHAPTER 8 (*Agriculture*)

1. Government reports containing statistical material include the *Census Reports* of the Republic, the quarterly *Irish Statistical Bulletin* and the annual *Statistical Abstract*. Two reports now of considerable historical value for the Republic are the *Agricultural Statistics* (*a*) *1847–1926*, Dublin, 1930; (*b*) *1927–1933*, Dublin, 1935; the former contains a classic account of Irish agriculture by John Hooper, and the latter has useful dot maps. Post-war trends are discussed in *Eire, Reports on Agricultural Policy*, Dublin, 1945. For Northern Ireland, see the *Marketing of Northern Ireland Produce*, 1932, the *Reports of the Agricultural Enquiry Committee, Cmd. 249*, Belfast, 1947, *Northern Ireland Agriculture* (the annual general reports of the Ministry of Agriculture).

2. *Books*. Stamp, L. D., *An Agricultural Atlas of Ireland*, London, 1931, and Stamp, L. D., and Beaver, S. H., *The British Isles*, London, 1933, contain distribution maps and comment based on the 1930 statistics: O'Donovan, J., *The Economic History of Livestock in Ireland*, Cork, 1940, gives a full account of the history of livestock and dairying, and is best used with Hooper's classic account (*v.s.*), and *British Agriculture*, an inquiry organized by Viscount Astor and B. Seebohm Rowntree, London, 1938. An admirable treatment of land tenure problems is given in Hooker, E. R., *Readjustments of Land Tenure in Ireland*, Chapel Hill, N. Carolina, 1938. See also the annual reports of the Land Commission of the Republic. Co-operation is discussed in *Agricultural Co-operation*, published by the Horace Plunkett Foundation, 1931, and in the periodic issues of the *Yearbook of Agricultural Co-operation*. Micks, W. L., *An Account of the Constitution, Administration and Dissolution of the Congested Districts Board from 1891 to 1923*, Dublin, 1925, is a review of conditions in the west of Ireland: Arensberg, C. M., and Kimball, S. T., *Family and Community in Ireland*, London, 1940, deals with social life among a farming people in Co. Clare and has some comments on the general agricultural organization.

3. *Articles*. For a consideration of one small area, see Henning, D. V.,

'The Demesne at Mitchelstown, Co. Cork', *Irish Geogr.*, vol. 1, no. 4, 1947, pp. 97–101, and also 'Mitchelstown and its demesne in the year 1841', ibid., vol. 2, no. 3, 1951, pp. 106–10: the varying fate of Irish demesnes offers a most promising field for research. General articles include, on Northern Ireland, Muskett, A. E., and Morrison, J., 'Agriculture', in *Belfast in its Regional Setting*, Belfast, 1952, pp. 139–48, and also Freeman, T. W. ,'Farming in Irish Life', *Geogr. J.*, vol. 110, 1947, pp. 38–56; 'The prospect for Irish agriculture', ibid., vol. 120, 1954, pp. 329–36. Interesting techniques are used in Gillmor, D. A. 'The agricultural regions of the Republic of Ireland', *Irish Geogr.*, vol. 5, no. 4, 1967, pp. 245–61. An article of considerable contemporary interest is Gillmor, D. A., 'Cattle marketing in Ireland', *Administration*, Vol. 15, 1967, pp. 308–27. Other articles include Haughton, J. P., 'The live stock fair in relation to Irish country towns', *Irish Geogr.*, vol. 3, no. 2, 1955, pp. 107–113, with useful maps: Proudfoot, V. B., 'Studies of conacre', ibid., vol. 3, no. 3, 1956, pp. 162–7: Jones Hughes, T., 'East Leinster in the mid-nineteenth century', ibid., vol. 3, no. 5, 1958, pp. 227–41, is an historical study of agriculture. The division between the area of Ireland dominated by the landlords and the rest is the main theme of Jones Hughes, T., 'Society and settlement in nineteenth-century Ireland', ibid., vol. 5, no. 2, 1965, pp. 79–96. Statistical information on conacre in the Republic is given on p. 84 on the *Irish Statistical Bulletin*, vol. 41, 1966. The question of the use of Irish acres in 1841 is raised in Bourke, P. M. A., 'The extent of the potato crop in Ireland at the time of the Famine', *Journal of the Statistical and Social Inquiry Society of Ireland*, vol. 20, part 3, 1959–60, pp. 1–35.

4. *Land Utilization.* The Land Utilization Survey of Northern Ireland (Hon. Director, D. A. Hill) has produced maps on the scale of 1 : 63,360 for the whole of the six counties: the first memoir, Part I, *The Belfast Region* (H.M.S.O., Belfast), appeared in 1948. Articles include Hill, D. A., 'Land Utilization in the Belfast area', *Geography*, vol. 32, 1947, pp. 20–33; 'The Land Utilization Survey in Northern Ireland', *Irish Geogr.*, vol. 1, no. 4, Dublin, 1947, pp. 101–5; 'Land use in East Ulster', ibid., vol. 2, no. 5, 1953, pp. 197–205; Symons, L. 'Hill land utilization in Ulster', *Journal of the Statistical and Social Inquiry Society of Ireland*, vol. 19, 1955–6, pp. 58–81; 'The economic conquest of the hills', *Scot. Geogr. Mag.*, vol. 75, 1959, pp. 18–29; 'Agricultural progress in Ulster', *Geography*, vol. 44, 1959, pp. 157–70. Attwood, E. A., 'Agriculture and economic growth in Western Ireland', *Journal of the Statistical and Social Inquiry Society of Ireland*, vol. 20, 1961–2, pp. 172–95, includes some useful statistics and an interesting discussion. A paper by Welsh, T., O'Hara P. J., and Quin, E.,

'The use of peatland in Irish agriculture', *Advanc. Sci.*, vol. 14, 1958, pp. 405–16, deals with the possible agricultural value of reclaimed peat bogs, obviously a subject of considerable importance in Ireland, though it may not be easy to find the agricultural pioneers who will be required. The work of the Northern Ireland Land Use Survey is admirably summarized in L. Symons (ed.), *Land Use in Northern Ireland*, London, 1963.

CHAPTER 9 (*Trade and Industry*)

Industries in the late eighteenth century are discussed in Young, A., *Travels in Ireland, 1776–1779*, ed. Hutton, 1892, and in the works of Maxwell, E. C., *Dublin under the Georges*, 1936, and *Country and Town in Ireland under the Georges*, 1940, and in the economic histories of G. O'Brien, listed under Chapter 5. A famous work, immediately pre-Famine, is Kane, Sir Robert, *Industrial Resources of Ireland*, Dublin, 1844 and 1845. A useful summary of industries is Riordan, E. J., *Modern Irish Trade and Industry*, London, 1920 (immediately pre-Treaty). See also the relevant sections of Stamp, L. D., and Beaver, S. H., *The British Isles*, London, 1954 (especially the admirable short section on the linen industry of Northern Ireland), and Fletcher, G., *Ireland*, 1922, with the books on each of the four provinces. On power, see McCarthy, A. J. P., 'The Irish national electrification scheme', *Geogr. Rev.*, vol. 47, 1957, pp. 539–54; but see a note in ibid., vol. 48, 1958, pp. 572–3, in which D. J. Dwyer suggests that A. J. P. McCarthy is unduly optimistic about the potential value of areas now peat-covered. Concentration of trade on the larger seaports appears to be characteristic, but there are others, as shown in Andrews, J. H., 'The pattern of trade in Ireland's smaller seaports', *Irish Geogr.*, vol. 4, no. 2, 1960, pp. 93–8. A survey of mineral resources is given in O'Brien, M. V., 'Economic geology of Ireland', *Advanc. Sci.*, vol. 14, 1958, pp. 349–52. Valuable recent papers include Dwyer, D. J., 'The leather industry of the Irish republic 1922–55: a study in industrial development and location', *Irish Geogr.*, vol. 4, no. 3, 1961, pp. 175–89; 'The development and location of the textile industries in the Irish republic', ibid., vol. 4, no. 6, 1963, pp. 415–31. Gillmor, D. A., 'Foreign participation in Irish manufacturing', ibid., vol. 5, no. 2, 1965, 97–102. Much valuable information is available in the publications of the Economic and Social Research Institute, 73, Lower Baggott Street, Dublin 2.

Agricultural industries are treated in various books, including those already named, and also in *Ireland Today*, London, 1915, which is based on a *Times* supplement of March 17, 1913; Smith-Gordon, L., and Staples, L. C., *Rural Reconstruction in Ireland*, London, 1917; *Agricultural*

Co-operation in Ireland, published by the Horace Plunkett Foundation, 1917; and O'Donovan, J., *Economic History of Livestock in Ireland*, Cork, 1940. The need for more agricultural industries is stressed in two official publications: *Reports on Agricultural Policy* (P. no. 7175), Dublin, 1945, and *Reports of the Agricultural Enquiry Committee, Cmd. 249*, Belfast, 1947 (the Babington report).

Abundant statistical material is available in the *Census* volumes, the quarterly *Irish Statistical Bulletin*, the annual *Statistical Abstract*, the annual *Census of Production*, the *Trade and Shipping Statistics*, and an excellent economic review, *Ireland, National Income and Expenditure, 1938–1944*, Dublin, 1946. A general review with useful maps is *The Times* (Trade and Engineering) with Eire section, vol. 44, no. 901, March, 1939: a paper which roused considerable interest at the time is Beddy, J. P., 'A Comparison of the Principal Economic Features of Eire and Denmark', *Journal of the Statistical and Social Inquiry Society of Ireland*, vol. 17, 1943–4, pp. 269–308. For minerals, an important summary pamphlet is Bishopp, D. W., *Irish Mineral Resources*, Dublin, 1943. There has recently been a resurgence of interest in base metal mining, on which see a note by D. A. Gillmor in *Irish Geogr.*, vol. 5, no. 2, 1966, pp. 220–5. The industries of the Republic are treated in the *Irish Industrial Yearbook*.

For Northern Ireland see the *Ulster Year Book*; an important book of the linen industry is Gill, C., *The Rise of the Irish Linen Industry*, Oxford, 1925, and, as a complement on the cotton phase, see Green, E. R. R., 'The cotton hand-loom weavers in the north-east of Ireland', *Ulster Journal of Archaeology*, vol. 7, 1944, pp. 30–42: Green, E. R. R., *The Lagan Valley, 1800–50*, London, 1949, is an excellent regional study. The 1902 Handbook of the British Association, 'Belfast and adjoining counties', is of considerable interest and in the 1952 Handbook, *Belfast in its Regional Setting*, pp. 149–58, there is a survey of current trends by K. S. Isles and N. Cuthbert. This has been followed by a detailed economic survey, Isles, K. S., and Cuthbert, N., *An Economic Survey of Northern Ireland*, H.M.S.O., Belfast, 1957. Some of the government publications are also of value, not least *Who makes what in Northern Ireland*, H.M.S.O., 1957.

CHAPTER 10 (*Communications*)

Several works dealing with social and economic conditions in Ireland include comments on transport: among them are Young, A., Lewis, S. (1837), Kane, R. (1844), Micks (1925), Maxwell (1940), whose works are noted under Chapters 5 and 9. Data on railways are available in Conroy, J. C., *A History of Railways in Ireland*, London, 1928; Murray, K., *The*

Great Northern Railway (Ireland), Dublin, 1944, and Fayle, H., *The Narrow Gauge Railways of Ireland*, London, 1946, and Barrie, D. S. M., *The Dundalk, Newry and Greenore Railway*, Lingfield, 1957. Later works on railways include Patterson, E. M., *The County Donegal Railways*, Dawlish and London, 1962; *The Great Northern Railway of Ireland*, Lingfield, 1962; *The Lough Swilly Railway*, Dawlish and London, 1964: see also Fisher, C. A., 'Evolution of the Irish Railway System', *Econ. Geogr.*, vol. 17, 1941, pp. 262–74. For Northern Ireland there is an official report (by Sir Felix J. C. Pole), *Transport Conditions, Cmd. 160*, Belfast, 1934, and the *Report of the Transport Tribunal, Cmd. 310*, Belfast, 1952. Recent data on the situation in the Republic comes from the *Report of the Committee of inquiry into internal transport*, Dublin Stationery Office, 1957. On various railways there are informative articles in the *Railway Magazine*: for example, the number for May 1956 (vol. 102, no. 661) is largely devoted to Irish railways. For some of the material in this chapter, the author is indebted to the central offices of the railways. Excellent material on ports is given in the Dublin Stationery Office publication which appears at intervals of a few years, the *Shipping Handbook*.

Three important papers by Herring, I. J., deal with the 'bians' (see pp. 239–41), in *Ulster Journal of Archaeology*, vol. 2, 3rd series, 1939, pp. 130–7; vol. 3, 1940, pp. 115–22, and vol. 4, 1941, pp. 1–11. See also Herring, I. J., 'Ulster roads on the eve of the railway age', *Irish Historical Studies*, vol. 2, 1940–1, pp. 160–88. The data for the map on p. 241 are from lists in O'Connell, Mary (Mrs M. J.), *Charles Bianconi, 1786–1875*, London, 1878.

On *canals* see McCutcheon, W. A., 'The Lagan navigation', *Irish Geogr.*, vol. 4, no. 4, 1962, pp. 244–55; 'The Newry navigation: the earliest inland canal in the British Isles', *Geogr. J.*, vol. 129, 1963, pp. 466–79.

Two important books on canals are McCutcheon, W. A., *The Canals of the north of Ireland*, Dawlish and London 1965; Delany, V. T. H., and D. R., *The Canals of the south of Ireland*, Newton Abbot 1966.

CHAPTER 11 (*Regional*)

There are no particular references for this chapter, which serves only as an introduction to the remaining chapters of the book.

CHAPTER 12 (*Dublin and the East Central Lowland*)

Few works deal specifically with this region: for morphology, see Cole, G. A. J., and Hallissy, T., *Handbook of the Geology of Ireland*, 1924, esp. pp.

29–30; Smyth, L. B., and others, *Proc. Geol. Ass. Lond.*, vol. 50, 1939, pp. 287–351, which includes articles by T. Hallissy, pp. 288–302, and A. Farrington, pp. 337–44. For glaciation, see Sollas, W. J., 'A map to show the distribution of eskers in Ireland', *Sci. Proc. R. Dublin Soc.*, vol. 5, series 2, 1896, pp. 785–822; Charlesworth, J. K., 'The glacial retreat from central and southern Ireland', *Quart. J. Geol. Soc. Lond.*, vol. 84, 1928, esp. pp. 312–15, 337–42; Flint, R. F., 'The origin of the Irish "eskers"', *Geogr. Rev.*, vol. 20, 1930, pp. 615–30; Synge, F. M., 'The glacial deposits around Trim, Co. Meath', *Proc. R. Irish Acad.*, vol. 53, B10, 1950, pp. 99–110. The *Geological Memoirs* for this area were published between 1858 and 1880, with a revised *Dublin Memoir* in 1903. On other aspects of the region, information is available only from standard or official sources, such as the *Agricultural Statistics, 1847–1926, 1927–1933*, the annual *Statistical Abstract*, the *Census* volumes, and the unofficial *Irish Industrial Yearbook*, annual, Dublin. There are comments of particular interest in Lewis, S., *Topographical Dictionary*, 1837: Jones Hughes, T., 'East Leinster in the mid-nineteenth century', *Irish Geogr.*, vol. 3, no. 5, 1958, pp. 227–41. A local study is Freeman, T. W., 'Tullamore and its environs, Co. Offaly', ibid., vol. 1, no. 5, 1948, pp. 133–50.

Dublin

Historical material includes Chart, D. A., *The Story of Dublin*, London, 1932, a most attractive volume in the Mediaeval Towns series; Joyce, W. St J., *The Neighbourhood of Dublin*, Dublin, 1912 and 1939, which has interesting information on historic sites in and around the city; the Georgian Society's *Records of Eighteenth Century Architecture and Decoration in Dublin*, 5 vols., Dublin, 1909–13; Maxwell, C. E., *Dublin under the Georges*, London, 1936, an admirable account of the city of the time; Curtis, E., 'Norse Dublin', *Dublin Historical Record*, vol. 4, 1932, pp. 96–108. The British Association's *Handbook for the Dublin District*, 1908, includes sections on history, architecture, the port and industries: there is some useful material on the city in the 1957 British Association volume, *A View of Ireland*. One aspect of historical geography is treated in Butlin, R. A., 'The population of Dublin in the late seventeenth century', *Irish Geogr.*, vol. 5, no. 2, 1965, pp. 51–66. Articles of a geographical character are Haughton, J. P., 'The Site of Dublin', *Irish Geogr.*, vol. 1, no. 3, 1946, pp. 53–6, and 'The Social Geography of Dublin', *Geogr. Rev.*, vol. 39, 1949, pp. 257–77. See also Hegarty, D. A., 'The port of Dublin', *Journal of the Statistical and Social Inquiry Society of Ireland*, vol. 20, 1954–5, pp. 88–100.

CHAPTER 13 (*Leinster Chain*)

There are a number of important papers on the geology and glaciation of the Leinster chain, with a useful summary account in Smyth, L. B., and others, 'The geology of southeast Ireland', *Proc. Geol. Ass. Lond.*, vol. 50, 1939, esp. pp. 287–96, 334–7; see also Cole, G. A. J., and Hallissy, T., *Handbook of the Geology of Ireland*, 1924, esp. pp. 9–10, 59–60, and the *Memoirs of the Geological Survey*, 1858–80, with the *Dublin Memoir*, 1903. Morphological papers include Farrington, A., 'The topographical features of the granite-schist junction in the Leinster Chain', *Proc. R. Irish Acad.*, vol. 37, B20, Dublin, 1927, pp. 181–92; Farrington, A., 'The pre-glacial topography of the Liffey basin', ibid., vol. 38, B9, Dublin, 1929, pp. 148–170; Haughton, J. P., 'Low-level surfaces of erosion in the Wicklow mountains', *Irish Geogr.*, vol. 1, no. 2, 1945, pp. 34–8; Synge, F. M., 'The Kilmacanoge valley, Co. Wicklow', ibid., vol. 2, no. 1, 1949, pp. 20–4. For glaciation, see Charlesworth, J. K., 'The glacial retreat from central and southern Ireland', *Quart. J. Geol. Soc. Lond.*, vol. 84, 1928, pp. 293–342; Farrington, A., 'The glaciation of the Wicklow mountains', *Proc. R. Irish Acad.*, vol. 42, B7, 1934, pp. 173–209; Charlesworth, J. K., 'A map of the glacier-lakes and the local glaciers of the Wicklow hills', ibid., vol. 44, B3, 1937, pp. 29–36; Farrington, A., 'The local glaciers of Mount Leinster and the Blackstairs mountains', ibid., vol. 45, B3, 1938, pp. 65–71; Farrington, A., 'The granite drift near Brittas', ibid., vol. 50, B6, 1944, pp. 279–291; Farrington, A., 'The glacial deposits of the district around Enniskerry, Co. Wicklow', ibid., vol. 50, B6, 1944, pp. 133–57; Synge, F. M., 'The Glendalough valley', *Irish Geogr.*, vol. 1, no. 3, 1946, pp. 58–61; Farrington, A., 'Glacial Lake Blessington', ibid., vol. 3, no. 4, 1957, pp. 216–22; Davies, G. L., 'Platforms developed in the boulder clay of the coastal margins of counties Wicklow and Wexford', ibid., vol. 4, no. 2, 1960, pp. 107–16.

A classic source on mineral deposits is Cole, G. A. J., 'Memoir and Map of Localities of Minerals of Economic Importance in Ireland', Geological Survey of Ireland, Dublin, 1922: see also Cole and Hallissy, op. cit. *supra*, esp. pp. 77–8; Bishopp, D. W., *A Short Review of Irish Mineral Resources*, Dublin, 1943, pp. 7–9, 13–14; and Bishopp, D. W., in *The Natural Resources of Ireland*, pub. Royal Dublin Society, 1943.

For land use and vegetation, see *Agricultural Statistics, 1847–1926* and *1927–1933*, and annual *Statistical Abstract*; Praeger, R. L., *The Botanist in Ireland*, Dublin, 1934, pp. 261–80, also Pethybridge, G. H., and Praeger, R. L., 'The vegetation of the district lying south of Dublin', *Proc. R. Irish*

Acad., vol. 25, B6, 1905, pp. 124–80; Finch, T. F., 'Cultivation limits in the Dublin mountains', *Irish Geogr.*, vol. 2, no. 5, 1953, pp. 206–10.

The towns of this area have been little studied though an excellent paper of local interest but widespread application is O'Farrell, P. N., 'The urban hinterlands of New Ross and Enniscorphy', *Irish Geogr.*, vol. 5, no. 2, 1965, 67–78.

For industries, see *Irish Industrial Yearbook.*

CHAPTER 14 (*Suir, Nore and Barrow*)

The *Memoirs of the Geological Survey* were mainly published from 1858 to 1862 and the coalfields are treated in a special *Memoir*, sheet 127 etc., 1891: see also Hallissy, T., 'The lower Palaeozoics and Old Red sandstones of southeast Ireland', *Proc. Geol. Ass. Lond.*, vol. 50, 1939, pp. 288–302; in the same number are articles by Smyth, L. B. (lower Carboniferous), Hallissy, T. (coalfields), Farrington, A. (glaciation). An important paper on river history is Miller, A. A., 'River development in southern Ireland', *Proc. R. Irish Acad.*, vol. 45, 1939, pp. 321–54. Criticism of some views in this paper appears in Farrington, A., 'The South Ireland Peneplane', *Irish Geogr.*, vol. 2, no. 5, 1953, pp. 211–17. Some vivid descriptive writing on this region will be found in Praeger, R. L., *The Way that I Went*, Dublin, 1937: for vegetation, see Praeger, R. L., *A Botanist in Ireland*, Dublin, 1934.

Agricultural information is derived from standard official sources (see p. 318): for industry see the annual *Irish Industrial Yearbook*, Dublin. Two articles have been published on the Graiguenamanagh basin: Freeman, T. W., in *Irish Geogr.*, vol. 1, no. 3, 1946, pp. 61–77; on strip farms, Richards-Orpen, J., ibid., pp. 77–80. The current state of various fair towns in south Tipperary is discussed in Haughton, J. P., 'The live stock fair in relation to Irish country towns', ibid., vol. 3, no. 2, 1955, pp. 107–13. There is an extremely entertaining 'Historical Account of Kilkenny', prepared for a British Association visit, and reprinted from the *Kilkenny Journal*, August 1878: a copy exists in the library of the Royal Dublin Society.

Dr Ingeborg Leister of Marburg has written a series of papers mainly on Co. Tipperary of which the last noted here is a comprehensive monograph in historical geography. The works are: 'Orchards in county Tipperary', *Irish Geogr.*, vol. 4, 1962, pp. 292–301; 'Wald und Forst in Irland', *Erdkunde*, vol. 17, 1963, pp. 58–76; Landwirtschaft und agrarräumliche Gliederung Irlands zur Zeit A. Young's', *Zeitschrift fur Agrargeschichte und Agrarsoziologie*, vol. 10, 1962, pp. 9–44; 'Das Werden der Agrarlandschaft in der Graftschaft Tipperary (Irland)', *Marburger Geographische Schriften*, Heft 18, 1963.

CHAPTER 15 (*Valleys and Uplands of Cork*)

The most recent geological memoir for the district is that for Cork and Cork harbour, 1905. Various notable contributions to geomorphology have been made by workers on this region: see Jukes, J. B., 'On the mode of formation of some of the river-valleys in the south of Ireland', *Quart. J. Geol. Soc. Lond.*, vol. 18, 1862, pp. 378–403; various papers summarized in Cole, G. A. J., and Hallissy, T., *Handbook of the Geology of Ireland*, 1924; Miller, A. A., 'River development in southern Ireland', *Proc. R. Irish Acad.*, vol. 45, B14, 1939, pp. 321–54. Notable contributions on glaciation include Wright, W. B., and Muff (Maufe), H. B., 'The pre-glacial raised beach of the south of Ireland', *Sci. Proc. R. Dublin Soc.*, vol. 10 (new series), part 2, 1904, pp. 250–324, with numerous annotated illustrations; Farrington, A., 'The glacial geology of south-eastern Ireland', *Proc. Geol. Ass. Lond.*, vol. 50, 1939, pp. 337–42; 'Unglaciated areas in southern Ireland', *Irish Geogr.*, vol. 1, no. 4, 1947, pp. 89–97; 'Notes on the geomorphology of the Kinsale district', ibid., vol. 2, no. 3, 1951, pp. 124–8.

Information on agriculture and forestry has been acquired from official sources (see pp. 514, 523); see also O'Donovan, J., *The Economic History of Livestock in Ireland*, Cork, 1940, and, for the creameries, *Agricultural Co-operation*, published by the Horace Plunkett Foundation, London, 1931. The factory at Mitchelstown is discussed in Henning, D. V., 'The demesne at Mitchelstown', *Irish Geogr.*, vol. 1, no. 4, 1947, pp. 97–101; see also 'Mitchelstown and its demesne in the year 1841', ibid., vol. 2, no. 3, 1951, pp. 106–10. Another local study is Haughton, J. P., 'Kinsale. An essay in regional geography', ibid., vol. 2, no. 3, 1951, pp. 11–123.

For the city of Cork, see two maps (cf. Fig. 64) in the *Geological Memoir*, 1905, and Hartnett, P. J., 'Cork city, its history and antiquities', *Journal of the Cork Historical and Archaeological Society*, vol. 48, pp. 1–105, esp. pp. 61–81; Robertson, M., *County Borough of Cork and Neighbourhood, Town Planning Report*, Cork, 1941. The *Irish Industrial Yearbook*, annual, Dublin, has been a valued source for industries, and articles on Cork normally appear in the 'Annual Review', published by the *Irish Times* on January 1. See also O'Sullivan, W., *The Economic History of Cork city from the earliest times to the Act of Union*, Cork and London, 1937. An article of great interest, especially for its historical approach, is Brookfield, H. C., 'Cobh and Passage West', *Irish Geogr.*, vol. 2, no. 4, 1952, pp. 156–67.

CHAPTER 16 (*Kerry and West Cork*)

On structure and landforms, see the *Memoirs of the Geological Survey*, 1859–64, and the more recent special memoir, Wright, W. B., and others, *The*

Geology of Killarney and Kenmare, Dublin, 1927, which includes some valuable material on glaciation: also useful is Cole, G. A. J., and Hallissy, T., *Handbook of the Geology of Ireland*, 1924; both these authors also helped to produce the Killarney memoir. The literature on glaciation is considerable: see the papers mentioned on pp. 510–12.

In the literature of vegetation, a book of particular value is Scully, R. W., *The Flora of Kerry*, Dublin, 1916, which also has some important information on climate and weather: there is also valuable information in Praeger, R. L., *The Botanist in Ireland*, Dublin, 1934, para. 300–30, and in Tansley, A. G., *The British Islands and their Vegetation*, Cambridge, 1939, pp. 327–40 (on the Killarney oakwoods). On land use, information is drawn mainly from official sources, *Agricultural Statistics, 1847–1926*, Dublin, 1930, and *Agricultural Statistics, 1927–1933*, Dublin, 1935, and the annual Statistical Abstracts.

A detailed local study is Haughton, J. P., 'The town of Skibbereen', *Irish Geogr.*, vol. 3, no. 4, 1957, pp. 181–91. Social and regional aspects are treated in the Base Line Reports of the Congested Districts Board, 1891 onwards, which are less full than those for Connacht or Donegal, but of value; Micks, W. L., *An Account of the Constitution, Administration and Dissolution of the Congested Districts Board for Ireland from 1891 to 1923*, Dublin, 1925; *Comisiun na Gaeltachta*, Dublin, 1926 (Irish-speaking districts, fishing and crafts); an account of the Blasket islands is given in Mason, T. H., *The Islands of Ireland*, London, 1936, pp. 89–106.

CHAPTER 17 (*Limerick and Clare*)

The main sources on the structure and landforms are the *Memoirs of the Geological Survey*, 1858–69, the *Limerick Memoir*, 1907, Cole, G. A. J., and Hallissy, T., *Handbook of the Geology of Ireland*, 1924. Papers on the Burren country include 'The Carboniferous limestones of County Clare', *Quart. J. Geol. Soc. Lond.*, vol. 65, 1909, pp. 538–86; Coleman, J. C., and Dunnington, N. J., 'The Pollnagollum cave, Co. Clare', *Proc. R. Irish Acad.*, vol. 50, B5, 1944, pp. 105–32; Hodson, F., 'The beds above the Carboniferous limestone in northwest County Clare, Eire', *Quart. J. Geol. Soc. Lond.*, vol. 109, 1953, pp. 259–79; Corbel, J., 'Une région karstique de l'Irlande, le Burren', *Revue de géographie de Lyon*, vol. 27, 1952, pp. 21–34; Sweeting, M. M., 'The enclosed depression of Carran, County Clare', *Irish Geogr.*, vol. 2, no. 5, 1953, pp. 218–24 (with a note on land use by J. P. Haughton, pp. 225–6); 'The landforms of north-west County Clare', *Transactions, Institute of British Geographers*, 21, 1955, pp. 33–49; Williams,

P. W., 'Limestone pavements with special reference to western Ireland', ibid., 40, 1966, 155–72.

A paper of great interest on the volcanic rocks is Ashby, D. F., 'The geological succession and petrology of the Lower Carboniferous volcanic area of Co. Limerick', *Proc. Geol. Ass. Lond.*, vol. 50, 1939, pp. 324–30. A local study of glaciation, on a little-known area, is Farrington, A., 'The glacial drifts of Slieve Bernagh', *Irish Geogr.*, vol. 2, no. 4, 1952, pp. 153–155. See also Farrington, A., 'The last glaciation in the Burren', *Proc. R. Irish Acad.*, vol. 64, B3, 1964–6, pp. 33–9. Excellent descriptions of Co. Clare appear in Praeger, R. L., *The Way that I Went*, Dublin, 1937, and with the addition of a full discussion of the vegetation in *The Botanist in Ireland*, 1934. There is also a summary treatment in Tansley, A. G., *The British Islands and their Vegetation*, Cambridge, 1939. Information on the woods has been acquired partly from these sources and also from McEvoy, T., 'Irish native woodlands', *Irish For.*, vol. 1, no. 2, 1944, pp. 27–35. Ivimey-Cook, R. B. and Proctor, M. C. F., 'The plant communities of the Burren, Co. Clare', *Proc. R. Irish Acad.*, vol. 64, B15 1964, pp. 211–302 is a new and comprehensive study.

Official sources have been used for the agriculture and population, and the *Irish Industrial Yearbook* has also been of value. For social factors, see Micks, W. L., op. cit., esp. pp. 120–7; Arensberg, C. M., *The Irish Countryman*, London, 1937; Arensberg, C. M., and Kimball, S. T., *Family and Community in Ireland*, Cambridge, Mass., 1940: both these works are based primarily on work in a rural area of Clare. A local study of particular interest is Haughton, J. P., 'Killaloe and district', *Irish Geogr.*, vol. 2, no. 4, 1952, pp. 141–52. The Shannon airport developments are described in the reports of the Shannon Free Development Company: see also Dwyer, D. J., *Geogr. Rev.*, vol. 51, 1961, pp. 583–5 and Soulsby, J. A., 'The Shannon Free Airport scheme: a new approach to industrial development', *Scot. Geogr. Mag.*, vol. 81, no. 2, 1965, 104–14.

The 'Shannon scheme' is described in *Engineering*, London, vol. 127, 1929, p. 141; vol. 129, 1930, pp. 731, 743, and 771; vol. 131, 1931, pp. 250, 340, 526, and in the annual reports of the Electricity Supply Board, Dublin, which includes a review of the river levels through each year: some of this data has been used in Chapter 3 (Climate and Weather).

CHAPTER 18 (*East Connacht*)

General sources on structure and landforms include the *Memoirs of the Geological Survey*, 1865–74: Cole, G. A. J., and Hallissy, T., *Handbook of the Geology of Ireland*, 1924. The eskers and drifts were the subject of a classic paper, Sollas, W. J., 'A map to show the distribution of eskers in Ireland', *Sci. Proc. R. Dublin Soc.*, vol. 5, series 2, 1896, pp. 785–822: for an interesting hypothesis, see Flint, R. F., 'The origin of the Irish "eskers"', *Geogr. Rev.*, vol. 20, 1930, pp. 615–30. For vegetation see Praeger, R. L., *The Botanist in Ireland*, 1934, and *The Way that I Went*, 1937 (some caustic comments on the Roscommon landscape). Official sources include the *Report of the Minister for Land on Forestry*, published every few years; *Agricultural Statistics, 1846–1926, 1927–1933*, Dublin, 1930 and 1935; and the annual *Statistical Abstracts*; the *Census Reports*. See also the distribution maps (especially of sheep) in Stamp, L. D., *Agricultural Atlas of Ireland*, London, 1931. O'Donovan, J., *Economic History of Livestock in Ireland*, Cork, 1940, has some general information on the region's agriculture and interesting historical data on the fairs of Ballinasloe. Economic and social factors are discussed in the Base Line Reports of the Congested Districts Board, 1891 onwards; Micks, W. L., op. cit., 1925; Freeman, T. W., op. cit., *Geogr. Rev.*, vol. 33, 1943, pp. 1–14; regional studies include 'The town and district of Roscommon', *Irish Geogr.*, vol. 2, no. 2, 1950, pp. 61–76 and 'Galway – the Key to West Connacht', ibid., vol. 3, no. 4, 1957, pp. 194–205. In this last article, the author has drawn freely on an excellent history, O'Sullivan, M. J. D., *Old Galway: the History of a Norman Colony in Ireland*, Cambridge, 1942

CHAPTER 19 (*East Connacht*)

Structure and Landforms. The *Memoirs of the Geological Survey* for this district were published in Dublin between 1869 and 1881: a later production is the *Memoir on Clare Island, Co. Mayo*, Dublin, 1914. Cole, G. A. J., and Hallissy, T., *Handbook on the Geology of Ireland*, London, 1924, is a valuable general source book: see esp. pp. 1–17, 52. Papers include (*a*) Kilroe, J. R., 'The Silurian and Metamorphic rocks of Mayo and North Galway', *Proc. R. Irish Acad.*, vol. 26, B10, 1907, pp. 129–60; (*b*) Wager, L. R., 'The geology of the Roundstone district', ibid., vol. 41, B5, 1932, pp. 46–72; (*c*) Wager, L. R., 'An outline of the geology of Connemara', *Proc. Geol. Ass. Lond.*, vol. 50, part 3, 1939, pp. 346–51. There are brief but pertinent comments on structure in Bailey, E. B., *Introduction to Geology*, London, 1939, p. 385. For the glaciation, see Kinahan, G. H., and Close, M. H., *The General Glaciation of Iar-Connaught*, Dublin, 1872; Charlesworth, J. K., 'The glacial geology of north Mayo and west Sligo', *Proc. R. Irish Acad.*, vol. 38, B6, 1928, pp. 110–15, and 'The glacial retreat in Iar Connaught', ibid., vol. 39, B2, 1929, pp. 95–106; Synge, F. M., 'The glaciation of the Nephin beg range, County Mayo', *Irish Geogr.*, vol. 4, no. 6, 1963, pp. 397–403.

Vegetation. The main sources are Praeger, R. L., *The Botanist in Ireland*, Dublin, 1934; Tansley, A. G., *The British Islands and their Vegetation*, London, 1939, pp. 340–3; Connolly, G., 'The vegetation of southern Connemara', *Proc. R. Irish Acad.*, vol. 39, B12, 1930, pp. 203–31.

Agriculture and Social Life. See Micks, W. L., op. cit.; The Base Line Reports of the Congested Districts Board, 1891 onwards, the 'Clare Island Survey' (team field work from 1908 to 1911), published in *Proc. R. Irish Acad.*, vol. 31, 1911–15; Freeman, T. W., op. cit., *Geogr. Rev.*, vol. 33, 1943, pp. 1–14; 'The changing distribution of population in Co. Mayo', *Journal of the Statistical and Social Inquiry Society of Ireland*, vol. 17, 1942-3, pp. 85–106. Arising from a field work visit to Inishbofin, there are three articles in *Irish Geogr.*, vol. 3, no. 3, 1956, pp. 123–38, by J. P. Haughton (general), T. Jones Hughes (townlands) and J. A. K. Grahame (fuel), with Freeman, T. W., 'Inishbofin – an Atlantic Island', *Econ. Geogr.*, vol. 34, 1958, pp. 202–9. Another island study, carried out by a group of students from the University of London Geographical Society, is Dwyer, D. J., 'Farming an outpost island: Clare island, county Mayo', *Geography*, vol. 48, 1963, pp. 255–67. On the Belmullet area, see Haughton, J. P., 'The Mullet of Mayo', *Irish Geogr.*, vol. 4, no. 1, 1959, pp. 1–15: Jones Hughes, T., 'Landlordism in the Mullet of Mayo', ibid., pp. 16–34. On the area west of Galway city, see Mac Aodha, B. S., 'Clachan settlement in Iar-Chonnacht', ibid., vol. 5. no. 2, 1965, pp. 20–8.

Aran Islands. Mason, T. H., *The Islands of Ireland*, London, 1936, has some interesting material: so has Praeger, R. L., op. cit. The *Memoirs of the Geological Survey*, sheets 104, 113, etc., by Kinahan, H. G., and others, was published in 1871. An old classic is Synge, J. M., *The Aran Islands*, 1910. There are valuable papers by Klimm, L. E., 'Inishmore: an outpost island', *Geogr. Rev.*, vol. 17, 1927, pp. 387–96; 'The rain tanks of Aran', *Bulletin of the Geographical Society of Philadelphia*, vol. 34, 1936, pp. 73–84. 'The relation between field patterns and jointing in the Aran islands', *Geogr. Rev.*, vol. 25, 1935, pp. 618–24. See also Gailey, R. A., 'Settlement and population in the Aran islands', *Irish Geogr.*, vol. 4, no. 1, 1959, pp. 65–78. See also the note on p. 424.

CHAPTER 20 (*Drumlin-Drift Belt*)

Cole, G. A. J., and Hallissy, T., *Handbook of the Geology of Ireland*, 1924, has valuable information on both the eastern and western sections of the drumlin-drift belt: the *Memoirs of the Geological Survey* are among the later publications of the series, having appeared between 1873 and 1887. There are two special memoirs, sheet 58 (around the town of Monaghan), by Hallissy, T., 2nd edition, 1914, and *Memoir on the Geological Features and Soils of the Agricultural Station at Ballyhaise, Co. Cavan*, by Kilroe, J. R., and others, Dublin, 1910. For the mineral resources, see *Memoir on the Coal-fields of Ireland*, Dublin, 1921; Bishopp, D. W., *A Short Review of Irish Mineral Resources*, Dublin, 1943, pp. 4–7. Praeger, R. L., *The Botanist in Ireland*, 1934, discusses Sligo and Leitrim in some detail in para. 416–31. For creameries, see *Agricultural Co-operation in Ireland*, a survey by the Horace Plunkett Foundation, London, 1931, pp. 193–200. Articles on the region are Freeman, T. W., op. cit., *Geogr. Rev.*, vol. 33, 1943, pp. 1–14; 'Population distribution in County Sligo', *Journal of the Statistical and Social Inquiry Society of Ireland*, vol. 17, 1943–4, pp. 254–68; 'The Castleblayney district, Co. Monaghan', *Irish Geogr.*, vol. 1, no. 4, 1947, pp. 106–26, and 'The town and district of Carrick-on-Shannon, Co. Leitrim', ibid, vol. 2, no. 1, 1949, pp. 1–19. Two papers deal with the Carlingford (Cooley) peninsula. Jones Hughes, T., 'Land holding and settlement in the Cooley peninsula of Louth', ibid., vol. 4, no. 3, 1961, pp. 149–74, is a study in historical geography showing the impact of the landed gentry in the peninsula: the pattern of farms was due to an economy in which the plains were farmed by privileged tenants employing labourers, producing crops and livestock for the British market. Elsewhere farming was on a small scale and the families were virtually self-sufficient. Aalen, F. H. A., 'Some historical aspects of landscape and rural life in Omeath, County Louth', ibid., vol. 4, no. 4, 1962, pp. 256–78, is a detailed study of a small area in the Carlingford peninsula. The port of Greenore is discussed by Gillmor, D. A., *Irish Geogr.*, vol. 5, no. 4, 1967, pp. 330–3.

CHAPTER 21 (*Donegal*)

Information on the structure and landforms is available in the *Geological Survey Memoirs*, 1887–91, and the more recent *Geology of the Country around Londonderry*, Dublin, 1908: Cole, G. A. J., and Hallissy, T., *Handbook of the Geology of Ireland*, 1924. The glaciation is discussed in Charlesworth, J. K., 'The glacial geology of northwest Ireland', *Proc. R. Irish Acad.*, vol. 36, B12, 1924, pp. 174–314. On land use, statistical sources include *Agricultural Statistics, 1847–1926, 1927–1933*, Dublin, 1930 and 1935, the annual 'Statistical Abstract': there is also valuable information in Praeger, R. L., *The Botanist in Ireland*, para. 440–52. 'Spray-level' vegetation is discussed in Tansley, A. G., *The British Islands and their Vegetation*, Cambridge, 1939, pp. 896–8; for woods, see Tansley, op. cit., pp. 342–3.

The main printed sources on social and regional aspects are the Base Line Reports of the Congested Districts Board, 1891 onwards; Micks, W. L., op. cit., which includes the admirable Base Line report on the Rosses, written by Mr Micks; *Comisiun na Gaeltachta, Report*, Dublin, 1926, which refers mainly to the Irish-speaking areas and also gives some details of fishing and home industries; Mason, T. H., *The Islands of Ireland*, London, 1936, has a description of Tory island. There is a discursive and entertaining account of co-operation in west Donegal in *My Story*, by Paddy the Cope (Patrick Gallagher), London, 1939. A list of industries is given in the annual *Irish Industrial Yearbook*. For the Dunlewy valley, see Evans, E. E., 'Some survivals of the Irish openfield system', *Geography*, London, vol. 24, 1939, pp. 24–36; see also Freeman, T. W., 'Buncrana and its environs', *Irish Geogr.*, vol. 2, no. 3, 1951, pp. 93–105; Williams, H. G., 'Tory island', ibid., vol. 2, no. 4, 1952, pp. 184–90. A very different area is treated in Freeman, T. W., 'North Inishowen, County Donegal', ibid., vol. 3, no. 2, 1955, pp. 59–68; see also Haughton, J. P., 'The town and district of Killybegs', ibid., vol. 4, no. 6, 1963, pp. 379–96. A notable contribution to historical geography is Macourt, D., 'The rundale system in Donegal: its distribution and decline', *Journal of the Donegal Historical Society*, vol. 3, 1955, pp. 47–60.

CHAPTER 22 (*Northern Ireland*)

1. *General Works*. Among geographical works, see Cole, G. A. J., *Ireland the Outpost*, Oxford, 1919; Fitzgerald, W., *Historical Geography of Early Ireland*, London, 1925; Stamp, L. D., and Beaver, S. H., *The British Isles*, London, 1954; Stamp, L. D., *An Agricultural Atlas of Ireland*, London, 1931; Evans, E. E., *Irish Heritage*, Dundalk, 1942; *The Mourne Country*, Dundalk, 1951; Mogey, J. M., *Rural Life in Northern Ireland*, Oxford, 1947. See also the *Ulster Year Book*, normally published at three-year intervals. The British Association volume, *Belfast in its Regional Setting*, 1952, contains a series of articles grouped under the headings 'physical background', 'human background', 'the region today', and 'Belfast'. For an economic study, see Isles, K. S., and Cuthbert, N., *An Economic Survey of Northern Ireland*, Belfast, 1957. For conditions *c.* 1918 and some economic history, see Riordan, E. J., *Modern Irish Trade and Industry*, London, 1920; a classic of its type is Gill, C., *The Rise of the Linen Industry*, Oxford, 1925.

2. *Physical Features*. For the general physical setting, see Cole, G. A. J., *Handbook of the Geology of Ireland*, 1924; *Memoirs of the Geological Survey*, 1871–91, with three special memoirs, *Belfast*, 1904 (by Lamplugh, G. W., and others), *Londonderry*, 1908 (by Wilkinson, S. B., and others), *The*

Ballycastle Coalfield, 1924 (by Wright, W. B., and others). Much of the important work done on the igneous rocks, with a general account of the geology of the northeast, is summarized in Charlesworth, J. K., and others, 'The geology of northeast Ireland', *Proc. Geol. Ass. Lond.*, vol. 46, 1935, pp. 441–502, and in Charlesworth, J. K., *The Geology of Ireland*, Edinburgh, 1953. On glaciation see Dwerryhouse, A. R., 'The glaciation of northeast Ireland', *Quart. J. Geol. Soc. Lond.*, vol. 79, 1923, pp. 352–421; Charlesworth, J. K., 'The glacial geology of the northwest of Ireland', *Proc. R. Irish Acad.*, vol. 36, B12, 1924, pp. 174–314, 'Some observations on the glaciation of northeast Ireland', ibid., vol. 45, 1939, B11, pp. 255–95 and Prior, D. B., 'Late glacial and post-glacial shorelines in north-east Antrim', *Irish Geogr.*, vol. 5, no. 3, 1966, pp. 173–87. Studies of a local character include Proudfoot, V. B., 'Erosion surfaces in the Mourne mountains', *Irish Geogr.*, vol. 3, no. 1, 1954, pp. 26–35; Stephens, M., and Symons, L., 'The lough Erne drainage scheme', *Geography*, vol. 41, 1956, pp. 123–6; McMillan, Nora F., 'Quaternary deposits around lough Foyle, North Ireland', *Proc. R. Irish Acad.*, vol. 58, B9, 1957, pp. 185–205; Colhoun, E. A., 'Slope profiles of the Mourne mountains', *Irish Geogr.*, vol. 5, no. 4, 1967, 311–19.

3. *Vegetation, Soils, Land Use, Agriculture.* For official statistics, see the *Ulster Year Book, Annual Reports upon the Agricultural Statistics*; see also *Marketing of Northern Ireland Produce*, 1932, and the *Reports of the Agricultural Enquiry Committee*, Belfast, *Cmd. 249*, Belfast, 1947. The agriculture of Northern Ireland is comprehensively treated in Symons, L. (ed.), *Land Use in Northern Ireland*, London, 1963. A short but valuable paper on forestry is Stewart, D., 'Afforestation in Northern Ireland', *Irish For.*, vol. 2, 1945, pp. 54–6. See also Thomas, D. E. L., 'Economic aspects of forestry in Northern Ireland', *Journal of the Statistical and Social Inquiry Society of Ireland*, vol. 20, part 5, 1961–2, pp. 196–219. The series of papers on the lough Neagh fens and the paper on the Mourne mountains by Armstrong, J. L., and others, are noted under Chapter 4: see the bibliography of Chapter 8 for a list of further papers by L. Symons and others on land use. A gap in knowledge of soils is common to all parts of Ireland, but two papers are welcome, Cruickshank, J. R., 'A reconnaissance soil survey of county Fermanagh', *Irish Geogr.*, vol. 4, no. 3, 1961, pp. 190–201, is largely a study of drainage conditions. Proudfoot, V. B., and Boal, F. W., 'Two soil maps of county Down', *Geogr. J.*, vol. 126, 1960, pp. 60–6, uses some earlier work with additional fieldwork to provide maps as a physical basis for a study of the distribution of early settlement, and modern land use.

4. *Population and Social Life*. The *Handbook of the Ulster Question*, Dublin, 1923, is of particular interest for this area: for co-operation, *Agricultural Co-operation*, a survey by the Horace Plunkett Foundation, 1931. A source of particular interest on Londonderry is the first (and last) Ordnance Survey memoir: *Memoir of the City and Northwestern Liberties of Londonderry* (parish of Templemore), vol. 1 of the Ordnance Survey of the County of Londonderry, Dublin, 1837; another important period work is Lewis, S., *Topographical Dictionary*, London, 1837, which has some interesting descriptions of towns. A notable study of history on the ground is Green, E. R. R., *The Industrial Archaeology of County Down*, H.M.S.O., Belfast, 1963. A recent attempt to provide guidance on the economic geography is reflected in two papers, Thomas, M. D., 'Manufacturing industry in Belfast', *Annals of the Association of American Geographers*, vol. 46, 1956, pp. 175–96, and 'Economic geography and the manufacturing industries of Northern Ireland', *Econ. Geogr.*, vol. 32, 1956, pp. 75–86. An excellent series of papers on Belfast includes Evans, E. E., 'The Site of Belfast', *Geography*, vol. 22, 1937, pp. 169–77: see also Evans, E. E., 'Belfast: the site and the city', *Ulster Journal of Archaeology*, vol. 7, 1944, pp. 7–29; Jones, Emrys, 'The social geography of Belfast', *Journal of the Statistical and Social Inquiry Society of Ireland*, vol. 19, 1953–4, pp. 1–18; Johnson, J. H., 'The Geography of a Belfast suburb', *Irish Geogr.*, vol. 3, no. 3, 1956, pp. 150–61. This work is developed further in Jones, E., *The Social Geography of Belfast*, London, 1960. See also Davies, G. L., 'The Town and coalfield of Ballycastle, Co. Antrim', *Irish Geogr.*, vol. 3, no. 4, 1957, pp. 206–15; Green, F. H. W., 'Town and country in Northern Ireland', *Geography*, vol. 34, 1949, pp. 89–96. A comparison between the eastern counties of Northern Ireland and the southwest of Scotland is given in Freeman, T. W., 'Two landscapes: a study of southwest Scotland and northeast Ireland', in *Essays in memory of Alan G. Ogilvie*, Edinburgh, 1959. Note that Northern Ireland is included in *The Atlas of Britain and Northern Ireland*, Oxford, 1963.

A note on the bibliography

As a whole, it looks impressive. It is not complete, but is intended to help other workers. But those who study it will realize that some of the books and papers enter more than once, like the stage army carrying different standards in 'Aida'. Nevertheless, the geographical material available on Ireland has vastly increased since this book was first written more than twenty years ago, especially on the physical aspects. To a great extent the credit for this change is due to the work of the Geographical Society of Ireland, though articles on various aspects of Irish geography have appeared in a wide range of journals. And Ireland has attracted workers from several countries. The hope for the future lies in the same growth, at increasing momentum: probably nobody realizes better than the present author how much more there is to do.

Index

In the making of this index, the reader wishing to find some information on a particular subject or area has been considered. But difficulties arise in listing counties, as some are divided between two or even three regional chapters: Tipperary, the largest, is treated mainly with the Suir, Nore and Barrow basins, but a small part belongs to the East Central Lowland and the Nenagh rural district has been included with Clare and Limerick so that the countryside flanking the Shannon from lough Derg to the estuary shall be considered in the same chapter. Mayo, similarly, is divided between three chapters. Those therefore who look for a systematic treatment of counties will not find it, at least not in the form that it is available in several excellent encyclopaedias. References to individual authors are given in the bibliography, pp. 507–40.

s*

For Product Safety Concerns and Information please contact our EU representative GPSR@taylorandfrancis.com Taylor & Francis Verlag GmbH, Kaufingerstraße 24, 80331 München, Germany